Essentials of
Systems Analysis
and Design

Essentials of Systems Analysis and Design

SECOND EDITION

Joseph S. Valacich
Washington State University

Joey F. George
Florida State University

Jeffrey A. Hoffer
University of Dayton

PEARSON
Prentice
Hall

Pearson Education International

Executive Editor: David Alexander
Publisher: Natalie E. Anderson
Project Manager (Editorial): Lori Cerreto
Editorial Assistant: Maat Van Uitert
Media Project Manager: Joan Waxman
Senior Marketing Manager: Sharon K. Turkovich
Marketing Assistant: Danielle Torio
Managing Editor (Production): Gail Steier de Acevedo
Production Editor: Vanessa Nuttry
Permissions Coordinator: Suzanne Grappi
Associate Director, Manufacturing: Vincent Scelta
Manufacturing Buyer: Natacha St. Hill Moore
Design Manager: Maria Lange
Art Director: Patricia Smythe
Interior Design: Judy Allan
Cover Design: Judy Allan
Cover Image: Salem Krieger
Line Art: ElectraGraphics, Inc.
Manager, Print Production: Christy Mahon
Full-Service Project Management/Composition: Carlisle Publishers Service
Cover Printer: Phoenix Color
Printer/Binder: Quebecor World-Dubuque

Credits and acknowledgments borrowed from other sources and reproduced, with permission, in this textbook appear on page 461.

Microsoft Excel, Solver, and Windows are registered trademarks of Microsoft Corporation in the U.S.A. and other countries. Screen shots and icons reprinted with permission from the Microsoft Corporation. This book is not sponsored or endorsed by or affiliated with Microsoft Corporation.

If you have purchased this book within the United States or Canada you should be aware that it has been wrongfully imported without the approval of the Publisher or the Author.

10 9 8 7 6 5 4 3 2
ISBN 0-13-121192-7

To my student and friend Jeremy Alexander.

—Joe

To Karen, Evan, and Caitlin, for your love, support, and patience.

—Joey

To Patty, for her sacrifices, encouragement, and support. To my students, for being receptive and critical, and for challenging me to be a better teacher.

—Jeff

Brief Contents

Contents

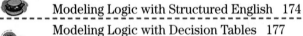

PART V   **SYSTEMS IMPLEMENTATION AND OPERATION 362**

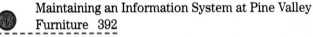

Preface

Our Approach

In today's information and technology-driven business world, students need to be aware of three key factors. First, it is more crucial than ever to know how to organize and access information strategically. Second, success often depends on the ability to work as part of a team. Third, the Internet will play an important part in their work lives. We developed *Essentials of Systems Analysis and Design, Second Edition*, to address these key factors.

We have over 50 years' combined teaching experience in systems analysis and design and have used that experience to create *Essentials of Systems Analysis and Design, Second Edition*, a text that emphasizes hands-on, experimental learning. We provide a clear presentation of the concepts, skills, and techniques students need to become effective systems analysts who work with others to create information systems for businesses. We use the systems development life cycle model as an organizing tool throughout the book to provide students with a strong conceptual and systematic framework.

Internet coverage is provided in each chapter via an integrated, extended illustrative case (Pine Valley WebStore), an end-of-chapter case (Broadway Entertainment Company, Inc.), and a margin feature (Net Search).

Many systems analysis and design courses involve lab work and outside reading. This means that lecture time can be limited. Based on market research and our own teaching experience, we understand the importance of using a book that combines depth of coverage with brevity. We have created a 10-chapter book that covers key systems analysis and design content without overwhelming students with unnecessary detail.

Essentials of Systems Analysis and Design, Second Edition, is characterized by the following themes:

1. *Systems development is firmly rooted in an organizational context.* The successful systems analyst requires a broad understanding of organizations, organizational culture, and operation.
2. *Systems development is a practical field.* Coverage of current practices as well as accepted concepts and principles are essential in a textbook.
3. *Systems development is a profession.* Standards of practice, a sense of continuing personal development, ethics, and a respect for and collaboration with the work of others are general themes in the textbook.
4. *Systems development has significantly changed with the explosive growth in databases, data-driven architecture for systems, and the Internet.* Systems development and database management can be and possibly should be taught in a highly coordinated fashion. The Internet has rapidly become a common development platform for database-driven electronic commerce systems.
5. *Success in systems analysis and design requires not only skills in methodologies and techniques but also in the management of time, resources, and risks.* Thus, learning systems analysis and design requires a thorough understanding of the process as well as the techniques and deliverables of the profession.

Given these themes, this textbook emphasizes the following:

- ◐ A business rather than a technology perspective
- ◐ The role, responsibilities, and mind-set of the systems analyst as well as the systems project manager rather than those of the programmer or business manager
- ◐ The methods and principles of systems development rather than the specific tools or tool-related skills of the field.

Audience

Many of you may be familiar with our other Prentice Hall book, *Modern Systems Analysis and Design, Third Edition*, which targets primarily upper-division undergraduates in Information Systems programs and majors in MS and MBA programs and provides a comprehensive examination of the systems analysis and design process. In this book, *Essentials of Systems Analysis and Design, Second Edition*, we provide a streamlined examination of the process, making this book useful for courses that either are more project based or take a more introductory focus.

The book is written assuming that students have taken an introductory course on computer systems and have experience writing programs in at least one programming language. We review basic system principles for those students who have not been exposed to the material on which systems development methods are based. We also assume that students have a solid background in computing literacy and a general understanding of the core elements of a business, including basic terms associated with the production, marketing, finance, and accounting functions.

Organization

The outline of the book follows the systems development life cycle, which allows for a logical progression of topics.

- ◐ Part I, "Foundations for Systems Development," gives an overview of systems development and previews the remainder of the book.
- ◐ Part II, "Systems Planning and Selection," covers how to assess project feasiblity and build the baseline project.
- ◐ Part III, "Systems Analysis," covers determining system requirements, process modeling, conceptual modeling, and determining the best design.
- ◐ Part IV, "Systems Design," covers how to design the human interface and databases.
- ◐ Part V, "System Implementation and Operation," covers system implementation, operation, closedown, and system maintenance.

Appendix A, "Object-Oriented Analysis and Design," and Appendix B, "Rapid Application Development and CASE Tools," can be skipped or treated as advanced topics at the end of the course.

Distinctive Features

Some of the distinctive features of *Essentials of Systems Analysis and Design, Second Edition*, are:

1. The grounding of systems development in the typical architecture for systems in modern organizations, including database management and Web-based systems.

2. A clear linkage of all dimensions of systems description and modeling—process, decision, and data modeling—into a comprehensive and compatible set of systems analysis and design approaches. Such a broad coverage is necessary for students in order to understand the advanced capabilities of many systems development methodologies and tools that are automatically generating a large percentage of code from design specifications.

3. Extensive coverage of oral and written communication skills including systems documentation, project management, team management, and a variety of systems development and acquisition strategies (e.g., life cycle, prototyping, rapid application development, object orientation, joint application development, participatory design, and systems reengineering).

4. Coverage of rules and principles of systems design, including decoupling, cohesion, modularity, and audits and controls.

5. A discussion of systems development and implementation within the context of management of change, conversion strategies, and organizational factors in systems acceptance.

6. Careful attention to human factors in systems design that emphasize usability in both character-based and graphical user interface situations.

Pedagogical Features

The pedagogical features of *Essentials of Systems Analysis and Design, Second Edition*, reinforce and apply the key content of the book.

SDLC Framework

Although there are several conceptual processes for guiding a systems development effort, each with strengths and weaknesses, the systems development life cycle (SDLC) is arguably the most widely applied method for designing contemporary information systems. We highlight four key SDLC steps (Figure P-1):

- Planning and selection
- Analysis
- Design
- Implementation and operation

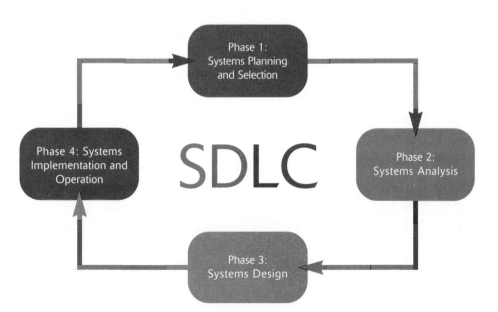

FIGURE P-1
SYSTEMS DEVELOPMENT LIFE CYCLE (SDLC)
Project management occurs throughout the systems development life cycle (SDLC).

We use the SDLC to frame the part and chapter organization of our book. Each chapter opens with an SDLC figure with various parts highlighted to show students how each chapter, and each step of the SDLC, systematically builds on the previous one.

Internet Coverage and Features

Pine Valley Furniture WebStore. A furniture company founded in 1980 has decided to explore electronic commerce as an avenue to increase its market share. Should this company sell its products online? How would a team of analysts work together to develop, propose, and implement a plan? Beginning in Chapter 3, we explore the step-by-step process.

Broadway Entertainment Company, Inc. This end-of-chapter fictional case illustrates how a video and music retailer develops a Web-based customer relationship management system. This case first appears at the end of Chapter 2 and concludes at the end of Chapter 10.

NET SEARCH
Visit http://www.prenhall.com/valacich

Net Search. Each chapter includes a margin feature entitled "Net Search." Students can access http://www.prenhall.com/valacich to link to a specific site related to the topic within the chapter and complete an exercise.

Three Illustrative Fictional Cases

Pine Valley Furniture (PVF). This case is introduced in Chapter 2 and revisited throughout the book. As key systems development life cycle concepts are presented, they are applied and illustrated with this illustrative case. For example, in Chapter 2, we explore how PVF implements the purchasing fulfillment system, and in Chapter 3, we explore how PVF implements a customer tracking system. A margin icon identifies the location of the case. A case problem related to PVF is included in the end-of-chapter material.

Hoosier Burger (HB). This second illustrative case is introduced in Chapter 5 and revisited throughout the book. Hoosier Burger is a fictional fast-food restaurant in Bloomington, Indiana. We use this case to illustrate how analysts would develop and implement an automated food ordering system. A margin icon identifies the location of these case segments. A case problem related to HB is included in the end-of-chapter material.

Broadway Entertainment Company, Inc. (BEC). This fictional video rental and music company is used as an extended case at the end of each chapter, beginning with Chapter 2. Designed to bring the chapter concepts to life, this case illustrates how a company initiates, plans, models, designs, and implements a Web-based customer relationship management system. Discussion questions are included to promote critical thinking and class participation. Suggested solutions to the discussion questions are provided in the Instructor's Resource CD-ROM.

End-of-Chapter Material

We have developed an extensive selection of end-of-chapter material designed to accommodate various learning and teaching styles.

Key Points Review. This repeats the learning objectives that appear at the opening of the chapter and summarizes the key points related to the objectives.

Key Terms Checkpoint. This is designed as a self-test feature. Students match each key term in the chapter with its definition.

Review Questions. These test students' understanding of key concepts.

Problems and Exercises. These test students' analytical skills and require them to apply key concepts.

Discussion Questions. These promote class participation and discussion.

Case Problems. These require students to apply the concepts of the chapter to three fictional cases from various industries. The two illustrative cases from the chapters are revisited—Pine Valley Furniture and Hoosier Burger. Other cases are from various fields such as medicine, agriculture, and technology. Solutions are provided on the Instructor's Resource CD-ROM.

Margin Term Definitions

Each key term and its definition appears in the margin. A glossary of terms appears at the back of the book.

References

Located at the end of the text, references organized by chapter list over 100 books and journals that can provide students and faculty with additional coverage of topics.

Software Packaging Options

- Visible Analyst
- Microsoft Visio
- Oracle 9i
- Microsoft Project
- System Architect

To enhance the hands-on learning process, Prentice Hall is planning to package this text with a choice of Visible Analyst, Microsoft Visio, Microsoft Project, System Architect, or Oracle 9i software. Your Prentice Hall sales representative can provide additional information on pricing and ordering.

The Supplement Package

A comprehensive and flexible technology support package is available to enhance the teaching experience.

Instructor's Resource CD-ROM

The Instructor's Resource CD-ROM features three key components:

- *Instructor's Resource Manual*, by Joseph S. Valacich, Joey F. George, Jeffrey A. Hoffer, and Lisa Miller, includes teaching suggestions and answers to all text review questions, problems, and exercises. Lecture notes on how to use the video series (described below) are also included.

- *Electronic Test Bank*, by Lisa Miller (University of Central Oklahoma), includes 1,500 test questions including multiple choice, matching, and essay questions. This computerized test bank is a comprehensive suite of tools for testing and assessment. The test bank is also available as Microsoft Word files on the Instructor's Resource CD-ROM.

◉ PowerPoint presentation slides, created by John Russo of the Wentworth Institute of Technology, feature lecture notes that highlight key text terms and concepts. Professors can customize the presentation by adding their own slides or editing the existing ones.

MyCompanion Website (http://www.prenhall.com/valacich)

MyCompanion Website to *Essentials of Systems Analysis and Design, Second Edition,* includes:

1. An interactive study guide with multiple choice, true/false, and essay questions. Students receive automatic feedback to their answers. Responses to the essay questions, and results from the multiple choice and true/false questions, can be e-mailed to the instructor after a student finishes a quiz.
2. Web-based exploratory exercises, referenced in the text margin as "Net Search" features, are developed on the site.
3. Destinations module (links) includes many useful Web links to help students explore systems analysis and design, CASE tools, and information systems on the Web.
4. PowerPoint presentations for each chapter are available in the student area of the site.
5. A full glossary is available along with a glossary of acronyms.
6. A secure, password-protected Instructor's area features downloads of the *Instructor's Resource Manual,* and the Test Item File in Microsoft Word, converted WebCT, and BlackBoard files.

Video Series

This video series, prepared in part by Electronic Data Systems (EDS) Corporation, consists of five video segments, each approximately fifteen minutes in length, that focus on systems analysis and design. Each includes an introduction and prologue from the text authors. Lecture notes and suggestions on how to use the videos are included in the *Instructor's Resource Manual.*

Online Courses

WebCT www.prenhall.com/webct. Gold Level Customer Support, available exclusively to adopters of Prentice Hall courses, is provided free-of-charge upon adoption and provides you with priority assistance, training discounts, and dedicated technical support.

BlackBoard www.prenhall.com/blackboard. Prentice Hall's abundant online content, combined with BlackBoard's popular tools and interface, results in robust Web-based courses that are easy to implement, manage, and use—taking your courses to new heights in student interaction and learning.

CourseCompass www.prenhall.com/coursecompass. CourseCompass is a dynamic, interactive online course management tool powered exclusively for Pearson Education by BlackBoard. This exciting product allows you to teach market-leading Pearson Education content in an easy-to-use customizable format.

Acknowledgments

The authors have been blessed by considerable assistance from many people on all aspects of preparation of this text and its supplements. We are, of course, responsible for what eventually appears between the covers, but the insights, corrections, contributions, and proddings of others have greatly improved our manuscript. The people we recognize here all have a strong commitment to students, to the IS field, and to excellence. Their contributions have stimulated us, and frequently rejuvenated us during periods of waning energy for this project.

We would like to recognize the efforts of the many faculty and practicing systems analysts who have been reviewers of the *Second Edition* and the *First Edition*. We have tried to deal with each reviewer comment, and although we did not always agree with specific points (within the approach we wanted to take with this book), all reviewers made us stop and think carefully about what and how we were writing. The reviewers were:

Richard Allen, Richland Community College

Bill Boroski, Trident Technical College

Rowland Brengle, Anne Arundel Community College

Veronica Echols-Noble, DeVry University-Chicago

Gerald Evans, University of Montana

Carol Grimm, Palm Beach Community College

Daniel Ivancevich, University of North Carolina-Wilmington

Jon Jasperson, University of Oklahoma

Len Jessup, Washington State University

James Scott Magruder, University of Southern Mississippi

Klara Nelson, University of Tampa

Lou Pierro, Indiana University

Mary Prescott, University of Tampa

Robert Saldarini, Bergen Community College

Elaine Seeman, Pitt Community College

Sultan Bhimjee, San Francisco State University

Dominic Thomas, University of Georgia

Merrill Warkentin, Northeastern University

Steven Zeltmann, University of Central Arkansas

We extend a special note of thanks to Jeremy Alexander of Web-X.com. Jeremy was instrumental in conceptualizing and writing the Pine Valley WebStore feature that appears in Chapters 3 through 10. The addition of this feature has helped make those chapters more applied and innovative. Jeremy also built the installation procedures on the Web site for Oracle, and Saonee Sarker of Washington State University developed the Oracle tutorial modules.

Lisa Miller from the University of Central Oklahoma has worked with us on several projects and has once again provided thoughtful and timely content that has improved the pedagogy of our book. Lisa wrote the end-of-chapter Case Problems, prepared an extensive test bank, and revised the *Instructor's Resource Manual* for this text.

We also wish to thank Atish Sinha of the University of Wisconsin-Milwaukee for writing Appendix A on object-oriented analysis and design. Dr. Sinha, who has been teaching this topic for several years to both undergraduates and MBA students, executed a challenging assignment with creativity and cooperation. We are also indebted to our undergraduate and MBA students at the University of Dayton, Florida State University, and Washington State University who have given us many helpful comments as they worked with drafts of this text.

Our unique supplement to this text is a series of five videotapes that illustrate common activities and situations encountered by systems analysts. We are very excited about the pedagogical value of these tapes and compliment EDS Corporation for its sizable commitment of human and financial resources to develop and produce four of these tapes for exclusive use with our book. Specifically, we thank Stu Bailey, Michael Cummings, Vern Olsen, Chris Ryan, and Terry Zuechow of EDS, Bob Tucker of Antares Alliance, and Bill Satterwhite of Whitecap Productions for all of their work on this project.

Thanks also go to Fred McFadden (University of Colorado, Colorado Springs) and Mary Prescott (University of Tampa) for their assistance in coordinating this text with its companion book—*Modern Database Management*, also by Prentice Hall.

Finally, we have been fortunate to work with a large number of creative and insightful people at Prentice Hall, who have added much to the development, format, and production of this text. We have been thoroughly impressed with their commitment to this text and to the IS education market. These people include David Alexander, executive editor; Sharon Turkovich, senior marketing manager; Lori Cerreto, project manager, who helped create a complete and comprehensive supplement package; Vanessa Nuttry, production editor; Pat Smythe, senior designer; Maat Van Uitert, editorial assistant; Joan Waxman, media project manager; and Danielle Torio, marketing assistant.

The writing of this text has involved thousands of hours of time from the authors and from all of the people listed above. Although our names will be visibly associated with this book, we know that much of the credit goes to the individuals and organizations listed here for any success this book might achieve. It is important for the reader to recognize all the individuals and organizations who have been committed to the preparation and production of this book.

About the Authors

Joseph S. Valacich is The Marian E. Smith Presidential Endowed Chair and The George and Carolyn Hubman Distinguished Professor in Information Systems for the College of Business and Economics at Washington State University, Pullman. He received a B.S. degree in computer science and M.B.A. from the University of Montana, and a Ph.D. degree in management information systems from the University of Arizona. He is a member of the Institute for Operations Research and Management Sciences (INFORMS), the Association for Computing Machinery (ACM), and is a charter member of the Association for Information Systems (AIS). Professor Valacich served on the national task forces to design IS '97: The Model Curriculum and Guidelines for Undergraduate Degree Programs in Information Systems (as well as the Executive Committee for designing the IS 2002 update to the Model) and MSIS 2000, the Master of Science in Information Systems Curriculum. He served on the Executive Committee, funded by the National Science Foundation, working to define IS Program Accreditation Standards and is on the Board of Directors for the Computing Sciences Accreditation Board (CSAB), representing the Association for Information Systems (AIS). He is the general conference co-chair for the 2003 International Conference on Information Systems (ICIS) that will be held in Seattle.

Prior to his academic career, Dr. Valacich worked in the information systems field as a programmer, systems analyst, and technical product manager. He has conducted numerous corporate training and executive development programs for organizations, including AT&T, Dow Chemical, EDS, Exxon, FedEx, General Motors, and Xerox.

Dr. Valacich serves on the editorial board of *Small Group Research*, *Information Systems Research* and was formerly an associate editor for *MIS Quarterly*. His research has appeared in publications such as *MIS Quarterly*, *Information Systems Research*, *Management Science*, and *Academy of Management Journal*. With Leonard M. Jessup, he coedited the book *Group Support Systems: New Perspectives* for Macmillan in 1993 and coauthored *Information Systems Foundations* for QUE Education and Training in 1999. Dr. Valacich is a coauthor with Jeffery A. Hoffer and Joey F. George of the best-selling *Modern Systems Analysis and Design, Third Edition*, published by Prentice Hall. A forthcoming book, *Information Systems Today*, is coauthored with Leonard M. Jessup and is published by Prentice Hall.

Joey F. George is professor and Thomas L. Williams Jr. Eminent Scholar in Information Systems in the College of Business at Florida State University. Dr. George earned his bachelor's degree at Stanford University in 1979 and his Ph.D. in management at the University of California at Irvine in 1986. He was previously the Edward G. Schlieder Chair of Information Systems in the E. J. Ourso College of Business Administration at Louisiana State University. He also served at Florida State University as Chair of the Department of Information and Management Sciences from 1995 to 1998.

Dr. George has published over three-dozen articles in such journals as *Information Systems Research, Communications of the ACM, MIS Quarterly, Journal of MIS,* and *Communication Research.* His research interests focus on the use of information systems in the workplace, including computer-based monitoring, computer-mediated deceptive communication, and group support systems.

Dr. George is coauthor of the textbook *Modern Systems Analysis and Design, Third Edition,* published in 2002 by Prentice Hall. He served as an associate editor for *MIS Quarterly* and *Information Systems Research.* He is a member of the editorial boards for *Information Technology and People* (since 1990), *Internet Research* (since 1998), and is senior editor for the new journal *eServices Journal.* Dr. George was the conference cochair for the 2001 ICIS, held in New Orleans, Louisiana.

Jeffrey Alan Hoffer is professor and chair of the Department of MIS, Operations Management, and Decision Sciences in the School of Business Administration at University of Dayton. He also taught at Indiana University and Case Western Reserve University. Dr. Hoffer earned his A.B. from Miami University in 1969 and his Ph.D. from Cornell University in 1975.

Dr. Hoffer has published three college textbooks: *Modern Systems Analysis and Design, Third Edition,* with George and Valacich; *Information Technology for Managers: What Managers Need to Know,* with DeHayes, Martin, and Perkins; and *Modern Database Management,* with Prescott and McFadden, all published by Prentice Hall. His research articles have appeared in numerous journals, including the *Journal of Database Management, Small Group Research, Communications of the ACM,* and *Sloan Management Review.* He has received research grants from IBM Corporation and the U.S. Department of the Navy.

Dr. Hoffer is cofounder of the International Conference on Information Systems and Association for Information Systems and has served as a guest lecturer at Catholic University of Chile, Santiago, and the Helsinki School of Economics and Business in Mikkeli, Finland.

Joseph S. Valacich, Pullman, Washington
Joey F. George, Tallahassee, Florida
Jeffrey A. Hoffer, Dayton, Ohio

Essentials of Systems Analysis and Design

1

The Systems Development Environment

➔ Objectives

*After studying this chapter,
you should be able to:*

- ➔ Define information systems analysis
and design.
- ➔ Discuss the modern approach to systems analysis and design that combines both process and data views of
systems.
- ➔ Describe the organizational role of
the systems analyst in information
systems development.
- ➔ Describe four types of information
systems: transaction processing systems, management information systems, decision support systems, and
expert systems.
- ➔ Describe the information systems
development life cycle (SDLC).
- ➔ List alternatives to the systems development life cycle.
- ➔ Explain briefly the role of computer-
aided software engineering (CASE)
tools in systems development.

Chapter Preview . . .

The key to success in business is the ability to gather, organize, and interpret information. Systems analysis and design is a proven methodology that helps both large and small businesses reap the rewards of utilizing information to its full capacity. As a **systems analyst,** the person in the organization most involved with systems analysis and design, you will enjoy a rich career path that will enhance both your computer and interpersonal skills.

The systems development life cycle (SDLC) is central to the development of an efficient information system. We will highlight four key SDLC steps: (1) planning and selection, (2) analysis, (3) design, and (4) implementation and operation. Be aware that these steps may vary in each organization depending on its goals. The SDLC is illustrated in Figure 1-1. Each chapter of this book includes an updated version of the SDLC highlighting which steps have been covered and which steps remain.

This text requires that you have a general understanding of computer-based information systems as provided in an introductory information systems course. This chapter previews systems analysis and lays the groundwork for the rest of the book.

FIGURE 1-1
The four steps of the systems development life cycle (SDLC): (1) planning and selection, (2) analysis, (3) design, and (4) implementation and operation.

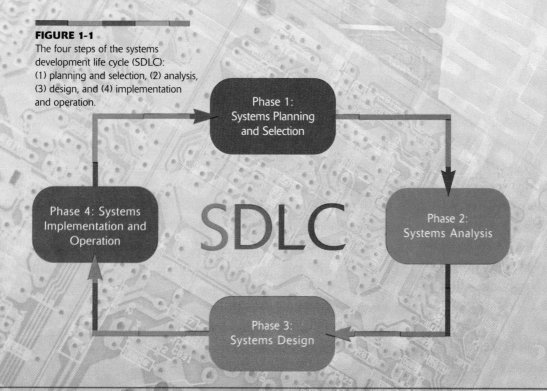

What Is Information Systems Analysis and Design?

Information systems analysis and design is a method used by companies ranging from IBM to Pepsico to Hasbro, Inc., to create and maintain information systems that perform basic business functions such as keeping track of customer names and addresses, processing orders, and paying employees. The main goal of systems analysis and design is to improve organizational systems, typically through applying software that can help employees accomplish key business tasks more easily and efficiently. As a systems analyst, you will be at the center of developing this software. The analysis and design of information systems are based on:

> **Information systems analysis and design**
> The process of developing and maintaining an information system.

- Your understanding of the organization's objectives, structure, and processes
- Your knowledge of how to exploit information technology for advantage

To be successful in this endeavor, you should follow a structured approach. The SDLC, shown in Figure 1-1, is a four-phased approach to identifying, analyzing, designing, and implementing an information system. Throughout this book, we use the SDLC to organize our discussion of the systems development process. Before we talk about the SDLC, we first describe what is meant by systems analysis and design.

> **NET SEARCH**
> *The number of new terms and words that appear each year related to information systems and new technologies is incredible. Visit http://www.prenhall.com/valacich to complete an exercise related to this topic.*

Systems Analysis and Design: Core Concepts

The major goal of systems analysis and design is to improve organizational systems. Often this involves developing or acquiring **application software** and training employees to use it. Application software, also called a *system*, is designed to support a specific organizational function or process, such as inventory management, payroll, or market analysis. The goal of application software is to turn data into information. For example, software developed for the inventory department at a bookstore may keep track of the number of books in stock for the latest bestseller. Software for the payroll department may keep track of the changing pay rates of employees. A variety of off-the-shelf application software can be purchased including WordPerfect, Excel, and PowerPoint. However, off-the-shelf software may not fit the needs of a particular organization, and so the organization must develop its own product.

> **Application Software**
> Software designed to process data and support users in an organization. Examples include spreadsheets, word processors, and database management systems.

In addition to application software, the information system includes:

- The hardware and systems software on which the application software runs. Note that the system software helps the computer function, whereas the application software helps the user perform tasks such as writing a paper, preparing a spreadsheet, and linking to the Internet.
- Documentation and training materials, which are materials created by the systems analyst to help employees use the software they've helped create.
- The specific job roles associated with the overall system, such as the people who run the computers and keep the software operating.
- Controls, which are parts of the software written to help prevent fraud and theft.
- The people who use the software in order to do their jobs.

The components of a computer-based information systems application are summarized in Figure 1-2. We address all the dimensions of the overall system,

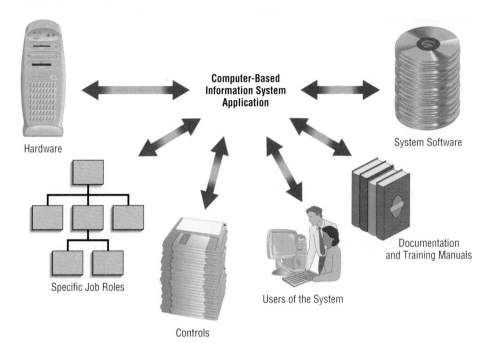

Hardware

Specific Job Roles

Controls

Users of the System

System Software

Documentation
and Training Manuals

FIGURE 1-2
COMPONENTS OF A COMPUTER-
BASED INFORMATION SYSTEM
APPLICATION

with particular emphasis on application software development—your primary responsibility as a systems analyst.

Our goal is to help you understand and follow the software engineering process that leads to the creation of information systems. As shown in Figure 1-3, proven methodologies, techniques, and tools are central to software engineering processes (and to this book).

Methodologies are a sequence of step-by-step approaches that help develop your final product: the information system. Most methodologies incorporate several development techniques, such as direct observations and interviews with users of the current system.

Techniques are processes that you, as an analyst, will follow to help ensure that your work is well thought-out, complete, and comprehensible to others on your project team. Techniques provide support for a wide range of tasks including conducting thorough interviews with current and future users of the information system to determine what your system should do, planning and managing the activities in a systems development project, diagramming how the system will function, and designing the reports, such as invoices, your system will generate for its users to perform their jobs.

Tools are computer programs, such as computer-aided software engineering (CASE) tools, that make it easy to use specific techniques. These three elements—methodologies, techniques, and tools—work together to form an organizational approach to systems analysis and design.

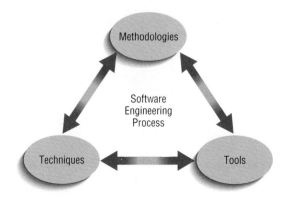

Methodologies

Software
Engineering
Process

Techniques

Tools

FIGURE 1-3
The software engineering process
uses methodologies, techniques,
and tools.

In the rest of this chapter, you will learn about approaches to systems development—the data- and process-oriented approaches. You will also identify the various people who develop systems and the different types of systems they develop. The chapter ends with a discussion of some of the methodologies, techniques, and tools created to support the systems development process. Before we talk more about computer-based information systems, let's briefly discuss what we mean by the word *system*.

Systems

The key term used most frequently in this book is *system*. Understanding systems and how they work is critical to understanding systems analysis and design.

Definition of a System and Its Parts

System
A group of interrelated procedures used for a business function, with an identifiable boundary, working together for some purpose.

A **system** is an interrelated set of business procedures (or components) used within one business unit, working together for some purpose. For example, a system in the payroll department keeps track of checks, whereas an inventory system keeps track of supplies. The two systems are separate. A system has nine characteristics, seven of which are shown in Figure 1-4. A detailed explanation of each characteristic follows, but from the figure you can see that a system exists within a larger world, an environment. A boundary separates the system from its environment. The system takes input from outside, processes it, and sends the resulting output back to its environment. The arrows in the figure show this interaction between the system and the world outside of it.

1. Components
2. Interrelated components
3. Boundary

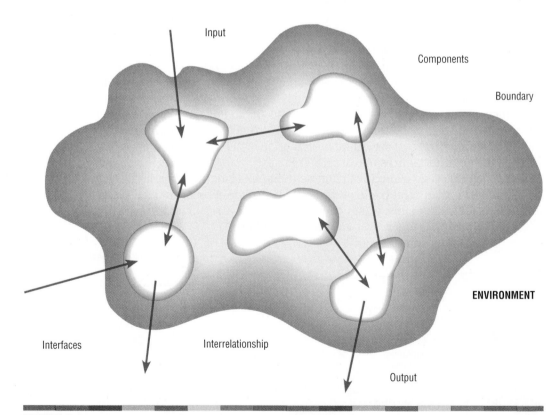

FIGURE 1-4
SEVEN CHARACTERISTICS OF A SYSTEM

4. Purpose
5. Environment
6. Interfaces
7. Input
8. Output
9. Constraints

A system is made up of components. A **component** is either an irreducible part or an aggregate of parts, also called a *subsystem*. The simple concept of a component is very powerful. For example, just as with an automobile or a stereo system, with proper design, we can repair or upgrade the system by changing individual components without having to make changes throughout the entire system. The components are **interrelated;** that is, the function of one is somehow tied to the functions of the others. For example, the work of one component, such as producing a daily report of customer orders received, may not progress successfully until the work of another component is finished, such as sorting customer orders by date of receipt. A system has a **boundary,** within which all of its components are contained and which establishes the limits of a system, separating it from other systems. Components within the boundary can be changed whereas systems outside the boundary cannot be changed. All of the components work together to achieve some overall **purpose** for the larger system: the system's reason for existing.

A system exists within an **environment**—everything outside the system's boundary that influences the system. For example, the environment of a state university includes prospective students, foundations and funding agencies, and the news media. Usually the system interacts with its environment. A university interacts with prospective students by having open houses and recruiting from local high schools. An information system interacts with its environment by receiving data (raw facts) and information (data processed in a useful format). Figure 1-5 shows how a university can be seen as a system. The points at which the system meets its environment are called **interfaces,** and there are also interfaces between subsystems.

A system must face **constraints** in its functioning because there are limits (in terms of capacity, speed, or capabilities) to what it can do and how it can achieve its purpose within its environment. Some of these constraints are imposed inside the system (e.g., a limited number of staff available), and others are imposed by the environment (e.g., due dates or regulations). A system takes input from its environment in order to function. People, for example, take in food, oxygen, and water from the environment as input. You are constrained from breathing fresh air if you're in an elevator with someone who is smoking. Finally, a system returns output to its environment as a result of its functioning and thus achieves its purpose. The system is constrained if electrical power is cut.

Important System Concepts

There are several other important systems concepts that systems analysts need to know:

- ◗ Decomposition
- ◗ Modularity
- ◗ Coupling
- ◗ Cohesion

Decomposition is the process of breaking down a system into its smaller components. These components may themselves be systems (subsystems) and can be broken down into their components as well. How does decomposition

---

Component
An irreducible part or aggregation of parts that makes up a system; also called a *subsystem.*

Interrelated
Dependence of one part of the system on one or more other system parts.

Boundary
The line that marks the inside and outside of a system and that sets off one system from other systems in the organization.

Purpose
The overall goal or function of a system.

Environment
Everything external to a system that interacts with the system.

Interface
Point of contact where a system meets its environment or where subsystems meet each other.

Constraint
A limit to what a system can accomplish.

NET SEARCH
Understanding the meaning of the word system is fundamental to becoming a systems analyst. Visit http:// www.prenhall.com/ valacich to complete an exercise related to this topic.

Decomposition
The process of breaking the description of a system down into small components; also known as *functional decomposition.*

FIGURE 1-5
A UNIVERSITY AS A SYSTEM
Source: Part of the University of Illinois at Urbana-Champaign, www.uiuc.edu/navigation/maps/kran block.html. Reprinted by permission of the University of Illinois at Urbana-Champaign.

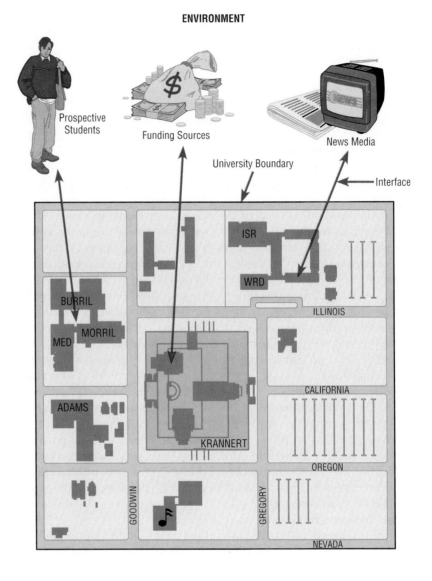

aid understanding of a system? It results in smaller and less complex pieces that are easier to understand than larger, complicated pieces. Decomposing a system also allows us to focus on one particular part of a system, making it easier to think of how to modify that one part independent of the entire system. Decomposition is a technique that allows the systems analyst to:

- Break a system into small, manageable, and understandable subsystems
- Focus attention on one area (subsystem) at a time, without interference from other areas
- Concentrate on the part of the system pertinent to a particular group of users, without confusing users with unnecessary details
- Build different parts of the system at independent times and have the help of different analysts

Figure 1-6 shows the decomposition of a portable compact disc (CD) player. This system accepts CDs and settings of volume and tone as input and produces music as output. Decomposing the system into subsystems reveals the system's inner workings. Separate systems read the digital signals

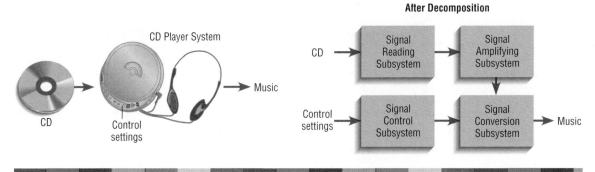

After Decomposition

FIGURE 1-6

A CD system is decomposed into four subsystems: signal reading subsystem, signal amplifying subsystem, signal control subsystem, and signal conversion subsystem.

from the CDs, amplify the signals, turn the signals into sound waves, and control the volume and tone of the sound. Breaking the subsystems down into their components reveals even more about the inner workings of the system and greatly enhances our understanding of how the overall system works.

Modularity is a direct result of decomposition. It refers to dividing a system into chunks or modules of a relatively uniform size. Modules can represent a system simply, making it easier to understand and easier to redesign and rebuild. For example, each of the separate subsystem modules for the CD player in Figure 1-6 shows how decomposition makes it easier to understand the overall system.

Coupling means that subsystems are dependent on each other. Subsystems should be as independent as possible. If one subsystem fails and other subsystems are highly dependent on it, the others will either fail themselves or have problems functioning. Looking at Figure 1-6, we would say the components of a portable CD player are tightly coupled. The amplifier and the unit that reads the CD signals are wired together in the same container, and the boundaries between these two subsystems may be difficult to draw clearly. If one subsystem fails, the entire CD player must be repaired. In a home stereo system, the components are loosely coupled because the subsystems, such as the speakers, the amplifier, the receiver, and the CD player, are all physically separate and function independently. If the amplifier in a home stereo system fails, only the amplifier needs to be repaired.

Cohesion is the extent to which a subsystem performs a single function. In the CD example, signal reading is a single function.

This brief discussion of systems should better prepare you to think about computer-based information systems and how they are built. Many of the same principles that apply to systems in general apply to information systems as well. In the next section, we review how the information systems development process and the tools that have supported it have changed over the decades.

Modularity
Dividing a system into chunks or modules of equal size.

Coupling
The extent to which subsystems depend on each other.

Cohesion
The extent to which a system or subsystem performs a single function.

A Modern Approach to Systems Analysis and Design

Today, systems development focuses on systems integration. Systems integration allows hardware and software from different vendors to work together in an application. It also enables existing systems developed in procedural languages to work with new systems built with visual programming environments. Developers use visual programming environments, such as PowerBuilder or Visual Basic, to design the user interfaces for systems that run on client/server

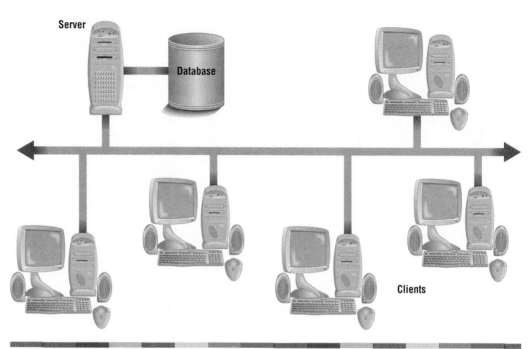

FIGURE 1-7
THE CLIENT/SERVER MODEL

platforms. In a client/server environment, some of the software runs on the server, a powerful computer designed to allow many people access to software and data stored on it, and some of the software runs on client machines. Client machines are the PCs you use at your desk at work. The database usually resides on the server. These relationships are shown in Figure 1-7. The Internet is also organized in a client/server format. With the browser software on your home PC, you can get files and applications from many different computers throughout the world. Your home PC is the client and all of the Internet computers are servers.

Alternatively, organizations may purchase an enterprise-wide system from a company such as SAP (Systems, Applications, and Products in Data Processing) or J. D. Edwards. Enterprise-wide systems are large, complex systems that consist of a series of independent system modules. Developers assemble systems by choosing and implementing specific modules. Enterprise-wide systems usually contain software to support many different tasks in an organization rather than only one or two functions. For example, an enterprise-wide system may handle all human resources management, payroll, benefits, and retirement functions within a single, integrated system.

Some basic principles that govern systems development have remained constant since the 1950s. One of those principles is the distinction between data, data flows, and processing logic, a distinction considered in the next section.

Separating Data and Processes That Handle Data

Every information system consists of three key components that anyone who analyzes and designs systems must understand. They are data, data flows, and processing logic. Figure 1-8 illustrates the differences among these three components. **Data** are raw facts that describe people, objects, and events in an organization. Examples include a customer's account number, the number of boxes of cereal bought, and whether someone is a Democrat or a Republican. Every information system uses data to produce information. **Information** is

Data
Raw facts about people, objects, and events in an organization.

Information
Data that have been processed and presented in a form suitable for human interpretation, often with the purpose of revealing trends or patterns.

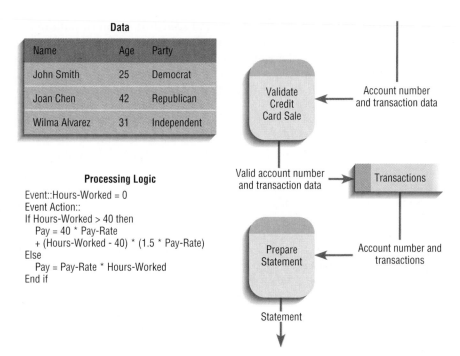

Data

Name	Age	Party
John Smith	25	Democrat
Joan Chen	42	Republican
Wilma Alvarez	31	Independent

Processing Logic

Event::Hours-Worked = 0
Event Action::
If Hours-Worked > 40 then
   Pay = 40 * Pay-Rate
   + (Hours-Worked - 40) * (1.5 * Pay-Rate)
Else
   Pay = Pay-Rate * Hours-Worked
End if

FIGURE 1-8
DIFFERENCES AMONG DATA, DATA FLOW, AND PROCESSING LOGIC
Note that data are facts about people, data flow illustrates how a sequence of activities moves data from one system to another, and processing logic is the steps that transform data from one form to another.

data organized in a form that humans can interpret. Systems developers must understand what kind of data a system uses and where the data originate. Data and the relationships among data may be described using various techniques. Figure 1-8 shows the structure of employee data as a simple table of rows (records about different employees) and columns (attributes describing each employee).

Data flows are groups of data that move and flow through a system. They include a description of the sources and destinations for each data flow. For example, a customer's account number may be captured when he or she uses a credit card to pay for a purchased item. The account number may then be stored in a file within the system until needed to compile a billing statement or prepare a mailing address for a sales circular. When needed, the account number can be extracted from storage and used to complete a system function. Figure 1-8 illustrates data flows with directional lines that connect rounded rectangles. These rounded rectangles represent the processing steps that accept input data flows and produce output data flows.

Processing logic, the third component, describes the steps that transform the data and the events that trigger these steps. For example, processing logic in a credit card application will explain how to compute available credit given the current credit balance and the amount of the current transaction. Processing logic will also indicate that the computation of the new credit balance will occur when a clerk presses a key on a credit card scanner to confirm the sales transaction. Figure 1-8, using an English-like language, illustrates the rules for calculating an employee's pay and the event (receipt of a new hours-worked value) that causes this calculation to be made.

Traditionally, an information system's design was based upon what the system was supposed to do, such as billing and inventory control: The focus was on output and processing logic. Although the data the system used as input were important, data were secondary to the application. The assumption was that we could anticipate all reports and the proper processing steps with their need for data. Each credit card application contained its own files and data storage capacity. The data had to match the specifications established in each application, and each application was considered separately.

Data flow
Data in motion, moving from one place in a system to another.

Processing logic
The steps by which data are transformed or moved and a description of the events that trigger these steps.

This concentration on the flow, use, and transformation of data in an information system typified the **process-oriented approach** to systems development. This approach involves creating graphical representations such as data flow diagrams and charts. The techniques and notations developed from this approach track the movement of data from their sources, through intermediate processing steps, and on to final destinations. Because various parts of an information system work on different schedules and at different speeds, the process-oriented approach also shows where data are temporarily stored until needed for processing. However, the natural structure of the data is not specified within the traditional process-oriented approach.

Systems analysts soon realized that there were problems with analyzing and designing systems using only a process-oriented approach. One problem was the existence of several files of data, each locked within different applications and programs. Many of the files in these different applications contained the same data elements. Figure 1-9 (A) illustrates the relationship between data and applications. Note that "personnel data" appears in two separate systems: the payroll system and the project management system. When a single data element changed, it had to be changed in each of these files. If, for example, such a system were in effect at your university and your address changed, it would have to be changed in the files of the library, the registrar's office, the financial aid office, and every other place your address was stored. It also became difficult to combine data files that had been created especially for specific applications into files that could be used more generally. Even if the specialized files contained the same data elements, each file might use a different name and format for the data. Because it was important to standardize how data elements were represented, data processing managers gradually came to separate the application programs and the data these programs used.

This focus on data typified the **data-oriented approach** to information systems development. The data-oriented approach depicts the ideal organization

FIGURE 1-9
**THE RELATIONSHIP BETWEEN DATA
AND APPLICATIONS**
(A) PROCESS-ORIENTED APPROACH
(B) DATA-ORIENTED APPROACH
Note that "personnel data" appears
in two separate systems: the payroll
system and the project
management system.

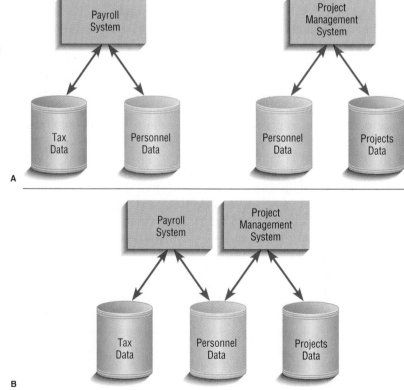

of data, independent of where and how data are used within a system. Figure 1-9 (B) illustrates the data-oriented approach. The techniques used for data orientation result in a *data model* that describes the kinds of data needed in systems and the business relationships among the data. A data model describes the rules and policies of a business. A business rule captures how an organization has chosen to capture and process data. For example, a university deals with many different vendors, ranging from suppliers of laboratory equipment, like test tubes, to suppliers of lightbulbs. If a university stops working with a vendor, perhaps because the vendor no longer supplies quality goods, the university might not want to delete all records of dealings with that vendor, at least not right away. Some offices in the university might need to access information about the vendor later. The university would need to have a business rule that indicated what happened to information about a vendor when that vendor was no longer being used. Business rules about vendors would all be shown in the data model for the university. Table 1-1 highlights the key differences between the process-oriented and the data-oriented approaches to systems development.

Separating Databases and Applications

A **database** is a shared collection of logically related data organized in ways that facilitate capture, storage, and retrieval for multiple users in an organization. For example, your name, address, and social security number are stored in your school's database. If you use an American Express credit card, your credit card usage is stored in that company's database. Databases group data so they can be centrally managed and easily accessible to people in various business units. Instead of a proliferation of separate and distinct data files, the database approach allows central databases to be the sole source of data. For example, your name, address, and social security number may be stored in a central location, but that information can be accessed by authorized people in the bursar's office, the office of the registrar, and the library.

> **Database**
> A shared collection of logically related data designed to meet the information needs of multiple users in an organization.

When you use the data-oriented approach to develop a system, databases are designed around subjects, such as customers, suppliers, and parts. Designing databases around subjects lets you use and revise databases for many different independent applications. This focus on subjects results in **application independence,** the separation of data and the definition of data from applications.

> **Application independence**
> The separation of data and the definition of data from the applications that use these data.

TABLE 1-1: Key Differences between the Process-Oriented and Data-Oriented Approaches to Systems Development

Characteristic	Process Orientation	Data Orientation
System focus	What the system is supposed to do and when	Data the system needs to operate
Design stability	Limited, because business processes and the applications that support them change constantly	More enduring, because the data needs of an organization do not change rapidly
Data organization	Data files designed for each individual application	Data files designed for the enterprise
State of the data	Much uncontrolled duplication	Limited, controlled duplication

Application independence means that data and applications are separate. For the data-oriented approach to be effective, organizations that store and manage all of their organizational data centrally must design new applications to work with existing databases. Organizations that currently store and manage their organizational data centrally must design databases that will support both current and future applications.

Your Role in Systems Development

Although many people in organizations are involved in systems analysis and design, the systems analyst has the primary responsibility. A career as a systems analyst will allow you to have a significant impact on how your organization operates. This is a fast-growing and rewarding position in both large and small companies. In 2002, a report released by the Information Technology Association of America predicted there would be 578,711 unfilled jobs in the information technology field by the end of the year. Information technology workers remain in demand.

The primary role of a systems analyst is to study the problems and needs of an organization in order to determine how people, methods, and information technology can best be combined to bring about improvements in the organization. A systems analyst helps system users and other business managers define their requirements for new or enhanced information services.

Let's consider two examples of the types of organizational problems you could face as a systems analyst. First, you work in the information systems department of a major magazine company. The company is having problems keeping an updated and accurate list of subscribers, and some customers are getting two magazines instead of one. The company will lose money and subscribers if this keeps up. To create a more efficient tracking system, the users of the current computer system as well as financial managers submit their problem to you and your colleagues in the information systems department. Second, you work in the information systems department at a university, where you are called upon to address an organizational problem such as the mailing of student grades to the wrong addresses.

When developing information systems to deal with problems such as these, an organization and its systems analysts have several options: They can use an in-house staff to develop the system, they can buy the system off-the-shelf, they can implement an enterprise-wide system from a company like SAP, they can hire a company to develop and run the software on its own computers, or they can go to a consulting company, such as Andersen Consulting or EDS, to have the system developed for them.

In an organization with its own in-house staff, several types of jobs are involved. In medium to large organizations, such as Procter & Gamble or Caterpillar, there is usually a separate Information Systems (IS) department. The IS department may be an independent department, reporting to the organization's top manager, or it may be part of another functional department, such as Finance. There may even be IS departments in several major business units. In any of these cases, the manager of an IS department is involved in systems development. If the department is large enough, there is a separate division for systems development, which would be home base for systems analysts, and another division for programming, where programmers would be based. Figure 1-10 highlights an organizational chart for a typical information systems department. Systems are designed to help people in functional departments do their jobs. These people are called **end users.** The end users often request new or modified software applications, test and approve applications, and may serve on project teams as business experts.

End users

Non-information-systems professionals in an organization. End users often request new or modified software applications, test and approve applications, and may serve on project teams as business experts.

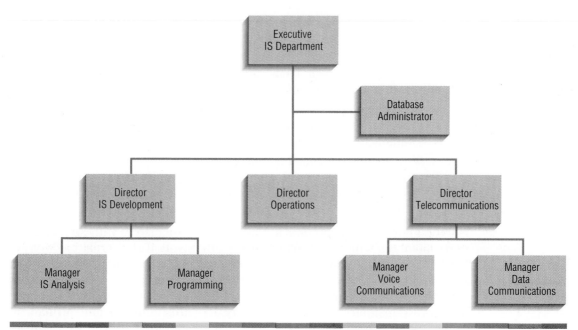

FIGURE 1-10
AN ORGANIZATIONAL CHART FOR A TYPICAL INFORMATION SYSTEMS DEPARTMENT

Following the data-oriented model, analysts are assigned and may report to functional departments and are able to learn more about the department they support. This approach results in better systems, because the analyst becomes an expert in both systems development and the business area.

Regardless of how an organization structures its information systems department, systems development is a team effort. Systems analysts work together in a team, usually organized on a project basis. A good team has certain characteristics, as also described in Table 1-2:

- ◉ A diverse team has representation from all the different groups interested in a system. For example, the team may consist of people from the database administration area and the group responsible for training as well as the marketing department (future system users). The representation of these groups on the team increases the likelihood of acceptance of the changes a new system will cause.

- ◉ Diversity exposes team members to new and different ideas, ideas they might never think of if all team members were from the same background, with the same skills and

TABLE 1-2:   Characteristics of Successful Teams

- • Diversity in backgrounds, skills, and goals
- • Tolerance of diversity, uncertainty, ambiguity
- • Clear and complete communication
- • Trust
- • Mutual respect and putting one's own views second to the team
- • Reward structure that promotes shared responsibility and accountability

goals. Someone in the data administration area would have a different view than someone in the marketing department. These different views could add new information.

- New and different ideas can help a team generate better solutions to its problems.
- Team members must be open to new ideas without being overly critical and without dismissing new ideas out of hand simply because they are new.
- Team members must be able to deal with ambiguous information as well as with complexity. They must learn to play a role on a team (and different roles on different teams) so that the talents of all team members can best be utilized.

A good team must communicate clearly and completely with its members. Team members will communicate more effectively if they trust each other. Trust, in turn, is built on mutual respect and an ability to place one's own goals and views secondary to the goals and views of the group. To help a team work well together, management needs to develop a reward structure that promotes shared responsibility and accountability. Rewards could include a financial bonus or special company acknowledgment. In addition to rewards for individual efforts, team members must be rewarded by IS managers for their work as members of an effective work unit.

Team success depends not only on how a team is assembled or the efforts of the group but also on the management of the team. In addition to a reward system, effective project management is another key element of successful teams. Project management includes:

- Devising a feasible and realistic work plan and schedule
- Monitoring progress against this schedule
- Coordinating the project with executives who sponsor the system
- Allocating resources to the project
- Sometimes even deciding whether and when a project should be terminated before completing the system

In general, a systems development team includes IS managers, systems analysts, programmers, end users, and business managers. The role you are preparing for, however, is that of systems analyst.

Systems Analysts in Systems Development

Systems analysts are key to the systems development process. To succeed as a systems analyst, you will need to develop four types of skills: analytical, technical, managerial, and interpersonal. Analytical skills enable you to understand the organization and its functions, to identify opportunities and problems, and to analyze and solve problems. One of the most important analytical skills you can develop is systems thinking, or the ability to see organizations and information systems as systems. Systems thinking provides a framework from which to see the important relationships among information systems, the organizations they exist in, and the environment in which the organizations themselves exist. Technical skills help you understand the potential and the limitations of information technology. As an analyst, you must be able to envision an information system that will help users solve problems and that will guide the system's design and development. You must also be able to work with programming languages such as C++ and Java, various operating systems such as Windows and Linux, and computer hardware platforms such as IBM and Mac. Management skills help you manage projects, resources, risk, and change.

Simon & Taylor, Inc., a candle manufacturer, has an immediate opening for a systems analyst in its Vermont-based office.

The ideal candidate will have:

1. A bachelor's degree in computer science and/or business administration

2. Two to three years, UNIX/RDBMS programming experience

3. Experience with the HP/UX Operating System, Informix, or VisualBasic

4. Familiarity with distribution and manufacturing concepts (allocation, replenishment, shop floor control, and production scheduling)

5. Working knowledge of project management and all phases of software development life cycle

6. Strong analytical and organization skills

We offer a competitive salary, a signing bonus, relocation assistance, and the challenges of working in a state-of-the-art IT environment.

E-mail your resume to www.human_resources@simontaylor.com with salary requirement.

Simon & Taylor, Inc., is an Equal Opportunity Employer

FIGURE 1-11
A JOB ADVERTISEMENT FOR A SYSTEMS ANALYST

Interpersonal skills help you work with end users as well as with other analysts and programmers. As a systems analyst, you will play a major role as a liaison among users, programmers, and other systems professionals. Effective written and oral communication, including competence in leading meetings, interviewing end users, and listening, is a key skill that analysts must master. Effective analysts successfully combine these four types of skills, as Figure 1-11, a typical advertisement for a systems analyst position, illustrates.

Types of Information Systems and Systems Development

Given the broad range of people and interests represented in organizations, it could take several different types of information systems to satisfy all of an organization's information system needs. Until now we have talked about information systems in generic terms, but there are actually several different types or classes of information systems. These classes are distinguished from each other on the basis of what the system does or by the technology used to construct the system. As a systems analyst, part of your job will be to determine which kind of system will best address the organizational problem or opportunity on which you are focusing. In addition, different classes of systems may require different methodologies, techniques, and tools for development.

As a systems analyst working as part of a team, you will work with at least four classes of information systems:

- Transaction processing systems
- Management information systems
- Decision support systems (for individuals, groups, and executives)
- Expert systems

These system types are represented graphically in Figure 1-12. Although each of these major system types is explained in more detail in the following sections, Figure 1-12 shows some contrasts between them. In the diagram for transaction processing systems, you can see that the major focus is capturing

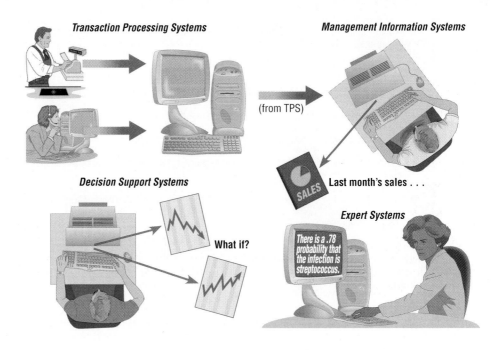

transaction data, which is then sent on to a computerized database of all transactions. The icons of the man with the cash register and the woman taking orders represent the capture of transaction data. The arrows from the icons to the computer represent moving the data to a database for storage. The picture in Figure 1-12 that illustrates management information systems shows a manager using transaction data to make a report about last month's sales. Management information systems are designed to process transaction data into standard reports. The next picture in Figure 1-12 shows a decision support system. Decision support systems help managers make decisions by analyzing data in different ways. Managers can make changes to their data, like changing interest rates, and see how those changes affect the parts of the business they manage. In the picture in Figure 1-12, the manager tries to determine what it takes to turn a downward trend into an upward trend. Finally, Figure 1-12 also shows an expert system. An expert system uses knowledge gathered from experts to make recommendations to managers. Some of the first such systems were designed to help doctors diagnose diseases, which is why the expert system depicted in Figure 1-12 shows a medical diagnosis on its screen. The system has determined, from the information provided by its operator, that the best diagnosis for the patient in question is streptococcus, the common bacteria that cause strep throat. The following sections briefly highlight how systems analysis and design methods differ across the four major types of systems.

Transaction Processing Systems

A *transaction processing system (TPS)* automates the handling of data about business activities or transactions. For example, a bank's TPS would capture information about withdrawals from and deposits to customer accounts. Data about each transaction are captured, transactions are verified and accepted or rejected, and validated transactions are stored. Reports may be produced immediately to provide summaries of transactions, and transactions may be moved from process to process in order to handle all aspects of the business activity.

The analysis and design of a TPS requires you to focus on the firm's current procedures for processing transactions. How does the organization track, capture, process, and output data? The goal of TPS development is to improve transaction processing by speeding it up, using fewer people, improving efficiency and accuracy, integrating it with other organizational information systems, or providing information not previously available.

Management Information Systems

A *management information system (MIS)* is a computer-based system that takes the raw data available through a TPS and converts them into a meaningful aggregated form. For example, whereas a transaction processing system keeps track of sales, a management information system can pinpoint which items are selling slowly and which are selling quickly. The MIS system can therefore direct the manufacturing department on what to produce and when. Developing an MIS calls for a good understanding of what kind of information managers require and how managers use information in their jobs. Sometimes managers themselves may not know precisely what they need or how they will use information. Thus, the analyst must also develop a good understanding of the business and the transaction processing systems that provide data for an MIS.

Management information systems often require data from several transaction processing systems (e.g., customer order processing, raw material purchasing, and employee timekeeping). Development of an MIS can, therefore, benefit from a data orientation, in which data are considered an organization resource separate from the TPS in which they are captured. Because it is important to be able to draw on data from various subject areas, developing a comprehensive and accurate model of data is essential in building an MIS.

Decision Support Systems

A *decision support system (DSS)* is designed to help decision makers with decisions. Whereas an MIS produces a report, a DSS provides an interactive environment in which decision makers can quickly manipulate data and models of business operations. A DSS has three parts. The first part is composed of a database (which may be extracted from a TPS or MIS). The second part consists of mathematical or graphical models of business processes. The third part is made up of a user interface (or dialogue module) that provides a way for the decision maker to communicate with the DSS. A DSS may use both historical data as well as judgments (or "what if" analysis) about alternative histories or possible futures. An *executive information system (EIS)* is a DSS that allows senior management to explore data starting at a high level of aggregation and selectively drill down into specific areas where more detailed information is required. A DSS is characterized by less structured and predictable use. DSS software supports certain decision-making activities (from problem finding to choosing a course of action).

The systems analysis and design for a DSS often concentrates on the three main DSS components: database, model base, and user dialogue. As with an MIS, a data orientation is most often used for understanding user requirements. The systems analysis and design project will carefully document the mathematical rules that define interrelationships among different data. These relationships are used to predict future data or to find the best solutions to decision problems. Decision logic must be carefully understood and documented. Also, because a decision maker typically interacts with a DSS, the design of easy-to-use yet thorough user dialogues and screens is important.

Expert Systems

An *expert system (ES)* is different from any of the other classes of systems we have discussed so far. The ES replicates the decision-making process rather than manipulating information. If-then-else rules or other knowledge representation forms describe the way an expert would approach situations in a specific domain of problems. Typically, users communicate with an ES through an interactive dialogue. The ES asks questions (which an expert would ask) and the end user supplies the answers. The answers are then used to determine which rules apply, and the ES provides a recommendation based on the rules.

The focus on developing an ES is acquiring the knowledge of the expert in the particular problem domain. Knowledge engineers perform knowledge acquisition; they are similar to systems analysts but are trained to use different techniques, as determining knowledge is considered more difficult than determining data.

Information Systems: An Overview

Many information systems you build or maintain will contain aspects of each of the four major types of information systems. As a systems analyst, you will likely use specific methodologies, techniques, and tools associated with each of the four information system types. Table 1-3 summarizes the general characteristics and development methods for each type.

We have concentrated on where an information system is developed and by whom. We have also seen the four types of information systems that exist in organizations. Now, we can turn to the process of developing information systems—the systems development life cycle.

TABLE 1-3: Systems Development for Different IS Types

IS Type	IS Characteristics	Systems Development Methods
Transaction processing system	High-volume, data capture focus; goals are efficiency of data movement and processing and interfacing different TPSs	Process orientation; concern with capturing, validating, and storing data and with moving data between each required step
Management information system	Draws on diverse yet predictable data resources to aggregate and summarize data; may involve forecasting future data from historical trends and business knowledge	Data orientation; concern with understanding relationships between data so data can be accessed and summarized in a variety of ways; builds a model of data that supports a variety of uses
Decision support system	Provides guidance in identifying problems, finding and evaluating alternative solutions, and selecting or comparing alternatives; potentially involves groups of decision makers; often involves semistructured problems and the need to access data at different levels of detail	Data and decision logic orientations; design of user dialogue; group communication may also be key and access to unpredictable data may be necessary; nature of systems requires iterative development and almost constant updating
Expert system	Provides expert advice by asking users a sequence of questions dependent on prior answers that lead to a conclusion or recommendation	A specialized decision logic orientation in which knowledge is elicited from experts and described by rules or other forms

Developing Information Systems and the Systems Development Life Cycle

Organizations use a standard set of steps, called a **systems development methodology,** to develop and support their information systems. Like many processes, the development of information systems often follows a life cycle. For example, a commercial product, such as a Nike sneaker or Honda car, follows a life cycle: It is created, tested, and introduced to the market. Its sales increase, peak, and decline. Finally, the product is removed from the market and replaced by something else. The **systems development life cycle (SDLC)** is a common methodology for systems development in many organizations. It marks the phases or steps of information systems development: Someone has an idea for an information system and what it should do. The organization that will use the system decides to devote the necessary resources to acquiring it. A careful study is done of how the organization currently handles the work the system will support. Professionals develop a strategy for designing the new system, which is then either built or purchased. Once complete, the system is installed in the organization, and after proper training, the users begin to incorporate the new system into their daily work. Every organization uses a slightly different life cycle model to model these steps, with anywhere from three to almost twenty identifiable phases. In this book, we highlight four SDLC steps: (1) planning and selection, (2) analysis, (3) design, and (4) implementation and operation.

Although any life cycle appears at first glance to be a sequentially ordered set of phases, it actually is not. Figure 1-13 highlights the four steps in the SDLC: (1) planning and selection, (2) analysis, (3) design, and (4) implementation and operation. The specific steps and their sequence are meant to be adapted as required for a project. For example, in any given SDLC phase, the project can return to an earlier phase if necessary. Similarly, if a commercial product does not perform well just after its introduction, it may be temporarily removed from the market and improved before being reintroduced. In the systems development life cycle, it is also possible to complete some activities in one phase in parallel with some activities of another phase. Sometimes the life cycle is iterative; that is, phases are repeated as required until an acceptable system is found. Some systems analysts consider the life cycle to be a spiral, in which we

Systems development methodology
A standard process followed in an organization to conduct all the steps necessary to analyze, design, implement, and maintain information systems.

Systems development life cycle (SDLC)
The series of steps used to mark the phases of development for an information system.

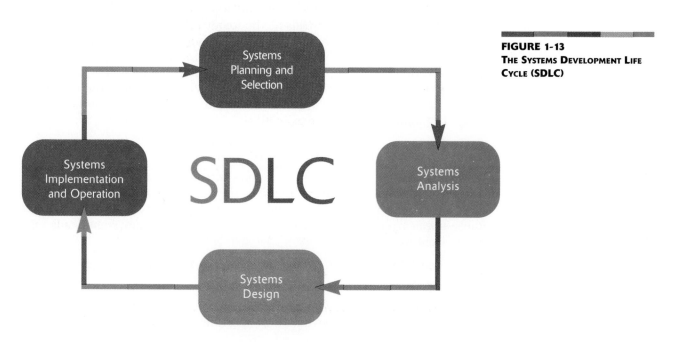

FIGURE 1-13
THE SYSTEMS DEVELOPMENT LIFE CYCLE (SDLC)

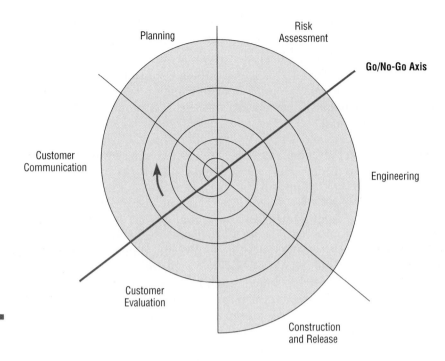

FIGURE 1-14
EVOLUTIONARY MODEL SDLC

constantly cycle through the phases at different levels of detail, as illustrated in Figure 1-14. The circular nature of the life cycle diagram in Figure 1-14 illustrates how the end of the useful life of one system leads to the beginning of another project that will replace the existing system altogether. However conceived, the systems development life cycle used in an organization is an orderly set of activities conducted and planned for each development project. The skills required of a systems analyst apply to all life cycle models.

Every medium to large corporation, such as Wal-Mart, and every custom software producer, such as SAP, will have its own specific, detailed life cycle or systems development methodology in place. Even if a particular methodology does not look like a cycle, many of the SDLC steps are performed and SDLC techniques and tools are used.

This book follows a generic SDLC model, as illustrated in Figure 1-13. We use this SDLC as an example methodology and a way to think about systems analysis and design. You can apply this methodology to almost any life cycle. As we describe this SDLC throughout the book, it becomes clear that each phase has specific outcomes and deliverables that feed important information to other phases. At the end of each phase (and sometimes within phases for intermediate steps), a systems development project reaches a milestone. Then, as deliverables are produced, they are often reviewed by parties outside the project team, including managers and executives.

Phase 1: Systems Planning and Selection

Systems planning and selection

The first phase of the SDLC in which an organization's total information system needs are analyzed and arranged, and in which a potential information systems project is identified and an argument for continuing or not continuing with the project is presented.

The first phase in the SDLC, **systems planning and selection,** has two primary activities. First, someone identifies the need for a new or enhanced system. Information needs of the organization are examined and projects to meet these needs are identified. The organization's information system needs may result from:

- ❿ Requests to deal with problems in current procedures
- ❿ The desire to perform additional tasks
- ❿ The realization that information technology could be used to capitalize on an existing opportunity

The systems analyst prioritizes and translates the needs into a written plan for the IS department, including a schedule for developing new major systems. Requests for new systems spring from users who need new or enhanced systems. During the systems planning and selection phase, an organization determines whether or not resources should be devoted to the development or enhancement of each information system under consideration. A *feasibility study* is conducted before the second phase of the SDLC to determine the economic and organizational impact of the system.

The second task in the systems planning and selection phase is to investigate the system and determine the proposed system's scope. The team of systems analysts then produces a specific plan for the proposed project for the team to follow. This baseline project plan customizes the standardized SDLC and specifies the time and resources needed for its execution. The formal definition of a project is based on the likelihood that the organization's IS department is able to develop a system that will solve the problem or exploit the opportunity and determine whether the costs of developing the system outweigh the possible benefits. The final presentation to the organization's management of the plan for proceeding with the subsequent project phases is usually made by the project leader and other team members.

Phase 2: Systems Analysis

The second phase of the systems development life cycle is **systems analysis.** During this phase, the analyst thoroughly studies the organization's current procedures and the information systems used to perform tasks such as general ledger, shipping, order entry, machine scheduling, and payroll. Analysis has several subphases. The first subphase involves determining the requirements of the system. In this subphase, you and other analysts work with users to determine what the users want from a proposed system. This subphase involves a careful study of any current systems, manual and computerized, that might be replaced or enhanced as part of this project. Next, you study the requirements and structure them according to their interrelationships, eliminating any redundancies. Third, you generate alternative initial designs to match the requirements. Then you compare these alternatives to determine which best meets the requirements within the cost, labor, and technical levels the organization is willing to commit to the development process. The output of the analysis phase is a description of the alternative solution recommended by the analysis team. Once the recommendation is accepted by the organization, you can make plans to acquire any hardware and system software necessary to build or operate the system as proposed.

Systems analysis
Phase of the SDLC in which the current system is studied and alternative replacement systems are proposed.

Phase 3: Systems Design

The third phase of the SDLC is called **systems design.** During systems design, analysts convert the description of the recommended alternative solution into logical and then physical system specifications. You must design all aspects of the system from input and output screens to reports, databases, and computer processes.

Logical design is not tied to any specific hardware and systems software platform. Theoretically, the system you design could be implemented on any hardware and systems software. Logical design concentrates on the business aspects of the system; that is, how the system will impact the functional units within the organization. Figure 1-15 shows both the logical design for a product and its physical design, side by side for comparison. You can see from the comparison that many specific decisions had to be made to move from the logical model to the physical product. The situation is very similar in information systems design.

In physical design, you turn the logical design into physical, or technical, specifications. For example, you must convert diagrams that map the origin, flow,

Systems design
Phase of the SDLC in which the system chosen for development in systems analysis is first described independent of any computer platform (logical design) and is then transformed into technology-specific details (physical design) from which all programming and system construction can be accomplished.

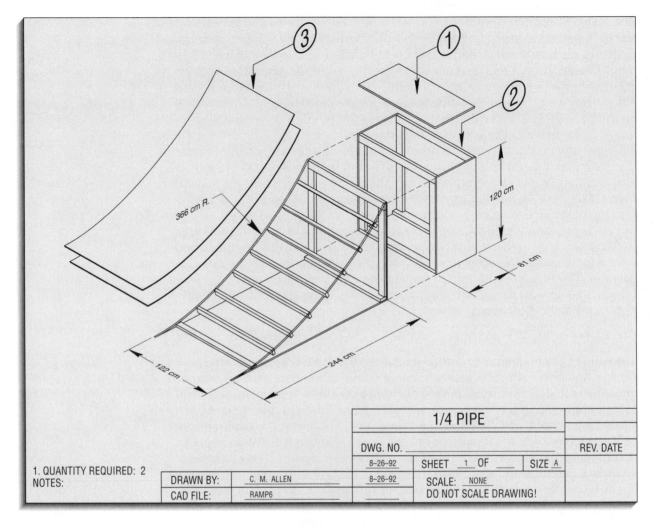

FIGURE 1-15

THE DIFFERENCE BETWEEN LOGICAL DESIGN AND PHYSICAL DESIGN (A) A SKATEBOARD RAMP BLUEPRINT (LOGICAL DESIGN) (B) A SKATEBOARD RAMP (PHYSICAL DESIGN)

Sources: www.tumyeto.com/tydu/skatebrd/organizations/plans/14pipe.jpg; www.tumyeto.com/tydu/skatebrd/organizations/iuscblue.html; accessed September 16, 1999. Reprinted by permission of the International Association of Skateboard Companies.

and processing of data in a system into a structured systems design that can then be broken down into smaller and smaller units for conversion to instructions written in a programming language. You design the various parts of the system to perform the physical operations necessary to facilitate data capture, processing, and information output. During physical design, the analyst team decides which programming languages the computer instructions will be written in, which database systems and file structures will be used for the data, and which hardware platform, operating system, and network environment the system will run under. These decisions finalize the hardware and software plans initiated at the end of the analysis phase. Now you can acquire any new technology not already present in the organization. The final product of the design phase is the physical system specifications, presented in a form, such as a diagram or written report, ready to be turned over to programmers and other system builders for construction.

Phase 4: Systems Implementation and Operation

The final phase of the SDLC is a two-step process: **systems implementation and operation.** During systems implementation and operation, you turn system specifications into a working system that is tested and then put into use. Implementation includes coding, testing, and installation. During coding, programmers write the programs that make up the system. During testing, programmers and analysts test individual programs and the entire system in order to find and correct errors. During installation, the new system becomes a part of the daily activities of the organization. Application software is installed, or loaded, on existing or new hardware; then users are introduced to the new system and trained. Begin planning for both testing and installation as early as the project planning and selection phase, because they both require extensive analysis in order to develop exactly the right approach.

Systems implementation activities also include initial user support such as the finalization of documentation, training programs, and ongoing user assistance. Note that documentation and training programs are finalized during implementation; documentation is produced throughout the life cycle, and training (and education) occurs from the inception of a project. Systems implementation can continue for as long as the system exists because ongoing user support is also part of implementation. Despite the best efforts of analysts, managers, and programmers, however, installation is not always a simple process. Many well-designed systems have failed because the installation process was faulty. Note that even a well-designed system can fail if implementation is not well managed. Because the management of systems implementation is usually done by the project team, we stress implementation issues throughout this book.

The second part of the fourth phase of the SDLC is operation. While a system is operating in an organization, users sometimes find problems with how it works and often think of improvements. During operation, programmers make the changes that users ask for and modify the system to reflect changing business conditions. These changes are necessary to keep the system running and useful. The amount of time and effort devoted to system enhancements during operation depends a great deal on the performance of the previous phases of the life cycle. There inevitably comes a time, however, when an information system is no longer performing as desired, when the costs of keeping a system running become prohibitive, or when an organization's needs have changed substantially. Such problems indicate that it is time to begin designing the system's replacement, thereby completing the loop and starting the life cycle over again.

The SDLC is a highly linked set of phases whose products feed the activities in subsequent phases. Table 1-4 summarizes the outputs or products of each phase based on the above descriptions. The subsequent chapters on the SDLC phases discuss the products of each phase and how they are developed.

Throughout the systems development life cycle, the systems development project itself needs to be carefully planned and managed. The larger the systems project, the greater the need for project management. Several project management techniques have been developed in the last quarter century and many have been improved through automation. Chapter 2 contains a more detailed treatment of project planning and management techniques.

Systems implementation and operation
Final phase of the SDLC, in which the information system is coded, tested, and installed in the organization, and in which the information system is systematically repaired and improved.

TABLE 1-4: Products of the SDLC Phases

Phase	Products, Outputs, or Deliverables
Systems planning and selection	Priorities for systems and projects
	Architecture for data, networks, hardware, and IS management
	Detailed work plan for selected project
	Specification of system scope
	System justification or business case
Systems analysis	Description of current system
	General recommendation on how to fix, enhance, or replace current system
	Explanation of alternative systems and justification for chosen alternative
Systems design	Detailed specifications of all system elements
	Acquisition plan for new technology
Systems implementation and operation	Code
	Documentation
	Training procedures and support capabilities
	New versions or releases of software with associated updates to documentation, training, and support

Approaches to Development

Prototyping, rapid application development (RAD), joint application design (JAD), and participatory design (PD) are four approaches that streamline and improve the systems analysis and design process.

Prototyping

Prototyping
Building a scaled-down version of the desired information system.

Designing and building a scaled-down but working version of a desired system is known as **prototyping.** A prototype can be developed with a computer-aided software engineering (CASE) tool, a software product that automates the systems development life cycle steps. CASE tools make prototyping easier and more creative by supporting the design of screens and reports and other parts of a system interface. CASE tools also support many of the diagramming techniques you will learn, such as data flow diagrams and entity-relationship diagrams.

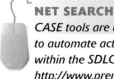

NET SEARCH
CASE tools are used to automate activities within the SDLC. Visit http://www.prenhall.com/valacich to complete an exercise related to this topic.

Figure 1-16 illustrates prototyping. The analyst works with users to determine the initial or basic requirements for the system. The analyst then quickly builds a prototype. When the prototype is completed, the users work with it and tell the analyst what they like and do not like about it. The analyst uses this feedback to improve the prototype and takes the new version back to the users. This iterative process continues until the users are relatively satisfied with what they have seen. The key advantages of the prototyping technique are (1) it involves the user in analysis and design, and (2) it captures requirements in concrete, rather than verbal or abstract, form. In addition to being used as a stand-alone, prototyping may also be used to augment the SDLC. For example, a prototype of the final system may be developed early in analysis to help the analysts identify what users want. Then the final system is developed based on the specifications of the prototype. We discuss prototyping in greater detail in Chapter 4 and use various prototyping tools in Chapter 8 to illustrate the design of system outputs.

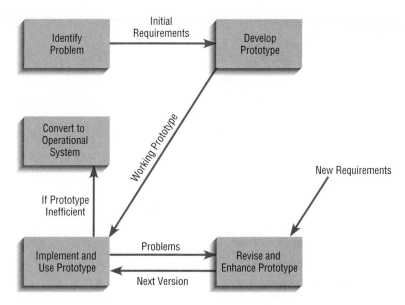

FIGURE 1-16
THE PROTOTYPING METHOD
Source: Adapted from J. D. Naumann and A. M. Jenkins, "Prototyping: The New Paradigm for Systems Development," *MIS Quarterly* 6, no. 3 (1982): 29–44.

Prototyping is a key tool that supports *rapid application development (RAD)*. The fundamental principle of any RAD methodology is to delay producing detailed system design documents until after user requirements are clear. The prototype serves as the working description of needs. RAD involves gaining user acceptance of the interface and developing key system capabilities as quickly as possible. RAD is widely used by consulting firms. It is also used as an in-house methodology by firms like Hughes Electronics Corporation. RAD sacrifices computer efficiency for gains in human efficiency in rapidly building and rebuilding working systems. On the other hand, RAD methodologies can overlook important systems development principles, which may result in problems with systems developed this way. Appendix B addresses RAD in detail.

Joint Application Design

In the late 1970s, systems development personnel at IBM developed a new process for collecting information system requirements and reviewing system designs. The process is called **Joint Application Design (JAD).** The idea behind JAD is to structure the requirements determination phase of analysis and the reviews that occur as part of design. Users, managers, and systems developers are brought together for a series of intensive structured meetings run by a JAD session leader. By gathering the people directly affected by an IS in one room at the same time to work together to agree on system requirements and design details, time and organizational resources are better managed. Group members are more likely to develop a shared understanding of what the IS is supposed to do. JAD has become common in certain industries such as insurance and in specific companies such as CIGNA. We discuss JAD in more detail in Chapter 4.

Joint Application Design (JAD)
A structured process in which users, managers, and analysts work together for several days in a series of intensive meetings to specify or review system requirements.

Participatory Design

Developed in northern Europe, **Participatory Design (PD)** represents a viable alternative approach to the SDLC. One of the best-known companies that has used this approach is StatOil, the Norwegian state-owned oil company. PD emphasizes the role of the user much more than traditional North American techniques such as structured analysis and structured design do. In some cases, PD may involve the entire user community in the development process. Each user has an equal voice in determining system requirements and in approving system design. In other cases, an elected group of users controls the

Participatory Design (PD)
A systems development approach that originated in northern Europe in which users and the improvement of their work lives are the central focus.

process. These users represent the larger community, much as a legislature represents the needs and wants of the electorate. Typically, under PD, systems analysts work for the users. The organization's management and outside consultants provide advice rather than control. PD is partly a result of the roles of labor and management in the northern European workplace where labor is more organized, carries more clout, and is more intimately involved with technological changes than is true in North America.

Key Points Review

1. **Define information systems analysis and design.**

   Systems analysis and design is the complex organizational process whereby computer-based information systems are developed and operated.

2. **Discuss the modern approach to systems analysis and design that combines both process and data views of systems.**

   Process orientation focuses on what the system is supposed to do whereas data orientation focuses on the data the system needs to operate. Process orientation provides a less stable design than does data orientation, because business processes change faster than do the data an organization uses. With process orientation, data files are designed for specific applications, whereas data files are designed for the whole enterprise with data orientation. Process orientation leads to much uncontrolled data redundancy whereas data redundancy is controlled under data orientation. Data orientation and application independence (the separation of data from the computer applications that use the data) frame the way you learn about systems analysis and design in this book.

3. **Describe the organizational role of the systems analyst in information systems development.**

   Systems analysts play a key organizational role in systems development. They act as liaisons between business users on one hand and technical personnel on the other. Analysts need to develop four sets of skills in order to succeed: analytical, technical, managerial, and interpersonal.

4. **Describe four types of information systems: transaction processing systems, management information systems, decision support systems, and expert systems.**

   There are many different kinds of information systems used in organizations. These include transaction processing systems, management information systems, decision support systems, and expert systems. Development techniques vary with system type.

5. **Describe the information systems development life cycle (SDLC).**

   The systems development life cycle used in this book has four major phases: (1) systems planning and selection, (2) systems analysis, (3) systems design, and (4) systems implementation and operation. In the first phase, which is planning and selection, analysts make detailed road maps of the system development project. In analysis, analysts work to solve the business problem being studied. In design, the solution to the problem is built. Finally, in the last phase, the system is given to users and kept running.

6. **List alternatives to the systems development life cycle.**

   The three alternative frameworks mentioned in this chapter are prototyping, Joint Application Design (JAD), and Participatory Design (PD). Using prototyping, analysts build a working model of the system. In JAD, analysts and users meet to solve problems and design systems. In PD, the emphasis is on the user community.

7. **Explain briefly the role of computer-aided software engineering (CASE) tools in systems development.**

   CASE tools represent the use of information technology to assist in the systems development process. They include diagramming tools, screen and report design tools, and other special-purpose tools. CASE tools help programmers and analysts do their jobs efficiently and effectively by automating routine tasks.

Key Terms Checkpoint

Here are the key terms from the chapter. The page where each term is first explained is in parentheses after the term.

1. Application independence (p. 13)
2. Application software (p. 4)
3. Boundary (p. 7)
4. Cohesion (p. 9)
5. Component (p. 7)
6. Constraints (p. 7)
7. Coupling (p. 9)
8. Data (p. 10)
9. Database (p. 13)
10. Data flow (p. 11)
11. Data-oriented approach (p. 12)
12. Decomposition (p. 7)
13. End users (p. 14)
14. Environment (p. 7)
15. Information (p. 10)
16. Information systems analysis and design (p. 4)
17. Interfaces (p. 7)
18. Interrelated components (p. 7)
19. Joint Application Design (JAD) (p. 27)
20. Modularity (p. 9)
21. Participatory Design (PD) (p. 27)
22. Processing logic (p. 11)
23. Process-oriented approach (p. 12)
24. Prototyping (p. 26)
25. Purpose (p. 7)
26. System (p. 6)
27. Systems analysis (p. 23)
28. Systems analyst (p. 3)
29. Systems design (p. 23)
30. Systems development life cycle (SDLC) (p. 21)
31. Systems development methodology (p. 21)
32. Systems implementation and operation (p. 25)
33. Systems planning and selection (p. 22)

Match each of the key terms above with the definition that best fits it.

_____ 1. The first phase of the SDLC, in which an organization's total information system needs are analyzed and arranged, and in which a potential information systems project is identified and an argument for continuing or not continuing with the project is presented.

_____ 2. The complex organizational process whereby computer-based information systems are developed and maintained.

_____ 3. A systems development approach that originated in northern Europe in which users and the improvement of their work lives are the central focus.

_____ 4. Computer software designed to support organizational functions or processes.

_____ 5. The organizational role most responsible for the analysis and design of information systems.

_____ 6. A structured process in which users, managers, and analysts work together for several days in a series of intensive meetings to specify or review system requirements.

_____ 7. An iterative process of systems development in which requirements are converted to a working system that is continually revised through close work between an analyst and users.

_____ 8. An interrelated set of components, with an identifiable boundary, working together for some purpose.

_____ 9. An irreducible part or aggregation of parts that make up a system, also called a subsystem.

_____ 10. Dependence of one part of the system on one or more other system parts.

_____ 11. The line that marks the inside and outside of a system, and that sets off the system from its environment.

_____ 12. The overall goal or function of a system.

_____ 13. Phase of the SDLC in which the system chosen for development in systems analysis is first described independent of any computer platform and is then transformed into technology-specific details from which all programming and system construction can be accomplished.

_____ 14. Phase of the SDLC in which the current system is studied and alternative replacement systems are proposed.

_____ 15. Everything external to a system that interacts with the system.

_____ 16. Point of contact where a system meets its environment or where subsystems meet each other.

_____ 17. A limit to what a system can accomplish.

_____ 18. The final phase of the SDLC in which the information system is coded, tested, and installed in the organization, and in which the information system is systematically repaired and improved.

_____ 19. A standard process followed in an organization to conduct all the steps necessary to analyze, design, implement, and maintain information systems.

_____ 20. The traditional methodology used to develop, maintain, and replace information systems.

_____ 21. Non-information-systems professionals in an organization who specify the business requirements for and use software applications.

_____ 22. An iterative process of breaking the description of a system down into finer and finer detail, which creates a set of charts in which one process on a given chart is explained in greater detail on another chart.

_____ 23. The separation of data and the definition of data from the applications that use these data.

_____ 24. A shared collection of logically related data designed to meet the information needs of multiple users in an organization.

_____ 25. Dividing a system up into chunks or modules of a relatively uniform size.

_____ 26. The extent to which subsystems depend on each other.

_____ 27. The extent to which a system or subsystem performs a single function.

_____ 28. Raw facts about people, objects, and events in an organization.

_____ 29. Data in motion, moving from one place in a system to another.

_____ 30. The steps by which data are transformed or moved and a description of the events that trigger these steps.

_____ 31. An overall strategy for information systems development that focuses on how and when data are moved through and changed by an information system.

_____ 32. An overall strategy of information systems development that focuses on the ideal organization of data rather than on where and how they are used.

_____ 33. Data that have been processed and presented in a form suitable for human interpretation, often with the purpose of revealing trends or patterns.

Review Questions

1. What is information systems analysis and design?
2. What is systems thinking? How is it useful for thinking about computer-based information systems?
3. What is decomposition? coupling? cohesion?
4. In what way are organizations systems?
5. Explain the traditional application-based approach to systems development. How is this different from the data-based approach?
6. List the different classes of information systems described in this chapter. How do they differ from each other?
7. List and explain the different phases in the systems development life cycle.
8. What is prototyping?
9. What is JAD? What is Participatory Design?
10. Explain how systems analysis and design has changed from 1950 to 2000.
11. What are the characteristics of successful systems development teams?

Problems and Exercises

1. Why is it important to use systems analysis and design methodologies when building a system? Why not just build the system in whatever way seems to be "quick and easy"? What value is provided by using an "engineering" approach?
2. Describe your university or college as a system. What is the input? the output? the boundary? the components? their interrelationships? the constraints? the purpose? the interfaces? the environment? Draw a diagram of this system.
3. A car is a system with several subsystems, including the braking subsystem, the electrical subsystem, the engine, the fuel subsystem, the climate control subsystem, and the passenger subsystem. Draw a diagram of a car as a system and label all of its system characteristics.
4. Your personal computer is a system. Draw and label a personal computer as a system as you did for a car in Problem and Exercise 3.

5. Choose a business transaction you undertake regularly, such as using an ATM machine, buying groceries at the supermarket, or buying a ticket for a university's basketball game. For this transaction, define the data, draw the data flow diagram, and describe processing logic.
6. How is the Joint Application Design (JAD) approach different from the Participatory Design (PD) approach developed in northern Europe? (You may have to do some digging at the library to answer this question adequately.) What are the benefits in using approaches like this in building information systems? What are the barriers?
7. How would you organize a project team of students to work with a small business client? How would you organize a project team if you were working for a professional consulting organization? How might these two methods of organization differ? Why?
8. How might prototyping be used as part of the SDLC?
9. Describe the difference in the role of a systems analyst in the SDLC versus prototyping.
10. Contrast process-oriented and data-oriented approaches to systems analysis and design. Why does this book make the point that these are complementary, not competing, approaches to systems development?
11. Compare Figures 1-14 and 1-16. What similarities and differences do you see?

Discussion Questions

1. If someone at a party asked you what a systems analyst was and why anyone would want to be one, what would you say? Support your answer with evidence from this chapter.
2. Explain how a computer-based information system designed to process payroll is a specific example of a system. Be sure to account for all nine components of any system in your explanation.
3. How does the Internet, and more specifically the World Wide Web, fit into the picture of systems analysis and systems development drawn in this chapter?
4. What do you think systems analysis and design will look like in the next decade? As you saw earlier in the chapter, changes in systems development have been pretty dramatic in the past. A computer programmer suddenly transported from the 1950s to the 2000s would have trouble recognizing the computing environment that had evolved just 50 years later. What dramatic changes might occur in the next 10 years?

Case Problems

1. Pine Valley Furniture
   Alex Schuster began Pine Valley Furniture as a hobby. Initially, Alex would build custom furniture in his garage for friends and family. As word spread about his quality craftsmanship, he began taking orders. The hobby has since evolved into a medium-size business, employing over 50 workers.
   Over the years, increased demand has forced Alex to relocate several times, increase his sales force, expand his product line, and renovate Pine Valley Furniture's information systems. As the company began to grow, Alex organized the company into functional areas—manufacturing, sales, orders, accounting, and purchasing. Originally, manual information systems were used; however, as the business began to expand rapidly, a minicomputer was installed to automate applications.

   In the beginning, a process-oriented approach was utilized. Each separate application had its own data files. The applications automated the manual systems on which they were modeled. In an effort to improve its information systems, PVF has recently renovated its information systems, resulting in a company-wide database and applications that work with this database. Pine Valley Furniture's computer-based applications are primarily in the accounting and financial areas. All applications have been built in-house, and when necessary, new information systems staff is hired to support Pine Valley Furniture's expanding information systems.

   a. How did Pine Valley Furniture go about developing its information systems? Why do you think the company chose this option? What other options were available?

b. One option available to Pine Valley Furniture was an enterprise-wide system. What features does an enterprise-wide system, such as SAP, provide? What is the primary advantage of an enterprise-wide system?

c. Pine Valley Furniture will be hiring two systems analysts next month. Your task is to develop a job advertisement for these positions. Locate several Web sites and/or newspapers that have job advertisements for systems analysts. What skills are required?

d. What types of information systems are currently utilized at Pine Valley Furniture? Provide an example of each.

2. Hoosier Burger

As college students in the 1970s, Bob and Thelma Mellankamp often dreamed of starting their own business. While on their way to an economics class, Bob and Thelma drove by Myrtle's Family Restaurant and noticed a "for sale" sign in the window. Bob and Thelma quickly made arrangements to purchase the business, and Hoosier Burger Restaurant was born. The restaurant is moderately sized, consisting of a kitchen, dining room, counter, storage area, and office. Currently, all paperwork is done by hand. Thelma and Bob have discussed the benefits of purchasing a computer system; however, Bob wants to investigate alternatives and hire a consultant to help them.

Perishable food items, such as beef patties, buns, and vegetables are delivered daily to the restaurant. Other items, such as napkins, straws, and cups, are ordered and delivered as needed. Bob Mellankamp receives deliveries at the restaurant's back door and then updates a stock log form. The stock log form helps Bob track inventory items. The stock log form is updated when deliveries are received and also nightly after daily sales have been tallied.

Customers place their orders at the counter and are called when their orders are ready. The orders are written on an order ticket, totaled on the cash register, and then passed to the kitchen where the orders are prepared. The cash register is not capable of capturing point-of-sale information. Once an order is prepared and delivered, the order ticket is placed in the order ticket box. Bob reviews these order tickets nightly and makes adjustments to inventory.

In the past several months, Bob has noticed several problems with Hoosier Burger's current information systems, especially with the inventory control, customer ordering, and management reporting systems. Because the inventory control and customer ordering systems are paper-based, errors occur frequently, often impacting delivery orders received from suppliers and the taking of customer orders. Bob has often wanted to have electronic access to forecasting information, inventory usage, and basic sales information. This access is impossible because of the paper-based system.

a. Apply the SDLC approach to Hoosier Burger.

b. Using the Hoosier Burger scenario, identify an example of each system characteristic.

c. Decompose Hoosier Burger into its major subsystems.

d. Briefly summarize the approaches to systems development discussed in this chapter. Which approach do you feel should be used by Hoosier Burger?

3. Natural Best Health Food Stores

Natural Best Health Food Stores is a chain of health food stores serving Oklahoma, Arkansas, and Texas. Garrett Davis opened his first Natural Best Health Food Store in 1975 and has since opened 15 stores in three states. Initially, he sold only herbal supplements, gourmet coffees and teas, and household products. In 1990, he expanded his product line to include personal care, pet care, and grocery items.

In the past several months, many of Mr. Davis's customers have requested the ability to purchase prepackaged meals, such as chicken, turkey, fish, and vegetarian, and have these prepackaged meals automatically delivered to their homes weekly, biweekly, or monthly. Mr. Davis feels that this is a viable option, because Natural Best has an automatic delivery system in place for its existing product lines.

With the current system, a customer can subscribe to the Natural Best Delivery Service (NBDS), and have personal care, pet care, gourmet products, and grocery items delivered on a weekly, biweekly, or monthly basis. The entire subscription process takes approximately five minutes. The salesclerk obtains the customer's name, mailing address, credit card number, desired delivery items and quantity, delivery frequency, and phone number. After the customer's subscription has been processed, delivery usually begins within a week. As customer orders are placed, inventory is automatically updated.

The NBDS system is a client/server system. Each store is equipped with a client computer that accesses a centralized database housed on a central server. The server tracks inventory, customer activity, delivery schedules, and individual store sales. Each week the NBDS generates sales summary reports, low-in-stock reports, and delivery schedule reports for each store. The information contained on each of these individual reports is then consolidated into master sales summary, low-in-stock, and forecasting reports. Information contained on these reports facilitates restocking, product delivery, and forecasting decisions. Mr. Davis has an Excel worksheet that he uses to consolidate sales information from each store. He then uses this worksheet to make forecasting decisions for each store.

a. Identify the different types of information systems used at Natural Best Health Food Stores. Provide an example of each. Is an expert system currently used? If not, how could Natural Best benefit from the use of such a system?

b. Figure 1-4 identifies seven characteristics of a system. Using the Natural Best Health Food Stores scenario, provide an example of each system characteristic.

c. What type of computing environment does Natural Health Food Stores have?

d. Figure 1-8 illustrates the differences among data, data flow, and processing logic. Using this figure as a guide, provide a similar example for Natural Health Food Stores.

2

Managing the Information Systems Project

◗ Objectives

After studying this chapter, you should be able to:

◗ Describe the skills required to be an effective project manager.

◗ List and describe the skills and activities of a project manager during project initiation, project planning, project execution, and project closedown.

◗ Explain what is meant by critical path scheduling and describe the process of creating Gantt charts and Network diagrams charts.

◗ Explain how commercial project management software packages can be used to assist in representing and managing project schedules.

Chapter Preview . . .

In Chapter 1, we introduced the four phases of the systems development life cycle (SDLC) and explained how an information system project moves through those four phases. In this chapter we focus on the systems analyst's role as project manager of information systems projects. Throughout the SDLC, the project manager is responsible for initiating, planning, executing, and closing down the systems development project. Figure 2-1 illustrates these four functions.

We use two fictional companies in this book—Pine Valley Furniture and Hoosier Burger—to help illustrate key SDLC concepts. Icons appear in the margins to make references to these companies easy to spot while you read. The next section gives you background on Pine Valley Furniture, a manufacturing company. (We will introduce Hoosier Burger in Chapter 3.) Next we describe the project manager's role and the project management process. The subsequent section examines techniques for reporting project plans using Gantt charts and Network digrams. At the end of the chapter, we discuss commercially available project management software that a systems analyst can use in a wide variety of project management activities.

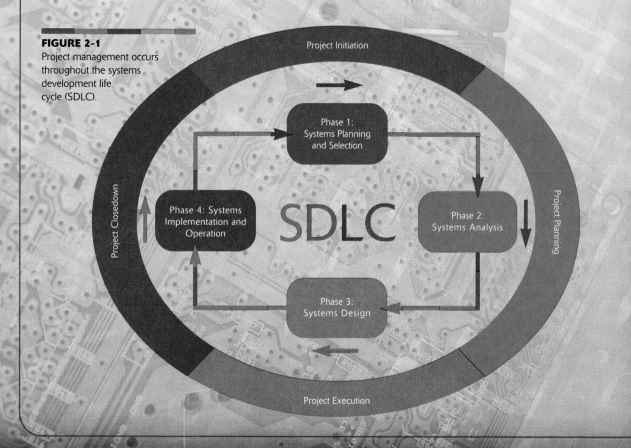

FIGURE 2-1
Project management occurs throughout the systems development life cycle (SDLC).

Pine Valley Furniture Company Background

Pine Valley Furniture (PVF) Company manufactures high-quality wood furniture and distributes it to retail stores within the United States. Its product lines include dinette sets, stereo cabinets, wall units, living room furniture, and bedroom furniture. In the early 1980s, PVF's founder, Alex Schuster, started to make and sell custom furniture in his garage. Alex managed invoices and kept track of customers by using file folders and a filing cabinet. By 1984, business expanded and Alex had to rent a warehouse and hire a part-time bookkeeper. PVF's product line had multiplied, sales volume had doubled, and staff had increased to 50 employees. By 1990, PVF moved into its third and present location. Due to the added complexity of the company's operations, Alex reorganized the company into the following functional areas:

- Manufacturing, which was further subdivided into three separate functions—Fabrication, Assembling, and Finishing
- Sales
- Orders
- Accounting
- Purchasing

Alex and the heads of the functional areas established manual information systems, such as accounting ledgers and file folders, which worked well for a time. Eventually, however, PVF selected and installed a minicomputer to automate invoicing, accounts receivable, and inventory control applications.

When the applications were first computerized, each separate application had its own individual data files tailored to the needs of each functional area. As is typical in such situations, the applications closely resembled the manual systems on which they were based. Three computer applications at PVF are depicted in Figure 2-2: order filling, invoicing, and payroll. In the late 1990s, PVF formed a task force to study the possibility of moving to a database approach. After a preliminary study, management decided to convert its information sys-

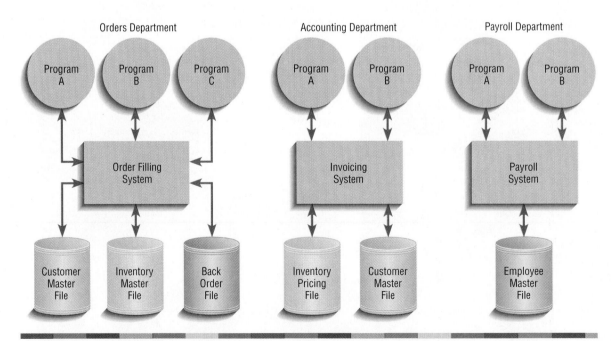

FIGURE 2-2
THREE COMPUTER APPLICATIONS AT PINE VALLEY FURNITURE: ORDER FILLING, INVOICING, AND PAYROLL
Source: Hoffer, Prescrott, and McFadden, 2002.

tems to such an approach. The company upgraded its minicomputer and implemented a database management system. By the time we catch up with PVF, it has successfully designed and populated a company-wide database and has converted its applications to work with the database. However, PVF is continuing to grow at a rapid rate, putting pressure on its current application systems.

The computer-based applications at PVF support its business processes. When customers order furniture, their orders must be processed appropriately: Furniture must be built and shipped to the right customer and the right invoice mailed to the right address. Employees have to be paid for their work. Given these tasks, most of PVF's computer-based applications are located in the accounting and financial areas. The applications include order filling, invoicing, accounts receivable, inventory control, accounts payable, payroll, and general ledger. At one time, each application had its own data files. For example, there was a customer master file, an inventory master file, a back order file, an inventory pricing file, and an employee master file. The order filling system used data from three files: customer master, inventory master, and back order. With PVF's new centralized database, data are organized around entities, or subjects, such as customers, invoices, and orders.

Pine Valley Furniture Company, like many firms, decided to develop its application software in-house; that is, it hired staff and bought computer hardware and software necessary to build application software suited to its own needs. (Other methods used to obtain application software are explained in Chapter 7.) Let's see how a project manager plays a key role in developing a new information system for PVF.

Managing the Information Systems Project

Project management is an important aspect of the development of information systems and a critical skill for a systems analyst. The focus of project management is to assure that system development projects meet customer expectations and are delivered within budget and time constraints.

The **project manager** is a systems analyst with a diverse set of skills—management, leadership, technical, conflict management, and customer relationship—who is responsible for initiating, planning, executing, and closing down a project. As a project manager, your environment is one of continual change and problem solving. In some organizations the project manager is a senior systems analyst who "has been around the block" a time or two. In others, both junior and senior analysts are expected to take on this role, managing parts of a project or actively supporting a more senior colleague who is assuming this role. Understanding the project management process is a critical skill for your future success.

Creating and implementing successful projects require managing resources, activities, and tasks needed to complete the information systems project. A **project** is a planned undertaking of a series of related activities to reach an objective that has a beginning and an end. The first question you might ask yourself is "Where do projects come from?" and, after considering all the different things that you could be asked to work on within an organization, "How do I know which projects to work on?" The ways in which each organization answers these vary.

In the rest of this section, we describe the process followed by Juanita Lopez and Chris Martin during the development of Pine Valley Furniture's Purchasing Fulfillment System. Juanita works in the order department, and Chris is a systems analyst.

Juanita observed problems with the way orders were processed and reported: Sales growth had increased the workload for the Manufacturing department, and the current systems no longer adequately supported the tracking of orders.

NET SEARCH
You may not be aware tht you can become a certified project manager. Visit http://www.prenhall.com/valacich to complete an exercise related to this topic.

Project manager
A systems analyst with a diverse set of skills—management, leadership, technical, conflict management, and customer relationship—who is responsible for initiating, planning, executing, and closing down a project.

Project
A planned undertaking of related activities to reach an objective that has a beginning and an end.

Pine Valley Furniture
System Service Request

REQUESTED BY   Juanita Lopez   DATE   November 1, 2003

DEPARTMENT   Purchasing, Manufacturing Support

LOCATION   Headquarters, 1-322

CONTACT   Tel: 4-3267   FAX: 4-3270   e-mail: jlopez

TYPE OF REQUEST   URGENCY

[X]   New System   [   ]   Immediate – Operations are impaired or opportunity lost

[   ]   System Enhancement   [   ]   Problems exist, but can be worked around

[   ]   System Error Correction   [X]   Business losses can be tolerated until new system installed

PROBLEM STATEMENT

Sales growth at PVF has caused greater volume of work for the manufacturing support unit within Purchasing. Further, more concentration on customer service has reduced manufacturing lead times, which puts more pressure on purchasing activities. In addition, cost-cutting measures force Purchasing to be more agressive in negotiating terms with vendors, improving delivery times, and lowering our investments in inventory. The current modest systems support for manufacturing purchasing is not responsive to these new business conditions. Data are not available, information cannot be summarized, supplier orders cannot be adequately tracked, and commodity buying is not well supported. PVF is spending too much on raw materials and not being responsive to manufacturing needs.

SERVICE REQUEST

I request a thorough analysis of our current operations with the intent to design and build a completely new information system. This system should handle all purchasing transactions, support display and reporting of critical purchasing data, and assist purchasing agents in commodity buying.

IS LIAISON   Chris Martin (Tel: 4-6204   FAX: 4-6200   e-mail: cmartin)

SPONSOR   Sal Divario, Director, Purchasing

------------------------- TO BE COMPLETED BY SYSTEMS PRIORITY BOARD -------------------------

[   ]   Request approved   Assigned to _____
                          Start date _____

[   ]   Recommend revision

[   ]   Suggest user development

[   ]   Reject for reason _____

FIGURE 2-3
SYSTEM SERVICE REQUEST FOR PURCHASING FULFILLMENT WITH NAME AND CONTACT INFORMATION OF THE PERSON REQUESTING THE SYSTEM, A STATEMENT OF THE PROBLEM, AND THE NAME AND CONTACT INFORMATION OF THE LIAISON AND SPONSOR

It was becoming more difficult to track orders and get the right furniture and invoice to the right customers. Juanita contacted Chris, and together they developed a system that corrected these Ordering department problems.

The first **deliverable,** or end product, produced by Chris and Juanita was a System Service Request (SSR), a standard form PVF uses for requesting systems development work. Figure 2-3 shows an SSR for purchasing a fulfillment system. The form includes the name and contact information of the person requesting the system, a statement of the problem, and the name and contact information of the liaison and sponsor.

This request was then evaluated by the Systems Priority Board of PVF. Because all organizations have limited time and resources, not all requests can be approved. The board evaluates development requests in relation to the business problems or opportunities the system will solve or create. It also considers how the proposed project fits within the organization's information systems architecture and long-range development plans. The review board selects those projects that best meet overall organizational goals (we learn more about organizational goals in Chapter 3). In the case of the Purchasing Fulfillment System request, the board found merit in the request and approved a more detailed **feasibility study.** A feasibility study, conducted by the project manager, involves determining if the information system makes sense for the organization from an economic and operational standpoint. The study takes place before the system is constructed. Figure 2-4 is a graphical view of the steps followed during the project initiation of the Purchasing Fulfillment System.

In summary, systems development projects are undertaken for two primary reasons: to take advantage of business opportunities and to solve business problems. Taking advantage of an opportunity might mean providing an innovative service to customers through the creation of a new system. For example, PVF may want to create a Web page so that customers can easily access its catalog and place orders at any time. Solving a business problem could involve modifying how an existing system processes data so that more accurate or

Deliverable
An end product in a phase of the SDLC.

Feasibility study
Determines if the information system makes sense for the organization from an economic and operational standpoint.

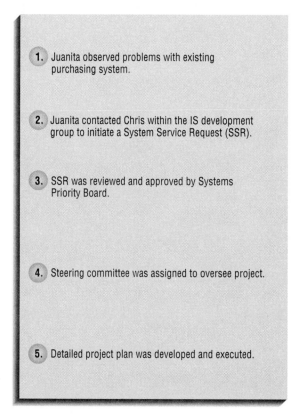

1. Juanita observed problems with existing purchasing system.
2. Juanita contacted Chris within the IS development group to initiate a System Service Request (SSR).
3. SSR was reviewed and approved by Systems Priority Board.
4. Steering committee was assigned to oversee project.
5. Detailed project plan was developed and executed.

FIGURE 2-4
A GRAPHICAL VIEW OF THE FIVE STEPS FOLLOWED DURING THE PROJECT INITIATION OF THE PURCHASING FULFILLMENT SYSTEM

timely information is provided to users. For example, a company such as PVF may create a password-protected intranet site that contains important announcements and budget information.

Projects are not always initiated for the rational reasons (taking advantage of business opportunities or solving business problems) stated above. For example, in some instances organizations and government undertake projects to spend resources, attain or pad budgets, keep people busy, or help train people and develop their skills. Our focus in this chapter is not on how and why organizations identify projects but on the management of projects once they have been identified.

Once a potential project has been identified, an organization must determine the resources required for its completion. This is done by analyzing the scope of the project and determining the probability of successful completion. After getting this information, the organization can then determine whether taking advantage of an opportunity or solving a particular problem is feasible within time and resource constraints. If deemed feasible, a more detailed project analysis is then conducted.

As you will see, determining the size, scope, and resource requirements for a project are just a few of the many skills that a project manager must possess. A project manager is often referred to as a juggler keeping aloft many balls, which reflect the various aspects of a project's development, as depicted in Figure 2-5.

FIGURE 2-5
A project manager juggles numerous activities.

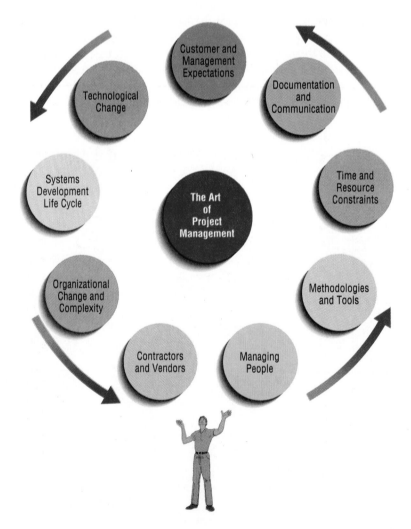

To successfully orchestrate the construction of a complex information system, a project manager must have interpersonal, leadership, and technical skills. Table 2-1 lists the project manager's common skills and activities. Note that many of the skills are related to personnel or general management, not simply technical skills. Table 2-1 shows that not only does an effective project manager have varied skills, but he or she is also the most instrumental person to the successful completion of any project.

The remainder of this chapter will focus on the **project management** process, which involves four phases:

1. Initiating the project
2. Planning the project
3. Executing the project
4. Closing down the project

Several activities must be performed during each of these four phases. Following this formal project management process greatly increases the likelihood of project success.

Initiating the Project

During **project initiation** the project manager performs several activities that assess the size, scope, and complexity of the project, and establishes procedures to support subsequent activities. Depending on the project, some initiation activities may be unnecessary and some may be very involved. The types

Project management
A controlled process of initiating, planning, executing, and closing down a project.

Project initiation
The first phase of the project management process in which activities are performed to assess the size, scope, and complexity of the project and to establish procedures to support later project activities.

TABLE 2-1: Common Activities and Skills of a Project Manager

Activity	Description	Skill
Leadership	Influencing the activities of others toward the attainment of a common goal through the use of intelligence, personality, and abilities	Communication; liaison between management, users, and developers; assigning activities; monitoring progress
Management	Getting projects completed through the effective utilization of resources	Defining and sequencing activities; communicating expectations; assigning resources to activities; monitoring outcomes
Customer relations	Working closely with customers to assure project deliverables meet expectations	Interpreting system requests and specifications; site preparation and user training; contact point for customers
Technical problem solving	Designing and sequencing activities to attain project goals	Interpreting system requests and specifications; defining activities and their sequence; making trade-offs between alternative solutions; designing solutions to problems
Conflict management	Managing conflict within a project team to assure that conflict is not too high or too low	Problem solving; smoothing out personality differences; compromising; goal setting
Team management	Managing the project team for effective team performance	Communication within and between teams; peer evaluations; conflict resolution; team building; self-management
Risk and change management	Identifying, assessing, and managing the risks and day-to-day changes that occur during a project	Environmental scanning; risk and opportunity identification and assessment; forecasting; resource redeployment

of activities you will perform when initiating a project are summarized in Figure 2-6 and described next.

1 *Establishing the project initiation team.* This activity involves organizing an initial core of project team members to assist in accomplishing the project initiation activities. For example, during the Purchasing Fulfillment System project at PVF, Chris Martin was assigned to support the Purchasing department. It is a PVF policy that all initiation teams consist of at least one user representative, in this case Juanita Lopez, and one member of the IS development group. Therefore, the project initiation team consisted of Chris and Juanita; Chris was the project manager.

2 *Establishing a relationship with the customer.* A thorough understanding of your customer builds stronger partnerships and higher levels of trust. At PVF, management has tried to foster strong working relationships between business units (like Purchasing) and the IS development group by assigning a specific individual to work as a liaison between both groups. Because Chris had been assigned to the Purchasing unit for some time, he was already aware of some of the problems with the existing purchasing systems. PVF's policy of assigning specific individuals to each business unit helped to assure that both Chris and Juanita were comfortable working together prior to the initiation of the project. Many organizations use a similar mechanism for establishing relationships with customers.

3 *Establishing the project initiation plan.* This step defines the activities required to organize the initiation team while it is working to define the scope of the project. Chris's role was to help Juanita translate her business requirements into a written request for an improved information system. This required the collection, analysis, organization, and transformation of a lot of information. Because Chris and Juanita were already familiar with each other and their roles within a development project, they next needed to define when and how they would communicate, define deliverables and project steps, and set deadlines. Their initiation plan included agendas for several meetings. These steps eventually led to the creation of their System Service Request (SSR) form.

4 *Establishing management procedures.* Successful projects require the development of effective management procedures. Within PVF, many of these management procedures had been established as standard operating procedures by the Systems Priority Board and the IS development group. For example, all project development work is charged back to the functional unit requesting the work. In other organizations, each project may have unique procedures tailored to its needs. Yet, in general when establishing procedures, you are concerned with developing team communication and reporting procedures, job assignments and roles, project change procedures, and determining how project funding and billing will be handled. It was fortunate for Chris and Juanita that most of these procedures were

FIGURE 2-6
FIVE PROJECT INITIATION ACTIVITIES

Project Initiation

1. Establishing the Project Initiation Team
2. Establishing a Relationship with the Customer
3. Establishing the Project Initiation Plan
4. Establishing Management Procedures
5. Establishing the Project Management Environment and Project Workbook

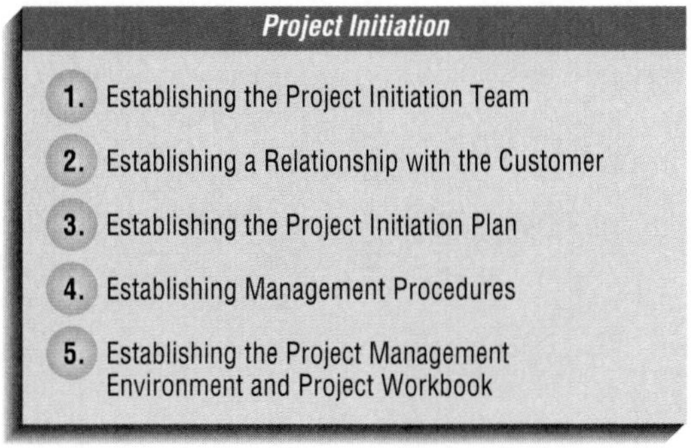

already established at PVF, allowing them to move quickly on to other project activities.

5 *Establishing the project management environment and project workbook.* The focus of this activity is to collect and organize the tools that you will use while managing the project and to construct the **project workbook.** For example, most diagrams, charts, and system descriptions provide much of the project workbook contents. Thus, the project workbook serves as a repository for all project correspondence, inputs, outputs, deliverables, procedures, and standards established by the project team. The project workbook can be stored as an online electronic document or in a large three-ring binder. The project workbook is used by all team members and is useful for project audits, orientation of new team members, communication with management and customers, identifying future projects, and performing postproject reviews. The establishment and diligent recording of all project information in the workbook are two of the most important activities you will perform as project manager.

Figure 2-7 shows the project workbook for the Purchasing Fulfillment System. It consists of both a large hard-copy binder and electronic diskettes where the system data dictionary, a catalog of data stored in the database, and diagrams are stored. For this system, all project documents can fit into a single binder. It is not unusual, however, for project documentation to be spread over several binders. As more information is captured and recorded electronically, however, fewer hard-copy binders may be needed. Many project teams keep their project workbooks on the Web. A Web site can be created so that all project members can easily access all project documents. This Web site can be a simple repository of documents or an elaborate site with password protection and security levels. The best feature of using the Web as your repository is that it allows all project members and customers to review a project's status and all related information continually.

Project initiation is complete once these five activities have been performed. Before moving on to the next phase of the project, the work performed during project initiation is reviewed at a meeting attended by management, customers, and project team members. An outcome of this meeting is a decision to continue

Project workbook
An online or hard-copy repository for all project correspondence, inputs, outputs, deliverables, procedures, and standards that is used for performing project audits, orientating new team members, communicating with management and customers, identifying future projects, and performing postproject reviews.

1. Project overview
2. Initiation plan and SSR
3. Project scope and risks
4. Management procedures
5. Data descriptions
6. Process descriptions
7. Team correspondence
8. Statement of work
9. Project schedule

Online copies of data dictionary, diagrams, schedules, reports, etc.

FIGURE 2-7
The project workbook for the Purchase Fulfillment System project contains nine key documents in both hard-copy and electronic form.

Pine Valley Furniture Information Systems Development Group

Purchasing Fulfillment System

Manager: Chris Martin

PFS Project Data Dictionary Diagrams

the project, modify it, or abandon it. In the case of the Purchasing Fulfillment System project at Pine Valley Furniture, the board accepted the SSR and selected a project steering committee to monitor project progress and to provide guidance to the team members during subsequent activities. If the scope of the project is modified, it may be necessary to return to project initiation activities and collect additional information. Once a decision is made to continue the project, a much more detailed project plan is developed during the project planning phase.

Planning the Project

Project planning
The second phase of the project management process, which focuses on defining clear, discrete activities and the work needed to complete each activity within a single project.

The next step in the project management process is **project planning.** Project planning involves defining clear, discrete activities and the work needed to complete each activity within a single project. It often requires you to make numerous assumptions about the availability of resources such as hardware, software, and personnel. It is much easier to plan nearer-term activities than those occurring in the future. In actual fact, you often have to construct longer-term plans that are more general in scope and nearer-term plans that are more detailed. The repetitive nature of the project management process requires that plans be constantly monitored throughout the project and periodically updated (usually after each phase) based upon the most recent information.

Figure 2-8 illustrates the principle that nearer-term plans are typically more specific and firmer than longer-term plans. For example, it is virtually impossible to rigorously plan activities late in the project without first completing earlier activities. Also, the outcome of activities performed earlier in the project are likely to have impact on later activities. This means that it is very difficult, and very likely inefficient, to try to plan detailed solutions for activities that will occur far into the future.

As with the project initiation process, varied and numerous activities must be performed during project planning. For example, during the Purchasing Fulfillment System project, Chris and Juanita developed a 10-page plan. However, project plans for very large systems may be several hundred pages in length. The types of activities that you can perform during project planning are summarized in Figure 2-9 and are described in the following list:

1 *Describing project scope, alternatives, and feasibility.* The purpose of this activity is to understand the content and complexity of the project. Within PVF's system development methodology, one of the first meetings must focus on defining a project's scope. Although project scope information was not included in the SSR developed by Chris and Juanita, it was important that both shared the same vision for the project before moving too far along. During this activity, you should reach agreement on the following questions:

   ◗ What problem or opportunity does the project address?
   ◗ What are the quantifiable results to be achieved?

FIGURE 2-8
Level of project planning detail should be high in the short term, with less detail as time goes on.

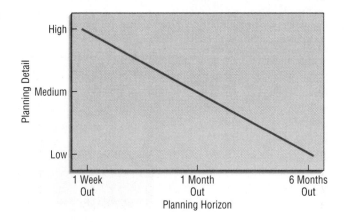

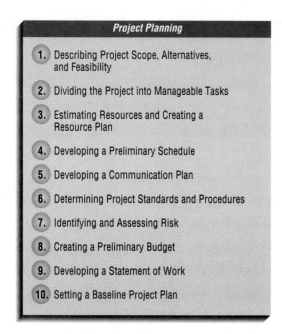

Project Planning

1. Describing Project Scope, Alternatives, and Feasibility
2. Dividing the Project into Manageable Tasks
3. Estimating Resources and Creating a Resource Plan
4. Developing a Preliminary Schedule
5. Developing a Communication Plan
6. Determining Project Standards and Procedures
7. Identifying and Assessing Risk
8. Creating a Preliminary Budget
9. Developing a Statement of Work
10. Setting a Baseline Project Plan

FIGURE 2-9
TEN PROJECT PLANNING ACTIVITIES

- What needs to be done?
- How will success be measured?
- How will we know when we are finished?

After defining the scope of the project, your next objective is to identify and document general alternative solutions for the current business problem or opportunity. You must then assess the feasibility of each alternative solution and choose which to consider during subsequent SDLC phases. In some instances, off-the-shelf software can be found. It is also important that any unique problems, constraints, and assumptions about the project be clearly stated.

2 *Dividing the project into manageable tasks.* This is a critical activity during the project planning process. Here, you must divide the entire project into manageable tasks and then logically order them to ensure a smooth evolution between tasks. The definition of tasks and their sequence is referred to as the **work breakdown structure.** Some tasks may be performed in parallel whereas others must follow one another sequentially. Task sequence depends on which tasks produce deliverables needed in other tasks, when critical resources are available, the constraints placed on the project by the client, and the process outlined in the SDLC.

For example, suppose that you are working on a new development project and need to collect system requirements by interviewing users of the new system and reviewing reports they currently use to do their job. A work breakdown for these activities is represented in a Gantt chart in Figure 2-10. A **Gantt chart** is a graphical representation of a project that shows each task as a horizontal bar whose length is proportional to its time for completion. Different colors, shades, or shapes can be used to highlight each kind of task. For example, those activities on the critical path (defined later in this chapter) may be in red and a summary task could have a special bar. Note that the black horizontal bars—rows 1, 2, and 8 in Figure 2-10—represent summary tasks. Planned versus actual times or progress for an activity can be compared by parallel bars of different colors, shades, or shapes. Gantt charts do not show how tasks must be ordered (precedence) but simply show when an activity should begin and end. In Figure 2-10, the task duration is shown in the second column by days, "d," and necessary prior tasks

Work breakdown structure
The process of dividing the project into manageable tasks and logically ordering them to ensure a smooth evolution between tasks.

Gantt chart
A graphical representation of a project that shows each task as a horizontal bar whose length is proportional to its time for completion.

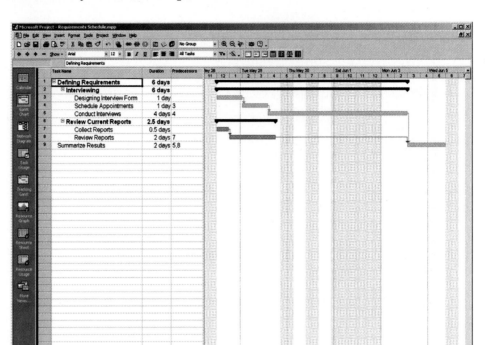

FIGURE 2-10
**GANTT CHART SHOWING PROJECT
TASKS, DURATION TIMES FOR
THOSE TASKS (D = DAYS), AND
PREDECESSORS.**

are noted in the third column as predecessors. Most project management software tools support a broad range of task durations including minutes, hours, days, weeks, and months. As you will learn in later chapters, the SDLC consists of several phases, which you need to break down into activities. Creating a work breakdown structure requires that you decompose phases into activities—summary tasks—and activities into specific tasks. For example, Figure 2-10 shows that the activity Interviewing consists of three tasks: design interview form, schedule appointments, and conduct interviews.

Defining tasks in too much detail will make the management of the project unnecessarily complex.

What are the characteristics of a "task"? A task—

   ● Can be done by one person or a well-defined group.

   ● Has a single and identifiable deliverable. (The task, however, is the process of creating the deliverable.)

   ● Has a known method or technique.

   ● Has well-accepted predecessor and successor steps.

   ● Is measurable so that percent completed can be determined.

You will develop the skill of discovering the optimal level of detail for representing tasks through experience. For example, it may be very difficult to list tasks that require less than one hour of time to complete in a final work breakdown structure. Alternatively, choosing tasks that are too large in scope (e.g., several weeks long) will not provide you with a clear sense of the status of the project or of the interdependencies between tasks.

3   *Estimating resources and creating a resource plan.* The goal of this activity is to estimate resource requirements for each project activity and use this information to create a project resource plan. The resource plan helps assemble and deploy resources in the most effective manner. For example,

you would not want to bring additional programmers onto the project at a rate faster than you could prepare work for them.

People are the most important, and expensive, part of project resource planning. Project time estimates for task completion and overall system quality are significantly influenced by the assignment of people to tasks. It is important to give people tasks that allow them to learn new skills. It is equally important to make sure that project members are not in "over their heads" or working on a task that is not well suited to their skills. Resource estimates may need to be revised based upon the skills of the actual person (or people) assigned to a particular activity. Figure 2-11 indicates the relative programming speed versus the relative programming quality of three programmers. The figure suggests that Carl should not be assigned tasks in which completion time is critical and that Brenda should be assigned to tasks in which high quality is most vital.

One approach to assigning tasks is to assign a single task type (or only a few task types) to each worker for the duration of the project. For example, you could assign one worker to create all computer displays and another to create all system reports. Such specialization ensures that both workers become efficient at their own particular tasks. A worker may become bored if the task is too specialized or is long in duration, so you could assign workers to a wider variety of tasks. However, this approach may lead to lowered task efficiency. A middle ground would be to make assignments with a balance of both specialization and task variety. Assignments depend upon the size of the development project and the skills of the project team. Regardless of the manner in which you assign tasks, make sure that each team member works only on one task at a time. Exceptions to this rule can occur when a task occupies only a small portion of a team member's time (e.g., testing the programs developed by another team member) or during an emergency.

4   *Developing a preliminary schedule.* During this activity, you use the information on tasks and resource availability to assign time estimates to each activity in the work breakdown structure. These time estimates will allow you to create target starting and ending dates for the project. Target dates can be revisited and modified until a schedule produced is acceptable to the customer. Determining an acceptable schedule may require that you find additional or different resources or that the scope of the project be changed. The schedule may be represented as a Gantt chart, as illustrated in Figure 2-10, or as a Network diagram, as illustrated in Figure 2-12. A **Network diagram** is a graphical depiction of project tasks and their interrelationships. As with a Gantt chart, each type of task can be highlighted by different features on the Network diagram. The distinguishing feature of a

Network diagram
A diagram that depicts project tasks and their interrelationships.

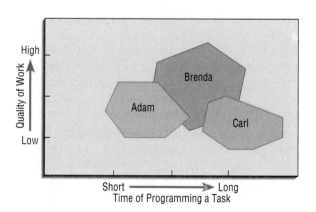

FIGURE 2-11
TRADE-OFFS BETWEEN THE QUALITY OF THE PROGRAM CODE VERSUS THE SPEED OF PROGRAMMING

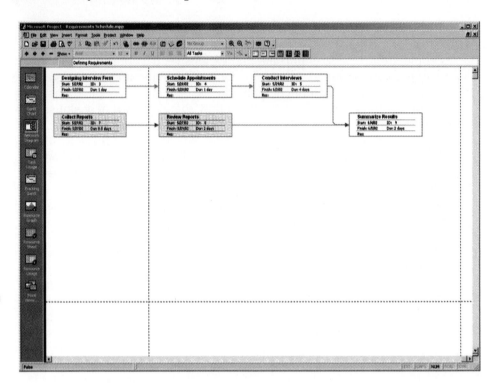

FIGURE 2-12

A Network diagram illustrates tasks with rectangles (or ovals) and the relationships and sequences of those activities with arrows.

Network diagram is that the ordering of tasks is shown by connecting tasks—depicted as rectangles or ovals—with its predecessor and successor tasks. However, the relative size of a node (representing a task) or a gap between nodes does not imply the task's duration. We describe both of these charts later in this chapter.

5 *Developing a communication plan.* The goal of this activity is to outline the communication procedures among management, project team members, and the customer. The communication plan includes when and how written and oral reports will be provided by the team, how team members will coordinate work, what messages will be sent to announce the project to interested parties, and what kinds of information will be shared with vendors and external contractors involved with the project. It is important that free and open communication occurs among all parties, with respect for proprietary information and confidentiality with the customer.

6 *Determining project standards and procedures.* During this activity, you specify how various deliverables are produced and tested by you and your project team. For example, the team must decide on which tools to use, how the standard SDLC might be modified, which SDLC methods will be used, documentation styles (e.g., type fonts and margins for user manuals), how team members will report the status of their assigned activities, and terminology. Setting project standards and procedures for work acceptance is a way to assure the development of a high-quality system. Also, it is much easier to train new team members when clear standards are in place. Organizational standards for project management and conduct make the determination of individual project standards easier and the interchange or sharing of personnel among different projects feasible.

7 *Identifying and assessing risk.* The goal of this activity is to identify sources of project risk and to estimate the consequences of those risks. Risks might arise from the use of new technology, prospective users' resistance to change, availability of critical resources, competitive reactions or changes in regulatory actions due to the construction of a system, or team

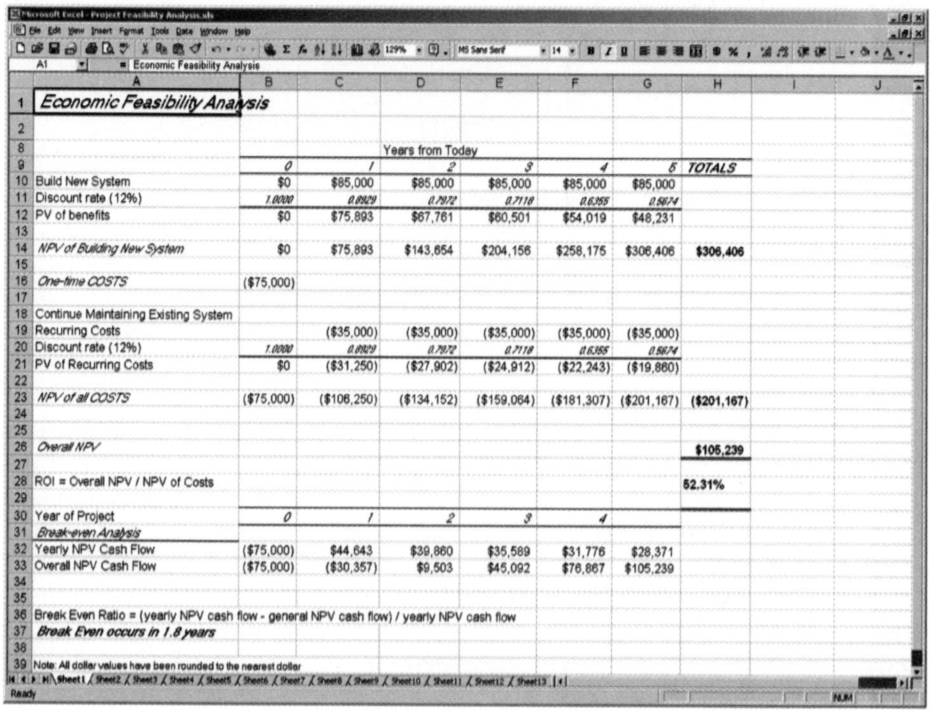

Economic Feasibility Analysis							
				Years from Today			
	0	*1*	*2*	*3*	*4*	*5*	*TOTALS*
Build New System	$0	$85,000	$85,000	$85,000	$85,000	$85,000	
Discount rate (12%)	*1.0000*	*0.8929*	*0.7972*	*0.7118*	*0.6355*	*0.5674*	
PV of benefits	$0	$75,893	$67,761	$60,501	$54,019	$48,231	
NPV of Building New System	$0	$75,893	$143,654	$204,156	$258,175	$306,406	**$306,406**
One-time COSTS	($75,000)						
Continue Maintaining Existing System							
Recurring Costs		($35,000)	($35,000)	($35,000)	($35,000)	($35,000)	
Discount rate (12%)	*1.0000*	*0.8929*	*0.7972*	*0.7118*	*0.6355*	*0.5674*	
PV of Recurring Costs	$0	($31,250)	($27,902)	($24,912)	($22,243)	($19,860)	
NPV of all COSTS	($75,000)	($106,250)	($134,152)	($159,064)	($181,307)	($201,167)	**($201,167)**
Overall NPV							**$105,239**
ROI = Overall NPV / NPV of Costs							**52.31%**
Year of Project	*0*	*1*	*2*	*3*	*4*		
Break-even Analysis							
Yearly NPV Cash Flow	($75,000)	$44,643	$39,860	$35,589	$31,776	$28,371	
Overall NPV Cash Flow	($75,000)	($30,357)	$9,503	$45,092	$76,867	$105,239	

Break Even Ratio = (yearly NPV cash flow - general NPV cash flow) / yearly NPV cash flow

Break Even occurs in 1.8 years

Note: All dollar values have been rounded to the nearest dollar

FIGURE 2-13
A FINANCIAL COST AND BENEFIT ANALYSIS FOR A SYSTEMS DEVELOPMENT PROJECT

member inexperience with technology or the business area. You should continually try to identify and assess project risk.

The identification of project risks is required to develop PVF's new Purchasing Fulfillment System. Chris and Juanita met to identify and describe possible negative outcomes of the project and their probabilities of occurrence. Although we list the identification of risks and the outline of project scope as two discrete activities, they are highly related and often concurrently discussed.

8 *Creating a preliminary budget.* During this phase, you need to create a preliminary budget that outlines the planned expenses and revenues associated with your project. The project justification will demonstrate that the benefits are worth these costs. Figure 2-13 shows a cost-benefit analysis for a new development project. This analysis shows net present value calculations of the project's benefits and costs as well as a return on investment and cash flow analysis. We discuss project budgets fully in Chapter 3.

9 *Developing a Statement of Work.* An important activity that occurs near the end of the project planning phase is the development of the Statement of Work. Developed primarily for the customer, this document outlines work that will be done and clearly describes what the project will deliver. The Statement of Work is useful to make sure that you, the customer, and other project team members have a clear understanding of the intended project size, duration, and outcomes.

10 *Setting a Baseline Project Plan.* Once all of the prior project planning activities have been completed, you will be able to develop a Baseline Project Plan. This baseline plan provides an estimate of the project's tasks and resource requirements and is used to guide the next project phase—execution. As new information is acquired during project execution, the baseline plan will continue to be updated.

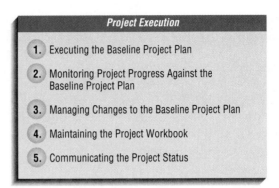

FIGURE 2-14
FIVE PROJECT EXECUTION
ACTIVITIES

At the end of the project planning phase, a review of the Baseline Project Plan is conducted to double-check all the information in the plan. As with the project initiation phase, it may be necessary to modify the plan, which means returning to prior project planning activities before proceeding. As with the Purchasing Fulfillment System project, you may submit the plan and make a brief presentation to the project steering committee at this time. The committee can endorse the plan, ask for modifications, or determine that it is not wise to continue the project as currently outlined.

Executing the Project

Project execution

The third phase of the project management process in which the plans created in the prior phases (project initiation and planning) are put into action.

Project execution puts the Baseline Project Plan into action. Within the context of the SDLC, project execution occurs primarily during the analysis, design, and implementation phases. During the development of the Purchasing Fulfillment System, Chris Martin was responsible for five key activities during project execution. These activities are summarized in Figure 2-14 and described in the remainder of this section:

1 *Executing the Baseline Project Plan.* As project manager, you oversee the execution of the baseline plan. This means that you initiate the execution of project activities, acquire and assign resources, orient and train new team members, keep the project on schedule, and assure the quality of project deliverables. This is a formidable task, but a task made much easier through the use of sound project management techniques. For example, as tasks are completed during a project, they can be "marked" as completed on the project schedule. In Figure 2-15, tasks 3 and 7 are marked as completed by showing 100 percent in the "% Complete" column. Members of the project team will come and go. You are responsible for initiating new team members by providing them with the resources they need and helping them assimilate into the team. You may want to plan social events, regular team project status meetings, team-level reviews of project deliverables, and other group events to mold the group into an effective team.

2 *Monitoring project progress against the Baseline Project Plan.* While you execute the Baseline Project Plan, you should monitor your progress. If the project gets ahead of (or behind) schedule, you may have to adjust resources, activities, and budgets. Monitoring project activities can result in modifications to the current plan. Measuring the time and effort expended on each activity helps you improve the accuracy of estimations for future projects. It is possible with project schedule charts, like Gantt, to show progress against a plan; and it is easy with Network diagrams to understand the ramifications of delays in an activity. Monitoring progress also means that the team leader must evaluate and appraise each team member, occasionally change work assignments or request changes in personnel, and provide feedback to the employee's supervisor.

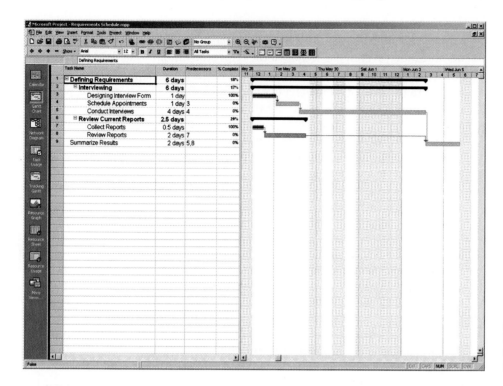

FIGURE 2-15
GANTT CHART WITH TASKS 3 AND 7
COMPLETED

3 *Managing changes to the Baseline Project Plan.* You will encounter pressure to make changes to the baseline plan. At PVF, policies dictate that only approved changes to the project specification can be made and all changes must be reflected in the baseline plan and project workbook, including all charts. For example, if Juanita suggests a significant change to the existing design of the Purchasing Fulfillment System, a formal change request must be approved by the steering committee. The request should explain why changes are desired and describe all possible impacts on prior and subsequent activities, project resources, and the overall project schedule. Chris would have to help Juanita develop such a request. This information allows the project steering committee to more easily evaluate the costs and benefits of a significant midcourse change.

In addition to changes occurring through formal request, changes may also occur from events outside your control. In fact, numerous events may initiate a change to the Baseline Project Plan, including the following possibilities:

- A slipped completion date for an activity
- A bungled activity that must be redone
- The identification of a new activity that becomes evident later in the project
- An unforeseen change in personnel due to sickness, resignation, or termination

When an event occurs that delays the completion of an activity, you typically have two choices: Devise a way to get back on schedule or revise the plan. Devising a way to get back on schedule is the preferred approach because no changes to the plan will have to be made. The ability to head off and smoothly work around problems is a critical skill that you need to master.

As you see later in the chapter, project schedule charts are very helpful in assessing the impact of change. Using such charts, you can quickly see if the completion time of other activities will be affected by changes in the duration of a given activity or if the whole project completion date will change. Often you will have to find a way to rearrange the activities because the ultimate project completion date may be rather fixed. There may be a penalty to the organization (even legal action) if the expected completion date is not met.

4 *Maintaining the project workbook.* As in all project phases, maintaining complete records of all project events is necessary. The workbook provides the documentation new team members require to assimilate project tasks quickly. It explains why design decisions were made and is a primary source of information for producing all project reports.

5 *Communicating the project status.* The project manager is responsible for keeping all team members—system developers, managers, and customers—abreast of the project status. Clear communication is required to create a shared understanding of the activities and goals of the project; such an understanding assures better coordination of activities. This means that the entire project plan should be shared with the entire project team and any revisions to the plan should be communicated to all interested parties so that everyone understands how the plan is evolving. Procedures for communicating project activities vary from formal meetings to informal hallway discussions. Some procedures are useful for informing others of project status, others for resolving issues, and others for keeping permanent records of information and events. Table 2-2 lists numerous communication procedures, their level of formality and most likely use. Whichever procedure you use, frequent communication helps to assure project success.

This section outlined your role as the project manager during the execution of the Baseline Project Plan. The ease with which the project can be managed is significantly influenced by the quality of prior project phases. If you develop a high-quality project plan, it is much more likely that the project will be successfully executed. The next section describes your role during project closedown, the final phase of the project management process.

Closing Down the Project

Project closedown
The final phase of the project management process which focuses on bringing a project to an end.

The focus of **project closedown** is to bring the project to an end. Projects can conclude with a natural or unnatural termination. A natural termination occurs when the requirements of the project have been met—the project has been completed and is a success. An unnatural termination occurs when the project

TABLE 2-2:   Project Team Communication Methods

Procedure	Formality	Use
Project workbook	High	Inform Permanent record
Meetings	Medium to high	Resolve issues
Seminars and workshops	Low to medium	Inform
Project newsletters	Medium to high	Inform
Status reports	High	Inform
Specification documents	High	Inform Permanent record
Minutes of meetings	High	Inform Permanent record
Bulletin boards	Low	Inform
Memos	Medium to high	Inform
Brown bag lunches	Low	Inform
Hallway discussions	Low	Inform Resolve issues

is stopped before completion. Several events can cause an unnatural termination of a project. For example, it may be learned that the assumption used to guide the project proved to be false or that the performance of the system or development group was somehow inadequate or that the requirements are no longer relevant or valid in the customer's business environment. The most likely reasons for the unnatural termination of a project relate to running out of time or money, or both. Regardless of the project termination outcome, several activities must be performed: closing down the project, conducting postproject reviews, and closing the customer contract. Within the context of the SDLC, project closedown occurs after the implementation phase. The system maintenance phase typically represents an ongoing series of projects, each needing to be individually managed. Figure 2-16 summarizes the project closedown activities which are described more fully in the remainder of this section:

1 *Closing down the project.* During closedown, you perform several diverse activities. For example, if you have several team members working with you, project completion may signify job and assignment changes for some members. You will likely be required to assess each team member and provide an appraisal for personnel files and salary determination. You may also want to provide career advice to team members, write letters to superiors praising special accomplishments of team members, and send thank-you letters to those who helped but were not team members. As project manager, you must be prepared to handle possible negative personnel issues such as job termination, especially if the project was not successful. When closing down the project, it is also important to notify all interested parties that the project has been completed and to finalize all project documentation and financial records so that a final review of the project can be conducted. You should also celebrate the accomplishments of the team. Some teams will hold a party, and each team member may receive memorabilia (e.g., a T-shirt with "I survived the X project"). The goal is to celebrate the team's effort in bringing a difficult task to a successful conclusion.

2 *Conducting postproject reviews.* Once you have closed down the project, final reviews of the project should be conducted with management and customers. The objective of these reviews is to determine the strengths and weaknesses of project deliverables, the processes used to create them, and the project management process. It is important that everyone understands what went right and what went wrong in order to improve the process for the next project. Remember, the systems development methodology adopted by an organization is a living guideline that must undergo continual improvement.

3 *Closing the customer contract.* The focus of this final activity is to ensure that all contractual terms of the project have been met. A project governed by a contractual agreement is typically not completed until agreed to by both parties, often in writing. Thus, it is paramount that you gain agreement from your customer that all contractual obligations have been met and that further work is either their responsibility or covered under another System Service Request or contract.

NET SEARCH
There is ample information available to help you become a better project manager. Visit http://www. prenhall.com/valacich to complete an exercise related to this topic.

Project Closedown

1. Closing Down the Project

2. Conducting Postproject Reviews

3. Closing the Customer Contract

FIGURE 2-16
THREE PROJECT CLOSEDOWN ACTIVITIES

Closedown is a very important activity. A project is not complete until it is closed, and it is at closedown that projects are deemed a success or failure. Completion also signifies the chance to begin a new project and apply what you have learned. Now that you have an understanding of the project management process, the next section describes specific techniques used in systems development for representing and scheduling activities and resources.

Representing and Scheduling Project Plans

A project manager has a wide variety of techniques available for depicting and documenting project plans. These planning documents can take the form of graphical or textual reports, although graphical reports have become most popular for depicting project plans. The most commonly used methods are Gantt charts and Network diagrams. Because Gantt charts do not show how tasks must be ordered (precedence) but simply show when a task should begin and when it should end, they are often more useful for depicting relatively simple projects or subparts of a larger project, the activities of a single worker, or for monitoring the progress of activities compared to scheduled completion dates (see Figure 2-17[A]). Recall that a Network diagram shows the ordering of activities by connecting a task to its predecessor and successor tasks (see Figure 2-17[B]). Sometimes a Network diagram is preferable; other times a Gantt chart more easily shows certain aspects of a project. Here are the key differences between these two representations.

- Gantt visually shows the duration of tasks whereas a Network diagram visually shows the sequence dependencies between tasks.
- Gantt visually shows the time overlap of tasks whereas a Network diagram does not show time overlap but does show which tasks could be done in parallel.
- Some forms of Gantt charts can visually show slack time available within an earliest start and latest finish duration. A Network diagram shows this by data within activity rectangles.

Project managers also use textual reports that depict resource utilization by tasks, complexity of the project, and cost distributions to control activities. For example, Figure 2-18 shows a screen from Microsoft Project for Windows that summarizes all project activities, their durations in weeks, and their scheduled starting and ending dates. Most project managers use computer-based systems to help develop their graphical and textual reports. Later in this chapter, we discuss these automated systems in more detail.

A project manager will periodically review the status of all ongoing project task activities to assess whether the activities will be completed early, on time, or late. If early or late, the duration of the activity, depicted in column 2 of Figure 2-18, can be updated. Once changed, the scheduled start and finish times of all subsequent tasks will also change. Making such a change will also alter a Gantt chart or Network diagram used to represent the project tasks. The ability to easily make changes to a project is a very powerful feature of most project management environments. It allows the project manager to determine easily how changes in task duration impact the project completion date. It is also useful for examining the impact of "what if" scenarios for adding or reducing resources, such as personnel, for an activity.

Resources
Any person, group of people, piece of equipment, or material used in accomplishing an activity.

Representing Project Plans

Project scheduling and management requires that time, costs, and resources be controlled. **Resources** are any person, group of people, piece of equipment, or

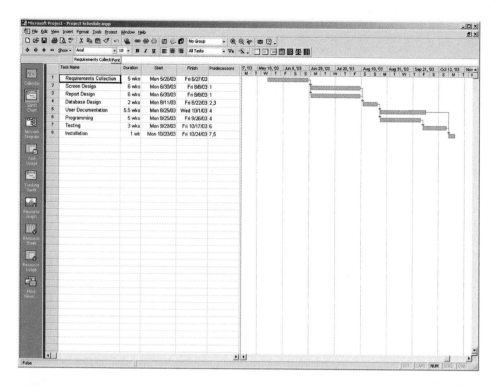

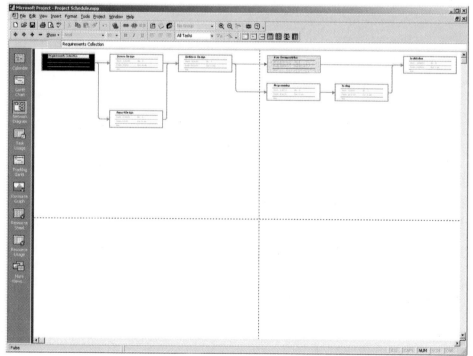

FIGURE 2-17
GRAPHICAL DIAGRAMS THAT
DEPICT PROJECT PLANS
(A) A GANTT CHART
(B) A NETWORK DIAGRAM

material used in accomplishing an activity. Network digramming is a **critical path scheduling** technique used for controlling resources. A critical path refers to a sequence of task activities whose order and durations directly affect the completion date of a project. A Network diagram is one of the most widely used and best-known scheduling methods. You would use a Network diagram when tasks:

> Are well-defined and have a clear beginning and endpoint

> Can be worked on independently of other tasks

Critical path scheduling
A scheduling technique whose order and duration of a sequence of task activities directly affect the completion date of a project.

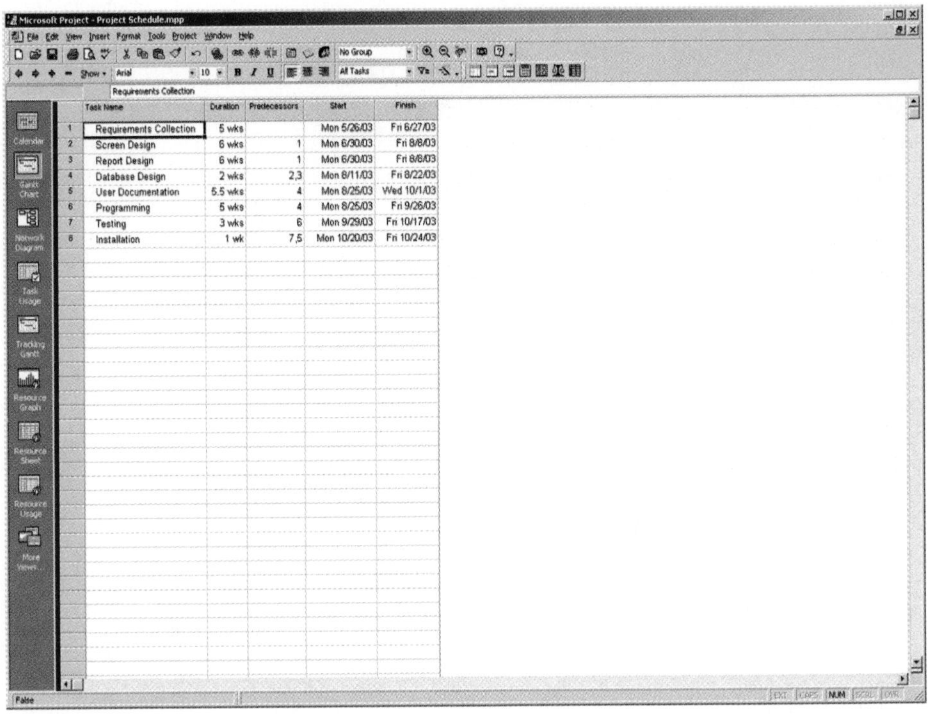

FIGURE 2-18

A screen from Microsoft Project for Windows summarizes all project activities, their durations in weeks, and their scheduled starting and ending dates.
Source: Reprinted with permission of Microsoft.

- Are ordered
- Serve the purpose of the project

A major strength of Network diagramming is its ability to represent how completion times vary for activities. Because of this, it is more often used than Gantt charts to manage projects such as information systems development where variability in the duration of activities is the norm. **Network diagrams** are composed of circles or rectangles representing activities and connecting arrows showing required work flows, as illustrated in Figure 2-19.

Calculating Expected Time Durations Using PERT

One of the most difficult and most error prone activities when constructing a project schedule is the determination of the time duration for each task within a work breakdown structure. It is particularly problematic to make these estimates when there is a high degree of complexity and uncertainty about a task. **PERT (Program Evaluation Review Technique)** is a technique that uses optimistic, pessimistic, and realistic time estimates to calculate the *expected time* for a particular task. This technique helps you obtain a better time estimate when there is some uncertainty as to how much time a task will require to be completed.

PERT (Program Evaluation Review Technique)
A technique that uses optimistic, pessimistic, and realistic time to calculate the expected time for a particular task.

FIGURE 2-19
A NETWORK DIAGRAM SHOWING ACTIVITIES (REPRESENTED BY CIRCLES) AND SEQUENCE OF THOSE ACTIVITIES (REPRESENTED BY ARROWS)

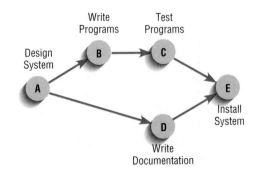

The optimistic (o) and pessimistic (p) times reflect the minimum and maximum possible periods of time for an activity to be completed. The realistic time (r), or most likely time, reflects the project manager's "best guess" of the amount of time the activity will require for completion. Once each of these estimates is made for an activity, an expected time (ET) can be calculated. Because the expected completion time should be closer to the realistic time (r), it is typically weighted four times more than the optimistic (o) and pessimistic (p) times. Once you add these values together, it must be divided by 6 to determine the ET. This equation is shown in the following formula.

$$ET = \frac{O + 4r + p}{6}$$

where

ET = expected time for the completion for an activity
o = optimistic completion time for an activity
r = realistic completion time for an activity
p = pessimistic completion time for an activity

For example, suppose that your instructor asked you to calculate an expected time for the completion of an upcoming programming assignment. For this assignment, you estimate an optimistic time of 2 hours, a pessimistic time of 8 hours, and a most likely time of 6 hours. Using PERT, the expected time for completing this assignment is 5.67 hours. Commercial project management software like Microsoft Project assists you in using PERT to make expected time calculations. Additionally, many commercial tools allow you to customize the weighing of optimistic, pessimistic, and realistic completion times.

Constructing a Gantt Chart and Network Diagram at Pine Valley Furniture

Although Pine Valley Furniture has historically been a manufacturing company, it has recently entered the direct sales market for selected target markets. One of the fastest growing of these markets is economically priced furniture suitable for college students. Management has requested that a new Sales Promotion Tracking System (SPTS) be developed. This project has already successfully moved through project initiation and is currently in the detailed project planning stage, which corresponds to the SDLC phase of project initiation and planning. The SPTS will be used to track the sales purchases by college students for the next fall semester. Students typically purchase low-priced beds, bookcases, desks, tables, chairs, and dressers. Because PVF does not normally stock a large quantity of lower-priced items, management feels that a tracking system will help provide information about the college student market that can be used for follow-up sales promotions (e.g., a midterm futon sale).

The project is to design, develop, and implement this information system before the start of the fall term in order to collect sales data at the next major buying period. This deadline gives the project team 24 weeks to develop and implement the system. The Systems Priority Board at PVF wants to make a decision this week based on the feasibility of completing the project within the 24-week deadline. Using PVF's project planning methodology, the project manager, Jim Woo, knows that the next step is to construct a Gantt chart and a Network diagram of the project to represent the Baseline Project Plan so that he can use these charts to estimate the likelihood of completing the project within 24 weeks. A major activity of project planning focuses on dividing the project into manageable activities, estimating times for each, and sequencing their order. Here are the steps Jim followed to do this.

1. *Identify each activity to be completed in the project.* After discussing the new Sales Promotion Tracking System with PVF's management, sales, and development staff, Jim identified the following major activities for the project:

   - Requirements collection
   - Screen design
   - Report design
   - Database construction
   - User documentation creation
   - Software programming
   - System testing
   - System installation

2. *Determine time estimates and calculate the expected completion time for each activity.* After identifying the major project activities, Jim established optimistic, realistic, and pessimistic time estimates for each activity. These numbers were then used to calculate the expected completion times for all project activities. Figure 2-20 shows the estimated time calculations for each activity of the Sales Promotion Tracking System project.

3. *Determine the sequence of the activities and precedence relationships among all activities by constructing a Gantt chart and Network diagram.* This step helps you understand how various activities are related. Jim starts by determining the order in which activities should take place. The results of this analysis for the SPTS project are shown in Figure 2-21. The first row of this figure shows that no activities precede requirements collection. Row 2 shows that screen design must be preceded by requirements

FIGURE 2-20
ESTIMATED TIME CALCULATIONS FOR THE SPTS PROJECT

ACTIVITY	TIME ESTIMATE (in weeks)			EXPECTED TIME (ET) $\dfrac{o + 4r + p}{6}$
	o	r	p	
1. Requirements Collection	1	5	9	5
2. Screen Design	5	6	7	6
3. Report Design	3	6	9	6
4. Database Design	1	2	3	2
5. User Documentation	3	6	7	5.5
6. Programming	4	5	6	5
7. Testing	1	3	5	3
8. Installation	1	1	1	1

FIGURE 2-21
SEQUENCE OF ACTIVITIES WITHIN THE SPTS PROJECT

ACTIVITY	PRECEDING ACTIVITY
1. Requirements Collection	—
2. Screen Design	1
3. Report Design	1
4. Database Design	2,3
5. User Documentation	4
6. Programming	4
7. Testing	6
8. Installation	5,7

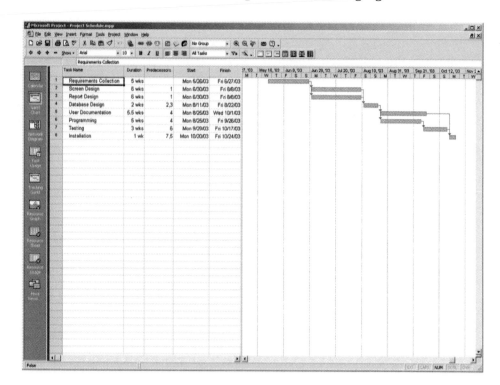

FIGURE 2-22
**GANTT CHART THAT ILLUSTRATES
THE SEQUENCE AND DURATION OF
EACH ACTIVITY OF THE SPTS
PROJECT**

collection. Row 4 shows that both screen and report design must precede database construction. Thus, activities may be preceded by zero, one, or more activities.

Using the estimated time and activity sequencing information from Figures 2-20 and 2-21, Jim can now construct a Gantt chart and Network diagram of the project's activities. To construct the Gantt chart, a horizontal bar is drawn for each activity that reflects its sequence and duration, as shown in Figure 2-22. The Gantt chart may not, however, show direct inter-relationships between activities. For example, just because the database design activity begins right after the screen design and report design bars finish does not imply that these two activities must finish before database design can begin. To show such precedence relationships, a Network diagram must be used. The Gantt chart in Figure 2-22 does, however, show precedence relationships.

Network diagrams have two major components: arrows and nodes. Arrows reflect the sequence of activities whereas nodes reflect activities that consume time and resources. A Network diagram for the SPTS project is shown in Figure 2-23. This diagram has eight nodes labeled 1 through 8.

4   *Determine the critical path.* The critical path of a Network diagram is represented by the sequence of connected activities that produces the longest

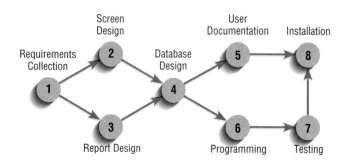

FIGURE 2-23
**A NETWORK DIAGRAM THAT
ILLUSTRATES THE ACTIVITIES
(CIRCLES) AND THE SEQUENCE
(ARROWS) OF THOSE ACTIVITIES**

Critical path
The shortest time in which a project can be completed.

Slack time
The amount of time that an activity can be delayed without delaying the project.

overall time period. All nodes and activities within this sequence are referred to as being "on" the **critical path.** The critical path represents the shortest time in which a project can be completed. In other words, any activity on the critical path that is delayed in completion delays the entire project. Nodes not on the critical path, however, can be delayed (for some amount of time) without delaying the final completion of the project. Nodes not on the critical path contain **slack time** and allow the project manager some flexibility in scheduling.

Figure 2-24 shows the Network diagram that Jim constructed to determine the critical path and expected completion time for the SPTS project. To determine the critical path, Jim calculated the earliest and latest expected completion time for each activity. He found each activity's earliest expected completion time (T_E) by summing the estimated time (ET) for each activity from left to right (i.e., in precedence order), starting at activity 1 and working toward activity 8. In this case, T_E for activity 8 is equal to 22 weeks. If two or more activities precede an activity, the largest expected completion time of these activities is used in calculating the new activity's expected completion time. For example, because activity 8 is preceded by both activities 5 and 7, the largest expected completion time between 5 and 7 is 21, so T_E for activity 8 is 21 = 1, or 22. The earliest expected completion time for the last activity of the project represents the amount of time the project should take to complete. Because the time of each activity can vary, however, the projected completion time represents only an estimate. The project may in fact require more or less time for completion.

The latest expected completion time (T_L) refers to the time in which an activity can be completed without delaying the project. To find the values for each activity's T_L, Jim started at activity 8 and set T_L equal to the final T_E (22 weeks). Next, he worked right to left toward activity 1 and subtracted the expected time for each activity. The slack time for each activity is equal to the difference between its latest and earliest expected completion times ($T_L - T_E$). Figure 2-25 shows the slack time calculations for all activities of the SPTS project. All activities with a slack time equal to zero are on the critical path. Thus, all activities except 5 are on the critical path. Part of the diagram in Figure 2-24 shows two critical paths, between activities 1-2-4 and 1-3-4, because both of these parallel activities have zero slack.

In addition to the possibility of having multiple critical paths, there are actually two possible types of slack. *Free slack* refers to the amount of time a task can be delayed without delaying the early start of any immediately following. *Total slack* refers to the amount of time a task can be delayed without delaying the completion of the project. Understanding free and total slack allows the project manager to better identify where tradeoffs can be

FIGURE 2-24
A nework diagram for the SPTS project showing estimated times for each activity and the earliest and latest expected completion time for each activity.

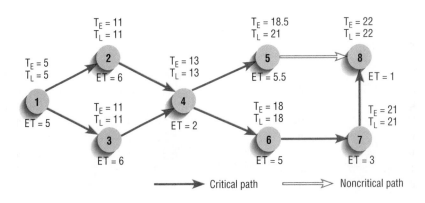

Critical path ⇒ Noncritical path

ACTIVITY	T_E	T_L	SLACK $T_L - T_E$	ON CRITICAL PATH
1	5	5	0	✓
2	11	11	0	✓
3	11	11	0	✓
4	13	13	0	✓
5	18.5	21	2.5	
6	18	18	0	✓
7	21	21	0	✓
8	22	22	0	✓

FIGURE 2-25
ACTIVITY SLACK TIME CALCULATIONS FOR THE SPTS PROJECT
All activities except number 5 are on the critical path.

made if changes to the project schedule need to be made. For more information of understanding slack and how it can be used to manage tasks see *Project Management for Business and Technology: Principles and Practice* (© 2001) by John Nicholas (Prentice Hall: Upper Saddle River, NJ).

Using Project Management Software

A wide variety of automated project management tools are available to help you manage a development project. New versions of these tools are continuously being developed and released by software vendors. Most of the available tools have a common set of features that include the ability to define and order tasks, assign resources to tasks, and easily modify tasks and resources. Project management tools are available to run on Windows-compatible personal computers, the Macintosh, and larger mainframe and workstation-based systems. These systems vary in the number of task activities supported, the complexity of relationships, system processing and storage requirements, and, of course, cost. Prices for these systems can range from a few hundred dollars for personal computer–based systems to more than $100,000 for large-scale multiproject systems. Yet a lot can be done with systems like Microsoft Project as well as public domain and shareware systems. For example, numerous shareware project management programs (e.g., MinuteMan, Delegator, or Project KickStart) can be downloaded from the World Wide Web (e.g., at www.download.com) and online bulletin boards. Because these systems are continuously changing, you should comparison shop before choosing a particular package.

We now illustrate the types of activities you would perform when using project management software. Microsoft Project for Windows is a project management system that has earned consistently high marks in computer publication reviews. When using this system to manage a project, you need to perform at least the following activities:

NET SEARCH
There is powerful software available to support the management of a project's activities. Visit http://www.prenhall. com/valacich to complete an exercise related to this topic.

- Establish a project starting or ending date
- Enter tasks and assign task relationships
- Select a scheduling method to review project reports

Establishing a Project Starting Date

Defining the general project information includes obtaining the name of the project and project manager and the starting or ending date of the project. Starting and ending dates are used to schedule future activities or backdate others (see below) based upon their duration and relationships to other activities. An example from Microsoft Project for Windows of the data entry screen for establishing a project starting or ending date is shown in Figure 2-26. This screen shows PVF's Purchasing Fulfillment System project. Here, the starting date for the project is Monday, November 3, 2003.

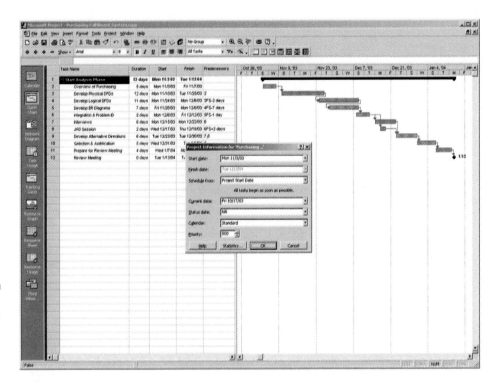

FIGURE 2-26
ESTABLISHING A PROJECT STARTING DATE IN MICROSOFT PROJECT FOR WINDOWS
Source: Reprinted with permission of Microsoft.

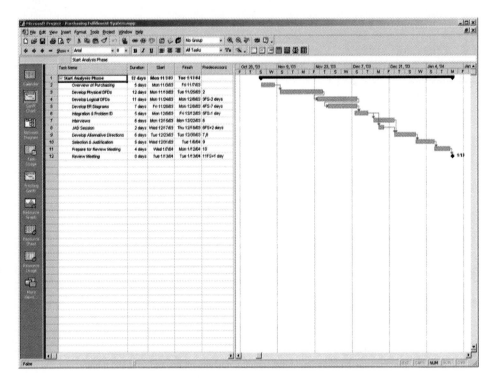

FIGURE 2-27
ENTERING TASKS AND ASSIGNING TASK RELATIONSHIPS IN MICROSOFT PROJECT FOR WINDOWS
Source: Reprinted with permission of Microsoft.

Entering Tasks and Assigning Task Relationships

The next step in defining a project is to define project tasks and their relationships. For the Purchasing Fulfillment System project, Chris defined 11 tasks to be completed when he performed the initial system analysis activities of the project (Task 1—Start Analysis Phase—is a summary task that is used to group related tasks). The task entry screen, shown in Figure 2-27, is similar to a financial spreadsheet program. The user moves the cursor to a cell with arrow keys or the mouse and then simply enters a textual Name and a numeric Duration for each activity. Scheduled Start and Scheduled Finish are automatically entered based upon the project start date and duration. To set an activity relationship, the ID number (or numbers) of the activity that must be completed before the start of the current activity is entered in the Predecessors column. Additional codes under this column make the precedence relationships more precise. For example, consider the Predecessor column for ID 6. The entry in this cell says that activity 6 cannot start until one day before the finish of activity 5. (Microsoft Project provides many different options for precedence and delays such as in this example, but discussion of these is beyond the scope of our coverage.) The project management software uses this information to construct Gantt, Network diagrams, and other project-related reports.

Selecting a Scheduling Method to Review Project Reports

Once information about all the activities for a project has been entered, it is very easy to review the information in a variety of graphical and textual formats using displays or printed reports. For example, Figure 2-27 shows the project information in a Gantt chart screen whereas Figure 2-28 shows the project information as a Network diagram. You can easily change how you view the information by making a selection from the View menu shown in Figure 2-28.

As mentioned in the chapter, interim project reports to management will often compare actual progress to plans. Figure 2-29 illustrates how Microsoft Project shows progress with a solid line within the activity bar. In this figure, task 2 is completed and task 3 is almost completed, but there remains a small percentage of work, as shown by the incomplete solid lines within the bar for this task. Assuming that this screen represents the status of the project on Friday, November 14, 2003, the third activity is approximately on schedule. Tabular reports can summarize the same information.

This brief introduction to project management software has only scratched the surface to show you the power and the features of these systems. Other features widely available and especially useful for multiperson projects relate to resource usage and utilization. Resource-related features allow you to define

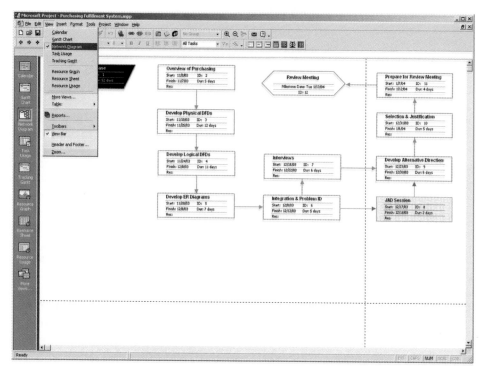

FIGURE 2-28
VIEWING PROJECT INFORMATION AS A NETWORK DIAGRAM IN MICROSOFT PROJECT FOR WINDOWS
Source: Reprinted with permission of Microsoft.

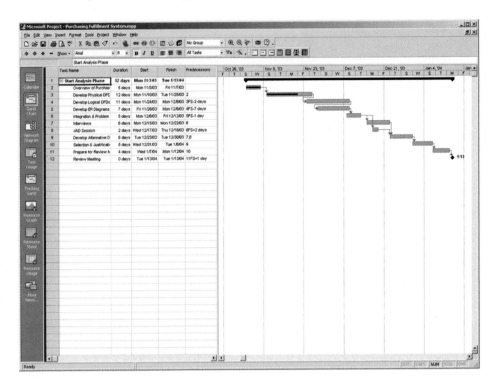

FIGURE 2-29
GANTT CHART SHOWING PROGRESS OF ACTIVITIES (RIGHT FRAME) VERSUS PLANNED ACTIVITIES (LEFT FRAME)

characteristics such as standard costing rates and daily availability via a calendar that records holidays, working hours, and vacations. These features are particularly useful for billing and estimating project costs. Often, resources are shared across multiple projects, which could significantly affect a project's schedule. Depending upon how projects are billed within an organization, assigning and billing resources to tasks is a very time-consuming activity for most project managers. The features provided in these powerful tools can greatly ease both the planning and managing of projects so that both project and management resources are effectively utilized.

Key Points Review

1. **Describe the skills required to be an effective project manager.**

   A project manager has both technical and managerial skills and is ultimately responsible for determining the size, scope, and resource requirements for a project. Once a project is deemed feasible by an organization, the project manager ensures that the project meets the customer's needs and is delivered within budget and time constraints.

2. **List and describe the skills and activities of a project manager during project initiation, project planning, project execution, and project closedown.**

   To manage the project, the project manager must execute four primary activities: project ini-

tiation, project planning, project execution, and project closedown. The focus of project initiation is on assessing the size, scope, and complexity of a project and establishing procedures to support later project activities. The focus of project planning is on defining clear, discrete activities and the work needed to complete each activity. The focus of project execution is on putting the plans developed in project initiation and planning into action. Project closedown focuses on bringing the project to an end.

3. **Explain what is meant by critical path scheduling and describe the process of creating Gantt charts and Network diagrams.**

   Critical path scheduling refers to planning methods whereby the order and duration of the project's activities directly affect the completion date of the project. Gantt charts and Network diagrams are powerful graphical techniques used in planning and controlling projects. Both Gantt and Network diagramming scheduling techniques require that a project have activities that can be defined as having a clear beginning and end, can be worked on independently of other activities, are ordered, and are such that their completion signifies the end of the project. Gantt charts use horizontal bars to represent the beginning, duration, and ending of an activity. Network diagramming is a critical path scheduling method that shows the interrelationships between activities. These charts show when activities can begin and end, which activities cannot be delayed without delaying the whole project, how much slack time each activity has, and progress against planned activities. Network diagramming's ability to use probability estimates in determining critical paths and deadlines makes it a widely used technique for very complex projects.

4. **Explain how commercial project management software packages can be used to assist in representing and managing project schedules.**

A wide variety of automated tools for assisting the project manager is available. Most tools have common features including the ability to define and order tasks, assign resources to tasks, and modify tasks and resources. Systems vary regarding the number of activities supported, the complexity of relationships, processing and storage requirements, and cost.

Key Terms Checkpoint

Here are the key terms from the chapter. The page where each term is first explained is in parentheses after the term.

1. **Critical path (p. 60)**
2. **Critical path scheduling (p. 55)**
3. **Deliverable (p. 39)**
4. **Feasibility study (p. 39)**
5. **Gantt chart (p. 45)**
6. **Network diagram (p. 47)**
7. **PERT (p. 56)**
8. **Project (p. 37)**
9. **Project closedown (p. 52)**
10. **Project execution (p. 50)**
11. **Project initiation (p. 41)**
12. **Project management (p. 41)**
13. **Project manager (p. 37)**
14. **Project planning (p. 44)**
15. **Project workbook (p. 43)**
16. **Resources (p. 54)**
17. **Slack time (p. 60)**
18. **Work breakdown structure (p. 45)**

Match each of the key terms above with the definition that best fits it.

_____ 1. An online or hard-copy repository for all project correspondence, inputs, outputs, deliverables, procedures, and standards that is used for performing project audits, orientation of new team members, communication with management and customers, scoping future projects, and performing post-project reviews.

_____ 2. An end product in a phase of the SDLC.

_____ 3. Determines if the information system makes sense for the organization from an economic and operational standpoint.

_____ 4. A controlled process of initiating, planning, executing, and closing down a project.

_____ 5. The third phase of the project management process in which the plans created in the prior phases (project initiation and planning) are put into action.

_____ 6. The first phase of the project management process in which activities are performed to assess the size, scope, and complexity of the project and to establish procedures to support later project activities.

_____ 7. A diagram that depicts project tasks and their interrelationships.

_____ 8. A planned undertaking of related activities to reach an objective that has a beginning and an end.

_____ 9. The amount of time that an activity can be delayed without delaying the project.

_____ 10. The process of dividing the project into manageable tasks and logically ordering them to ensure a smooth evolution between tasks.

_____ 11. The final phase of the project management process that focuses on bringing a project to an end.

_____ 12. A graphical representation of a project that shows each task activity as a horizontal bar whose length is proportional to its time for completion.

_____ 13. Any person, group of people, piece of equipment, or material used in accomplishing an activity.

_____ 14. A scheduling technique where the order and duration of a sequence of activities directly affect the completion date of a project.

_____ 15. An individual with a diverse set of skills—management, leadership, technical, conflict management, and customer relationship—who is responsible for initiating, planning, executing, and closing down a project.

_____ 16. The second phase of the project management process, which focuses on defining clear, discrete activities and the work needed to complete each activity within a single project.

_____ 17. The shortest time in which a project can be completed.

_____ 18. A technique that uses optimistic, pessimistic, and realistic time estimates to calculate the expected time for a particular task.

Review Questions

1. Discuss the reasons why organizations undertake information system projects.

2. List and describe the common skills and activities of a project manager. Which skill do you think is most important? Why?

3. Describe the activities performed by the project manager during project initiation.

4. Describe the activities performed by the project manager during project planning.

5. Describe the activities performed by the project manager during project execution.

6. List various project team communication methods, and describe an example of the type of information that might be shared among team members using each method.

7. Describe the activities performed by the project manager during project closedown.

8. What characteristics must a project have in order for critical path scheduling to be applicable?

9. Describe the steps involved in making a Gantt chart.

10. Describe the steps involved in making a Network diagram.

11. In which phase of the systems development life cycle does project planning typically occur? In which phase does project management occur?

12. What are some reasons why one activity may have to precede another activity before the second activity can begin? In other words, what causes precedence relationships between project activities?

Problems and Exercises

1. Which of the four phases of the project management process do you feel is most challenging? Why?

2. What are some sources of risk in a systems analysis and design project, and how does a project manager cope with risk during the stages of project management?

3. Search computer magazines for recent reviews of project management software. Which packages seem to be most popular? What are the relative strengths and weaknesses of each packaged software? What advice would you give to someone intending to buy project management software for his or her PC? Why?

4. How are information system projects similar to other types of projects? How are they different?

Are the project management packages you evaluated in Problem and Exercise 3 suited for all types of projects or for particular types of projects? Which package is best suited for information systems projects? Why?

5. If given the chance, would you become the manager of an information systems project? If so, why? Prepare a list of the strengths that you would bring to the project as its manager. If not, why not? What would it take for you to feel more comfortable managing an information systems project? Prepare a list and timetable for the necessary training you would need to feel more comfortable about managing an information systems project.

6. Calculate the expected time for the following tasks.

Task	Optimistic Time	Most Likely Time	Pessimistic Time	Expected Time
A	3	7	11	
B	5	9	13	
C	1	2	9	
D	2	3	16	
E	2	4	18	
F	3	4	11	
G	1	4	7	
H	3	4	5	
I	2	4	12	
J	4	7	9	

7. A project has been defined to contain the following list of activities along with their required times for completion.

Activity No.	Activity	Time (weeks)	Immediate Predecessors
1	Collect requirements	2	—
2	Analyze processes	3	1
3	Analyze data	3	2
4	Design processes	7	2
5	Design data	6	2
6	Design screens	1	3,4
7	Design reports	5	4,5
8	Program	4	6,7
9	Test and document	8	7
10	Install	2	8,9

a. Draw a Network diagram for the activities.
b. Calculate the earliest expected completion time.
c. Show the critical path.
d. What would happen if activity 6 were revised to take six weeks instead of one week?

8. Construct a Gantt chart for the project defined in Problem and Exercise 7 above.

9. Look again at the activities outlined in Problem and Exercise 7. Assume that your team is in its first week of the project and has discovered that each of the activity duration estimates is wrong. Activity 2 will take only two weeks to complete. Activities 4 and 7 will each take three times longer than anticipated. All other activities will take twice as long to complete as previously estimated. In addition, a new activity, number 11, has been added. It will take one week to complete and its immediate predecessors are activities 10 and 9. Adjust the Network diagram and recalculate the earliest expected completion times.

10. Construct a Gantt chart and Network diagram for a project you are or will be involved in. Choose a project of sufficient depth at either work, home, or school. Identify the activities to be completed, determine the sequence of the activities, and construct a diagram reflecting the starting, ending, duration, and precedence (Network diagram only) relationships among all activities. For your Network diagram, use the procedure in this chapter to determine time estimates for each activity and calculate the expected time for each activity. Now determine the critical path and the early and late starting and finishing times for each activity. Which activities have slack time?

11. For the project you described in Problem and Exercise 10, assume that the worst has happened. A key team member has dropped out of the project and has been assigned to another project in another part of the country. The remaining team members are having personality clashes. Key deliverables for the project are now due much earlier than expected. In addition, you have just determined that a key phase in the early life of the project will now take much longer than you had originally expected. To make matters worse, your boss absolutely will not accept that this project cannot be completed by this new deadline. What will you do to account for these project changes and problems? Begin by reconstructing your Gantt chart and Network diagram and determining a strategy for dealing with the specific changes and problems described above. If new resources are needed to meet the new deadline, outline the rationale that you will use to convince your boss that these additional resources are critical to the success of the project.

12. Assume you have a project with seven activities labeled A–G (below). Derive the earliest completion time (or early finish—EF), latest completion time (or late finish—LF), and slack for each of the following tasks (begin at time = 0). Which tasks are on the critical path? Draw a Gantt chart for these tasks.

Task	Preceding Event	Expected Duration	EF	LF	Slack	Critical Path?
A	—	5				
B	A	3				
C	A	4				
D	C	6				
E	B,C	4				
F	D	1				
G	D,E,F	5				

13. Draw a Network diagram for the tasks shown in Problem and Exercise 12. Highlight the critical path.

14. Assume you have a project with 10 activities labeled A–J. Derive the earliest completion time (or early finish—EF), latest completion time (or late finish—LF), and slack for each of the following tasks (begin at time = 0). Which tasks are on

the critical path? Highlight the critical path on your Network diagram.

Activity	Preceding Event	Expected Duration	EF	LF	Slack	Critical Path?
A	–	4				
B	A	5				
C	A	6				
D	A	7				
E	A,D	6				
F	C,E	5				
G	D,E	4				
H	E	3				
I	F,G	4				
J	H,I	5				

15. Draw a Gantt chart for the tasks shown in Problem and Exercise 14.
16. Assume you have a project with 11 activities labeled A–K (below). Derive the earliest completion time (or early finish—EF), latest completion time (or late finish—LF), and slack for each of the following tasks (begin at time = 0). Which tasks are on the critical path? Draw both Gantt chart and Network diagram for these tasks, and

make sure you highlight the critical path on your Network diagram.

Activity	Preceding Event	Expected Duration	EF	LF	Slack	Critical Path?
A	–	2				
B	A	3				
C	B	4				
D	C	5				
E	C	4				
F	D,E	3				
G	F	4				
H	F	6				
I	G,H	5				
J	G	2				
K	I,J	4				

17. Make a list of the tasks that you performed when designing your schedule of classes for this term. Develop a table showing each task, its duration, preceding event(s), and expected duration. Develop a Network diagram for these tasks. Highlight the critical path on your Network diagram.

Discussion Questions

1. You interview for a job and the employer asks you if the project management process for systems development should be a structured, formal process. What will your answer be?
2. Do you agree that breaking projects down into small, manageable tasks is an important part of managing a project? What are the pros and cons of doing this?
3. What are the strengths and weaknesses of using a Gantt chart for representing a project plan? When using a Network diagram? Is one method "better" than the other?
4. When completing a project some tasks are independent of others, whereas others are interdependent on others. What does this mean in regard to slack? How are slack and the critical path related?

Case Problems

1. Pine Valley Furniture

In an effort to better serve the various departments at Pine Valley Furniture, the PVF Information Systems department assigns one of its systems analysts to serve as a liaison to a particular business unit. Chris Martin is currently the liaison to the Purchasing department.

After graduating from Valley State University, Chris began working at Pine Valley Furniture. He began his career at Pine Valley Furniture as a Programmer/Analyst I. This job assignment required him to code and maintain financial application systems in COBOL. In the six years he

has been at PVF, he has been promoted several times; his most recent promotion was to a junior systems analyst position. During his tenure at PVF, Chris has worked on several important projects, including serving as a team member on a project that developed a five-year plan that would renovate the manufacturing information systems.

Chris enjoys his work at Pine Valley Furniture and wishes to continue moving up the information systems ladder. Over the last three years, Chris has often thought about becoming certified by the Project Management Institute. In the last three years, he has taken several courses toward his MBA, has attended three technology-related seminars, and has helped the local Feed the Hungry chapter develop, implement, and maintain its computerized information system.

a. While eating lunch one day, Juanita asked Chris about the benefits of becoming a project management professional. Briefly make a case for becoming a project management professional.
b. What are the project management professional eligibility criteria for Chris? What documentation must he provide?
c. Assume Chris has obtained his certification. What are PDUs, and how many must Chris acquire over a three-year period?
d. Several activity categories are listed as qualifying for PDUs on the Project Management Institute's Web site. Identify these categories. In which categories would you place Chris's experience?

2. Hoosier Burger

Bob and Thelma Mellankamp have come to realize that the current problems with their inventory control, customer ordering, and management reporting systems are seriously impacting Hoosier Burger's day-to-day operations. At the close of business one evening, Bob and Thelma decide to hire the Build a Better System (BBS) consulting firm. Harold Parker and Lucy Chen, two of BBS's owners, are frequent Hoosier Burger customers. Bob and Thelma are aware of the excellent consulting service BBS is providing to the Bloomington area.

Build a Better System is a medium-size consulting firm based in Bloomington, Indiana. Six months ago BBS hired you as a junior systems analyst for the firm. Harold and Lucy were impressed with your résumé, course work, and systems analysis and design internship. During your six months with BBS, you have had the opportunity to work alongside several senior systems analysts and observe the project management process.

On a Friday afternoon, you learn that you have been assigned to the Hoosier Burger project and that the lead analyst on the project is Juan Rodriquez. A short while later, Juan stops by your desk and mentions that you will be participating in the project management process. Mr. Rodriquez has scheduled a meeting with you for 10:00 A.M. on Monday to review the project management process with you. You know from your brief discussion with Mr. Rodriquez that you will be asked to prepare various planning documents, particularly a Gantt chart and a Network diagram.

a. In an effort to learn more about project management, you decide to research this topic over the weekend. Locate an article(s) that discusses project management. Summarize your findings.
b. At your meeting on Monday, Mr. Rodriquez asks you to prepare a Gantt chart for the Hoosier Burger project. Using the following information, prepare a Gantt chart.

Activity No.	Activity	Time (weeks)	Immediate Predecessors
1	Requirements collection	1	—
2	Requirements structuring	2	1
3	Alternative generation	1	2
4	Logical design	2	3
5	Physical design	3	4
6	Implementation	2	5

c. Using the information provided in part b, prepare a Network diagram.
d. After reviewing the Gantt chart and a Network diagram, Mr. Rodriquez feels that alternative generation should take only one-half week and that implementation may take three weeks. Modify your charts to reflect these changes.

3. Lilly Langley's Baking Goods Company

In 1919 Lionel Langley opened his first bakery store, which he named after his wife, Lilly. Initially he sold only breads, cakes, and flour to his customers. Through the years, the business expanded rapidly by opening additional bakeries, acquiring flour mills, and acquiring food-processing companies. After 81 years in business, the company is now a well-known, highly reputable, international corporation. Lilly Langley's Baking Goods Company (LLBGC) has over 15,000 employees, operates in 50 countries, and offers a wide variety of products.

Frederica Frampton, LLBGC's chief information officer, has just returned from a meeting with Chung Lau, LLBGC's director of operations. They discussed the many problems the company is having with getting supplies and distributing products. In essence, the end users of the current operations/manufacturing systems are demanding information that the current system just cannot provide. The current information systems are inflexible.

● Combining data housed in separate plant databases is difficult, if not impossible.

● End users have difficulty generating ad hoc reports.

• Scheduling the production lines is becoming quite tedious.

Costs to enhance the systems are becoming unwieldy, so it is now time to consider renovating these systems. Due to a top management directive, the systems must be operational within nine months.

Frederica Frampton recognizes the importance of the LLBGC operations/manufacturing systems renovation. She decides to assemble a team of her best systems analysts to develop new operations/manufacturing systems for LLBGC. You are assigned as a member of this team.

a. Lorraine Banderez, the project manager, has asked you to investigate how other companies have used project management software, particularly Microsoft Project. Investigate two companies and provide a brief summary of how each has used project management software.

b. Part of your responsibility is to assist in the preparation of the planning documents. Using the following information, prepare a Gantt chart.

Activity No.	Activity	Time (weeks)	Immediate Predecessors
1	Requirements collection	3	—
2	Requirements structuring	4	1
3	Process analysis	3	2
4	Data analysis	3	2
5	Logical design	5	3,4
6	Physical design	5	5
7	Implementation	6	6

c. Using the information from part b, prepare a Network diagram. Identify the critical path.

d. After reviewing your planning documents, Lorraine decides to modify several of the activity times. Revise both your Gantt chart and Network diagram to reflect these modifications.

Activity	Time (weeks)
Requirements collection	4
Requirements structuring	3
Process analysis	4
Data analysis	4.5
Logical design	5
Physical design	5.5
Implementation	7

CASE: BROADWAY ENTERTAINMENT COMPANY, INC.

Company Background

Case Introduction

Broadway Entertainment Company, Inc. (BEC) is a fictional company in the video rental and recorded music retail industry, but its size, strategies, and business problems (and opportunities) are comparable to those of real businesses in this fast-growing industry.

In this section, we introduce you to the company, the people who work for it, and the company's information systems. At the end of each subsequent chapter, we revisit BEC to illustrate the phase of the life cycle discussed in that chapter. Our aim is to provide you with a realistic case example of how the systems development life cycle moves through its phases and how analysts, managers, and users work together to develop an information system. Through this example, you practice working on tasks and discussing issues related to each phase in an ongoing systems development project.

The Company

As of January 2003, Broadway Entertainment Company, or BEC, owned 2,403 outlets across the United States, Canada, Mexico, and Costa Rica. There is at least one BEC outlet in every state (except Montana) and in each Canadian province. There are 58 Canadian stores, 10 in Mexico, and 5 in Costa Rica. The company is currently struggling to open a retail outlet in Japan and plans to expand into the European Union (EU) within two years. In the United States, Broadway is headquartered in Spartanburg, South Carolina, Canadian operations are headquartered in Vancouver, British Columbia, and Latin American operations are based in Mexico City.

Each BEC outlet offers for sale two product lines, recorded music (on CDs and cassette tapes) and video games. Each outlet also rents two product lines, recorded videos (on VHS tape and DVDs) and video games. In calendar year 2000, music sales and video rentals together accounted for 85 percent of Broadway's U.S. revenues (see BEC Table 2-1). Foreign operations added another $21,500,000 to company revenues.

The home video and music retail industries are strong and growing, both domestically and internationally. For several years, home video has generated more revenue than either theatrical box office or movie pay-per-view.

TABLE 2-1:   BEC Domestic Revenue by Category, Calendar Year 2000

Category	Revenue (in $000s)	Percent
Music sales	572,020	37
Compact discs	386,500	25
Cassettes	185,520	12
Video game sales	92,760	6
Video game rentals	139,140	9
Video rentals	742,080	48
Videotapes	664,780	43
DVDs	77,300	5
Total	1,546,000	100

To get a good idea of the industry in which Broadway competes, we look at five key elements of the home video and music retail industries:

1. Suppliers—all of the major distributors of recorded music (Sony, Matsushita, Time Warner), video games (Nintendo, Sega), and recorded videos (CBS, Fox, Viacom)
2. Buyers—individual consumers
3. Substitutes—television (broadcast, cable, satellite), first-run movies, Internet-based multimedia, theater, radio, concerts, and sporting events
4. Barriers to entry—few barriers and many threats, including alliances between telecommunications and entertainment companies to create cable television and WebTV, which lets consumers choose from a large number and variety of videos, music, and other home entertainment products from a computerized menu system in their homes
5. Rivalries among competing firms—large music chains (such as Musicland and Tower Records, all smaller than BEC) and large video chains (such as Blockbuster Entertainment, which is larger and more globally competitive than BEC)

Company History

The first BEC outlet opened in the Westgate Mall in Spartanburg, South Carolina, in 1977 as a music (record) sales store. The first store exclusively sold recorded music, primarily in vinyl format, but also stocked cassette tapes. Broadway's founder and current chairman of the board, Nigel Broad, had immigrated to South Carolina from his native Great Britain in 1968. After nine years of playing in a band in jazz clubs, Nigel used the money he had been left by his mother, formed Broadway Entertainment Company, Inc., and opened the first BEC outlet.

Sales were steady and profits increased. Soon Nigel was able to open a second outlet and then a third. Predicting that his BEC stores had already met Spartanburg's demand for recorded music, Nigel decided to open his fourth store in nearby Greenville in 1981. At about the same time, he added a new product line—Atari video game cartridges. Atari's release of its Space Invaders game cartridge resulted in huge profits for Nigel. The company continued to grow and Broadway expanded beyond South Carolina into neighboring states.

In the early 1980s, Nigel saw the potential in videotapes. A few video rental outlets had opened in some of Broadway's markets, but they were all small independent operations. Nigel saw the opportunity to combine video rentals with music sales in one place. He also decided that he could rent more videos to customers if he changed some of the typical video store rules such as eliminating the heavy membership fee and allowing customers to keep videos more than one night. Nigel also wanted to offer the best selection of videos anywhere.

Nigel opened his first joint music and video store at the original BEC outlet in Spartanburg in 1985. Customer response was overwhelming. In 1986, Nigel decided to turn all 17 BEC outlets into joint music and video stores. To move into the video rental business in a big way, Nigel and his chief financial officer, Bill Patton, decided to have a public offering. They were happily surprised when all 1 million shares sold at $7 per share. The proceeds also allowed Broadway to revive the dying video game line by dropping Atari and adding the newly released Nintendo game cartridges.

Profits from BEC outlets continued to grow throughout the 1980s, and Broadway further expanded by acquiring existing music and video store chains including Music World. From 1987 through 1993, the number of BEC outlets roughly doubled each year. The decision to go international, made in 1991, resulted in 12 Canadian stores that year. The initial 3 Latin American stores were opened in mid-1994. From its beginnings in 1977, with 10 employees and $398,000 in revenues, Broadway Entertainment Company, Inc., grew to 24,225 employees and worldwide revenues of $1,567,500,000 by January 1, 2003.

Company Organization

In 1992, when the company opened its one-thousandth store, Nigel decided that he no longer wanted to be chief executive officer of the company. Nigel decided to fill only the position of chairman, and he promoted his close friend Ira Abramowitz to the offices of president and CEO (see BEC Figure 2-1).

Most of Broadway's other senior officers have also been promoted from within. Bill Patton, the chief financial officer, started as the fledgling company's

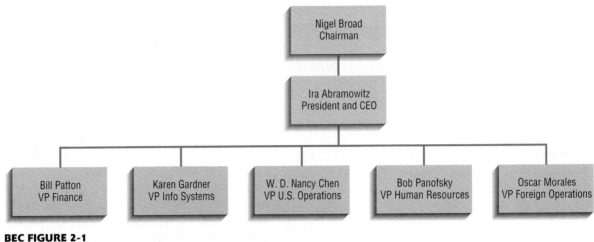

BEC FIGURE 2-1
BROADWAY ENTERTAINMENT COMPANY, INC., ORGANIZATION CHART

first bookkeeper and accountant. Karen Gardner had been part of the outside consulting team that built Broadway's first information system in 1986 and 1987. She became the vice president in charge of IS for BEC in 1990. Bob Panofsky, the vice president for human resources, had been with the company since 1981. An exception to the promote-from-within tendency, W. D. Nancy Chen, the vice president for domestic operations, had been recruited from Music World in 1991 shortly before the chain was purchased by Broadway. Oscar Morales had been hired in 1992 from Blockbuster Entertainment, where he had been in charge of Latin American expansion.

Development of Information Systems

Broadway Entertainment Company operated from 1977 until 1984 without any computer-based information systems support. As the company grew, ledgers, files, and customer account information became unruly. As with many businesses this size, the owner did not have the expertise or the capital for developing the company's own information systems. For example, Bill Patton managed inventory by hand until he bought an IBM AT in 1984. Computerizing the company made the expansion to 10 stores in 1984 much easier.

In 1985 BEC had nobody trained in information systems on staff, and all the BEC managers were quite busy coping with the business expansion. Nigel and Bill considered hiring a small staff of experienced IS professionals, but they did not know how to manage such a group, how to select quality staff, or what to expect from such employees. Nigel and Bill realized that computer software could be quite complicated, and building systems for a rapidly changing organization could be quite a challenge. Nigel and Bill also knew that building information systems required discipline. So Nigel, after talking

with leaders of several other South Carolina businesses, contacted the information consulting firm of Fitzgerald McNally, Inc., about designing and building a custom computer-based information system for Broadway. In 1985, no prewritten programs were available to help run the still relatively new business of video and music rental and sales stores.

Nigel and Bill wanted the new system to perform accounting, payroll, and inventory control. Nigel wanted the system to be readily expandable as he was planning for Broadway's rapid growth. At the operational level, Nigel realized that the video rental business would require unique features in its information system. For one thing, rental customers would not only be taking product from the store, they would also be returning it at the end of the rental period. Further, customers would be required to register with Broadway and attach some kind of deposit to their account in order to help ensure that videos would be returned.

At a managerial level, Nigel wanted the movement of videos in and out of the stores and all customer accounts computerized. Nigel also wanted to be able to search through the data on Broadway's customers describing their rental habits. He wanted to know which videos were the most popular, and he wanted to know who Broadway's most frequent customers were, not only in South Carolina but also in every location where Broadway did business.

Fitzgerald McNally, Inc., was happy to get Broadway's account. It assigned Karen Gardner to head the development team. Karen led a team of her own staff of analysts and programmers, along with several BEC managers, in a thorough analysis and design study. The methodology applied in this study provided the discipline needed for such a major systems development effort. The methodology began with information planning and continued through all phases of the systems development life cycle.

Karen and her team delivered and installed the system at the end of the two-year project. The system was centralized, with an IBM 4381 mainframe installed at headquarters in Spartanburg and three terminals, three light pens, and three dot-matrix printers installed in each BEC outlet. The light pens recorded, for example, when the tapes were rented and when they were returned by reading the bar code on the cassette. The light pens were also used to read the customer's account number, which was recorded in a bar code on the customer's BEC account card. The printers generated receipts. In addition, the system included a small personal computer and printer to handle a few office functions such as the ordering and receiving of goods. The software monitored and updated inventory levels. Another software product generated and updated the customer database, whereas other parts of the final software package were designed for accounting and payroll.

In 1990, Karen Gardner left Fitzgerald McNally and joined Broadway as the head of its information systems group. Karen led the effort to expand and enhance Broadway's information systems as the company grew to over 2,000 company-owned stores in 1995. Broadway now uses a client/server network of computers at headquarters and in-store point-of-sale (POS) computer systems to handle the transaction volume generated by millions of customers at all BEC outlets.

Information Systems at BEC Today

BEC has two systems development and support groups, one for in-store applications and the other for corporate, regional, and country-specific applications. The corporate development group has liaison staff with the in-store group, because data in many corporate systems feed or are fed by in-store applications (e.g., market analysis systems depend on transaction data collected by the in-store systems). BEC creates both one-year and three-year IS plans that encompass both store and corporate functions.

The functions of the original in-store systems at BEC have changed very little since they were installed in 1987—for example, customer and inventory tracking are still done by pen-based, bar code scanning of product labels and membership cards. Rentals and returns, sales, and other changes in inventory as well as employee time in and out are all captured at the store in electronic form via a local POS computer system. These data are transmitted in batches at night using modems and regular telephone connections to corporate headquarters, where all records are stored in a network of IBM AS/400 computers (see BEC Figure 2-2).

As shown in BEC Figure 2-2, each BEC store has an NCR computer that serves as a host for a number of POS terminals at checkout counters and is used for generating reports. Some managers have also learned how to use spreadsheet, word processing, and other packages to handle functions not supported by systems provided by BEC. The front-end communications processor offloads traffic from the IBM AS/400 network so that the servers can concentrate on data processing applications. BEC's communication protocol is SNA (System Network Architecture), an IBM standard. Corporate databases are managed by IBM's relational DBMS DB2. BEC uses a variety of programming environments, including C, COBOL, SQL (as part of DB2), and code generators.

Inventory control and purchasing are done centrally and employees are paid by the corporation. Each store has electronic records of only its own activity, including inventory and personnel. Profit and loss, balance sheets, and other financial statements are produced for each store by centralized systems. In the following sections we review the applications that exist in the stores and at the corporate level.

In-Store Systems

BEC Table 2-2 lists the application systems installed in each store. BEC has developed a turnkey package of hardware and software (called Entertainment Tracker—ET), which is installed in each store worldwide. Besides English, the system also works in Spanish and French.

As you can see from BEC Table 2-2, all of these applications are transaction processing systems. In fact, there is a master screen on the POS terminals from which each ET application is activated. These systems work off a local decentralized database, and there is a similarly structured database for each store. Various batched data transfers occur between corporate and store systems at night (store transactions, price and membership data updates, etc.). The local database contains data on members, products, sales, rentals, returns, employees, and work assignments. The database contains only current data—the history of customer sales and rentals is retained in a corporate database. Thus, local stores do not retain any customer sales and rental activity (except for open rentals).

Data for those members who have had no activity at a local store for more than one year are purged from the local database. When members use a BEC membership card and no member record exists in the local database, members are asked to provide a local address and phone number where they can be contacted.

All store employees, except the store manager who is on salary, are paid on an hourly basis, so clock-in and -out times are entered as a transaction,

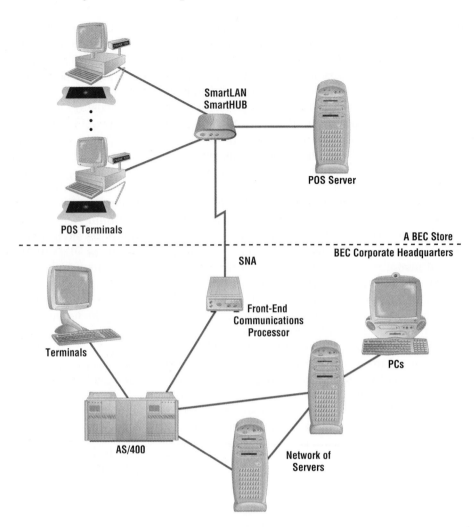

BEC FIGURE 2-2
BEC HARDWARE AND NETWORK ARCHITECTURE

TABLE 2-2: List of BEC In-Store (Entertainment Tracker) Applications

System Name	Description
Membership	Supports enrollment of new members, issuing membership cards, reinstatement of inactive members, and local data management for transient members
Rental	Supports rentals and returns of all products and outstanding rental reports
Sales	Supports sales and returns of all products (including videos, music, snack food, BEC apparel, and gift certificates)
Inventory control	Supports all changes in rental and sales inventory that are not sales based (e.g., receipt of a new tape for rental, rejection of goods damaged in shipment, and transfer of an item from rental to sales categories)
Employee	Supports hiring and terminating hourly employees, as well as all time-reporting activities

using employee badges with a bar code strip, on the same POS terminal used for member transactions. Paychecks are delivered by express mail twice a month. Employee reports (e.g., attendance, payroll, and productivity) are produced by corporate systems and sent to store managers.

All other store record keeping is manual and corporation offices handle accounts receivables and payables. The local store manager is responsible for contacting via phone or mail members who are late in returning rented items. Each night a file of delinquent members is transmitted to each store and, if a member tries to use a delinquent membership, the member is asked to return all outstanding rentals before renting any more items and the current transaction is invalidated. When terminated members try to use their cards, a BEC store clerk keeps the membership card and members are given a printed form that explains their rights at that point. Stolen membership cards are handled similarly, except that the store manager deals personally with people using cards that have been reported stolen.

Corporate Systems

Corporate systems run on IBM servers using IBM's DB2 relational database management system, although some run on PCs. Application software is written in COBOL, C, SQL (a database processing language), and several 4GLs, and all systems are developed by BEC. Clerks and managers use PCs for interactive access into corporate systems as well as for stand-alone, end-user applications such as word processing, spreadsheets, specialized databases, and business graphics.

There are more than 20 major corporate systems with over 350 programs and approximately 500,000 lines of code. There are many more specialized systems, often developed for individual managers, projects, or special events. BEC Table 2-3 lists some of the most active and largest of the major corporate systems.

One interesting aspect of the Banking application is that because stores have no financial responsibilities, BEC uses a local bank only for daily deposits and getting change. BEC's corporate bank, NCNB, arranges correspondent banking relationships for BEC so that local deposits are electronically transferred to BEC's corporate accounts with NCNB.

BEC's applications are still expanding, and they are under constant revision. For example, in cooperation with several hotel and motel chains that provide VCRs for rental, BEC is undertaking a new marketing campaign aimed at frequent travelers. At any one time, there are approximately 10 major system changes or new systems under development for corporate applications with over 250 change requests received annually covering requirements from minor bug fixes to reformatting or creating new reports to whole new systems.

TABLE 2-3:   List of BEC Corporate Applications

System Name	Description
Human resources	Supports all employee functions, including payroll, benefits, employment and evaluation history, training and career development (including a college scholarship for employees and dependents)
Accounts receivable	Supports notification of overdue fees and collection of payment from delinquent customers
Banking	Supports interactions with banking institutions, including account management and electronic funds transfers
Accounts payable, purchasing, and shipping	Supports ordering products and all other purchased items used internally and resold/rented, distribution of products to stores, and payment to vendors
General ledger and financial accounting	Supports all financial statement and reporting functions
Property management	Supports the purchasing, rental, and management of all properties and real estate used by BEC
Member tracking	Supports record keeping on all BEC members and transmits and receives member data between corporate and in-store systems
Inventory management	Supports tracking inventory of items in stores and elsewhere and reordering those items that must be replenished
Sales tracking and analysis	Supports a variety of sales analysis activities for marketing and product purchasing functions based on sales and rental transaction data transmitted nightly from stores
Store contact	Supports transmittal of data between corporate headquarters and stores nightly, and the transfer of data to and from corporate and store systems
Fraud	Supports monitoring abuse of membership privileges
Shareholder services	Supports all shareholder activities, including recording stock purchases and transfers, disbursement of dividends, and reporting
Store and site analysis	Supports the activity and profit analysis of stores and the analysis of potential sites for stores

Status of Systems

A rapidly expanding business, BEC has created significant growth for the information systems group managers. Karen Gardner is considering reorganizing her staff to provide more focused attention to the international area. BEC still uses the services of Fitzgerald McNally when Karen's resources are fully committed. Karen's department includes 33 developers (programmers, analysts, and other specialists in database, networking, etc.) plus data center staff, which is now large and technically skilled enough to handle almost all requests.

Karen's current challenge in managing the IS group is keeping her staff current in the skills they need to successfully support the systems in a rapidly changing and competitive business environment. In addition, the members of Karen's staff need to be excellent project managers, to understand the business completely, and to exhibit excellent communication with clients and each other. Karen is also concerned about information systems literacy among BEC management and that technology is not being as thoroughly exploited as it could be.

To deal with this situation, Karen is considering several initiatives. First, she has requested a sizable increase in her training budget, including expanding the benefits of the college tuition reimbursement program. Second, Karen is considering instituting a development program that will better develop junior staff members and will involve user departments. As part of this program, BEC personnel will rotate in and out of the IS group as part of normal career progression. This program should greatly improve relationships with user departments and increase end-user understanding of technology. The development of this set of technical, managerial, business, and interpersonal skills in and outside IS is a critical success factor for Karen's group in responding to the significant demands and opportunities of the IS area.

Case Summary

Broadway Entertainment Company is a $1.66 billion international chain of music, video, and game rental and sales outlets. BEC started with one store in Spartanburg, South Carolina in 1977 and has grown through astute management of expansion and acquisitions into over 2,000 stores in four countries.

BEC's hardware and software environment is similar to that used by many national retail chains. Each store has a computer system with point-of-sale terminals that run mainly sales and rental transaction processing applications, such as product sales and rental, membership, store-level inventory, and employee pay activities. Corporate systems are executed on a network of computers at a corporate data center. Corporate systems handle all accounting, banking, property, sales and member tracking, and other applications that involve data from all stores.

BEC is a rapidly growing business with significant demand for information services. To build and maintain systems, BEC has divided its staff into functional area groups for both domestic and international needs. BEC uses modern database management and programming language technologies. The BEC IS organization is challenged by keeping current in both business and technology areas. We will see in case studies in subsequent chapters how BEC responds to a request for a new system within this business and technology environment.

Case Questions

1. What qualities have led to BEC's success so far?
2. Is the IS organization at BEC poised to undertake significant systems development in the near future?
3. What specific management skills do systems analysts at BEC need?
4. What specific communication skills do systems analysts at BEC need?
5. What specific areas of organizational knowledge do systems analysts at BEC need beyond the information provided in this case?
6. Do corporate and in-store systems seem to be tightly or loosely related at BEC? Why do you think this is so?
7. What challenges and limitations will affect what and how systems are developed in the future at BEC?
8. Chapter 1 of this book identified roles associated with systems development. This BEC case lists the types of jobs held by people in the IS organization at BEC. Do you see any IS development roles missing at BEC?
9. Is BEC ready to use the Internet to deliver information systems to employees and customers? Explain.

3

Systems Planning and Selection

> **Objectives**

After studying this chapter, you should be able to:

- Describe the steps involved when identifying and selecting projects and initiating and planning projects.
- Explain the need for and the contents of a Statement of Work and Baseline Project Plan.
- List and describe various methods for assessing project feasibility.
- Describe the differences between tangible and intangible benefits and costs and the differences between one-time and recurring costs.
- Perform cost-benefit analysis and describe what is meant by the time value of money, present value, discount rate, net present value, return on investment, and break-even analysis.
- Describe the activities and participant roles within a structured walkthrough.
- Describe the three classes of Internet electronic commerce applications: Internet, intranets, and extranets.

Chapter Preview . . .

The acquisition, development, and maintenance of information systems consume substantial resources for most organizations. Organizations can benefit from following a formal process for identifying, selecting, initiating, and planning projects. The first phase of the systems development life cycle—systems planning and selection—deals with this issue. As you can see in Figure 3-1, there are two primary activities in this phase. In the next section, you learn about the first activity, a general method for identifying and selecting projects and the deliverables and outcomes from this process. Next, we review the second activity, project initiation and planning, and present several techniques for assessing project feasibility. The information uncovered during feasibility analysis is organized into a document called a Baseline Project Plan. Once this plan is developed, a formal review of the project can be conducted. The process of building this plan is discussed next. Before the project can evolve to the next phase of the systems development life cycle—systems analysis—the project plan must be reviewed and accepted. In the final major section of the chapter, we provide an overview of the project review process.

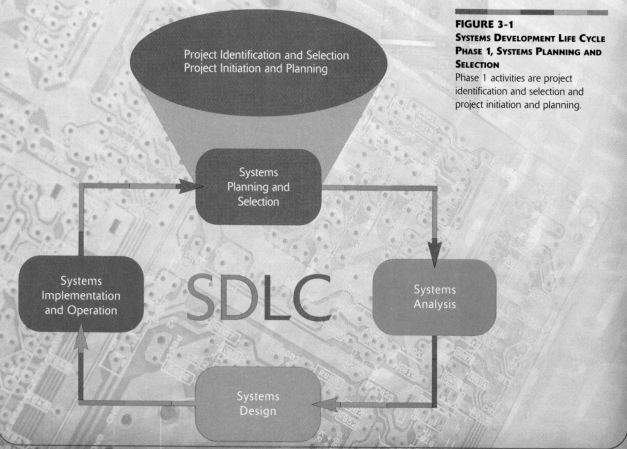

FIGURE 3-1
SYSTEMS DEVELOPMENT LIFE CYCLE PHASE 1, SYSTEMS PLANNING AND SELECTION
Phase 1 activities are project identification and selection and project initiation and planning.

Identifying and Selecting Projects

The first activity of the systems planning and selection phase of the SDLC is project identification and selection. During this activity, a senior manager, a business group, an IS manager, or a steering committee identifies and assesses all possible systems development projects that a business unit could undertake. Next, those projects deemed most likely to yield significant organizational benefits, given available resources, are selected. Organizations vary in their approach to identifying and selecting projects. In some organizations, project identification and selection is a very formal process in which projects are outcomes of a larger overall planning process. For example, a large organization may follow a formal project identification process that involves rigorously comparing all competing projects. Alternatively, a small organization may use informal project selection processes that allow the highest-ranking IS manager to select projects independently or allow individual business units to decide on projects after agreeing on funding.

Requests for information systems development can come from three key sources, as depicted in Figure 3-2:

1. Managers and business units who want to replace or extend an existing system in order to gain needed information or to provide a new service to customers
2. Managers who want to make a system more efficient, less costly to operate, or want to move a system to a new operating environment
3. Formal planning groups that want to improve an existing system in order to help the organization meet its corporate objectives, such as providing better customer service

Regardless of how an organization executes the project identification and selection process, a common sequence of activities occurs. In the following sections, we describe a general process for identifying and selecting projects and producing the deliverables and outcomes of this process.

The Process of Identifying and Selecting Information Systems Development Projects

Project identification and selection consists of three primary activities: identifying potential development projects, classifying and ranking projects, and selecting projects for development. Each of these activities is described next.

1. *Identifying potential development projects.* Organizations vary as to how they identify projects. This process can be performed by:

FIGURE 3-2
THREE KEY SOURCES FOR
INFORMATION SYSTEMS PROJECTS

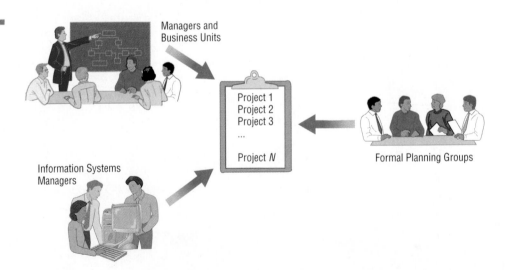

◗ A key member of top management, either the CEO of a
small- or medium-size organization or a senior executive in
a larger organization

◗ A steering committee, composed of a cross-section of managers with an interest in systems

◗ User departments, in which either the head of the requesting unit or a committee from the requesting department
decides which projects to submit (as a systems analyst, you
will help users prepare such requests)

◗ The development group or a senior IS manager

Each identification method has strengths and weaknesses. For example,
projects identified by top management have a strategic organizational
focus. Alternatively, projects identified by steering committees reflect the
diversity of the committee and therefore have a cross-functional focus.
Projects identified by individual departments or business units have a narrow, tactical focus. The development group identifies projects based on the
ease with which existing hardware and systems will integrate with the proposed project. Other factors, such as project cost, duration, complexity, and
risk, also influence the people who identify a project. Table 3-1 summarizes
the characteristics of each selection method.

Of all the possible project sources, those identified by top management
and steering committees most often reflect the broader needs of the organization. These groups have a better understanding of overall business
objectives and constraints. Projects identified by top management or by a
diverse steering committee are therefore referred to as coming from a top-
down source.

Projects identified by a functional manager, a business unit, or the information systems development group are often designed for a particular business need within a given business unit and may not reflect the overall objectives of the organization. This does not mean that projects identified by
individual managers, business units, or the IS development group are defi-
cient, only that they may not consider broader organizational issues.
Project initiatives stemming from managers, business units, or the development group are referred to as coming from a bottom-up source. As a systems analyst, you provide ongoing support for users of these types of projects and are involved early in the life cycle. You help managers describe
their information needs and the reasons for doing the project. These
descriptions are evaluated in selecting which projects will be approved to
move into the project initiation and planning activities.

In sum, projects are identified by both top-down and bottom-up initiatives. The formality of identifying and selecting projects can vary substantially across organizations. Because limited resources preclude the

TABLE 3-1: Common Characteristics of Alternative Methods for Making Information Systems Identification and Selection Decisions

Project Source	Cost	Duration	Complexity	System Size	Focus
Top management	Highest	Longest	Highest	Largest	Strategic
Steering committee	High	Long	High	Large	Cross-functional
User department	Low	Short	Low	Small	Departmental
Development group	Low–high	Short–long	Low–high	Small–large	Integration with existing systems

development of all proposed systems, most organizations have some process of classifying and ranking each project's merit. Those projects deemed to be inconsistent with overall organizational objectives, redundant in functionality to some existing system, or unnecessary will not be considered.

2. *Classifying and ranking IS development projects.* Assessing the merit of potential projects is the second major activity in the project identification and selection phase. As with project identification, classifying and ranking projects can be performed by top managers, a steering committee, business units, or the IS development group. The criteria used to assign the merit of a given project can vary based on the size of the organization. Table 3-2 summarizes the criteria commonly used to evaluate projects. In any given organization, one or several criteria might be used during the classifying and ranking process.

   As with project identification, the criteria used to evaluate projects will vary by organization. If, for example, an organization uses a steering committee, it may choose to meet monthly or quarterly to review projects and use a wide variety of evaluation criteria. At these meetings, new project requests are reviewed relative to projects already identified, and ongoing projects are monitored. The relative ratings of projects are used to guide the final activity of this identification process—project selection.

3. *Selecting IS development projects.* The selection of projects is the final activity in the project identification and selection phase. The short- and long-term projects most likely to achieve business objectives are considered. As business conditions change over time, the relative importance of any single project may substantially change. Thus, the identification and selection of projects is a very important and ongoing activity.

   Numerous factors must be considered when selecting a project, as illustrated in Figure 3-3. These factors are:

   ◗ Perceived needs of the organization
   ◗ Existing systems and ongoing projects
   ◗ Resource availability
   ◗ Evaluation criteria
   ◗ Current business conditions
   ◗ Perspectives of the decision makers

TABLE 3-2:   Possible Evaluation Criteria When Classifying and Ranking Projects

Evaluation Criteria	Description
Value chain analysis	Extent to which activities add value and costs when developing products and/or services; information systems projects providing the greatest overall benefits will be given priority over those with fewer benefits
Strategic alignment	Extent to which the project is viewed as helping the organization achieve its strategic objectives and long-term goals
Potential benefits	Extent to which the project is viewed as improving profits, customer service, etc. and the duration of these benefits
Resource availability	Amount and type of resources the project requires and their availability
Project size/duration	Number of individuals and the length of time needed to complete the project
Technical difficulty/risks	Level of technical difficulty to complete the project successfully within given time and resource constraints

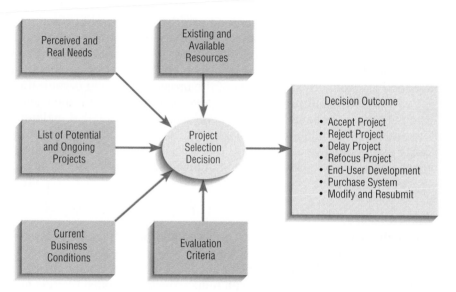

FIGURE 3-3
Numerous factors must be
considered when selecting a project.
Decisions can result in one of seven
outcomes.

This decision-making process can lead to numerous outcomes. Of course, projects can be accepted or rejected. Acceptance of a project usually means that funding to conduct the next SDLC activity has been approved. Rejection means that the project will no longer be considered for development. However, projects may also be conditionally accepted; projects may be accepted pending the approval or availability of needed resources or the demonstration that a particularly difficult aspect of the system can be developed. Projects may also be returned to the original requesters who are told to develop or purchase the requested system themselves. Finally, the requesters of a project may be asked to modify and resubmit their request after making suggested changes or clarifications.

Deliverables and Outcomes

The primary deliverable, or end product, from the project identification and selection phase is a schedule of specific IS development projects. These projects come from both top-down and bottom-up sources, and once selected they move into the second activity within this SDLC phase—project initiation and planning. This sequence of events is illustrated in Figure 3-4. An outcome of this activity is the assurance that people in the organization gave careful

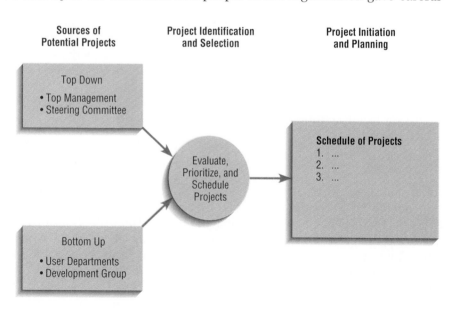

FIGURE 3-4
Information systems development
projects come from both top-down
and bottom-up initiatives.

consideration to project selection and clearly understood how each project could help the organization reach its objectives. Due to the principle of incremental commitment, a selected project does not necessarily result in a working system. **Incremental commitment** means that after each subsequent SDLC activity, you, other members of the project team, and organization officials will reassess your project. This reassessment will determine whether the business conditions have changed or whether a more detailed understanding of a system's costs, benefits, and risks would suggest that the project is not as worthy as previously thought. In the next section, we discuss several techniques for gaining a thorough understanding of your development project.

Incremental commitment
A strategy in systems analysis and design in which the project is reviewed after each phase and continuation of the project is rejustified in each of these reviews.

Initiating and Planning Systems Development Projects

Many activities performed during initiation and planning could also be completed during the next phase of the SDLC—systems analysis. Proper and insightful project initiation and planning, including determining project scope and identifying project activities, can reduce the time needed to complete later project phases, including systems analysis. For example, a careful feasibility analysis conducted during initiation and planning could lead to rejecting a project and saving a considerable expenditure of resources. The actual amount of time expended will be affected by the size and complexity of the project as well as by the experience of your organization in building similar systems. A rule of thumb is that between 10 and 20 percent of the entire development effort should be expended on initiation and planning. In other words, you should not be reluctant to spend considerable time and energy early in the project's life in order to fully understand the motivation for the requested system.

Most organizations assign an experienced systems analyst, or team of analysts for large projects, to perform project initiation and planning. The analyst will work with the proposed customers—managers and users in a business unit—of the system and other technical development staff in preparing the final plan. Experienced analysts working with customers who well understand their information services needs should be able to perform a detailed analysis with relatively little effort. Less experienced analysts with customers who only vaguely understand their needs will likely expend more effort in order to be certain that the project scope and work plan are feasible.

The objective of project initiation and planning is to transform a vague system request document into a tangible project description, as illustrated in Figure 3-5. Effective communication among the systems analyst, users, and management is crucial to the creation of a meaningful project plan. Getting all parties to agree on the direction of a project may be difficult for cross-department projects when different parties have different business objectives. Projects at large, complex organizations require systems analysts to take more time to analyze both the current and proposed systems.

In the remainder of this chapter, we describe how a systems analyst develops a clear project description.

The Process of Initiating and Planning Systems Development Projects

As its name implies, two major activities occur during project initiation and project planning. Project initiation focuses on activities that will help organize a team to conduct project planning. During initiation, one or more analysts are assigned to work with a customer to establish work standards and communication procedures. Table 3-3 summarizes five activities performed during project initiation.

The second activity, project planning, focuses on defining clear, discrete tasks and the work needed to complete each task. The objective of the project

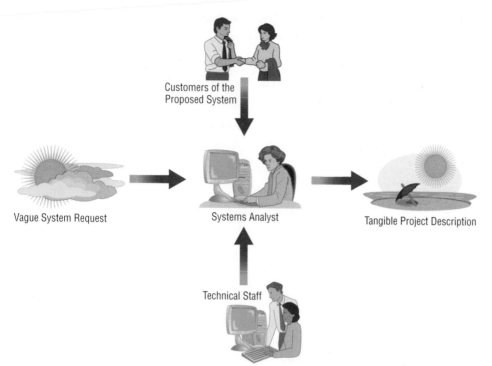

Customers of the
Proposed System

Vague System Request

Systems Analyst

Tangible Project Description

Technical Staff

FIGURE 3-5
The systems analyst transforms a
vague systems request into a
tangible project description during
systems planning and selection.

planning process is to produce two documents: a Baseline Project Plan (BPP)
and the Statement of Work (SOW). The BPP becomes the foundation for the
remainder of the development project. It is an internal document used by the
development team but not shared with customers. The SOW outlines the objectives and constraints of the project for the customer. Both the BPP and SOW
are discussed below. As with the project initiation process, the size, scope, and
complexity of a project dictate the comprehensiveness of the project planning
process and resulting documents. Further, numerous assumptions about
resource availability and potential problems will have to be made. Analysis of
these assumptions and system costs and benefits forms a **business case.** Table
3-4 lists the activities performed during project planning.

Deliverables and Outcomes

The major outcomes and deliverables from project initiation and planning are the
Baseline Project Plan and the Statement of Work. The **Baseline Project Plan
(BPP)** contains all information collected and analyzed during the project initiation and planning activity. The plan reflects the best estimate of the project's
scope, benefits, costs, risks, and resource requirements given the current understanding of the project. The BPP specifies detailed project activities for the next
life cycle phase—systems analysis—and less detail for subsequent project
phases (because these depend on the results of the analysis phase). Similarly,

Business case
A written report that outlines
the justification for an
information system. The report
highlights economic benefits
and costs and the technical and
organizational feasibility of the
proposed system.

Baseline Project Plan (BPP)
The major outcome and
deliverable from the project
initiation and planning phase. It
contains an estimate of the
project's scope, benefits, costs,
risks, and resource
requirements.

TABLE 3-3:   Types of Activities Performed during Project Initiation

- Establishing the project initiation team

- Establishing a relationship with the customer

- Establishing the project initiation plan

- Establishing management procedures

- Establishing the project management environment and project workbook

TABLE 3-4:   Activities Performed during Project Planning

- Describing the project scope, alternatives, and feasibility
- Dividing the project into manageable tasks
- Estimating resources and creating a resource plan
- Developing a preliminary schedule
- Developing a communication plan
- Determining project standards and procedures
- Identifying and assessing risk
- Creating a preliminary budget
- Developing a Statement of Work
- Setting a Baseline Project Plan

benefits, costs, risks, and resource requirements will become more specific and quantifiable as the project progresses. The project selection committee uses the BPP to help decide whether to continue, redirect, or cancel a project. If selected, the BPP becomes the foundation document for all subsequent SDLC activities; however, it is updated as new information is learned during subsequent SDLC activities. We explain how to construct the BPP later in the chapter.

Statement of Work (SOW)
A document prepared for the customer during project initiation and planning that describes what the project will deliver and outlines generally at a high level all work required to complete the project.

The **Statement of Work (SOW)** is a short document prepared for the customer that describes what the project will deliver and outlines all work required to complete the project. The SOW is a useful communication tool that assures that both you and your customer have a common understanding of the project. The SOW is an easy document to create because it typically consists of a high-level summary of the BPP information (described later). A sample SOW is shown in Figure 3-6. Depending upon your relationship with your customer, the role of the SOW may vary. At one extreme, the SOW can be used as the basis of a formal contractual agreement outlining firm deadlines, costs, and specifications. At the other extreme, the SOW can simply be used as a communication vehicle to outline the current estimates of what the project will deliver, when it will be completed, and the resources it may consume. A contract programming or consulting firm, for example, may establish a very formal relationship with a customer and use an extensive and formal SOW. Alternatively, an internal development group may develop a brief SOW that is intended to inform customers rather than to set contractual obligations and deadlines.

Assessing Project Feasibility

NET SEARCH
Building a high-performance development team relates directly to the speed of development and quality of the new system. Visit http://www.prenhall.com/valacich to complete an exercise related to this topic.

Most information systems projects have budgets and deadlines. Assessing project feasibility is a required task that can be a large undertaking because it requires you, as a systems analyst, to evaluate a wide range of factors. Although the specifics of a given project will dictate which factors are most important, most feasibility factors fall into the following six categories:

- Economic
- Operational
- Technical
- Schedule
- Legal and contractual
- Political

Pine Valley Furniture   Prepared: 9/20/2003
Statement of Work

Project Name: Customer Tracking Systems
PVF Project Manager: Jim Woo

Customer: Marketing
Project Sponsor: Jackie Judson

Project Start/End (projected):   10/1/03–2/1/04

PVF Development Staff Estimates (labor-months):
  Programmers: 2.0
  Jr. Analysts: 1.5
  Sr. Analysts: 0.3
  Supervisors: 0.1
  Consultants: 0.0
  Librarian: 0.1

  TOTAL: **4.0**

Project Description

Goal
   This project will implement a customer tracking system for the
   marketing department. The purpose of this system is to
   automate the . . . to save employee time, reduce errors, have
   more timely information, . . .

Objectives
  • minimize data entry errors
  • provide more timely information
  • . . .

Phases of Work
   The following tasks and deliverables reflect the current
   understanding of the project:
      In Analysis, . . .
      In Design, . . .
      In Implementation, . . .

FIGURE 3-6
**STATEMENT OF WORK (SOW) FOR
THE CUSTOMER TRACKING SYSTEM
AT PINE VALLEY FURNITURE**
The SOW outlines the staff needed
for the project and describes the
goal, objectives, and phases of the
project.

The analysis of these six factors forms the business case that justifies the
expenditure of resources on the project. In the remainder of this section, we
examine various feasibility studies, beginning with economic feasibility.

To help you better understand the feasibility assessment process, we examine a project at Pine Valley Furniture. Jackie Judson, Pine Valley Furniture's
(PVF) vice president of Marketing, prepares a System Service Request (SSR),
illustrated in Figure 3-7, to develop a Customer Tracking System. Jackie feels
that this system would allow PVF's marketing group to better track customer
purchase activity and sales trends. She also feels that, if implemented, the
Customer Tracking System (CTS) would help improve revenue, a tangible benefit, and improve employee morale, an intangible benefit. PVF's Systems
Priority Board selected this project for an initiation and planning study. The
board assigned senior systems analyst Jim Woo to work with Jackie to initiate
and plan the project. At this point in the project, all project initiation activities
have been completed: Jackie prepared an SSR, the selection board reviewed
the SSR, and Jim Woo was assigned to work on the project. Jackie and Jim can

Pine Valley Furniture
System Service Request

REQUESTED BY _____Jackie Judson_____ DATE: _March 23, 2003_____

DEPARTMENT _____Marketing_____

LOCATION _____Headquarters, 570c_____

CONTACT _____Tel: 4-3290   FAX: 4-3270   e-mail: jjudson_____

TYPE OF REQUEST   URGENCY

[X] New System   [   ] Immediate—Operations are impaired or opportunity lost
[   ] System Enhancement   [   ] Problems exist, but can be worked around
[   ] System Error Correction   [X] Business losses can be tolerated until new system installed

PROBLEM STATEMENT

Sales growth at PVF has caused a greater volume of work for the marketing department. This volume of work has greatly increased the volume and complexity of the data we need to deal with and understand. We are currently using manual methods and a complex PC-based electronic spreadsheet to track and forecast customer buying patterns. This method of analysis has many problems: (1) We are slow to catch buying trends as there is often a week or more delay before data can be taken from the point-of-sale system and manually enter it into our spreadsheet; (2) the process of manual data entry is prone to errors (which makes the results of our subsequent analysis suspect); and (3) the volume of data and the complexity of analyses conducted in the system seem to be overwhelming our current system—sometimes the program starts recalculating and never returns anything or it returns information that we know cannot be correct.

SERVICE REQUEST

I request a thorough analysis of our current method of tracking and analysis of customer purchasing activity with the intent to design and build a completely new information system. This system should handle all customer purchasing activity, support display and reporting of critical sales information, and assist marketing personnel in understanding the increasingly complex and competitive business environment. I feel that such a system will improve the competitiveness of PVF, particularly in our ability to better serve our customers.

IS LIAISON   Jim Woo   (Tel: 4-6207   FAX: 4-6200   e-mail: jwoo)

SPONSOR   Jackie Judson, Vice President, Marketing

---------------------- TO BE COMPLETED BY SYSTEMS PRIORITY BOARD ----------------------

[   ] Request approved   Assigned to _____
                              Start date _____
[   ] Recommend revision
[   ] Suggest user development
[   ] Reject for reason _____
_____

FIGURE 3-7
SYSTEM SERVICE REQUEST (SSR) FOR A CUSTOMER TRACKING SYSTEM AT PINE VALLEY FURNITURE
The SSR includes contact information, a problem statement, service request statement, and liaison contact information.

now focus on project planning activities, which will lead to the Baseline Project Plan.

Assessing Economic Feasibility

A study of economic feasibility is required for the Baseline Project Plan. The purpose for assessing **economic feasibility** is to identify the financial benefits and costs associated with the development project. Economic feasibility is often referred to as *cost-benefit analysis*. During project initiation and planning, it will be impossible for you to define precisely all benefits and costs related to a particular project. Yet, it is important that you identify and quantify benefits and costs or it will be impossible for you to conduct a sound economic analysis and determine if one project is more feasible than another. Next, we review worksheets you can use to record costs and benefits, and techniques for making cost-benefit calculations. These worksheets and techniques are used after each SDLC phase to decide whether to continue, redirect, or kill a project.

Determining Project Benefits An information system can provide many benefits to an organization. For example, a new or renovated IS can automate monotonous jobs, reduce errors, provide innovative services to customers and suppliers, and improve organizational efficiency, speed, flexibility, and morale. These benefits are both tangible and intangible. A **tangible benefit** is an item that can be measured in dollars and with certainty. Examples of tangible benefits include reduced personnel expenses, lower transaction costs, or higher profit margins. It is important to note that not all tangible benefits can be easily quantified. For example, a tangible benefit that allows a company to perform a task 50 percent of the time may be difficult to quantify in terms of hard dollar savings. Most tangible benefits fit in one or more of the following categories:

- Cost reduction and avoidance
- Error reduction
- Increased flexibility
- Increased speed of activity
- Improvement of management planning and control
- Opening new markets and increasing sales opportunities

Jim and Jackie identified several tangible benefits of the Customer Tracking System at PVF and summarized them in a worksheet, shown in Figure 3-8. Jackie and Jim collected information from users of the current

Economic feasibility
A process of identifying the financial benefits and costs associated with a development project.

Tangible benefit
A benefit derived from the creation of an information system that can be measured in dollars and with certainty.

TANGIBLE BENEFITS WORKSHEET *Customer Tracking System Project*	
	Year 1 through 5
A. Cost reduction or avoidance	$ 4,500
B. Error reduction	2,500
C. Increased flexibility	7,500
D. Increased speed of activity	10,500
E. Improvement in management planning or control	25,000
F. Other _____	0
TOTAL tangible benefits	**$50,000**

FIGURE 3-8
TANGIBLE BENEFITS WORKSHEET FOR THE CUSTOMER TRACKING SYSTEM AT PINE VALLEY FURNITURE

customer tracking system in order to create the worksheet. They first interviewed the person responsible for collecting, entering, and analyzing the correctness of the current customer tracking data. This person estimated that he spent 10 percent of his time correcting data entry errors. This person's salary is $25,000, so Jackie and Jim estimated an error reduction benefit of $2,500 (10 percent of $25,000). Jackie and Jim also interviewed managers who used the current customer tracking reports to estimate other tangible benefits. They learned that cost reduction or avoidance benefits could be gained with better inventory management. Also, increased flexibility would likely occur from a reduction in the time normally taken to reorganize data manually for different purposes. Further, improvements in management planning or control should result from a broader range of analyses in the new system. This analysis forecasts that benefits from the system would be approximately $50,000 per year.

Jim and Jackie also identified several intangible benefits of the system. Although they could not quantify these benefits, they will still be described in the final BPP. **Intangible benefits** refer to items that cannot be easily measured in dollars or with certainty. Intangible benefits may have direct organizational benefits, such as the improvement of employee morale, or they may have broader societal implications, such as the reduction of waste creation or resource consumption. Potential tangible benefits may have to be considered intangible during project initiation and planning because you may not be able to quantify them in dollars or with certainty at this stage in the life cycle. During later stages, such intangibles can become tangible benefits as you better understand the ramifications of the system you are designing. Intangible benefits include:

- Competitive necessity
- Increased organizational flexibility
- Increased employee morale
- Promotion of organizational learning and understanding
- More timely information

After determining project benefits, project costs must be identified.

Determining Project Costs An information system can have both tangible and intangible costs. A **tangible cost** refers to an item that you can easily measure in dollars and with certainty. From a systems development perspective, tangible costs include items such as hardware costs, labor costs, and operational costs from employee training and building renovations. Alternatively, an **intangible cost** refers to an item that you cannot easily measure in terms of dollars or with certainty. Intangible costs can include loss of customer goodwill, employee morale, or operational inefficiency.

Besides tangible and intangible costs, you can distinguish system-related development costs as either one time or recurring. A **one-time cost** refers to a cost associated with project initiation and development and the start-up of the system. These costs typically encompass the following activities:

- System development
- New hardware and software purchases
- User training
- Site preparation
- Data or system conversion

When conducting an economic cost-benefit analysis, you should create a worksheet for capturing these expenses. This worksheet can be a two-column document or a multicolumn spreadsheet. For very large projects, one-time

Intangible benefit
A benefit derived from the creation of an information system that cannot be easily measured in dollars or with certainty.

Tangible cost
A cost associated with an information system that can be easily measured in dollars and with certainty.

Intangible cost
A cost associated with an information system that cannot be easily measured in terms of dollars or with certainty.

One-time cost
A cost associated with project start-up and development, or system start-up.

costs may be staged over one or more years. In these cases, a separate one-time cost worksheet should be created for each year. This separation would make it easier to perform present value calculations (see below). A **recurring cost** refers to a cost resulting from the ongoing evolution and use of the system. Examples of these costs typically include:

- Application software maintenance
- Incremental data storage expense
- Incremental communications
- New software and hardware leases
- Consumable supplies and other expenses (e.g., paper, forms, data center personnel)

Both one-time and recurring costs can consist of items that are fixed or variable in nature. Fixed costs refer to costs that are billed or incurred at a regular interval and usually at a fixed rate. A facility lease payment is an example of a one-time cost. Variable costs refer to items that vary in relation to usage. Long-distance phone charges are variable costs.

Jim and Jackie identified both one-time and recurring costs for the Customer Tracking System project. Figure 3-9 shows that this project will incur a one-time cost of $42,500. Figure 3-10 shows a recurring cost of $28,500 per year. One-time costs were established by discussing the system with Jim's boss, who

Recurring cost
A cost resulting from the ongoing evolution and use of a system.

ONE-TIME COSTS WORKSHEET *Customer Tracking System Project*	
	Year 0
A. Development costs	$20,000
B. New hardware	15,000
C. New (purchased) software, if any	
1. Packaged applications software	5,000
2. Other _____	0
D. User training	2,500
E. Site preparation	0
F. Other _____	0
TOTAL one-time costs	**$42,500**

FIGURE 3-9
ONE-TIME COSTS WORKSHEET FOR THE CUSTOMER TRACKING SYSTEM AT PINE VALLEY FURNITURE

RECURRING COSTS WORKSHEET *Customer Tracking System Project*	
	Year 1 through 5
A. Application software maintenance	$25,000
B. Incremental data storage required: 20 MB 2 $50. (estimated cost/MB = $50)	1,000
C. Incremental communications (lines, messages, . . .)	2,000
D. New software or hardware leases	0
E. Supplies	500
F. Other _____	0
TOTAL recurring costs	**$28,500**

FIGURE 3-10
RECURRING COSTS WORKSHEET FOR THE CUSTOMER TRACKING SYSTEM AT PINE VALLEY FURNITURE

felt that the system would require approximately four months to develop (at $5,000 per month). To run the new system effectively, the Marketing department would need to upgrade at least five of its current workstations (at $3,000 each). Additionally, software licenses for each workstation (at $1,000 each) and modest user training fees (10 users at $250 each) would be necessary.

As you can see from Figure 3-10, Jim and Jackie estimate that the proposed system will require, on average, five months of annual maintenance, primarily for enhancements that users will expect from the system. Other ongoing expenses such as increased data storage, communications equipment, and supplies should also be expected.

You should now have an understanding of the types of benefit and cost categories associated with an information systems project. In the next section, we address the relationship between time and money.

The Time Value of Money   Most techniques used to determine economic feasibility encompass the concept of the **time value of money (TVM)**. TVM refers to comparing present cash outlays to future expected returns. As we've seen, the development of an information system has both one-time and recurring costs. Furthermore, benefits from systems development will likely occur sometime in the future. Because many projects may be competing for the same investment dollars and may have different useful life expectancies, all costs and benefits must be viewed in relation to their present rather than future value when comparing investment options.

A simple example will help you understand the concept of TVM. Suppose you want to buy a used car from an acquaintance, and she asks that you make three payments of $1,500 for three years, beginning next year, for a total of $4,500. If she would agree to a single lump sum payment at the time of sale (and if you had the money!), what amount do you think she would agree to? Should the single payment be $4,500? Should it be more or less? To answer this question, we must consider the time value of money. Most of us would gladly accept $4,500 today rather than three payments of $1,500, because a dollar today (or $4,500 for that matter) is worth more than a dollar tomorrow or next year, because money can be invested. The interest rate at which money can be borrowed or invested, the cost of capital, is called the **discount rate** for TVM calculations. Let's suppose that the seller could put the money received for the sale of the car in the bank and receive a 10 percent return on her investment. A simple formula can be used when figuring out the **present value** of the three $1,500 payments:

$$PV_n = Y = \frac{1}{(1+i)^n}$$

where PV_n is the present value of Y dollars n years from now when i is the discount rate.

From our example, the present value of the three payments of $1,500 can be calculated as

$$PV_1 = 1500 \times \frac{1}{(1+.10)^1} \, 1,500 \times .9091 = 1,363.65$$

$$PV_2 = 1500 \times \frac{1}{(1+.10)^2} \, 1,500 \times .8264 = 1,239.60$$

$$PV_3 = 1500 \times \frac{1}{(1+.10)^3} \, 1,500 \times .7513 = 1,126.95$$

where PV_1, PV_2, and PV_3 reflect the present value of each $1,500 payment in year 1, 2, and 3, respectively.

Time value of money (TVM)
The process of comparing present cash outlays to future expected returns.

Discount rate
The interest rate used to compute the present value of future cash flows.

Present value
The current value of a future cash flow.

To calculate the net present value (NPV) of the three $1,500 payments, simply add the present values calculated ($NPV = PV_1 + PV_2 + PV_3 = 1{,}363.65 + 1{,}239.60 + 1{,}126.95 = \$3{,}730.20$). In other words, the seller could accept a lump sum payment of $3,730.20 as equivalent to the three payments of $1,500, given a discount rate of 10 percent.

Now that we know the relationship between time and money, the next step in performing the economic analysis is to create a summary worksheet that reflects the present values of all benefits and costs. PVF's Systems Priority Board feels that the useful life of many information systems may not exceed five years. Therefore, all cost-benefit analysis calculations will be made using a five-year time horizon as the upper boundary on all time-related analyses. In addition, the management of PVF has set its cost of capital to be 12 percent (i.e., PVF's discount rate). The worksheet constructed by Jim is shown in Figure 3-11.

Cell H11 of the worksheet displayed in Figure 3-11 summarizes the NPV of the total tangible benefits from the project over five years ($180,239). Cell H19 summarizes the NPV of the total costs from the project. The NPV for the project, indicated in cell H22 ($35,003), shows that benefits from the project exceed costs.

The overall return on investment (ROI) for the project is also shown on the worksheet in cell H25 (.24). Because alternative projects will likely have different benefit and cost values and, possibly, different life expectancies, the overall ROI value is very useful for making project comparisons on an economic basis. Of course, this example shows ROI for the overall project over five years. An ROI analysis could be calculated for each year of the project.

The last analysis shown in Figure 3-11, on line 34, is a **break-even analysis.** The objective of the break-even analysis is to discover at what point (if ever) cumulative benefits equal costs (i.e., when break-even occurs). To conduct this analysis, the NPV of the yearly cash flows is determined. Here, the yearly cash flows are calculated by subtracting both the one-time cost and the present values of the recurring costs from the present value of the yearly benefits. The

Break-even analysis
A type of cost-benefit analysis to identify at what point (if ever) benefits equal costs.

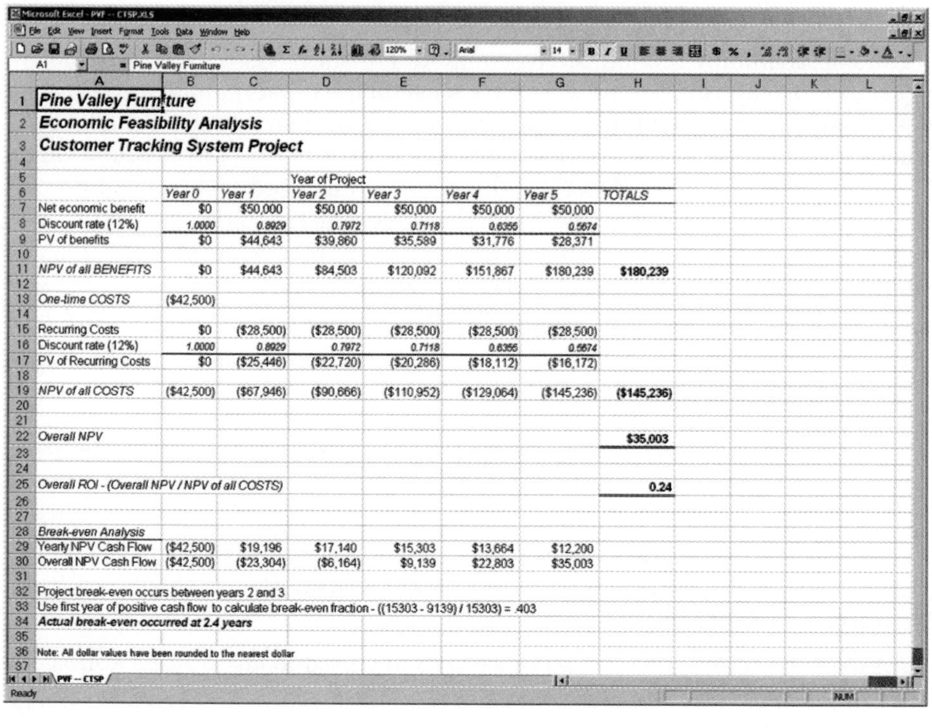

	A	B	C	D	E	F	G	H
1	*Pine Valley Furniture*							
2	*Economic Feasibility Analysis*							
3	*Customer Tracking System Project*							
4								
5				Year of Project				
6		Year 0	Year 1	Year 2	Year 3	Year 4	Year 5	TOTALS
7	Net economic benefit	$0	$50,000	$50,000	$50,000	$50,000	$50,000	
8	Discount rate (12%)	1.0000	0.8929	0.7972	0.7118	0.6355	0.5674	
9	PV of benefits	$0	$44,643	$39,860	$35,589	$31,776	$28,371	
10								
11	*NPV of all BENEFITS*	$0	$44,643	$84,503	$120,092	$151,867	$180,239	**$180,239**
12								
13	*One-time COSTS*	($42,500)						
14								
15	Recurring Costs	$0	($28,500)	($28,500)	($28,500)	($28,500)	($28,500)	
16	Discount rate (12%)	1.0000	0.8929	0.7972	0.7118	0.6355	0.5674	
17	PV of Recurring Costs	$0	($25,446)	($22,720)	($20,286)	($18,112)	($16,172)	
18								
19	*NPV of all COSTS*	($42,500)	($67,946)	($90,666)	($110,952)	($129,064)	($145,236)	**($145,236)**
20								
21								
22	*Overall NPV*							**$35,003**
23								
24								
25	*Overall ROI - (Overall NPV / NPV of all COSTS)*							**0.24**
26								
27								
28	*Break-even Analysis*							
29	Yearly NPV Cash Flow	($42,500)	$19,196	$17,140	$15,303	$13,664	$12,200	
30	Overall NPV Cash Flow	($42,500)	($23,304)	($6,164)	$9,139	$22,803	$35,003	
31								
32	Project break-even occurs between years 2 and 3							
33	Use first year of positive cash flow to calculate break-even fraction - ((15303 - 9139) / 15303) = .403							
34	***Actual break-even occurred at 2.4 years***							
35								
36	Note: All dollar values have been rounded to the nearest dollar							
37								

FIGURE 3-11
WORKSHEET REFLECTING THE PRESENT VALUE CALCULATIONS OF ALL BENEFITS AND COSTS FOR THE CUSTOMER TRACKING SYSTEM AT PINE VALLEY FURNITURE
This worksheet indicates that benefits from the project over five years exceed its costs by $35,003.

FIGURE 3-12
BREAK-EVEN ANALYSIS FOR THE
CUSTOMER TRACKING SYSTEM AT
PINE VALLEY FURNITURE

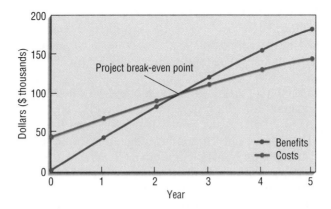

overall NPV of the cash flows reflects the total cash flows for all preceding years. If you examine line 30 of the worksheet, you'll see that break-even occurs between years two and three. Because year three is the first year in which the overall NPV cash flows figure is nonnegative, identifying the point when break-even occurs can be derived as follows:

$$\text{Break-Even Ratio} = \frac{\text{Yearly NPV Cash Flow } - \text{ Overall NPV Cash Flow}}{\text{Yearly NPV Cash Flow}}$$

Using data from Figure 3-11,

$$\text{Break-Even Ratio} = \frac{15,303 - 9,139}{15,303} = .403$$

Project break-even occurs at approximately 2.4 years. A graphical representation of this analysis is shown in Figure 3-12. Using the information from the economic analysis, PVF's Systems Priority Board will be in a much better position to understand the potential economic impact of the Customer Tracking System. Without this information, it would be virtually impossible to know the cost-benefits of a proposed system and impossible to make an informed decision on approving or rejecting the service request.

You can use many techniques to compute a project's economic feasibility. Because most information systems have a useful life of more than one year and will provide benefits and incur expenses for more than one year, most techniques for analyzing economic feasibility employ the concept of the time value of money, TVM. Table 3-5 describes three commonly used techniques for conducting economic feasibility analysis. (For a more detailed discussion of TVM

TABLE 3-5: Commonly Used Economic Cost-Benefit Analysis Techniques:
Net Present Value, Return on Investment, and Break-Even Analysis

Analysis Technique	Description
Net present value (NPV)	NPV uses a discount rate determined from the company's cost of capital to establish the present value of a project. The discount rate is used to determine the present value of both cash receipts and outlays.
Return on investment (ROI)	ROI is the ratio of the net cash receipts of the project divided by the cash outlays of the project. Trade-off analysis can be made among projects competing for investment by comparing their representative ROI ratios.
Break-even analysis (BEA)	BEA finds the amount of time required for the cumulative cash flow from a project to equal its initial and ongoing investment.

or cost-benefit analysis techniques in general, the interested reader is encouraged to review an introductory finance or managerial accounting textbook.)

To be approved for continuation, a systems project may not have to achieve break-even or have an ROI greater than estimated during project initiation and planning. Because you may not be able to quantify many benefits or costs at this point in a project, such financial hurdles for a project may be unattainable. In this case, simply doing as thorough an economic analysis as possible, including producing a long list of intangibles, may be sufficient for the project to progress. One other option is to run the type of economic analysis shown in Figure 3-11 using pessimistic, optimistic, and expected benefit and cost estimates during project initiation and planning. This range of possible outcomes, along with the list of intangible benefits and the support of the requesting business unit, will often be enough to allow the project to continue to the analysis phase. You must, however, be as precise as you can with the economic analysis, especially when investment capital is scarce. In this case, it may be necessary to conduct some typical analysis phase activities during project initiation and planning in order to clearly identify inefficiencies and shortcomings with the existing system and to explain how a new system will overcome these problems.

Assessing Other Feasibility Concerns

You may need to consider other feasibility studies when formulating the business case for a system during project planning. **Operational feasibility** is the process of examining the likelihood that the project will attain its desired objectives. The goal of this study is to understand the degree to which the proposed system will likely solve the business problems or take advantage of the opportunities outlined in the System Service Request or project identification study. In other words, assessing operational feasibility requires that you gain a clear understanding of how an IS will fit into the current day-to-day operations of the organization.

The goal of **technical feasibility** is to understand the organization's ability to construct the proposed system. This analysis should include an assessment of the development group's understanding of the possible target hardware, software, and operating environments to be used as well as system size, complexity, and the group's experience with similar systems. **Schedule feasibility** considers the likelihood that all potential time frames and completion date schedules can be met and that meeting these dates will be sufficient for dealing with the needs of the organization. For example, a system may have to be operational by a government-imposed deadline, by a particular point in the business cycle (such as the beginning of the season when new products are introduced), or at least by the time a competitor is expected to introduce a similar system.

Assessing **legal and contractual feasibility** requires that you gain an understanding of any potential legal and contractual ramifications due to the construction of the system. Considerations might include copyright or nondisclosure infringements, labor laws, antitrust legislation (which might limit the creation of systems to share data with other organizations), foreign trade regulations (for example, some countries limit access to employee data by foreign corporations), and financial reporting standards as well as current or pending contractual obligations. Typically, legal and contractual feasibility is a greater consideration if your organization has historically used an outside organization for specific systems or services that you now are considering handling yourself. Assessing **political feasibility** involves understanding how key stakeholders within the organization view the proposed system. Because an information system may affect the distribution of information within the organization, and thus the distribution of power, the construction of an IS can have political ramifications. Those stakeholders not supporting the project may take steps to block, disrupt, or change the project's intended focus.

Operational feasibility
The process of assessing the degree to which a proposed system solves business problems or takes advantage of business opportunities.

Technical feasibility
The process of assessing the development organization's ability to construct a proposed system.

Schedule feasibility
The process of assessing the degree to which the potential time frame and completion dates for all major activities within a project meet organizational deadlines and constraints for affecting change.

Legal and contractual feasibility
The process of assessing potential legal and contractual ramifications due to the construction of a system.

Political feasibility
The process of evaluating how key stakeholders within the organization view the proposed system.

In summary, numerous feasibility issues must be considered when planning a project. This analysis should consider economic, operational, technical, schedule, legal, contractual, and political issues related to the project. In addition to these considerations, project selection by an organization may be influenced by issues beyond those discussed here. For example, projects may be selected for construction given high project costs and high technical risk if the system is viewed as a strategic necessity; that is, a project viewed by the organization as being critical to its survival. Alternatively, projects may be selected because they are deemed to require few resources and have little risk. Projects may also be selected due to the power or persuasiveness of the manager proposing the system. This means that project selection may be influenced by factors beyond those discussed here and beyond items that can be analyzed. Your role as a systems analyst is to provide a thorough examination of the items that can be assessed so that a project review committee can make informed decisions. In the next section, we discuss how project plans are typically constructed.

Building the Baseline Project Plan

All the information collected during project initiation and planning is collected and organized into a document called the Baseline Project Plan. Once the BPP is completed, a formal review of the project can be conducted with customers. This presentation, a walkthrough, is discussed later in the chapter. The focus of the walkthrough is to verify all information and assumptions in the baseline plan before moving ahead with the project. An outline of a Baseline Project Plan, shown in Figure 3-13, contains four major sections:

1. Introduction
2. System description
3. Feasibility assessment
4. Management issues

The purpose of the *introduction* is to provide a brief overview of the entire document and outline a recommended course of action for the project. The introduction is often limited to only a few pages. Although it is sequenced as the first section of the BPP, it is often the final section to be written. It is only after performing most of the project planning activities that a clear overview and recommendation can be created. One initial activity that should be performed is the definition of project scope, its range, which is an important part of the BPP's introduction section.

When defining scope for the Customer Tracking System within PVF, Jim Woo first needed to gain a clear understanding of the project's objectives. Jim interviewed Jackie Judson and several of her colleagues to gain a good idea of their needs. He also reviewed the existing system's functionality, processes, and data use requirements for performing customer tracking activities. These activities provided him with the information needed to define the project scope and to identify possible alternative solutions. Alternative system solutions can relate to different system scopes, platforms for deployment, or approaches to acquiring the system. We elaborate on the idea of alternative solutions, called design strategies, when we discuss the systems analysis phase of the life cycle in Chapter 4. During project initiation and planning, the most crucial element of the design strategy is the system's scope. Scope depends on the answers to these questions:

- Which organizational units (business functions and divisions) might be affected by or use the proposed system or system change?
- With which current systems might the proposed system need to interact or be consistent, or which current systems might be changed due to a replacement system?

    ◉ Who inside and outside the requesting organization (or the organization as a whole) might care about the proposed system?

    ◉ What range of potential system capabilities is to be considered?

    The statement of project scope for the Customer Tracking System project at PVF is shown in Figure 3-14.

    For the Customer Tracking System (CTS), project scope was defined using only textual information. It is not uncommon, however, to define project scope

---

BASELINE PROJECT PLAN REPORT

1.0 Introduction
    A. Project Overview—Provides an executive summary that specifies the project's scope, feasibility, justification, resource requirements, and schedules. Additionally, a brief statement of the problem, the environment in which the system is to be implemented, and constraints that affect the project are provided.
    B. Recommendation—Provides a summary of important findings from the planning process and recommendations for subsequent activities.

2.0 System Description
    A. Alternatives—Provides a brief presentation of alternative system configurations.
    B. System Description—Provides a description of the selected configuration and a narrative of input information, tasks performed, and resultant information.

3.0 Feasibility Assessment
    A. Economic Analysis—Provides an economic justification for the system using cost-benefit analysis.
    B. Technical Analysis—Provides a discussion of relevant technical risk factors and an overall risk rating of the project.
    C. Operational Analysis—Provides an analysis of how the proposed system solves business problems or takes advantage of business opportunities in addition to an assessment of how current day-to-day activities will be changed by the system.
    D. Legal and Contractual Analysis—Provides a description of any legal or contractual risks related to the project (e.g., copyright or nondisclosure issues, data capture or transferring, and so on).
    E. Political Analysis—Provides a description of how key stakeholders within the organization view the proposed system.
    F. Schedules, Time Line, and Resource Analysis—Provides a description of potential time frame and completion date scenarios using various resource allocation schemes.

4.0 Management Issues
    A. Team Configuration and Management—Provides a description of the team member roles and reporting relationships.
    B. Communication Plan—Provides a description of the communication procedures to be followed by management, team members, and the customer.
    C. Project Standards and Procedures—Provides a description of how deliverables will be evaluated and accepted by the customer.
    D. Other Project-Specific Topics—Provides a description of any other relevant issues related to the project uncovered during planning.

FIGURE 3-13

An outline of a Baseline Project Plan contains four major sections: introduction, system description, feasibility assessment, and management issues.

Pine Valley Furniture *Statement of Project Scope*	Prepared by: Jim Woo Date: April 18, 2003

General Project Information
- **Project Name:** Customer Tracking System
- **Sponsor:** Jackie Judson, VP Marketing
- **Project Manager:** Jim Woo

Problem/Opportunity Statement:
Sales growth has outpaced the marketing department's ability to track and forecast customer buying trends accurately. An improved method for performing this process must be found in order to reach company objectives.

Project Objectives:
To enable the marketing department to track and forecast customer buying patterns accurately in order to better serve customers with the best mix of products. This will also enable PVF to identify the proper application of production and material resources.

Project Description:
A new information system will be constructed that will collect all customer purchasing activity, support display and reporting of sales information, aggregate data and show trends in order to assist marketing personnel in understanding dynamic market conditions. The project will follow PVF's systems development life cycle.

Business Benefits:
Improved understanding of customer buying patterns
Improved utilization of marketing and sales personnel
Improved utilization of production and materials

Project Deliverables:
Customer tracking system analysis and design
Customer tracking system programs
Customer tracking documentation
Training procedures

Estimated Project Duration:
5 months

FIGURE 3-14
STATEMENT OF PROJECT SCOPE FOR THE CUSTOMER TRACKING SYSTEM AT PINE VALLEY FURNITURE

using tools such as data flow diagrams and entity-relationship models. For example, Figure 3-15 shows a context-level data flow diagram used to define system scope for PVF's Purchasing Fulfillment System. As shown in Figure 3-15, the Purchasing Fulfillment System interacts with the "production schedulers," "suppliers," and "engineering." You will learn much more about data flow diagrams in Chapter 5. The other items in the introduction section of the BPP are simply executive summaries of the other sections of the document.

The second section of the BPP is the *system description*, in which you outline possible alternative solutions to the one deemed most appropriate for the given situation. Note that this description is at a very high level, mostly narrative in form. Alternatives may be stated as simply as this:

1. Web-based online system
2. Mainframe with central database
3. Local area network with decentralized databases

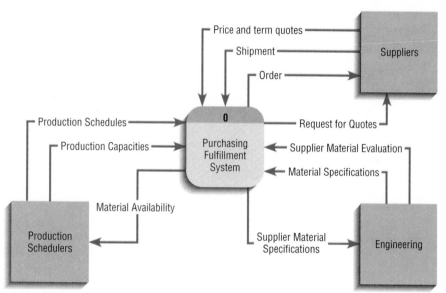

FIGURE 3-15
CONTEXT-LEVEL DATA FLOW DIAGRAM SHOWING PROJECT SCOPE FOR THE PURCHASING FULFILLMENT SYSTEM AT PINE VALLEY FURNITURE

4. Batch data input with online retrieval
5. Purchasing of a prewritten package

If the project is approved for construction or purchase, you will need to collect and structure information in a more detailed and rigorous manner during the systems analysis phase and evaluate in greater depth these and other alternatives for the system.

When Jim and Jackie were considering system alternatives for the CTS, they focused on two primary issues. First, they discussed how the system would be acquired and considered three options: purchase the system if one could be found that met PVF's needs, outsource the development of the system to an outside organization, or build the system within PVF. Next, Jim and Jackie defined the comprehensiveness of the system's functionality. To complete this task, Jackie wrote a series of statements listing the types of tasks that she thought marketing personnel would be able to accomplish when using the CTS. This list became the basis of the system description and was instrumental in helping them make the acquisition decision. After considering the unique needs of the marketing group, they decided that the best decision was to build the system within PVF.

In the third section of the BPP, *feasibility assessment*, the systems analyst outlines project costs and benefits and technical difficulties. This is also the section where high-level project schedules are specified using Network diagrams and Gantt charts. Recall from Chapter 2 that this process is referred to as a work breakdown structure. During project initiation and planning, task and activity estimates are generally not detailed. An accurate work breakdown can be done only for the next one or two life cycle activities—systems analysis and systems design. After defining the primary tasks for the project, an estimate of the resource requirements can be made. As with defining tasks and activities, this activity involves obtaining estimates of the human resource requirements, because people are typically the most expensive resource element of a project. Once you define the major tasks and resource requirements, a preliminary schedule can be developed. Defining an acceptable schedule may require that you find additional or different resources or that the scope of the project be changed. The greatest amount of project planning effort is typically expended on feasibility assessment activities.

The final section of the BPP, *management issues*, outlines the concerns that management has about the project. This will be a very short section if the proposed project is going to be conducted exactly as prescribed by the organization's standard systems development methodology. Most projects, however, have some unique characteristics that require minor to major deviation from the standard methodology. In the team configuration and management portion, you identify the types of people to work on the project, who will be responsible for which tasks, and how work will be supervised and reviewed. In the communications plan portion, you explain how the user will be kept informed about project progress, such as periodic review meetings or even a newsletter, and which mechanisms will be used to foster sharing of ideas among team members, such as a computer-based conference facility. An example of the type of information contained in the project standards and procedures portion would be procedures for submitting and approving project change requests and any other issues deemed important for the project's success.

You should now have a feel for how a BPP is constructed and the types of information it contains. Its creation is not meant to be a project in and of itself but rather a step in the overall systems development process. Developing the BPP has two primary objectives. First, it helps to assure that the customer and development group share a common understanding of the project. Second, it helps to provide the sponsoring organization with a clear idea of the scope, benefits, and duration of the project. Meeting these objectives creates the foundation for a successful project.

Reviewing the Baseline Project Plan

Before phase 2 of the SDLC analysis can begin, the users, management, and development group must review and approve the Baseline Project Plan. This review takes place before the BPP is submitted or presented to some project approval body, such as an IS steering committee or the person who must fund the project. The objective of this review is to assure that the proposed system conforms to organizational standards and to make sure that all relevant parties understand and agree with the information contained in the Baseline Project Plan. A common method for performing this review (as well as reviews during subsequent life cycle phases) is called a **walkthrough.** Walkthroughs, also called *structured walkthroughs*, are peer group reviews of any product created during the systems development process. They are widely used by professional development organizations, such as IBM, Xerox, and the U.S. government, and have proven very effective in ensuring the quality of an information system. As a systems analyst, you frequently will be involved in walkthroughs.

Although walkthroughs are not rigidly formal or exceedingly long in duration, they have a specific agenda that highlights what is to be covered and the expected completion time. Individuals attending the meeting have specific roles. These roles can include the following:

> - *Coordinator.* This person plans the meeting and facilitates discussions. This person may be the project leader or a lead analyst responsible for the current life cycle step.
> - *Presenter.* This person describes the work product to the group. The presenter is usually an analyst who has done all or some of the work being presented.
> - *User.* This person (or group) makes sure that the work product meets the needs of the project's customers. This user would usually be someone not on the project team.
> - *Secretary.* This person takes notes and records decisions or recommendations made by the group. This may be a clerk assigned to the project team or one of the analysts on the team.
> - *Standard-bearer.* This person ensures that the work product adheres to organizational technical standards. Many

Walkthrough
A peer group review of any product created during the systems development process; also called a *structured walkthrough.*

larger organizations have staff groups within the unit responsible for establishing standard procedures, methods, and documentation formats. For example, within Microsoft, user interface standards are developed and rigorously enforced on all development projects. As a result, all systems have the same look and feel to users. These standard-bearers validate the work so that it can be used by others in the development organization.

> ● *Maintenance oracle.* This person reviews the work product in terms of future maintenance activities. The goal is to make the system and its documentation easy to maintain.

After Jim and Jackie completed their BPP for the Customer Tracking System, Jim approached his boss and requested that a walkthrough meeting be scheduled and a walkthrough coordinator be assigned to the project. PVF provides the coordinator with a Walkthrough Review Form, shown in Figure 3-16. Using this form, the coordinator can more easily make sure that a qualified individual is assigned to each walkthrough role; that each member has been given a copy of the review materials; and that each member knows the agenda, date, time, and location of the meeting. At the meeting, Jim presented the BPP and Jackie added comments from a user perspective. Once the walkthrough presentation was completed, the coordinator polled each representative for his or her recommendation concerning the work product. The results of this voting may result in validation of the work product, validation pending changes suggested during the meeting, or a suggestion that the work product requires major revision before being presented for approval. In this latter case, substantial changes to the work product are usually requested after which another walkthrough must be scheduled before the project can be proposed to the Systems Priority Board (steering committee). In the case of the Customer Tracking System, the BPP was supported by the walkthrough panel pending some minor changes to the duration estimates of the schedule. These suggested changes were recorded by the secretary on a Walkthrough Action List, shown in Figure 3-17, and given to Jim to incorporate into a final version of the baseline plan presented to the steering committee.

Walkthrough meetings are a common occurrence in most systems development groups. In addition to reviewing the BPP, these meetings can be used for the following activities:

> ● System specifications
> ● Logical and physical designs
> ● Code or program segments
> ● Test procedures and results
> ● Manuals and documentation

One of the key advantages to using a structured review process is to ensure that formal review points occur during the project. At each subsequent phase of the project, a formal review should be conducted (and shown on the project schedule) to make sure that all aspects of the project are satisfactorily accomplished before assigning additional resources to the project. This conservative approach of reviewing each major project activity with continuation contingent on successful completion of the prior phase is called *incremental commitment*. It is much easier to stop or redirect a project at any point when using this approach.

Phase 1:
Systems Planning and Selection

NET SEARCH
Managing a structured walkthrough with a very large group can be a difficult task. Visit http://www.prenhall. com/valacich to complete an exercise related to this topic.

Pine Valley Furniture
Walkthrough Review Form

Session Coordinator:

Project/Segment:

Coordinator's Checklist:

1. Confirmation with producer(s) that material is ready and stable: _____
2. Issue invitations, assign responsibilities, distribute materials: [ ] Y   [ ] N
3. Set date, time, and location for meeting:

   Date: ___ / ___ / ___   Time: _____ A.M. / P.M. (circle one)

   Location: _____

Responsibilities	Participants	Can Attend	Received Materials
Coordinator	_____	[ ] Y [ ] N	[ ] Y [ ] N
Presenter	_____	[ ] Y [ ] N	[ ] Y [ ] N
User	_____	[ ] Y [ ] N	[ ] Y [ ] N
Secretary	_____	[ ] Y [ ] N	[ ] Y [ ] N
Standards	_____	[ ] Y [ ] N	[ ] Y [ ] N
Maintenance	_____	[ ] Y [ ] N	[ ] Y [ ] N

Agenda:
____ 1. All participants agree to follow PVF's Rules of a Walkthrough
____ 2. New material: Walkthrough of all material
____ 3. Old material: Item-by-item checkoff of previous action list
____ 4. Creation of new action list (contribution by each participant)
____ 5. Group decision (see below)
____ 6. Deliver copy of this form to the project control manager

Group Decision:
_____ Accept product as-is
_____ Revise (no further walkthrough)
_____ Review and schedule another walkthrough

Signatures		

FIGURE 3-16
WALKTHROUGH REVIEW FORM FOR THE CUSTOMER TRACKING SYSTEM AT PINE VALLEY FURNITURE

	Pine Valley Furniture *Walkthrough Action List*
Session Coordinator:	
Project/Segment:	
Date and Time of Walkthrough: Date: ___ / ___ / ___ Time: _____ A.M. / P.M. (circle one)	
Fixed (✓)	*Issues raised in review:*

FIGURE 3-17
WALKTHROUGH ACTION LIST FOR PINE VALLEY FURNITURE

Electronic Commerce Application: Systems Planning and Selection

Most businesses have discovered the power of Internet-based electronic commerce as a means to communicate efficiently with customers and to extend their marketing reach. As a systems analyst, you and a project team may be asked by your employer to help determine if an Internet-based electronic commerce application fits the goals of the company and, if so, how that application should be implemented.

The systems planning and selection process for an Internet-based electronic commerce application is no different than the process followed for other applications. Nonetheless, you should take into account special issues when developing an Internet-based application. In this section, we highlight those issues.

Internet Basics

The term *Internet* is derived from the term *internetworking*. The **Internet** is a global network comprised of thousands of interconnected individual networks that communicate with each other through *TCP/IP* (transmission control protocol/Internet protocol). The Internet refers to both the global computing network and to business-to-consumer electronic commerce applications. The interconnected networks include LINUX, Microsoft, UNIX, IBM, Novell, and Apple. Using the Internet and other technologies to support day-to-day business activities, such as communicating with customers and selling goods and services online, is referred to as **electronic commerce (EC).** Note that EC can also refer to the use of non-Internet technologies such as telephone voice messaging systems that route and process customer requests and inquiries. Nonetheless, for our purposes, we will use *EC* to mean Internet-enabled business. There are three classes of Internet EC applications: Internet, intranets, and extranets, as illustrated in Figure 3-18. Internet-based EC is transactions between individuals and businesses. **Intranet** refers to the use of the Internet within the same business. **Extranet** refers to the use of the Internet between firms.

Intranets and extranets are examples of two ways organizations communicate via technology. Having an intranet is a lot like having a "global" local area network. A company may create an intranet to house commonly used forms, up-to-date information on sales, and human resource information so that employees can access them easily and at any time. Organizations that have intranets dictate (1) what applications will run over the intranet—such as electronic mail or an inventory control system—as well as (2) the speed and quality of the hardware connected to the intranet. Intranets are a new way of using information systems to support business activities within a single organization. Extranets are another new way of using an established computing model, **electronic data interchange (EDI).** EDI refers to the use of telecommunications technologies to transfer business documents directly between organizations. Using EDI, trading partners—suppliers, manufacturers, and customers—establish computer-to-computer links that allow them to exchange data electronically. For example, a car manufacturer using EDI may send an electronic purchase order to a steel or tire supplier instead of a paper request. The paper order may take several days

---

Internet
A large worldwide network of networks that use a common protocol to communicate with each other; a global computing network to support business-to-consumer electronic commerce.

Electronic commerce (EC)
Internet-based communication to support day-to-day business activities.

Intranet
Internet-based communication to support business activities within a single organization.

Extranet
Internet-based communication to support business-to-business activities.

Electronic data interchange (EDI)
The use of telecommunications technologies to transfer business documents directly between organizations.

FIGURE 3-18
THREE POSSIBLE MODES OF ELECTRONIC COMMERCE

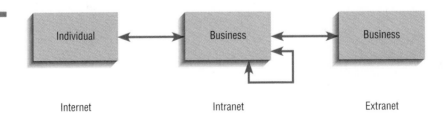

to arrive at the supplier, whereas an EDI purchase order will take only a few seconds. EDI is fast becoming the standard by which organizations will communicate with each other in the world of electronic commerce.

When developing either an intranet or an extranet, developers know who the users are, what applications will be used, the speed of the network connection, and the type of communication devices (e.g., Web browsers like Netscape or Internet Explorer, personal digital assistants like a Palm Pilot, or a Web-enabled cellular phone). On the other hand, when developing an Internet EC application (hereafter, simply EC), developers have to discern countless unknowns in order to build a useful system. Table 3-6 lists several unknowns you and your project team may deal with when designing and building an EC application. These unknowns may result in making trade-offs based on a careful analysis of who the users are likely to be, where they are likely to be located, and how they are likely to be connected to the Internet. Even with all these difficulties to contend with, there is no shortage of Internet EC applications springing up all across the world. One company that has decided to get onto the Web with its own EC site is Pine Valley Furniture.

Pine Valley Furniture WebStore

The PVF board of directors has requested that a project team be created to explore the opportunity to develop an EC system. Specifically, market research has found a good opportunity for online furniture purchases, especially in the areas of:

- Corporate furniture buying
- Home office furniture purchasing
- Student furniture purchasing

The board wants to incorporate all three target markets into its long-term EC plan but wants to focus initially on the corporate furniture buying system. The board feels that this segment has the greatest potential to provide an adequate return on investment and would be a good building block for moving into the customer-based markets. Because the corporate furniture buying system will be specifically targeted to the business furniture market, it will be easier to define the system's operational requirements. Additionally, this EC system should integrate nicely with two currently existing systems, Purchasing Fulfillment and Customer Tracking. Together, these attributes make it an ideal candidate for initiating PVF's Web strategy.

Initiating and Planning PVF's E-Commerce System Given the high priority of this project, Jackie Judson, vice president of Marketing, and senior systems analyst Jim Woo were assigned to work on this project. As for

TABLE 3-6: Unknowns That Must Be Dealt with When Designing and Building Internet Applications

User	Concern: Who is the user? Examples: Where is the user located? What is their expertise, education, or expectations?
Connection Speed	Concern: What is the speed of the connection and what information can be effectively displayed? Examples: modem, cable modem, satellite, broadband, cellular
Access Method	Concern: What is the method of accessing the Internet? Examples: Web browser, personal digital assistant (PDA), Web-enabled cellular phone, Web-enabled television

TABLE 3-7: Web-Based System Costs

Cost Category	Examples
Platform costs	Web-hosting service
	Web server
	Server software
	Software plug-ins
	Firewall server
	Router
	Internet connection
Content and service	Creative design and development
	Ongoing design fees
	Web project manager
	Technical site manager
	Content staff
	Graphics staff
	Support staff
	Site enhancement funds
	Fees to license outside content
	Programming, consulting, and research
	Training and travel
Marketing	Direct mail
	Launch and ongoing public relations
	Print advertisement
	Paid links to other Web sites
	Promotions
	Marketing staff
	Advertising sales staff

TABLE 3-8: PVF WebStore Project Benefits and Costs

Tangible Benefits	Intangible Benefits
Lower per-transaction overhead cost	First to market
Repeat business	Foundation for complete Web-based IS
Tangible Costs (one-time)	Simplicity for customers
Internet service setup fee	**Intangible Costs**
Hardware	No face-to-face interaction
Development cost	Not all customers use Internet
Data entry	
Tangible Costs (recurring)	
Internet service hosting fee	
Software	
Support	
Maintenance	
Decreased sales via traditional channels	

TABLE 3-9:   PVF WebStore: Feasibility Concerns

Feasibility Concern	Description
Operational	Online store open 24/7/365 Returns/customer support
Technical	New skill set for development, maintenance, and operation
Schedule	Must be open for business by Q3
Legal	Credit card fraud
Political	Traditional distribution channel loses business

the Customer Tracking System described earlier in the chapter, their first activity was to begin the project's initiation and planning activity. Over the next few days, Jim and Jackie met several times to initiate and plan the proposed system. At the first meeting they agreed that "WebStore" would be the proposed system project name. Next, they worked on identifying potential benefits, costs, and feasibility concerns. Jim developed a list of potential costs the company would incur to develop Web-based systems that he shared with Jackie and the other project team members (see Table 3-7).

WebStore Project Walkthrough   After meeting with the project team, Jim and Jackie established an initial list of benefits and costs (see Table 3-8) as well as several feasibility concerns (see Table 3-9). Next, Jim worked with several of PVF's technical specialists to develop an initial project schedule. Figure 3-19 shows the Gantt chart for this 84-day schedule. Finally, Jim and Jackie presented their initial project plans in a walkthrough to PVF's board of directors and senior management. All were excited about the project plan and approval was given to move the WebStore project on to the analysis phase.

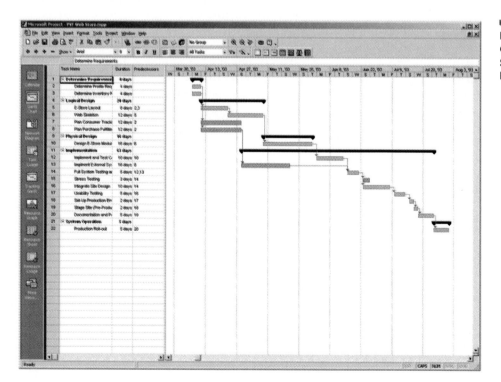

FIGURE 3-19
GANTT CHART SHOWING THE
SCHEDULE FOR THE WEBSTORE
PROJECT

Key Points Review

1. **Describe the steps involved when identifying and selecting projects and initiating and planning projects.**

   Project identification and selection consists of three primary activities: identifying potential development projects, classifying and ranking projects, and selecting projects for development. A variety of organizational members or units can be assigned to perform this process including top management, a diverse steering committee, business units and functional managers, the development group, or the most senior IS executive. Potential projects can be evaluated and selected using a broad range of criteria such as value chain analysis, alignment with business strategy, potential benefits, resource availability and requirements, and risks. Project initiation and planning is a critical activity in the life of a project. At this point projects are accepted for development, rejected as infeasible, or redirected. The objective of this process is to transform a vague system request into a tangible system description, clearly outlining the objectives, feasibility issues, benefits, costs, and time schedules for the project. Project initiation includes forming the project initiation team, establishing customer relationships, developing a plan to get the project started, setting project management procedures, and creating an overall project management environment. After project initiation, project planning focuses on assessing numerous feasibility issues associated with the project in order to create a clear statement of work and project plan.

2. **Explain the need for and the contents of a Statement of Work and a Baseline Project Plan.**

   A Statement of Work and a Baseline Project Plan are created during project initiation and planning. The Statement of Work is a short document prepared for the customer that describes what the project will deliver and outlines all work required to complete the project; it assures that both you and your customer gain a common understanding of the project. The Baseline Project Plan contains an introduction, a high-level description of the proposed system or system change, an outline of the various feasibilities, and an overview of management issues specific to the project. Before the development of an information system can begin, the users, management, and development group must review and agree on this specification.

3. **List and describe various methods for assessing project feasibility.**

   Assessing project feasibility can include an examination of economic, operational, technical, schedule, legal and contractual, and political aspects of the project. This assessment is influenced by the project size, the type of system proposed, and the collective experience of the development group and potential customers of the system. High project costs and risks are not necessarily bad; rather it is more important that the organization understands the costs and risks associated with a project and with the portfolio of active projects before proceeding.

4. **Describe the differences between tangible and intangible benefits and costs and the differences between one-time and recurring costs.**

   Tangible benefits can be easily measured in dollars and with certainty. Intangible benefits cannot be easily measured in dollars or with certainty. Tangible costs can be easily measured in dollars and with certainty. Intangible costs cannot be easily measured in terms of cost or with certainty. One-time costs are associated with project start-up and development. Recurring costs result from the ongoing evolution and use of a system.

5. **Perform cost-benefit analysis and describe what is meant by the time value of money, present value, discount rate, net present value, return on investment, and break-even analysis.**

   The time value of money refers to comparing present cash outlays to future expected returns. Thus, the present value represents the current value of a future cash flow. The discount rate refers to the rate of return used to compute the present value of future cash flows. The net present value uses a discount rate to gain the present value of a project's overall benefits and costs. The return on investment is the ratio of the cash benefits of a project divided by the cash costs; trade-off analysis can be made among projects by comparing their representative ROI ratios. Break-even analysis finds the amount of time required for the cumulative incoming cash flow (the benefits) from a project to equal its initial and ongoing investment (the costs).

6. **Describe the activities and participant roles within a structured walkthrough.**

   A walkthrough assesses the merits of the project and assures that the project, if accepted for development, conforms to organizational standards and goals. An objective of this process is also to make sure that all relevant parties understand and agree with the information contained in the project before subsequent development activities begin. Several individuals participate in a walkthrough, including the coordinator, presenter, user, secretary, standards bearer, and maintenance oracle. Each plays a specific role to make sure that the walkthrough is a success. Walkthroughs are used to assess all types of project deliverables, including system specifications, logical and physical designs, code and program segments, test procedures and results, and manuals and documentation.

7. **Describe the three classes of Internet electronic commerce applications: Internet, intranets, and extranets.**

The Internet is a global network consisting of thousands of interconnected individual networks that communicate with each other using TCP/IP. Electronic commerce refers to the use of the Internet to support day-to-day business activities. Internet-based EC refers to transactions between individuals and businesses. Intranet refers to the use of the Internet within the same business. Extranet refers to the use of the Internet between firms.

Key Terms Checkpoint

Here are the key terms from the chapter. The page where each term is first explained is in parentheses after the term.

1. **Baseline Project Plan (BPP) (p. 85)**
2. **Break-even analysis (p. 93)**
3. **Business case (p. 85)**
4. **Discount rate (p. 92)**
5. **Economic feasibility (p. 89)**
6. **Electronic commerce (EC) (p. 104)**
7. **Electronic data interchange (EDI) (p. 104)**
8. **Extranet (p. 104)**
9. **Incremental commitment (p. 84)**
10. **Intangible benefit (p. 90)**
11. **Intangible cost (p. 90)**
12. **Internet (p. 104)**
13. **Intranet (p. 104)**
14. **Legal and contractual feasibility (p. 95)**
15. **One-time cost (p. 90)**
16. **Operational feasibility (p. 95)**
17. **Political feasibility (p. 95)**
18. **Present value (p. 92)**
19. **Recurring cost (p. 91)**
20. **Schedule feasibility (p. 95)**
21. **Statement of Work (SOW) (p. 86)**
22. **Tangible benefit (p. 89)**
23. **Tangible cost (p. 90)**
24. **Technical feasibility (p. 95)**
25. **Time value of money (p. 92)**
26. **Walkthrough (p. 100)**

Match each of the key terms above with the definition that best fits it.

_____ 1. The process of evaluating how key stakeholders within the organization view the proposed system.

_____ 2. A document prepared for the customer during project initiation and planning that describes what the project will deliver and outlines generally at a high level all work required to complete the project.

_____ 3. The justification for an information system, presented in terms of the economic benefits and costs, and the technical and organizational feasibility of the proposed system.

_____ 4. A process of identifying the financial benefits and costs associated with a development project.

_____ 5. A strategy in systems analysis and design in which the project is reviewed after each phase and continuation of the project is rejustified in each of these reviews.

_____ 6. A cost resulting from the ongoing evolution and use of a system.

_____ 7. The interest rate used to compute the present value of future cash flows.

_____ 8. A benefit derived from the creation of an information system that cannot be easily measured in dollars or with certainty.

_____ 9. A large worldwide network of networks that use a common protocol to communicate with each other; a global computing network to support business-to-consumer electronic commerce.

_____ 10. The process of assessing the degree to which the potential time frame and completion dates for all major activities within a project meet organizational deadlines and constraints for affecting change.

_____ 11. A cost associated with an information system that can be easily measured in dollars and with certainty.

_____ 12. Internet-based communication to support day-to-day business activities.

_____ 13. A peer group review of any product created during the systems development process.

_____ 14. A process of assessing the development organization's ability to construct a proposed system.

_____ 15. A cost associated with project start-up and development, or system start-up.

_____ 16. The current value of a future cash flow.

_____ 17. Internet-based communication to support business activities within a single organization.

_____ 18. A benefit derived from the creation of an information system that can be measured in dollars and with certainty.

_____ 19. The process of assessing potential legal and contractual ramifications due to the construction of a system.

_____ 20. A cost associated with an information system that cannot be easily measured in terms of dollars or with certainty.

_____ 21. The major outcome and deliverable from the project initiation and planning phase and contains the best estimate of the project's scope, benefits, costs, risks, and resource requirements.

_____ 22. The process of assessing the degree to which a proposed system solves business problems or takes advantage of business opportunities.

_____ 23. The process of comparing present cash outlays to future expected returns.

_____ 24. A type of cost-benefit analysis to identify at what point (if ever) benefits equal costs.

_____ 25. Internet-based communication to support business-to-business activities.

_____ 26. The use of telecommunications technologies to transfer business documents directly between organizations.

Review Questions

1. Describe the project identification and selection process.
2. Describe several project evaluation criteria.
3. List and describe the steps in the project initiation and planning process.
4. What is contained in a Baseline Project Plan? Are the content and format of all baseline plans the same? Why or why not?
5. Describe three commonly used methods for performing economic cost-benefit analysis.
6. List and discuss the different types of project feasibility factors. Is any factor most important? Why or why not?
7. What are the potential consequences of not assessing the technical risks associated with an information systems development project?
8. What are the types or categories of benefits from an IS project?
9. What intangible benefits might an organization obtain from the development of an IS?
10. Describe the concept of the time value of money. How does the discount rate affect the value of $1 today versus one year from today?
11. Describe the structured walkthrough process. What roles need to be performed during a walkthrough?

Problems and Exercises

1. The economic analysis carried out during project identification and selection is rather superficial. Why is this? Consequently, what factors do you think tend to be most important for a potential project to survive this first phase of the life cycle?
2. Consider your use of a PC at either home or work and list tangible benefits from an information system. Based on this list, does your use of a PC seem to be beneficial? Why or why not?
3. Consider, as an example, buying a network of PCs for a department at your workplace, or alternatively, consider outfitting a laboratory of PCs for students at a university. Make sure you estimate both the one-time and recurring costs.
4. Assuming monetary benefits of an information system at $85,000 per year, one-time costs of $75,000, recurring costs of $35,000 per year, a discount rate of 12 percent, and a five-year time horizon, calculate the net present value of these costs and benefits of an information system. Also calculate the overall return on investment of the project and then present a break-even analysis. At what point does break-even occur?
5. Choose as an example one of the information systems you described in Problem and Exercise 3 above, either buying a network of PCs for a department at your workplace or outfitting a laboratory of PCs for students at a university.

Estimate the costs and benefits for your system and calculate the net present value and return on investment, and present a break-even analysis. Assume a discount rate of 12 percent and a five-year time horizon.

6. Use the outline for the Baseline Project Plan provided in Figure 3-13 to present the system specifications for the information system you chose for Problems and Exercises 3 and 5.

7. Change the discount rate for Problem and Exercise 4 to 10 percent and redo the analysis.

8. Change the recurring costs in Problem and Exercise 4 to $40,000 and redo the analysis.

9. Change the time horizon in Problem and Exercise 4 to three years and redo the analysis.

10. Assume monetary benefits of an information system of $50,000 the first year and increasing benefits of $5,000 a year for the next four years (year 1 = 50,000; year 2 = 55,000; year 3 = 60,000; year 4 = 65,000; year 5 = 70,000). One-time development costs were $90,000 and recurring costs beginning in year 1 were $40,000 over the duration of the system's life. The discount rate for the company was 10 percent. Using a five-year horizon, calculate the net present value of these costs and benefits. Also calculate the overall return on investment of the project and then present a break-even analysis. At what point does break-even occur?

11. Change the discount rate for Problem and Exercise 10 to 12 percent and redo the analysis.

12. Change the recurring costs in Problem and Exercise 10 to $60,000 and redo the analysis.

13. For the system you chose for Problems and Exercises 3 and 5, complete section 1.0.A, the project overview, of the Baseline Project Plan report. How important is it that this initial section of the Baseline Project Plan report be done well? What could go wrong if this section is incomplete or incorrect?

14. For the system you chose for Problems and Exercises 3 and 5, complete section 2.0.A, the alternatives, of the Baseline Project Plan report. Without conducting a full-blown feasibility analysis, what is your gut feeling as to the feasibility of this system?

15. For the system you chose for Problems and Exercises 3 and 5, complete section 3.0.A–F, the feasibility analysis, of the Baseline Project Plan report. How does this feasibility analysis compare with your gut feeling from the previous question? What might go wrong if you rely on your gut feeling in determining system feasibility?

16. For the system you chose for Problems and Exercises 3 and 5, complete section 4.0.A–C, management issues, of the Baseline Project Plan report. Why might people sometimes feel that these additional steps in the project plan are a waste of time? What could you say to convince them that these steps are important?

Discussion Questions

1. Imagine that you are the chief information officer (CIO) of a company and are responsible for making all technology investment decisions. Would you ever agree to build an information system that had a negative net present value? If so, why? If not, why not? How would you justify your decision?

2. Imagine that you are interviewing for a job when the interviewer asks you which cost-benefit analysis technique is best for assessing a project's economic feasibility. What would your response be?

3. Imagine that you are working at a company and a new project idea has been assigned to you. After getting this assignment, you have a conversation with your customer who says, "This systems planning and selection stuff takes too much time... let's get on with it and start building the system!" What would your response be?

4. Of the six methods for assessing project feasibility, which is the most important? In which situation is each method more or less important?

Case Problems

1. Pine Valley Furniture

Pine Valley Furniture has recently implemented a new internship program and has begun recruiting interns from nearby university campuses. As part of this program, interns have the opportunity to work alongside a systems analyst. This shadowing opportunity provides invaluable insights into the systems analysis and design process. Recently you were selected for a six-month internship at Pine Valley Furniture, and Jim Woo has been assigned as your supervisor.

At an initial meeting with Jim Woo, he explains that Pine Valley Furniture is currently involved with two important systems development projects, the Customer Tracking System and WebStore. The purpose of the Customer Tracking System is to enable the PVF marketing group to track customer purchase activity and sales trends better. The WebStore project will help move the company into the twenty-first century by facilitating online furniture purchases, with an initial focus on corporate furniture buying. During your meeting with Mr. Woo, he reviews the documentation assembled for both systems. Mr. Woo hands you a copy of the Customer Tracking System's economic feasibility analysis. He mentions that he would like to modify the spreadsheet to reflect the information provided in the following table. Because you are very familiar with a spreadsheet product, you volunteer to make the modifications for him.

	Year 0	Year 1	Year 2	Year 3	Year 4	Year 5
Net economic benefit	$ 0	$50,000	$55,000	$55,000	$60,000	$60,000
One-time costs		$47,500				
Recurring costs		$30,000	$30,000	$30,000	$30,000	$30,000

a. How were Pine Valley Furniture's projects initiated? What is the focus for each of the new systems?

b. Modify the Customer Tracking System's economic feasibility analysis to reflect the modifications mentioned in this case problem. Use a discount rate of 10 percent. After the changes are made, what are the new overall NPV, ROI, and BEP?

c. Modify the worksheet created in part b using discount rates of 12 and 14 percent. What impact do these values have on the overall NPV, ROI, and BEP?

d. Jim Woo would like to investigate how other online stores are targeting the business furniture market. Identify and evaluate two online stores that sell business furniture. Briefly summarize your findings.

2. Hoosier Burger

The Hoosier Burger project development team has met several times with Bob and Thelma Mellankamp. During these meetings Bob has stressed the importance of improving Hoosier Burger's inventory control, customer ordering, and management reporting systems. Demand for Hoosier Burger food is at an all-time high, and this increased demand is creating problems for

Hoosier Burger's staff, creating stock-out problems, and impacting sales.

During rush periods, customers sometimes wait 15 minutes to place an order and may wait an additional 25 minutes to receive their order. Low-in-stock inventory items are often not reordered in a timely fashion, thus creating problems with the food preparation. For instance, vanilla ice cream is used to prepare vanilla malts, an item that accompanies the Hoosier Burger Special. Last week, Bob did not order enough vanilla ice cream, resulting in a last-minute dash to the grocery store.

Bob and Thelma have expressed their feelings that a new information system will be beneficial in the areas of inventory management, marketing, customer service, and food preparation. Additionally, the project team discussed with Bob and Thelma the possibility of implementing a point-of-sale system as an alternative design strategy.

a. How was the Hoosier Burger project identified and selected? What focus will the new system have?

b. Identify the Hoosier Burger project's scope.

c. Using the six feasibility factors presented in the chapter, assess the Hoosier Burger project's feasibility.

d. Using Figure 3-6 as a guide, develop a Statement of Work for the Hoosier Burger project.

3. Golden Age Retirement Center

The Golden Age Retirement Center is a retirement village designed for adults over age 60 who want to "get away from it all." Golden Age leases apartments, sells condominiums, and provides housekeeping, basic utilities, cable television, and recreational activities for its residents. The retirement village is locally owned and managed; however, a residents' advisory board has significant input when changes or recommendations to the retirement village are contemplated.

Golden Age Retirement Center's manager, Mary Lou Tobias, has recently approached you for help with the retirement center's outdated information system. Currently, the retirement office has five employees, including Ms. Tobias. She explains that all data concerning residents, financial matters, suppliers, employees, and recreational activities are kept manually. The management office does have a Pentium II computer running Windows 95 and Office 97 software. Currently the computer is used only to prepare a weekly newsletter sent to current residents. Ms. Tobias would like to have a system that automates the areas mentioned above. She would also like to establish a network where any employee can access information. This new system must be implemented within six months.

After an initial analysis, you make the following estimations. You will use these data as part of your initial feasibility assessment.

	Year 0	Year 1	Year 2	Year 3	Year 4	Year 5
Net economic benefit		$25,000	$25,000	$25,000	$25,000	$25,000
One-time costs	$40,000					
Recurring costs		$15,000	$15,000	$15,000	$15,000	$15,000

a. Identify several benefits and costs associated with implementing this new system.
b. Using the feasibility factors identified in this chapter, assess the new system's feasibility.
c. Using Figure 3-11 as a guide, prepare an economic feasibility analysis worksheet for Ms. Tobias. Using a discount rate of 10 percent, what is the overall NPV and ROI? When will break-even occur?
d. Modify the spreadsheet developed for question c to reflect discount rates of 11 and 14 percent. What impact will these new rates have on the economic analysis?

CASE: BROADWAY ENTERTAINMENT COMPANY, INC.

Initiating and Planning a Web-Based Customer Relationship Management System

Case Introduction

Carrie Douglass graduated from St. Claire Community College with an associates degree in business marketing. Among the courses Carrie took at St. Claire were several on information technology in marketing, including one on electronic commerce. While at St. Claire, Carrie worked part-time as an assistant manager at the Broadway Entertainment Company (BEC) store in Centerville, Ohio, a suburb of Dayton. After graduation, Carrie was recruited by BEC for a full-time position because of her excellent job experience at BEC and her outstanding record in classes and student organizations at St. Claire. Carrie immediately entered the BEC Manager Development Program, which consisted of three months of training, observation of experienced managers at several stores, and work experience.

The first week of training was held at the BEC regional headquarters in Columbus, Ohio. Carrie learned about company procedures and policies, trends in the home entertainment industry, and personnel practices used in BEC stores. It was during this week that Carrie was introduced to the BEC Blueprint for the Decade, a vision statement for the firm, as shown in BEC Figure 3-1.

The Blueprint, as it is called, seemed rather abstract to Carrie while in training. Carrie saw a video in which Nigel Broad, BEC's chairman, explained the importance of the Blueprint. Nigel was very sincere and clearly passionate about BEC's future hinging on every employee finding innovative ways for BEC to achieve the vision outlined in the Blueprint.

After the three-month development program was over, Carrie was surprised to be appointed manager of the Centerville store. The previous manager was promoted to a marketing position in Columbus, which created this opportunity. Carrie started her job with enthusiasm, wanting to apply what she had learned at St. Claire and in the Management Development Program.

The Idea for a New System

Although confident in her skills, Carrie believes that learning never stops. So, she logged onto the Amazon.com Web site one night from her home computer to look for some books on trends in retail marketing. While on the Web site, Carrie saw that Amazon.com was selling some of the same products BEC sells and rents in its stores. She had visited the BEC Web site often. Although a rich source of information about the company (she had found her first job with BEC from a job posting on the company's Web site), BEC was not engaged in electronic commerce with customers. All of a sudden, the words of the BEC Blueprint for the Decade started to come to life for Carrie. The Blueprint said that "BEC will be a leader in all areas of our business—human resources, technology, operations, and marketing." And, "BEC will be innovative in the use of technology . . . to provide better service to our customers." These statements caused Carrie to recall a conversation she had in the store just that day with a mother of several young children.

The mother, a frequent BEC customer, had complimented Carrie on the cleanliness of the store and efficiency of checkout. The mother added, however, that she wished BEC better understood all her needs. For example, she allowed her children to pick

---

BLUEPRINT FOR THE DECADE

FOREWORD

This blueprint provides guidance to Broadway Entertainment Corporation (BEC) for this decade. It shows our vision for the firm—our mission, objectives, and strategy fit together—and provides direction for all individuals and decisions of the firm.

OUR MISSION

BEC is a publicly held, for-profit organization focusing on the home entertainment industry that has a global focus for operations. BEC exists to serve customers with a primary goal of enhancing shareholders' investment through the pursuit of excellence in everything we do. BEC will operate under the highest ethical standards; will respect the dignity, rights, and contributions of all employees; and will strive to better society.

OUR OBJECTIVES

1. BEC will strive to increase market share and profitability (prime objective).
2. BEC will be a leader in all areas of our business—human resources, technology, operations, and marketing.
3. BEC will be cost-effective in the use of all resources.
4. BEC will rank among industry leaders in both profitability and growth.
5. BEC will be innovative in the use of technology to help bring new products and services to market faster than our competition and to provide better service to our customers.
6. BEC will create an environment that values diversity in gender, race, values, and culture among employees, suppliers, and customers.

OUR STRATEGY

BEC will be a *global* provider of home entertainment products and services by providing the highest-quality *customer service*, the *broadest range of products and services*, at the *lowest possible price*.

BEC FIGURE 3-1
BROADWAY ENTERTAINMENT COMPANY'S MISSION, OBJECTIVES, AND STRATEGY

out movies and games, but she found that the industry rating system was not always consistent with her wishes. It would be great if she and other parents could submit and view comments about videos and games. This way, parents would be more aware of the content of the products and the reactions of other children to these products. Carrie wondered why this kind of information couldn't be placed on a Web site for anyone to use. Probably the comments made by parents shopping at the Centerville store would be different from those of parents shopping at other stores, so it seemed to make sense that this information service should be a part of local store operations.

One of the books Carrie found on Amazon.com discussed customer relationship marketing. This seemed like exactly what the mother wanted from BEC. The mother didn't want just products and services; rather she wanted a store that understood and supported all of her needs for home entertainment. She wanted the store to relate to her, not just sell and rent products to her and her children.

As a new store manager, Carrie was quite busy, but she was excited to do something about her idea. She still did not understand how all aspects of BEC worked (e.g., the Manager Development Program had not discussed how to work with BEC's IS organization), and she especially felt that without a more thorough plan for her idea about a customer information service, there was no way she could get BEC management to pay attention to it. Carrie knew a way, however, to better develop her idea while still giving all the attention she needed to her new job. All she needed to do was to make one phone call, and she thought her idea could take shape.

Requesting the Project

Carrie's call was to Professor Martha Tann, head of the computer information systems (CIS) program at St. Claire Community College. Carrie had taken Professor Tann's course on business information systems required of all business students at St. Claire. Professor Tann also teaches a two-quarter capstone course for CIS majors in which student teams work in local organizations to analyze and structure the requirements for a new or replacement information system. Carrie's idea was to have a CIS student team develop a prototype of the system and use this prototype to sell the concept of the system to BEC management.

Over the next few weeks Carrie and Professor Tann discussed Carrie's idea and how projects are conducted by CIS students. Students in the course indicate which projects they want to work on among a set of projects submitted for the course by local organizations. There are always more requests submitted by local organizations than can be handled by the course, just like most organizations have more demand for information systems than can be satisfied by the available resources. Projects are presented to the students via a System Service Request form, typical of what would be used inside an organization for a user to request the IS group to undertake a systems development project. Once a group of students is assigned by Professor Tann to a project of their choice, the student team proceeds as if they were a group of systems analysts employed by the sponsoring organization. Within any limitations imposed by the sponsoring organization, the students may conduct the project using any methodology or techniques appropriate for the situation.

The initial System Service Request that Carrie submitted for review by Professor Tann appears in BEC Figure 3-2. This request appears in a standard format used for all project submissions for the CIS project course at St. Claire Community College. Professor Tann reviews initial requests for understandability by the students and gives submitters guidance on how to make the project more appealing to students.

When selecting among final System Service Requests, the students look for the projects that will give them the best opportunity to learn and integrate the skills needed to manage and conduct a systems analysis and design project. Professor Tann also asks the students to pretend to be a steering committee (sometimes called a Systems Priority Board) to select projects that appear to be well justified and of value to the sponsoring organization. So, Carrie knows that she would have to make the case for the project succinctly and persuasively, even before a preliminary study of the situation could be conducted. Her project idea would have to compete with other submissions, just as it will when she proposes it later within BEC. At least by then, she will have the experience from the prototype to prove the value of her ideas—if the students at St. Claire accept her request.

Case Summary

Ideas for new or improved information systems come from a variety of sources, including the need to fix a broken system, the need to improve the performance of an existing system, competitive pressures or new/changed government regulations, requirements generated from top-down organizational initiatives, and creative ideas by individual managers. The request for a Web-based customer information system submitted by Carrie Douglass is an example of this common, last category. Often an organization is overwhelmed by such requests. An organization must determine which ideas are the most worthy and what action should be taken in response to each request.

Carrie's proposal creates an opportunity for students at St. Claire Community College to engage in an actual systems development project. Although Carrie is not expecting a final, professional, and complete system, a working prototype that will be used by actual customers can serve as an example of the type of system that could be built by Broadway Entertainment. The project, as proposed, requests that all the typical steps in the analysis and design of an information system be conducted. Carrie Douglass could be rewarded for her creativity if the system proves to be worthwhile, or her idea could flop. The success of her idea depends on the quality of the work done by students at St. Claire.

Case Questions

1. The System Service Request (SSR) submitted by Carrie Douglass (BEC Figure 3-2) has not been reviewed by Professor Tann. If you were Professor Tann, would you ask for any changes to the request as submitted? If so, what changes, and if no changes, why? Remember, an SSR is a call for a preliminary study, not a thorough problem statement.

2. If you were a student in Professor Tann's class, would you want to work on this project? Why or why not?

3. If you were a member of BEC's steering committee, what action would you recommend for this project request? Justify your answer.

4. If you were assigned to a team of students responsible for this project, identify a preliminary list of tangible and intangible costs you think would occur for this project and ultimately for the system. At this point, no tangible benefits have been computed, so all potential benefits are intangible. What intangible benefits do you anticipate for this system?

5. What do you consider to be the risks of the project as you currently understand it? Is this a

System Service Request
St. Claire Community College
Capstone CIS Project Course

REQUESTED BY _____Carrie Douglass_____ DATE _____August 12, 2002_____

DEPARTMENT _____Broadway Entertainment Company, Store OH-84_____

LOCATION _____4600 So. Main Street_____

CONTACT _____Tel: 422-7700   FAX: 422-7760   e-mail: CarrieDoug@aol.com_____

TYPE OF REQUEST   URGENCY

[X] New System   [   ] Immediate – Operations are impaired or opportunity lost
[   ] System Enhancement   [   ] Problems exist, but can be worked around
[   ] System Error Correction   [X] Business losses can be tolerated until new system installed

PROBLEM STATEMENT

Today, Broadway Entertainment Company (BEC) sells and rents videos, music, and games to customers. BEC is profitable and growing. Increased competition from existing and emerging competitors requires BEC constantly to consider better ways to meet the needs of its customers. Increasingly, customers want information services as well as products as part of the relationship with our store. Customers want us to be aware of their likes, dislikes, and preferences, and want us to create a sense of community for the exchange of information among customers. The vision of BEC is to be a market leader in the use of technology to provide the highest-quality customer service with the broadest range of products and services. Even though providing information services as part of our relationship with our customers is consistent with this vision, no such services are provided today. The purpose of the proposed project is to prove (or disprove) that such customer information services will improve customer satisfaction and lead to increased revenue and potentially increased market share. A sustainable competitive advantage would be desirable, but is not necessary at this stage.

Specifically, the proposed system will provide information services such as (1) ability for customers to submit unstructured and structured comments about movies, music, and games they have bought or rented; (2) submit requests for new products for sale and rent; (3) check on due dates for a customer's outstanding rentals; (4) extend a rental without penalty for a minor fee to be applied when the item is returned; (5) review the inventory of items carried in the store; (6) parents can monitor (see a list of) items rented or purchased by their children. This project should conduct a thorough analysis of such information services desired by customers, design a Web-based system to provide such services, and implement and test a prototype of this system.

SERVICE REQUEST

I request a thorough analysis of this idea be conducted. I need a working prototype of the system that could be tested with a selected group of actual customers. The prototype should include major system functions. A survey of users should be conducted to gather evidence to support (or possibly not support) my subsequent request to BEC to build such a system for all stores.

IS LIAISON _____Student team leader, assigned when a team is selected for this project_____

SPONSOR _____Carrie Douglass, Manager BEC Store OH-84_____

---------------------- TO BE COMPLETED BY SYSTEMS PRIORITY BOARD ----------------------

[   ] Request approved   Assigned to _____
                          Start date _____
[   ] Recommend revision
[   ] Suggest user development
[   ] Reject for reason _____
       _____

BEC FIGURE 3-2
SYSTEM SERVICE REQUEST FROM CARRIE DOUGLASS

low-, medium-, or high-risk project? Justify your answer. From your position as a member of a student team conducting this project, would you have any particular risks? From the position of Carrie Douglass, what risks does she have given that a team of students is conducting this project?

6. If you were assigned to a team of students responsible for this project, how would you utilize the concept of incremental commitment in the design of the Baseline Project Plan?

7. If you were assigned to a team of students responsible for this project, when in the project schedule (in what phase or after which activities are completed) do you think you could develop an economic analysis of the proposed system? What economic feasibility factors do you think would be relevant?

8. If you were assigned to a team of students responsible for this project, what activities would you conduct in order to prepare the details for the Baseline Project Plan? Explain the purpose of each activity and show a time line or schedule for these activities.

9. If you were an account representative with a small consulting firm that had received a request for proposal from Carrie Douglass to conduct the project she outlines, what would be your response? Is the System Service Request sufficient as a request for proposal? If so, why? If not, what is missing?

10. In Case Question 5, you analyze the risks associated with this project. Once deployed, what are the potential operational risks of the proposed systems? How do you factor operation risks into a systems development project?

Determining System Requirements

⊙ Objectives

After studying this chapter, you should be able to:

- ⊙ Describe options for designing and conducting interviews and develop a plan for conducting an interview to determine system requirements.
- ⊙ Design, distribute, and analyze questionnaires to determine system requirements.
- ⊙ Explain the advantages and pitfalls of observing workers and analyzing business documents to determine system requirements.
- ⊙ Participate in and help plan a Joint Application Design session.
- ⊙ Use prototyping during requirements determination.
- ⊙ Select the appropriate methods to elicit system requirements.
- ⊙ Explain business process redesign and how it affects requirements determination.
- ⊙ Understand how requirements determination techniques apply to development for Internet applications.

Chapter Preview . . .

Systems analysis is the part of the systems development life cycle in which you determine how a current information system in an organization functions. Then you assess what users would like to see in a new system. As you learned in Chapter 1, there are three parts to analysis: determining requirements, structuring requirements, and selecting the best alternative design strategy. Figure 4-1 illustrates these three parts and highlights our focus in this chapter—determining system requirements.

Techniques used in requirements determination have become more structured over time. As we see in this chapter, current methods increasingly rely on computers for support. We first study the more traditional requirements determination methods, which include interviewing, using questionnaires, observing users in their work environment, and collecting procedures and other written documents. We then discuss modern methods for collecting system requirements. The first of these methods is Joint Application Design (JAD), which you first read about in Chapter 1. Next, you read about how analysts rely more and more on information systems to help them perform analysis. You learn how prototyping can be used as a key tool for some requirements determination efforts. We end the chapter with a discussion of how requirements determination continues to be a major part of systems analysis and design, even when organizational change is radical, as with business process reengineering, and new, as with developing Internet applications.

FIGURE 4-1
There are three parts to analysis: determining requirements, structuring requirements, and selecting the best alternative design strategy.

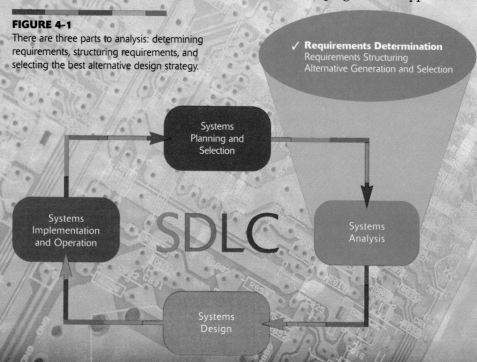

✓ **Requirements Determination**
Requirements Structuring
Alternative Generation and Selection

Systems Planning and Selection

Systems Implementation and Operation

SDLC

Systems Analysis

Systems Design

Performing Requirements Determination

As stated earlier and shown in Figure 4-1, there are three parts to systems analysis: determining requirements, structuring requirements, and selecting the best alternative design strategy. We address these as three separate steps, but you should consider these steps as somewhat parallel and repetitive. For example, as you determine some aspects of the current and desired system(s), you begin to structure these requirements or to build prototypes to show users how a system might behave. Inconsistencies and deficiencies discovered through structuring and prototyping lead you to explore further the operation of current system(s) and the future needs of the organization. Eventually your ideas and discoveries meet on a thorough and accurate depiction of current operations and the requirements for the new system. In the next section, we discuss how to begin the requirements determination process.

The Process of Determining Requirements

At the end of the systems planning and selection phase of the SDLC, management can grant permission to pursue development of a new system. A project is initiated and planned (as described in Chapter 3), and you begin determining what the new system should do. During requirements determination, you and other analysts gather information on what the system should do from as many sources as possible. Such sources include users of the current system, reports, forms, and procedures. All of the system requirements are carefully documented and made ready for structuring. Structuring means taking the system requirements you find during requirements determination and ordering them into tables, diagrams, and other formats that make them easier to translate into technical system specifications. We discuss structuring in detail in Chapters 5 and 6.

In many ways, gathering system requirements is like conducting any investigation. Have you read any of the Sherlock Holmes or similar mystery stories? Do you enjoy solving puzzles? The characteristics you need to enjoy solving mysteries and puzzles are the same ones you need to be a good systems analyst during requirements determination. These characteristics include:

- *Impertinence.* You should question everything. Ask such questions as "Are all transactions processed the same way?" "Could anyone be charged something other than the standard price?" "Might we someday want to allow and encourage employees to work for more than one department?"

- *Impartiality.* Your role is to find the best solution to a business problem or opportunity. It is not, for example, to find a way to justify the purchase of new hardware or to insist on incorporating what users think they want into the new system requirements. You must consider issues raised by all parties and try to find the best organizational solution.

- *Relaxing of constraints.* Assume anything is possible and eliminate the infeasible. For example, do not accept this statement: "We've always done it that way, so we have to continue the practice." Traditions are different from rules and policies. Traditions probably started for a good reason, but as the organization and its environment change, they may turn into habits rather than sensible procedures.

- *Attention to details.* Every fact must fit with every other fact. One element out of place means that the ultimate system will fail at some time. For example, an imprecise definition of who a customer is may mean that you purge customer data when a customer has no active orders; yet these past customers may be vital contacts for future sales.

 ◑ *Reframing.* Analysis is, in part, a creative process. You must challenge yourself to look at the organization in new ways. Consider how each user views his or her requirements. Be careful not to jump to this conclusion: "I worked on a system like that once—this new system must work the same way as the one I built before."

Deliverables and Outcomes

The primary deliverables from requirements determination are the types of information gathered during the determination process. The information can take many forms: transcripts of interviews; notes from observation and analysis of documents; analyzed responses from questionnaires; sets of forms, reports, job descriptions, and other documents; and computer-generated output such as system prototypes. In short, anything that the analysis team collects as part of determining system requirements is included in these deliverables. Table 4-1 lists examples of some specific information that might be gathered at this time.

   The deliverables summarized in Table 4-1 contain the information you need for systems analysis. In addition, you need to understand the following components of an organization:

 ◑ The business objectives that drive what and how work is done
 ◑ The information people need to do their jobs
 ◑ The data handled within the organization to support the jobs
 ◑ When, how, and by whom or what the data are moved, transformed, and stored
 ◑ The sequence and other dependencies among different data-handling activities
 ◑ The rules governing how data are handled and processed
 ◑ Policies and guidelines that describe the nature of the business and the market and environment in which it operates
 ◑ Key events affecting data values and when these events occur

TABLE 4·1:   Deliverables for Requirements Determination

Types of Deliverables	Specific Deliverables
Information collected from conversations with users	Interview transcripts
	Questionnaire responses
	Notes from observations
	Meeting notes
Existing documents and files	Business mission and strategy statement
	Sample business forms and reports and computer displays
	Procedure manuals
	Job descriptions
	Training manuals
	Flowcharts and documentation of existing systems
	Consultant reports
Computer-based information	Results from Joint Application Design sessions
	CASE repository contents and reports of existing systems
	Displays and reports from system prototypes

Such a large amount of information must be organized in order to be useful. This is the purpose of the next part of systems analysis—requirements structuring.

Requirements Structuring

The amount of information gathered during requirements determination could be huge, especially if the scope of the system under development is broad. The time required to collect and structure a great deal of information can be extensive and, because it involves so much human effort, quite expensive. Too much analysis is not productive, and the term *analysis paralysis* has been coined to describe a project that has bogged down in an abundance of analysis work. Because of the dangers of excessive analysis, today's systems analysts focus more on the system to be developed than on the current system. Later in the chapter, you learn about Joint Application Design (JAD) and prototyping, techniques developed to keep the analysis effort at a minimum yet still effective. Other processes have been developed to limit the analysis effort even more, providing an alternative to the SDLC. One of these is Rapid Application Development (RAD) (see Appendix B). Before you can fully appreciate alternative approaches, you need to learn traditional fact-gathering techniques.

Traditional Methods for Determining Requirements

Collection of information is at the core of systems analysis. At the outset, you must collect information about the information systems that are currently in use. You need to find out how users would like to improve the current systems and organizational operations with new or replacement information systems. One of the best ways to get this information is to talk to those directly or indirectly involved in the different parts of the organization affected by the possible system changes. Another way is to gather copies of documentation relevant to current systems and business processes. In this chapter, you learn about traditional ways to get information directly from those who have the information you need: interviews, questionnaires, and direct observation. You learn about collecting documentation on the current system and organizational operation in the form of written procedures, forms, reports, and other hard copy. These traditional methods of collecting system requirements are listed in Table 4-2.

TABLE 4-2: Traditional Methods of Collecting System Requirements

Traditional Method	Activities Involved
Interviews with individuals	Interview individuals informed about the operation and issues of the current system and needs for systems in future organizational activities.
Questionnaires	Survey people via questionnaires to discover issues and requirements.
Observations of workers	Observe workers at selected times to see how data are handled and what information people need to do their jobs.
Business documents	Study business documents to discover reported issues, policies, rules, and directions as well as concrete examples of the use of data and information in the organization.

Interviewing and Listening

Interviewing is one of the primary ways analysts gather information about an information systems project. Early in a project, an analyst may spend a large amount of time interviewing people about their work, the information they use to do it, and the types of information processing that might supplement their work. Others are interviewed to understand organizational direction, policies, and expectations that managers have on the units they supervise. During interviewing, you gather facts, opinions, and speculation and observe body language, emotions, and other signs of what people want and how they assess current systems.

There are many ways to interview someone effectively, and no one method is necessarily better than another. Some guidelines to keep in mind when you interview are summarized in Table 4-3 and discussed next.

First, prepare thoroughly before the interview. Set up an appointment at a time and for a duration convenient for the interviewee. The general nature of the interview should be explained to the interviewee in advance. You may ask the interviewee to think about specific questions or issues or to review certain documentation to prepare for the interview. Spend some time thinking about what you need to find out and write down your questions. Do not assume that you can anticipate all possible questions. You want the interview to be natural and, to some degree, you want to direct the interview spontaneously as you discover what expertise the interviewee brings to the session.

Prepare an interview guide or checklist so that you know in which sequence to ask your questions and how much time to spend in each area of the interview. The checklist might include some probing questions to ask as follow-up if you receive certain anticipated responses. You can, to some degree, integrate your interview guide with the notes you take during the interview, as depicted in a sample guide in Figure 4-2. This same guide can serve as an outline for a summary of what you discover during an interview.

The first page of the sample interview guide contains a general outline of the interview. Besides basic information on who is being interviewed and when, list major objectives for the interview. These objectives typically cover the most important data you need to collect, a list of issues on which you need to seek agreement (e.g., content for certain system reports), and which areas you need to explore. Also include reminder notes to yourself on key information about the interviewee (e.g., job history, known positions taken on issues, and

TABLE 4-3:   Guidelines for Effective Interviewing

Guidelines	What Is Involved
Plan the interview	Prepare interviewee by making an appointment and explaining the purpose of the interview.
	Prepare a checklist, an agenda, and questions.
Be neutral	Avoid asking leading questions.
Listen and take notes	Give your undivided attention to the interviewee and take notes and/or tape-record the interview (if permission is granted).
Review notes	Review your notes within 48 hours of the meeting. If you discover follow-up questions or need additional information, contact the interviewee.
Seek diverse views	Interview a wide range of people, including potential users and managers.

Interview Outline

Interviewee: *Name of person being interviewed*	Interviewer: *Name of person leading interview*

Location/Medium: *Office, conference room, or phone number*	Appointment Date: Start Time: End Time:

Objectives: *What data to collect* *On what to gain agreement* *What areas to explore*	Reminders: *Background/experience of interviewee* *Known opinions of interviewee*

Agenda:	Approximate Time:
Introduction	1 minute
Background on Project	2 minutes
Overview of Interview	
Topics to be Covered	1 minute
Permission to Tape Record	
Topic 1 Questions	5 minutes
Topic 2 Questions	7 minutes
...	...
Summary of Major Points	2 minutes
Questions from Interviewee	5 minutes
Closing	1 minute

General Observations:

    Interviewee seemed busy—probably need to call in a few days for follow-up questions because he gave only short answers. PC was turned off—probably not a regular PC user.

Unresolved Issues, Topics Not Covered:

    He needs to look up sales figures from 1998. He raised the issue of how to handle returned goods, but we did not have time to discuss.

(continues on next page)

FIGURE 4-2
A TYPICAL INTERVIEW GUIDE

role with current system). This information helps you to be personal, shows that you consider the interviewee important, and may assist in interpreting some answers. Also included is an agenda with approximate time limits for different sections of the interview. You may not follow the time limits precisely, but the schedule helps you cover all areas during the time the interviewee is available. Space is also allotted for general observations that do not fit under specific questions and for notes taken during the interview about topics skipped or issues raised that could not be resolved.

On subsequent pages you list specific questions. The sample form in Figure 4-2 includes space for taking notes on these questions. Because the interviewee may provide information you were not expecting, you may not follow the guide in sequence. You can, however, check off questions you have asked and write reminders to yourself to return to or skip other questions as the interview takes place.

Choosing Interview Questions You need to decide on the mix and sequence of open-ended and closed-ended questions to use. **Open-ended questions** are usually used to probe for information when you cannot anticipate all possible responses or when you do not know the precise question to ask. The person being interviewed is encouraged to talk about whatever interests him or her within the general bounds of the question. An example is "What would you say is the best thing about the information system you currently use to do your job?" or "List the three most frequently used menu options." You must react quickly to answers and determine whether or not any follow-up questions are needed for clarification or elaboration. Sometimes body language will suggest that a user has given an incomplete answer or is reluctant to provide certain information. This is where a follow-up question might result in more information. One advantage of open-ended questions is that previously unknown information can surface. You can then continue exploring along unexpected lines of inquiry to reveal even more new information. Open-ended questions also often put the interviewees at ease because they are able to respond in their own words using their own structure. Open-ended questions give interviewees more of a sense of involvement and control in the interview. A major disadvantage of open-ended questions is the length of time it can take for the questions to be answered. They also can be difficult to summarize.

Phase 2:
Systems
Analysis

Open-ended questions
Questions in interviews and on questionnaires that have no prespecified answers.

Interviewee:	Date:
Questions:	Notes:

When to ask question, if conditional
Question number: 1
 	
    Have you used the current sales tracking system? If so, how often?
	Answer
	      Yes, I ask for a report on my product line weekly.
	
	Observations
	      Seemed anxious — may be overestimating usage frequency

If yes, go to Question 2	

Question: 2
 	
    What do you like least about this system?
	Answer
	      Sales are shown in units, not dollars.
	
	Observations
	      System can show sales in dollars, but user does not know this.

Closed-ended questions
Questions in interviews and on questionnaires that ask those responding to choose from among a set of specified responses.

Closed-ended questions provide a range of answers from which the interviewee may choose. Here is an example:

Which of the following would you say is the one best things about the information system you currently use to do your job (pick only one)?
a. Having easy access to all of the data you need
b. The system's response time
c. The ability to run the system concurrently with other applications

Closed-ended questions work well when the major answers to questions are well known. Another plus is that interviews based on closed-ended questions do not necessarily require a large time commitment—more topics can be covered. Closed-ended questions can also be an easy way to begin an interview and to determine which line of open-ended questions to pursue. You can include an "other" option to encourage the interviewee to add unexpected responses. A major disadvantage of closed-ended questions is that useful information that does not quite fit the defined answers may be overlooked as the respondent tries to make a choice instead of providing his or her best answer.

Closed-ended questions, like objective questions on an examination, can follow several forms, including these choices:

- True or false
- Multiple choice (with only one response or selecting all relevant choices)
- Rating a response or idea on some scale, say from bad to good or strongly agree to strongly disagree. Each point on the scale should have a clear and consistent meaning to each person and there is usually a neutral point in the middle of the scale.
- Ranking items in order of importance

Interview Guidelines First, with either open- or closed-ended questions, do not phrase a question in a way that implies a right or wrong answer. Respondents must feel free to state their true opinions and perspectives and trust that their ideas will be considered. Avoid questions such as "Should the system continue to provide the ability to override the default value, even though most users now do not like the feature?" because such wording predefines a socially acceptable answer.

Second, listen very carefully to what is being said. Take careful notes or, if possible, record the interview on a tape recorder (be sure to ask permission first!). The answers may contain extremely important information for the project. Also, this may be your only chance to get information from this particular person. If you run out of time and still need more information from the person you are talking to, ask to schedule a follow-up interview.

Third, once the interview is over, go back to your office and key in your notes within 48 hours with a word processing program such as Microsoft Word. For numerical data, you can use a spreadsheet program such as Lotus 1-2-3 or Microsoft Excel. If you recorded the interview, use the recording to verify your notes. After 48 hours, your memory of the interview will fade quickly. As you type and organize your notes, write down any additional questions that might arise from lapses in your notes or ambiguous information. Separate facts from your opinions and interpretations. Make a list of unclear points that need clarification. Call the person you interviewed and get answers to these new questions. Use the phone call as an opportunity to verify the accuracy of your notes. You may also want to send a written copy of your notes to the person you interviewed to check your notes for accuracy. Finally, make sure to thank the person for his or her time. You may need to talk to your respondent again. If the interviewee will be a user of your system or is involved in some other way in the system's success, you want to leave a good impression.

Fourth, be careful during the interview not to set expectations about the new or replacement system unless you are sure these features will be part of the delivered system. Let the interviewee know that there are many steps to the project. Many people will have to be interviewed. Choices will have to be made from among many technically possible alternatives. Let respondents know that their ideas will be carefully considered. Due to the repetitive nature of the systems development process, however, it is premature to say now exactly what the ultimate system will or will not do.

Fifth, seek a variety of perspectives from the interviews. Talk to lots of different people: potential users of the system, users of other systems that might be affected by this new system, managers and superiors, information systems staff, and others. Encourage people to think about current problems and opportunities and what new information services might better serve the organization. You want to understand all possible perspectives so that later you will have information on which to base a recommendation or design decision that everyone can accept.

NET SEARCH
There are commercially available services that can help you conduct interviews over the phone. Visit http:// www.prenhall.com/ valacich to complete an exercise related to this topic.

Administering Questionnaires

Interviews are very effective ways of communicating with people and obtaining important information from them. However, they also are very expensive and time-consuming to conduct. Therefore, only a limited number of questions can be covered and people contacted. In contrast, questionnaires are passive and often yield less rich information than interviews, but questionnaires are not as expensive to administer per respondent. Also, questionnaires have the advantage of gathering information from many people in a relatively short time. There is also less bias involved in interpreting their results.

Choosing Questionnaire Respondents Sometimes there are more people to survey than you can handle. You must decide which questionnaire to send to which group of people. Whichever group of respondents you choose, it should be representative of all users. In general, you can achieve a representative sample by any one or a combination of these four methods:

1. *Those convenient to sample.* These may be people at a local site, those willing to be surveyed, or those most motivated to respond.
2. *A random group.* If you get a list of all users of the current system, simply choose every *n*th person on the list. Or you could select people by skipping names on the list based on numbers from a random number table.
3. *A purposeful sample.* Here you may specify only people who satisfy certain criteria, such as users of the system for more than two years or those who use the system most often.
4. *A stratified sample.* In this case, you have several categories of people you definitely want to include—choose a random set from each category (e.g., users, managers, foreign business unit users).

Samples that combine characteristics of several approaches are also common. In any case, once the questionnaires are returned, you should check for nonresponse bias; that is, a systematic bias in the results because those who responded are different from those who did not respond. You can refer to books on survey research to find out how to determine if your results are confounded by nonresponse bias.

Designing Questionnaires Questionnaires are usually administered on paper, although they can be administered in person (resembling a structured interview), over the phone (computer-assisted telephone interviewing), or even over the Internet or company intranet. Questionnaires are less expensive if they do not require a person to administer them directly; that is, if the people answering the questions can complete the questionnaire without help. Also, answers can be provided at the convenience of the respondent, as long as they are returned by a specific date.

Questionnaires typically include closed-ended questions, more than can be effectively asked in an interview, and sometimes contain open-ended questions as well. Closed-ended questions are preferable because they are easier to complete, and they define the exact coverage required. A few open-ended questions give the person being surveyed an opportunity to add insights not anticipated by the designer of the questionnaire. In general, questionnaires take less time to complete than interviews structured to obtain the same information. In addition, questionnaires are given to many people simultaneously, whereas interviews are usually limited to one person at a time.

Questionnaires are generally less rich in information content than interviews however, because they provide no direct means by which to ask follow-up questions. Also, because questionnaires are written, there is no opportunity to judge the accuracy of the responses. In an interview, you can sometimes determine if people are answering truthfully or fully by the words they use, whether they make direct eye contact, the tone of voice they use, or their body language.

The ability to create good questionnaires is a skill that improves with practice and experience. The person completing a questionnaire has only the written questions to interpret and answer. Hence, the questions must be extremely clear in meaning and logical in sequence. For example, what if a closed-ended question were phrased in this way:

How often do you back up your computer files?
a. Frequently
b. Sometimes
c. Hardly at all
d. Never

There are at least two sources of ambiguity in the wording of the question. The first source of ambiguity is the categories offered for the answer: The only nonambiguous answer is "never." "Hardly at all" could mean anything from once per year to once per month, depending on who is answering the question. "Sometimes" could cover the same range of possibilities as "Hardly at all." "Frequently" could be anything from once per hour to once per week.

The second source of ambiguity is in the question itself. Does the term *computer files* pertain only to those on my hard disk? Or does it also mean the files I have stored on other media? What if I have more than one PC in my office? And what about the public files I have stored on a file server? With no questioner present to explain the ambiguities, the respondent is at a loss and must try to answer the question in the best way he or she knows how. Whether the respondent's interpretation is the same as other respondents' is anyone's guess. The respondent cannot be there when the data are analyzed to tell exactly what was meant.

A less ambiguous way to phrase the question and its response categories would be something like this:

How often do you back up the computer files stored on the hard disk on the PC you use for the majority of your work time?
a. Frequently (at least once per week)
b. Sometimes (from one to three times per month)
c. Hardly at all (once per month or less)
d. Never

As you can see, the phrasing of the question is a bit awkward, but it avoids ambiguity. You may want to break up a single question into multiple questions, or a set of questions and statements, to avoid awkward phrasing. Notice also that the possible responses are much clearer now that they have been specifically defined. They cover the full range of possibilities, from never to at least once per week with no overlapping time periods.

Obviously, care must be taken in composing closed-ended and open-ended questions. Further, you should be as careful in composing questions for interviews as for questionnaires. Sloppily worded questions cannot be identified every time in an interview unless the interviewee asks for clarification. For both interviews and questionnaires, it is wise to pretest your questions. Pose the questions in a simulated interview, and ask the interviewee to rephrase each question as he or she interprets the question. Check responses for reasonableness. You can even ask the same question in several different ways to see if you receive a materially different response. Use this feedback to adjust the questions to make them less ambiguous.

Questionnaires are most useful in the requirements determination process when used for very specific purposes rather than for more general information gathering. For example, one useful application of questionnaires is to measure levels of user satisfaction with a system or with particular aspects of it. Another useful application is to have several users choose from among a list of system features available in many off-the-shelf software packages. You could ask users to choose the features they most want and quickly tabulate the results to find out which features are most in demand. You could then recommend a system solution based on a particular software package to meet the demands of most of the users.

Choosing between Interviews and Questionnaires

You have seen that interviews are good tools for collecting rich, detailed information and that interviews allow exploration and follow-up. On the other hand, interviews are quite time intensive and expensive. In comparison, questionnaires are inexpensive and take less time, as specific information can be gathered from many people at once without the personal intervention of an interviewer. The information collected from a questionnaire is less rich, however, and is potentially ambiguous if questions are not phrased precisely. In addition, follow-up to a questionnaire is more difficult as it often involves interviews or phone calls, adding to the expense of the process. Table 4-4 compares the characteristics of interviews and questionnaires.

These differences are important to remember during the analysis phase. Deciding which method to use and what strategy to employ to gather information will vary with the system being studied and its organizational context. For

TABLE 4-4: **Comparison of Interviews and Questionnaires**

Characteristic	Interviews	Questionnaires
Information richness	High (many channels).	Medium to low (only responses).
Time required	Can be extensive.	Low to moderate.
Expense	Can be high.	Moderate.
Chance for follow-up and probing	Good: Probing and clarification questions can be asked by either interviewer or interviewee.	Limited: Probing and follow-up done after original data collection.
Confidentiality	Interviewee is known to interviewer.	Respondent can be unknown.
Involvement of subject	Interviewee is involved and committed.	Respondent is passive, no clear commitment.
Potential audience	Limited numbers, but complete responses from those interviewed.	Can be quite large, but lack of response from some can bias results.

example, if the organization is large and the system being studied is vast and complex, there will probably be dozens of affected users. If you know little about the system or the organization, a good strategy is to identify key users and others and interview them. You then use the information gathered to create a questionnaire to distribute to a large number of users. You can schedule follow-up interviews with a few users. At the other extreme, if the system and organization are small and you understand them well, the best strategy may be to interview only one or two key users or stakeholders.

Directly Observing Users

All the methods of collecting information that we have discussed involve getting people to recall and convey information they have about organizational processes and the information systems that support them. People, however, are not always very reliable, even when they try to be and say what they think is the truth. As odd as it may sound, people often do not have a completely accurate appreciation of what they do or how they do it. This is especially true concerning infrequent events, issues from the past, or issues for which people have considerable passion. Because people cannot always be trusted to interpret and report their own actions reliably, you can supplement what people tell you by watching what they do in work situations.

For example, one possible view of how a hypothetical manager does her job is that a manager carefully plans her activities, works long and consistently on solving problems, and controls the pace of her work. A manager might tell you that is how she spends her day. Several studies have shown, however, that a manager's day is actually punctuated by many, many interruptions. Managers work in a fragmented manner, focusing on a problem or a communication for only a short time before they are interrupted by phone calls or visits from subordinates and other managers. An information system designed to fit the work environment described by our hypothetical manager would not effectively support the actual work environment in which that manager finds herself.

As another example, consider the difference between what another employee might tell you about how much he uses electronic mail and how much electronic mail use you might discover through more objective means. An employee might tell you he is swamped with e-mail messages and spends a significant proportion of time responding to e-mail messages. However, if you were able to check electronic mail records, you might find that this employee receives only three e-mail messages per day on average, and that the most messages he has ever received during one eight-hour period is ten. In this case, you were able to obtain an accurate behavioral measure of how much e-mail this employee copes with without having to watch him read his e-mail.

The intent behind obtaining system records and direct observation is the same, however, and that is to obtain more firsthand and objective measures of employee interaction with information systems. In some cases, behavioral measures will more accurately reflect reality than what employees themselves believe. In other cases, the behavioral information will substantiate what employees have told you directly. Although observation and obtaining objective measures are desirable ways to collect pertinent information, such methods are not always possible in real organizational settings. Thus, these methods are not totally unbiased, just as no other one data-gathering method is unbiased.

For example, observation can cause people to change their normal operating behavior. Employees who know they are being observed may be nervous and make more mistakes than normal. On the other hand, employees under observation may follow exact procedures more carefully than they typically do. They may work faster or slower than normal. Because observation typically cannot be continuous, you receive only a snapshot image of the person or task you observe. Such a view may not include important events or activities. Due to time constraints, you observe for only a limited time, a limited number of people, and a limited number of sites. Observation yields only a small segment of data from a possibly vast variety of data sources. Exactly which people or sites to observe is a difficult selection problem. You want to pick both typical and atypical people and sites and observe during normal and abnormal conditions and times to receive the richest possible data from observation.

Analyzing Procedures and Other Documents

As noted above, interviewing people who use a system every day or who have an interest in a system is an effective way to gather information about current and future systems. Observing current system users is a more direct way of seeing how an existing system operates. Both interviewing and observing have limitations. Methods for determining system requirements can be enhanced by examining system and organizational documentation to discover more details about current systems and the organization they support.

We discuss several important types of documents that are useful in understanding system requirements, but our discussion is not necessarily exhaustive. In addition to the few specific documents we mention, there are other important documents to locate and consider. These include organizational mission statements, business plans, organization charts, business policy manuals, job descriptions, internal and external correspondence, and reports from prior organizational studies.

What can the analysis of documents tell you about the requirements for a new system? In documents you can find information about:

> Problems with existing systems (e.g., missing information or redundant steps)

> Opportunities to meet new needs if only certain information or information processing were available (e.g., analysis of sales based on customer type)

> Organizational direction that can influence information system requirements (e.g., trying to link customers and suppliers more closely to the organization)

> Titles and names of key individuals who have an interest in relevant existing systems (e.g., the name of a sales manager who has led a study of buying behavior of key customers)

> Values of the organization or individuals who can help determine priorities for different capabilities desired by different users (e.g., maintaining market share even if it means lower short-term profits)

> Special information processing circumstances that occur irregularly that may not be identified by any other requirements determination technique (e.g., special handling needed for a few very large-volume customers that require use of customized customer ordering procedures)

> The reason why current systems are designed as they are, which can suggest features left out of current software that may now be feasible and desirable (e.g., data about a customer's purchase of competitors' products not available when the current system was designed; these data now available from several sources)

Phase 2:
Systems
Analysis

● Data, rules for processing data, and principles by which the organization operates that must be enforced by the information system (e.g., each customer assigned exactly one sales department staff member as primary contact if customer has any questions)

One type of useful document is a written work procedure for an individual or a work group. The procedure describes how a particular job or task is performed, including data and information used and created in the process of performing the job. For example, the procedure shown in Figure 4-3 includes data

---

GUIDE FOR PREPARATION OF INVENTION DISCLOSURE
(See FACULTY and STAFF MANUALS for detailed Patent Policy and routing procedures.)

(1) DISCLOSE ONLY ONE INVENTION PER FORM.

(2) PREPARE COMPLETE DISCLOSURE.

   The disclosure of your invention is adequate for patent purposes ONLY if it enables a person skilled in the art to understand the invention.

(3) CONSIDER THE FOLLOWING IN PREPARING A COMPLETE DISCLOSURE:

   (a) All essential elements of the invention, their relationship to one another, and their mode of operation

   (b) Equivalents that can be substituted for any elements

   (c) List of features believed to be new

   (d) Advantages this invention has over the prior art

   (e) Whether the invention has been built and/or tested

(4) PROVIDE APPROPRIATE ADDITIONAL MATERIAL.

   Drawings and descriptive material should be provided as needed to clarify the disclosure. Each page of this material must be signed and dated by each inventor and properly witnessed. A copy of any current and/or planned publication relating to the invention should be included.

(5) INDICATE PRIOR KNOWLEDGE AND INFORMATION.

   Pertinent publications, patents or previous devices, and related research or engineering activities should be identified.

(6) HAVE DISCLOSURE WITNESSED.

   Persons other than co-inventors should serve as witnesses and should sign each sheet of the disclosure only after reading and understanding the disclosure.

(7) FORWARD ORIGINAL PLUS ONE COPY (two copies if supported by grant/contract) TO VICE PRESIDENT FOR RESEARCH VIA DEPARTMENT HEAD AND DEAN.

FIGURE 4-3
EXAMPLE OF A WRITTEN WORK PROCEDURE FOR AN INVENTION DISCLOSURE

(list of features and advantages, drawings, inventor name, and witness names) required to prepare an invention disclosure. It also indicates that besides the inventor, the vice president for research and department head and dean must review the material, and that a witness is required for any filing of an invention disclosure. These insights clearly affect what data must be kept, to whom information must be sent, and the rules that govern valid forms.

Procedures are not trouble-free sources of information, however. Sometimes your analysis of several written procedures reveals a duplication of effort in two or more jobs. You should call such duplication to the attention of management as an issue to be resolved before system design can proceed. That is, it may be necessary to redesign the organization before the redesign of an information system can achieve its full benefits. Another problem you may encounter is a missing procedure. Again, it is not your job to create a document for a missing procedure—that is up to management. A third and common problem happens when the procedure is out of date. You may realize this in your interview of the person responsible for performing the task described in the procedure. Once again, the decision to rewrite the procedure so that it matches reality is made by management, but you may make suggestions based upon your understanding of the organization. A fourth problem often encountered is that the formal procedures may contradict information you collected from interviews, questionnaires, and observation about how the organization operates and what information is required. As in the other cases, resolution rests with management.

All of these problems illustrate the difference between formal systems and informal systems. A **formal system** is one an organization has documented; an **informal system** is the way in which the organization actually works. Informal systems develop because of inadequacies of formal procedures and individual work habits, preferences, and resistance to control. It is important to understand both formal and informal systems because each provides insight into information requirements and what is necessary to convert from present to future systems.

A second type of document useful to systems analysts is a business form, illustrated in Figure 4-4. Forms are used for all types of business functions, from recording an order to acknowledging the payment of a bill to indicating what goods have been shipped. Forms are important for understanding a system because they explicitly indicate what data flow in or out of a system. In the sample invoice form in Figure 4-4, we see space for data such as the customer identification code, the "ship to" address, the quantity of items ordered, their descriptions, discounts, and unit prices.

A printed form may correspond to a computer display that the system will generate for someone to enter and maintain data or to display data to online users. The most useful forms contain actual organizational data, as this allows you to determine the data characteristics actually used by the application. The ways in which people use forms change over time, and data that were needed when a form was designed may no longer be required.

A third type of useful document is a report generated by current systems. As the primary output for some types of systems, a report enables you to work backward from the information on the report to the data that must have been necessary to generate it. Figure 4-5 presents an example of a common financial accounting report, the consolidated balance sheet. Every number listed on the balance sheet is actually an aggregated amount, based on the accumulation of millions of individual business transactions. You analyze such reports to determine which data need to be captured over what time period and what manipulation of these raw data is necessary to produce each field on the report.

If the current system is computer based, a fourth set of useful documents is those that describe the current information systems—how they were designed and how they work. A lot of different types of documents fit this description,

Formal system
The official way a system works as described in organizational documentation.

Informal system
The way a system actually works.

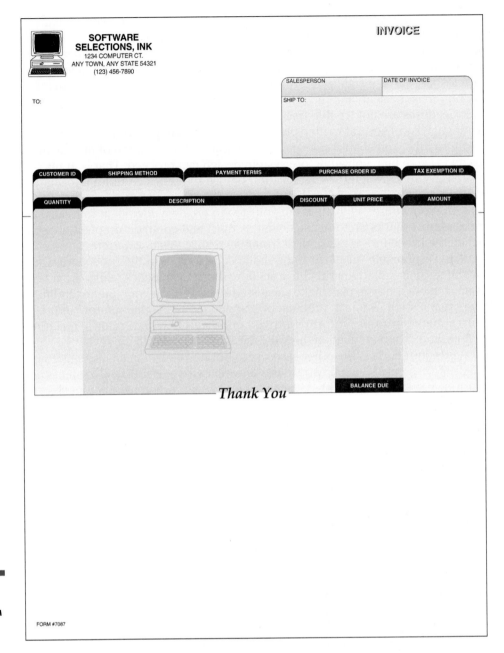

everything from flowcharts to data dictionaries to user manuals. An analyst who has access to such documents is fortunate because many in-house-developed information systems lack complete documentation.

Analysis of organizational documents and observation, along with interviewing and distributing questionnaires, are the methods used most for gathering system requirements. In Table 4-4 we summarize the comparative features of interviews and questionnaires. Table 4-5 summarizes the comparative features of observation and analysis of organizational documents.

Modern Methods for Determining System Requirements

Even though we called interviews, questionnaires, observation, and document analysis traditional methods for determining a system's requirements, all of

MICROSOFT BALANCE SHEETS
(In millions)

June 30	2000	2001
Assets		
Current assets:		
Cash and equivalents	$ 4,846	$ 3,922
Short-term investments	18,952	27,678
Total cash and short-term investments	23,798	31,600
Accounts receivable	3,250	3,671
Deferred income taxes	1,708	1,949
Other	1,552	2,417
Total current assets	30,308	39,637
Property and equipment, net	1,903	2,309
Equity and other investments	17,726	14,141
Other assets	2,213	3,170
Total assets	$52,150	$59,257
Liabilities and stockholders' equity		
Current liabilities:		
Accounts payable	$ 1,083	$ 1,188
Accrued compensation	557	742
Income taxes	585	1,468
Unearned revenue	4,816	5,614
Other	2,714	2,120
Total current liabilities	9,755	11,132
Deferred income taxes	1,027	836
Commitments and contigencies		
Stockholders' equity:		
Common stock and paid-in capital— shares authorized 12,000; shares issued and outstanding 5,283 and 5,383	23,195	28,390
Retained earnings, including accumulated other comprehensive income of $1,527 and $587	18,173	18,899
Total stockholders' equity	41,368	47,289
Total liabilities and stockholders' equity	$52,150	$59,257

FIGURE 4-5
AN EXAMPLE OF A REPORT—AN ACCOUNTING BALANCE SHEET Printed from www.microsoft.com Reprinted by permission of Microsoft Corp.

these methods are still very much used by analysts to collect important information. Today, however, there are additional techniques to collect information about the current system, the organizational area requesting the new system, and what the new system should be like. In this section, you learn about two modern information-gathering techniques for analysis: Joint Application Design (JAD) and prototyping. These techniques can support effective information collection and structuring while reducing the amount of time required for analysis.

Joint Application Design

You were introduced to Joint Application Design, or JAD, in Chapter 1. There you learned JAD started in the late 1970s at IBM as a means to bring together the key users, managers, and systems analysts involved in the analysis of a current system. Since the 1970s, JAD has spread throughout many companies and industries. For example, it is quite popular in the insurance industry. The

TABLE 4-5: Comparison of Observation and Document Analysis

Characteristic	Observation	Document Analysis
Information richness	High (many channels).	Low (passive) and old.
Time required	Can be extensive.	Low to moderate.
Expense	Can be high.	Low to moderate.
Chance for follow-up and probing	Good: Probing and clarification questions can be asked during or after observation.	Limited: Probing is possible only if original author is available.
Confidentiality	Observee is known to interviewer; observee may change behavior when observed.	Depends on nature of document; does not change simply by being read.
Involvement of subject	Interviewees may or may not be involved and committed depending on whether they know they are being observed.	None, no clear commitment.
Potential audience	Limited numbers and limited time (snapshot) of each.	Potentially biased by which documents were kept or because document not created for this purpose.

primary purpose of using JAD in the analysis phase is to collect systems requirements simultaneously from the key people involved with the system. The result is an intense and structured, but highly effective, process. Having all the key people together in one place at one time allows analysts to see the areas of agreement and the areas of conflict. Meeting with all these important people for over a week of intense sessions allows you the opportunity to resolve conflicts or at least to understand why a conflict may not be simple to resolve.

JAD sessions are usually conducted in a location away from where the people involved normally work. This is to keep participants away from as many distractions as possible so that they can concentrate on systems analysis. A JAD may last anywhere from four hours to an entire week and may consist of several sessions. A JAD employs thousands of dollars of corporate resources, the most expensive of which is the time of the people involved. Other expenses include the costs associated with flying people to a remote site and putting them up in hotels and feeding them for several days.

The following is a list of typical JAD participants:

JAD session leader
The trained individual who plans and leads Joint Application Design sessions.

- ◗ ***JAD session leader.*** The JAD leader organizes and runs the JAD. This person has been trained in group management and facilitation as well as in systems analysis. The JAD leader sets the agenda and sees that it is met. He or she remains neutral on issues and does not contribute ideas or opinions but rather concentrates on keeping the group on the agenda, resolving conflicts and disagreements, and soliciting all ideas.

- ◗ *Users.* The key users of the system under consideration are vital participants in a JAD. They are the only ones who clearly understand what it means to use the system on a daily basis.

- ◗ *Managers.* Managers of the work groups who use the system in question provide insight into new organizational directions, motivations for and organizational impacts of systems, and support for requirements determined during the JAD.

- *Sponsor.* As a major undertaking, due to its expense, a JAD must be sponsored by someone at a relatively high level in the company such as a vice president or chief executive officer. If the sponsor attends any sessions, it is usually only at the very beginning or the end.

- *Systems analysts.* Members of the systems analysis team attend the JAD although their actual participation may be limited. Analysts are there to learn from users and managers, not to run or dominate the process.

- **Scribe.** The scribe takes notes during the JAD sessions, usually on a personal computer or laptop.

- *IS staff.* Besides systems analysts, other IS staff, such as programmers, database analysts, IS planners, and data center personnel, may attend the session. Their purpose is to learn from the discussion and possibly to contribute their ideas on the technical feasibility of proposed ideas or on the technical limitations of current systems.

Scribe
The person who makes detailed notes of the happenings at a Joint Application Design session.

JAD sessions are usually held in special-purpose rooms where participants sit around horseshoe-shaped tables, as in Figure 4-6. These rooms are typically equipped with whiteboards (possibly electronic, with a printer to make copies of what is written on the board). Other audiovisual tools may be used, such as transparencies and overhead projectors, magnetic symbols that can be easily rearranged on a whiteboard, flip charts, and computer-generated displays. Flip

FIGURE 4-6
A TYPICAL ROOM LAYOUT FOR A JAD SESSION
Source: Adapted from Wood and Silver, 1989.

chart paper is typically used for keeping track of issues that cannot be resolved during the JAD or for those issues requiring additional information that can be gathered during breaks in the proceedings. Computers may be used to create and display form or report designs or to diagram existing or replacement systems. In general, however, most JADs do not benefit much from computer support.

The end result of a completed JAD is a set of documents that detail the workings of the current system and the features of a replacement system. Depending on the exact purpose of the JAD, analysts may gain detailed information on what is desired of the replacement system.

Taking Part in a JAD   Imagine that you are a systems analyst taking part in your first JAD. What might participating in a JAD be like? Typically, JADs are held off site, in comfortable conference facilities. On the first morning of the JAD, you and your fellow analysts walk into a room that looks much like the one depicted in Figure 4-6. The JAD facilitator is already there. She is finishing writing the day's agenda on a flip chart. The scribe is seated in a corner with a laptop, preparing to take notes on the day's activities. Users and managers begin to enter in groups and seat themselves around the U-shaped table. You and the other analysts review your notes describing what you have learned so far about the information system you are all there to discuss. The session leader opens the meeting with a welcome and a brief rundown of the agenda. The first day will be devoted to a general overview of the current system and major problems associated with it. The next two days will be devoted to an analysis of current system screens. The last two days will be devoted to analysis of reports.

The session leader introduces the corporate sponsor, who talks about the organizational unit and current system related to the systems analysis study and the importance of upgrading the current system to meet changing business conditions. He leaves and the JAD session leader takes over. She yields the floor to the senior analyst, who begins a presentation on key problems with the system that have already been identified. After the presentation, the session leader opens the discussion to the users and managers in the room.

After a few minutes of talk, a heated discussion begins between two users from different corporate locations. One user, who represents the office that served as the model for the original systems design, argues that the system's perceived lack of flexibility is really an asset, not a problem. The other user, who represents an office that was part of another company before a merger, argues that the current system is so inflexible as to be virtually unusable. The session leader intervenes and tries to help the users isolate particular aspects of the system that may contribute to the system's perceived lack of flexibility.

Questions arise about the intent of the original developers. The session leader asks the analysis team about their impressions of the original system design. Because these questions cannot be answered during this meeting, as none of the original designers are present nor are the original design documents readily available, the session leader assigns the question about intent to the "to-do" list. This becomes the first question on a flip chart sheet of to-do items, and the session leader gives you the assignment of finding out about the intent of the original designers. She writes your name next to the to-do item on the list and continues with the session. Before the end of the JAD, you must get an answer to this question.

The JAD will continue like this for its duration. Analysts will make presentations, help lead discussions of form and report design, answer questions from users and managers, and take notes on what is being said. After each meeting, the analysis team will meet, usually informally, to discuss what has occurred that day and to consolidate what they have learned. Users will continue to contribute during the meetings, and the session leader will facilitate, intervening in conflicts, seeing that the group follows the agenda. When the JAD is over, the

NET SEARCH
There is additional information available on the Web on Joint Application Design. Visit http://www. prenhall.com/valacich to complete an exercise related to this topic.

session leader and her assistants must prepare a report that documents the findings in the JAD and circulate it among users and analysts.

Using Prototyping during Requirements Determination

You were introduced to prototyping in Chapter 1 (see Figure 1-17 for an overview of prototyping). There you learned that prototyping is a repetitive process in which analysts and users build a rudimentary version of an information system based on user feedback. You also learned that prototyping could replace the systems development life cycle or augment it. In this section, we see how prototyping can augment the requirements determination process.

To establish requirements for prototyping, you still have to interview users and collect documentation. Prototyping, however, allows you quickly to convert basic requirements into a working, though limited, version of the desired information system. The user then views and tests the prototype. Typically, seeing verbal descriptions of requirements converted into a physical system prompts the user to modify existing requirements and generate new ones. For example, in the initial interviews, a user might have said he wanted all relevant utility billing information on a single computer display form, such as the client's name and address, the service record, and payment history. Once the same user sees how crowded and confusing such a design would be in the prototype, he might change his mind and instead ask for the information to be organized on several screens but with easy transitions from one screen to another. He might also be reminded of some important requirements (data, calculations, etc.) that had not surfaced during the initial interviews.

You would then redesign the prototype to incorporate the suggested changes. Once modified, users would again view and test the prototype. Once again, you would incorporate their suggestions for change. Through such a repetitive process, the chances are good that you will be able to better capture a system's requirements. The goal with using prototyping to support requirements determination is to develop concrete specifications for the ultimate system, not to build the ultimate system.

Prototyping is most useful for requirements determination when:

- ⊙ User requirements are not clear or well understood, which is often the case for totally new systems or systems that support decision making.

- ⊙ One or a few users and other stakeholders are involved with the system.

- ⊙ Possible designs are complex and require concrete form to evaluate fully.

- ⊙ Communication problems have existed in the past between users and analysts, and both parties want to be sure that system requirements are as specific as possible.

- ⊙ Tools (such as form and report generators) and data are readily available to rapidly build working systems.

Prototyping also has some drawbacks as a tool for requirements determination. These include:

- ⊙ A tendency to avoid creating formal documentation of system requirements, which can then make the system more difficult to develop into a fully working system.

- ⊙ Prototypes can become very idiosyncratic to the initial user and difficult to diffuse or adapt to other potential users.

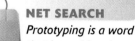

NET SEARCH
Prototyping is a word that is used in a broad variety of contexts. Visit http://www.prenhall. com/valacich to complete an exercise related to this topic.

> Prototypes are often built as stand-alone systems, thus ignoring issues of sharing data and interactions with other existing systems.

> Checks in the SDLC are bypassed so that some more subtle, but still important, system requirements might be forgotten (e.g., security, some data entry controls, or standardization of data across systems).

Radical Methods for Determining System Requirements

Whether traditional or modern, the methods for determining system requirements that you have read about in this chapter apply to any requirements determination effort, regardless of its motivation. Yet most of what you have learned has traditionally been applied to systems development projects that involve automating existing processes. Analysts use system requirements determination to understand current problems and opportunities, as well as what is needed and desired in future systems. Typically, the current way of doing things has a large impact on the new system. In some organizations, though, management is looking for new ways to perform current tasks. These ways may be radically different from how things are done now, but the payoffs may be enormous: Fewer people may be needed to do the same work, relationships with customers may improve dramatically, and processes may become much more efficient and effective, all of which can result in increased profits. The overall process by which current methods are replaced with radically new methods is referred to as **business process reengineering (BPR).**

To better understand BPR, consider the following analogy. Suppose you are a successful European golfer who has tuned your game to fit the style of golf courses and weather in Europe. You have learned how to control the flight of the ball in heavy winds, roll the ball on wide-open greens, putt on large and undulating greens, and aim at a target without the aid of the landscaping common on North American courses. When you come to the United States to make your fortune on the U.S. tour, you discover that improving your putting, driving accuracy, and sand shots will help, but the new competitive environment is simply not suited to your playing style. You need to reengineer your whole approach, learning how to aim at targets, spin and stop a ball on the green, and manage the distractions of crowds and press. If you are good enough, you may survive, but without reengineering, you will never become a winner.

Just as the competitiveness of golf forces good players to adapt their games to changing conditions, the competitiveness of our global economy has driven most companies into a mode of continuously improving the quality of their products and services. Organizations realize that creatively using information technologies can significantly improve most business processes. The idea behind BPR is not just to improve each business process but, in a systems modeling sense, to reorganize the complete flow of data in major sections of an organization to eliminate unnecessary steps, combine previously separate steps, and become more responsive to future changes. Companies such as IBM, Procter & Gamble, Wal-Mart, and Ford have had great success in actively pursuing BPR efforts. Yet, many other companies have found difficulty in applying BPR principles. Nonetheless, BPR concepts are actively applied in both corporate strategic planning and information systems planning as a way to improve business processes radically (as described in Chapter 5).

BPR advocates suggest that radical increases in the quality of business processes can be achieved through creatively applying information technologies. BPR advocates also suggest that radical improvement cannot be achieved

Business process reengineering (BPR)
The search for, and implementation of, radical change in business processes to achieve breakthrough improvements in products and services.

by making minor changes in existing processes but rather by using a clean sheet of paper and asking, "If we were a new organization, how would we accomplish this activity?" Changing the way work is performed also changes the way information is shared and stored, which means that the results of many BPR efforts are the development of information system maintenance requests or requests for system replacement. You likely have encountered or will encounter BPR initiatives in your own organization. A recent survey of IS executives found that they view BPR to be a top IS priority for the coming years.

Identifying Processes to Reengineer

A first step in any BPR effort is to understand what processes need to change. To do this, you must first understand what are the **key business processes** for the organization. Key business processes are the structured set of measurable activities designed to produce a specific output for a particular customer or market. The important aspect of this definition is that key processes are focused on some type of organizational outcome such as the creation of a product or the delivery of a service. Key business processes are also customer focused. In other words, key business processes would include all activities used to design, build, deliver, support, and service a particular product for a particular customer. BPR, therefore, requires you first to understand those activities that are part of the organization's key business processes and then to alter the sequence and structure of activities to achieve radical improvements in speed, quality, and customer satisfaction. The same techniques you learned to use for system requirements determination can be applied to discovering and understanding key business processes. These include interviewing key individuals, observing activities, reading and studying organizational documents, and conducting JADs.

After identifying key business processes, the next step is to identify specific activities that can be radically improved through reengineering. Michael Hammer and James Champy, two academics who coined the term *BPR*, suggest systems analysts ask three questions to identify activities for radical change:

1. How important is the activity to delivering an outcome?
2. How feasible is changing the activity?
3. How dysfunctional is the activity?

The answers to these questions provide guidance for selecting which activities to change. Those activities deemed important, changeable, yet dysfunctional, are primary candidates for alteration. To identify dysfunctional activities, Hammer and Champy suggest you look for activities that involve excessive information exchanges between individuals, information that is redundantly recorded or needs to be rekeyed, excessive inventory buffers or inspections, and a lot of rework or complexity. An example of a dysfunctional process and how BPR is used to change it is presented at the end of Chapter 5.

Disruptive Technologies

Once key business processes and activities have been identified, information technologies must be applied to improve business processes radically. To do this, Hammer and Champy suggest that organizations think "inductively" about information technology. Induction is the process of reasoning from the specific to the general, which means that managers must learn about the power of new technologies and think of innovative ways to alter the way work is done. This is contrary to deductive thinking, in which problems are first identified and solutions then formulated.

Hammer and Champy suggest that managers especially consider **disruptive technologies** when applying deductive thinking. Disruptive technologies are those that enable the breaking of long-held business rules that

Phase 2: Systems Analysis

Key business processes The structured, measured set of activities designed to produce a specific output for a particular customer or market.

Disruptive technologies Technologies that enable the breaking of long-held business rules that inhibit organizations from making radical business changes.

inhibit organizations from making radical business changes. For example, Saturn is using production schedule databases and electronic data interchange (EDI)—an information system that allows companies to link their computers directly to suppliers—to work with its suppliers as if they and Saturn were one company. Suppliers do not wait until Saturn sends them a purchase order for more parts but simply monitor inventory levels and automatically send shipments as needed. Table 4-6 shows several long-held business rules and beliefs that constrain organizations from making radical process improvements. For example, the first rule suggests that information can appear in only one place at a time. However, the advent of distributed databases, which allow business units to share a common database, has "disrupted" this long-held business belief.

ELECTRONIC COMMERCE APPLICATION: DETERMINING SYSTEM REQUIREMENTS

Determining system requirements for an Internet-based electronic commerce application is no different than the process followed for other applications. In the last chapter, you read how Pine Valley Furniture's management began the WebStore project—to sell furniture products over the Internet. Here we examine the process followed by PVF to determine system requirements and highlight some of the issues and capabilities that you may want to consider when developing your own Internet-based application.

Determining System Requirements for Pine Valley Furniture's WebStore

To collect system requirements as quickly as possible, Jim Woo and Jackie Judson decided to hold a three-day JAD session. In order to get the most out of these sessions, they invited a broad range of people, including representatives from Sales and Marketing, Operations, and Information Systems. Additionally,

TABLE 4-6: Long-Held Organizational Rules That Are Being Eliminated through Disruptive Technologies

Rule	Disruptive Technology
Information can appear in only one place at a time.	Distributed databases allow the sharing of information.
Only experts can perform complex work.	Expert systems can aid nonexperts.
Businesses must choose between centralization and decentralization.	Advanced telecommunications networks can support dynamic organizational structures.
Managers must make all decisions.	Decision support tools can aid nonmanagers.
Field personnel need offices where they can receive, store, retrieve, and transmit information.	Wireless data communication and portable computers provide a "virtual" office for workers.
The best contact with a potential buyer is personal contact.	Interactive communication technologies allow complex messaging capabilities.
You have to find out where things are.	Automatic identification and tracking technology know where things are.
Plans get revised periodically.	High-performance computing can provide real-time updating.

Phase 2:
Systems
Analysis

they asked an experienced JAD facilitator, Cheri Morris, to conduct the session. Together with Cheri, Jim and Jackie developed a very ambitious and detailed agenda for the session. Their goal was to collect requirements on the following items:

- ◗ System layout and navigation characteristics
- ◗ WebStore and site management system capabilities
- ◗ Customer and inventory information
- ◗ System prototype evolution

In the remainder of this section, we briefly highlight the outcomes of the JAD session.

System Layout and Navigation Characteristics As part of the process of preparing for the JAD session, all participants were asked to visit several established retail Web sites, including www.amazon.com, www.landsend.com, www.sony.com, and www.pier1.com. At the JAD session, participants were asked to identify characteristics of these sites that they found appealing and those they found cumbersome; this allowed participants to identify and discuss those features that they wanted the WebStore to possess. The outcomes of this activity are summarized in Table 4-7.

WebStore and Site Management System Capabilities After agreeing to the general layout and navigational characteristics of the WebStore, the session then turned its focus to the basic system capabilities. To assist in this process, systems analysts from the Information Systems department developed a draft skeleton of the WebStore based on the types of screens and capabilities of popular retail Web sites. For example, many retail Web sites have a "shopping cart" feature that allows customers to accumulate multiple items before checking out rather than buying a single item at a time. After some discussion, the participants agreed that the system structure shown in Table 4-8 would form the foundation for the WebStore system.

In addition to the WebStore capabilities, members of the Sales and Marketing department described several reports that would be necessary to manage customer accounts and sales transactions effectively. In addition, the department wants to be able to conduct detailed analyses of site visitors, sales tracking, and so on. Members of the Operations department expressed a need to update the product catalog easily. These collective requests and activities were organized into a system design structure, called the Site Management System, summarized in Table 4-8. The structures of both the WebStore and Site

TABLE 4-7:   Desired Layout and Navigation Feature of Webstore

Layout and Design

Navigation menu and logo placement should remain consistent throughout the entire site (this allows users to maintain familiarity while using the site and minimizes users who get "lost" in the site).

Graphics should be lightweight to allow for quick page display.

Text should be used over graphics whenever possible.

Navigation

Any section of the store should be accessible from any other section via the navigation menu.

Users should always be aware of what section they are currently in.

**TABLE 4-8:   System Structure of the WebStore
and Site Management Systems**

WebStore System	Site Management System
Main Page	User profile manager
Product line (catalog)	Order maintenance manager
• Desks	Content (catalog) manager
• Chairs	Reports
• Tables	Total hits
• File cabinets	Most frequent page views
Shopping cart	Users/time of day
Checkout	Users/day of week
Account profile	Shoppers not purchasing (used shopping cart—did not check out)
Order status/history	Feedback analysis
Customer comments	
Company information	
Feedback	
Contact information	

TABLE 4-9:   Customer and Inventory Information for WebStore

Corporate Customer	Home Office Customer	Student Customer	Inventory Information
Company name	Name	Name	SKU
Company address	Doing business as (company name)	School	Name
Company phone		Address	Description
Company fax	Address	Phone	Finished product size
Company preferred shipping method	Phone	E-mail	Finished product weight
	Fax		Available materials
Buyer name	E-mail		Available colors
Buyer phone			Price
Buyer e-mail			Lead time

Management systems will be given to the Information Systems department as the baseline for further analysis and design activities.

Customer and Inventory Information The WebStore will be designed to support the furniture purchases of three distinct types of customers:

- Corporate customers
- Home office customers
- Student customers

To track the sales to these different types of customers effectively, the system must capture and store distinct information. Table 4-9 summarizes this information for each customer type identified during the JAD session. Orders reflect the range of product information that must be specified to execute a sales transaction. Thus, in addition to capturing the customer information, product and sales data must also be captured and stored; Table 4-9 lists the results of this analysis.

System Prototype Evolution As a final activity, the JAD participants discussed, along with extensive input from the Information Systems staff, how the system implementation should evolve. After completing analysis and design activities, they agreed that the system implementation should progress in three main stages so that requirements changes could be more easily identified and implemented. Table 4-10 summarizes these stages and the functionality incorporated at each one.

At the conclusion of the JAD session, all the participants felt good about the progress that had been made and the clear requirements that had been identified. With these requirements in hand, Jim and the Information Systems staff could begin to turn these lists of requirements into formal analysis and design specifications. To show how information flows through the WebStore, Jim and his staff will produce data flow diagrams (Chapter 5). To show a conceptual model of the data used within the WebStore, they will generate an entity-relationship diagram (Chapter 6). Both of these analysis documents become the foundation for detailed system design and implementation.

As we saw in Chapter 1, there are three parts to the systems analysis phase of the systems development life cycle: determining requirements, structuring requirements, and selecting the best alternative design strategy. Chapter 4 focuses on requirements determination, the gathering of information about current systems and the need for replacement systems. Chapters 5 and 6 address techniques for structuring the information discovered during requirements determination. Chapter 7 closes Part III of the book by explaining how analysts generate alternative design strategies for replacement systems and choose the best one.

TABLE 4-10: Stages of System Implementation of WebStore

Stage 1 (Basic Functionality)

Simple catalog navigation; 2 products per section—limited attribute set

25 sample users

Simulated credit card transaction

Full shopping cart functionality

Stage 2 (Look and Feel)

Full product attribute set and media (images, video)—commonly referred to as "product data catalog"

Full site layout

Simulated integration with Purchasing Fulfillment and Customer Tracking systems

Stage 3 (Staging/Preproduction)

Full integration with Purchasing Fulfillment and Customer Tracking systems

Full credit card processing integration

Full product data catalog

Key Points Review

1. **Describe options for designing and conducting interviews and develop a plan for conducting an interview to determine system requirements.**

   Interviews can involve open-ended and closed-ended questions. In either case, you must be very precise in formulating a question in order to avoid ambiguity and to ensure a proper response. Making a list of questions is just one activity necessary to prepare for an interview. You must also create a general interview guide (see Figure 4-2) and schedule the interview.

2. **Design, distribute, and analyze questionnaires to determine system requirements.**

   Open-ended and closed-ended questions can be used on questionnaires, just as with interviews. Questionnaires tend to be much more specific than interviews. They also have to be carefully designed and tested because their designer will typically not be there when the questionnaire is answered by a respondent. There are also several methods for distributing questionnaires and for dealing with the bias introduced through nonresponse.

3. **Explain the advantages and pitfalls of observing workers and analyzing business documents to determine system requirements.**

   During observation you must try not to intrude or interfere with normal business activities so that the people being observed do not modify their activities from normal processes. Observation can be expensive because it is so labor intensive. Analyzing documents may be much less expensive, but any insights gained will be limited to what is available, based on the reader's interpretation. Often the creator of the document is not there to answer questions.

4. **Participate in and help plan a Joint Application Design session.**

   Joint Application Design (JAD) begins with the idea of the group interview and adds structure and a JAD session leader to it. Typical JAD participants include the session leader, a scribe, key users, managers, a sponsor, systems analysts, and IS staff members. JAD sessions are usually held off site and may last as long as one week.

5. **Use prototyping during requirements determination.**

   You read how information systems can support requirements determination with prototyping. As part of the prototyping process, users and analysts work closely together to determine requirements that the analyst then builds into a model. The analyst and user then work together on revising the model until it is close to what the user desires.

6. **Select the appropriate methods to elicit system requirements.**

   For requirements determination, the traditional sources of information about a system include interviews, questionnaires, observation, and procedures, forms, and other useful documents. Often many or even all of these sources are used to gather perspectives on the adequacy of current systems and the requirements for replacement systems. Each form of information collection has its advantages and disadvantages, which were summarized in Tables 4-4 and 4-5. Selecting the methods to use depends on the need for rich or thorough information, the time and budget available, the need to probe deeper once initial information is collected, the need for confidentiality for those providing assessments of system requirements, the desire to get people involved and committed to a project, and the potential audience from which requirements should be collected.

7. **Explain business process redesign and how it affects requirements determination.**

   Business process reengineering (BPR) is an approach to changing business processes radically. BPR efforts are a source of new information requirements. Information systems and technologies often enable BPR by allowing an organization to eliminate or relax constraints on traditional business rules.

8. **Understand how requirements determination techniques apply to development for Internet applications.**

   Most of the same techniques used for requirements determination for traditional systems can also be fruitfully applied to the development of Internet applications. Accurately capturing requirements in a timely manner for Internet applications is just as important as for more traditional systems.

Key Terms Checkpoint

Here are the key terms from the chapter. The page where each term is first explained is in parentheses after the term.

1. **Business process reengineering (BPR) (p. 140)**
2. **Closed-ended questions (p. 126)**
3. **Disruptive technologies (p. 141)**
4. **Formal system (p. 133)**
5. **Informal system (p. 133)**
6. **JAD session leader (p. 136)**
7. **Key business processes (p. 141)**
8. **Open-ended questions (p. 125)**
9. **Scribe (p. 137)**

Match each of the key terms above with the definition that best fits it.

_____ 1. The search for, and implementation of, radical change in business processes to achieve breakthrough improvements in products and services.

_____ 2. The person who makes detailed notes of the happenings at a Joint Application Design session.

_____ 3. Technologies that enable the breaking of long-held business rules that inhibit organizations from making radical business changes.

_____ 4. The way a system actually works.

_____ 5. The official way a system works as described in organizational documentation.

_____ 6. The structured, measured set of activities designed to produce a specific output for a particular customer or market.

_____ 7. Questions in interviews and on questionnaires that ask those responding to choose from among a set of specified responses.

_____ 8. Questions in interviews and on questionnaires that have no prespecified answers.

_____ 9. The trained individual who plans and leads Joint Application Design sessions.

Review Questions

1. Describe systems analysis and the major activities that occur during this phase of the systems development life cycle.
2. What are some useful character traits for an analyst involved in requirements determination?
3. Describe four traditional techniques for collecting information during analysis. When might one be better than another?
4. What are the general guidelines for conducting interviews?
5. What are the general guidelines for designing questionnaires?
6. Compare collecting information by interview and by questionnaire. Describe a hypothetical situation in which each of these methods would be an effective way to collect information system requirements.
7. What are the general guidelines for collecting data through observing workers?
8. What are the general guidelines for collecting data through analyzing documents?
9. Compare collecting information through observation and through document analysis. Describe a hypothetical situation in which each of these methods would be an effective way to collect information system requirements.
10. What is JAD? How is it better than traditional information-gathering techniques? What are its weaknesses?
11. How has computing been used to support requirements determination?
12. Describe how prototyping can be used during requirements determination. How is it better or worse than traditional methods?
13. When conducting a business process reengineering study, what should you look for when trying to identify business processes to change? Why?
14. What are disruptive technologies and how do they enable organizations to change their business processes radically?
15. What are the similarities in the requirements determination process for Internet applications and for traditional applications? What are the key differences?

Problems and Exercises

1. One of the potential problems mentioned in the chapter with gathering information requirements by observing potential system users is that people may change their behavior when observed. What could you do to overcome this potentially confounding factor in accurately determining information requirements?
2. Summarize the problems with the reliability and usefulness of analyzing business documents as a method for gathering information requirements. How could you cope with these problems to use business documents effectively as a source of insights on system requirements?
3. Suppose you were asked to lead a JAD session. List 10 guidelines you would follow in playing the proper role of a JAD session leader.
4. Prepare a plan, similar to Figure 4-2, for an interview with your academic adviser to determine which courses you should take to develop the skills you need to be hired as a programmer/analyst.
5. Write at least three closed-ended questions to use on a questionnaire that would be sent to users of a word processing package in order to develop ideas for the next version of the package. Test these questions by asking a friend to answer them; then interview your friend to determine why she responded as she did. From this interview, determine if she misunderstood any of your questions and, if so, rewrite the questions to be less ambiguous.
6. An interview lends itself easily to asking probing questions or asking different questions, depending on the answers provided by the interviewee. It is possible to use probing and alternative questions in a questionnaire. Discuss how you could include probing or alternative sets of questions in a questionnaire.
7. Figure 4-2 shows part of a guide for an interview. How might an interview guide differ when a group interview is to be conducted?
8. JADs are very powerful ways to collect system requirements, but special problems arise during group requirements collection sessions. Summarize these special interviewing and group problems, and suggest ways that you, as a group facilitator, might deal with them.

Discussion Questions

1. All of the methods of data collection discussed in this chapter take a lot of time. What are some ways analysts can still collect the information they need for systems analysis but also save time? What methods can you think of that would improve upon both traditional and newer techniques?
2. Some of the key problems with information systems that show up later in the systems development life cycle can be traced back to inadequate work during requirements determination. How might this be avoided?
3. Survey the literature on JAD in the academic and popular press and determine the "state of the art." How is JAD being used to help determine system requirements? Is using JAD for this process beneficial? Why or why not? Present your analysis to the IS manager at your work or at your university. Does your analysis of JAD fit with his or her perception? Why or why not? Is he or she currently using JAD, or a JAD-like method, for determining system requirements? Why or why not?
4. Is business process reengineering a business fad or is there more to it? Explain and justify your answer.

Case Problems

1. Pine Valley Furniture
   Jackie Judson, vice president of Marketing, and Jim Woo, a senior systems analyst, have been involved with Pine Valley Furniture's Customer Tracking System since the beginning of the project. After receiving project approval from the Systems Priority Board, Jim and his project development team turned their attention toward analyzing the Customer Tracking System.
   During a Wednesday afternoon meeting, Jim and his project team members decide to utilize several requirements determination methods.

Because the Customer Tracking System will facilitate the tracking of customer purchasing activity and help identify sales trends, various levels of end users will benefit from the new system. Therefore, the project team feels it is necessary to collect requirements from these potential end users. The project team will use interviews, observations, questionnaires, and JAD sessions as data-gathering tools.

Jim assigns you the task of interviewing Stacie Walker, a middle manager in the Marketing department; Pauline McBride, a sales representative; and Tom Percy, assistant vice president of Marketing. Tom is responsible for preparing the sales forecasts. In addition, Jim assigns Pete Polovich, a project team member, the task of organizing the upcoming JAD sessions.

a. Because this is Pete Polovich's first time organizing a JAD session, he would like to locate additional information about organizing and conducting a JAD session. Visit one of the Web sites recommended in the textbook or locate a site on your own. After visiting this site, provide Pete with several recommendations for conducting and organizing a JAD session.

b. When conducting your interviews, what guidelines should you follow?

c. As part of the requirements determination process, what business documents should be reviewed?

d. Is prototyping an appropriate requirements determination method for this project?

2. Hoosier Burger

Juan Rodriquez has assigned you the task of requirements determination for the Hoosier Burger project. You are looking forward to this opportunity because it will allow you to meet and interact with Hoosier Burger employees. Besides interviewing Bob and Thelma Mellankamp, you decide to collect information from Hoosier Burger's waiters, cooks, and customers.

Mr. Rodriquez suggests that you formally interview Bob and Thelma Mellankamp and perhaps observe them performing their daily management tasks. You decide that the best way to collect requirements from the waiters and cooks is to interview and observe them. You realize that discussing the order-taking process with Hoosier Burger employees and then observing them in action will provide you with a better idea of where potential system improvements can be made. You also decide to prepare a questionnaire

to distribute to Hoosier Burger customers. Because Hoosier Burger has a large customer base, it would be impossible to interview every customer; therefore, you feel that a customer satisfaction survey will suffice.

a. Assume you are preparing the customer satisfaction questionnaire. What types of questions would you include? Prepare five questions that you would ask.

b. What types of questions would you ask the waiters? What types of questions would you ask the cooks? Prepare five questions that you would ask each group.

c. What types of documents are you likely to obtain for further study? What types of documents will most likely not be available? Why?

d. What modern requirements determination methods are appropriate for this project?

3. Clothing Shack

The Clothing Shack is an online retailer of men's, women's, and children's clothing. The company has been in business for four years and makes a modest profit from its online sales. However, in an effort to compete successfully against online retailing heavyweights, the Clothing Shack's marketing director, Makaya O'Neil, has determined that the Clothing Shack's marketing information systems need improvement.

Ms. O'Neil feels that the Clothing Shack should begin sending out catalogs to its customers, keep better track of its customer's buying habits, perform target marketing, and provide a more personalized shopping experience for its customers. Several months ago, Ms. O'Neil submitted a Systems Service Request to the Clothing Shack's steering committee. The committee unanimously approved this project. You were assigned to the project at that time and have since helped your project team successfully complete the project initiation and planning phase. Your team is now ready to move into the analysis phase and begin identifying requirements for the new system.

a. Whom would you interview? Why?

b. What requirements determination methods are appropriate for this project?

c. Based on the answers provided for part b, which requirements determination methods are appropriate for the individuals identified in part a?

d. Identify the requirements determination deliverables that will likely result from this project.

CASE: BROADWAY ENTERTAINMENT COMPANY, INC.

Determining Requirements for the Web-Based Customer Relationship Management System

Case Introduction

Carrie Douglass, manager of the Broadway Entertainment Company store in Centerville, Ohio, was pleased when the Computer Information System (CIS) students at St. Claire Community College accepted her request to design a customer information system. The students saw the development of this system as a unique opportunity. This system deals with one of the hottest topics in business today—customer relationship management—and is a simple form of one of the most active areas of information systems development—electronic commerce. Many of the CIS students wanted to work on this project, but Professor Tann limits each team to four members. Professor Tann selected a team of Tracey Wesley, John Whitman, Missi Davies, and Aaron Sharp to work on the BEC project.

The BEC student team had never worked together before; in fact, the members did not know each other. This is not uncommon at St. Claire, because many students attend part-time and take classes around work and family obligations. Also, none of the team had ever worked in a store like those operated by BEC, although all of them were regular BEC customers. Tracey, John, and Missi are parents of children and, hence, have some personal interest in an information system such as Carrie has proposed. Professor Tann selected Tracey, John, Missi, and Aaron, in part, for their diversity. Tracey is a full-time COBOL computer programmer, who spent several busy years working on Y2K conversions for the local electric utility company. John works full-time for the Dayton Public Schools as a computer applications trainer. Missi wants to return to the workforce after her first child enters first grade; Missi worked in customer service at a local department store before her six-year break from working outside the home. Aaron recently graduated from the local high school Tech Prep program and actually began taking CIS classes at St. Claire while in this high school program. Aaron was in charge of the Web site for his high school and has done Web site consulting with several small businesses in the Dayton area. Missi and Aaron are full-time St. Claire students.

Getting Started on Requirements Determination

The first step for the student team was to meet Carrie at a project kickoff meeting (see BEC Figure 4-1). BEC stores do not have a meeting room, so the team visited Carrie's store for a tour and to meet a few store employees, and then the team and Carrie got a table at a nearby restaurant for their discussion. The team shared with Carrie information about their project course requirements, including a tentative schedule of when the team was expected to submit deliverables for each system development phase to Professor Tann. Each team member explained his or her background and skills, and stated personal goals for the project. Carrie explained her background. The team members were surprised to discover that Carrie had only recently graduated from St. Claire; however, they felt that they had found a kindred spirit for a project sponsor.

This preliminary meeting at the start of the project was actually a transition meeting from the project initiation and planning phase to the analysis phase. The System Service Request (see BEC Figure 3-2 at the end of Chapter 3) provided the team with a basic background. Before developing a plan for the detailed steps of the analysis phase, the team wanted to determine if anything had changed in Carrie's mind since she submitted the request.

Carrie, not too surprisingly, had been very busy since she submitted the request and had only a few new ideas. First, Carrie had become even more excited about the system she proposed. She explained that once the project started she would probably generate new ideas every day. The team explained that although this would be helpful, at some point the system requirements would have to be fixed so that a detailed design could be completed and the system prototype built. Further ideas could be incorporated in later enhancements. Second, Carrie raised a concern about what would happen once the course was over and an initial system was built. How would she get any further help? The team suggested that it would consider how follow-up might be handled. Missi offered the option that if Carrie thought that the final product of the project was good enough, it might be time to involve BEC corporate IS people near the end of the project in a handoff meeting. Carrie suggested that this would be an alternative to readdress later in the project.

Third, Carrie also asked how the team would interact with her during the project. The team responded

Interview Outline	
Interviewee: *Carrie Douglass* *Manager, BEC Store OH-84* *Centerville, Ohio*	**Interviewers:** *Missi Davies* *Aaron Sharp* *Tracey Wesley* *John Whitman*
Location/Medium: *Centerville BEC store* *and nearby restaurant*	**Appointment Date:** Start Time: *3:00* P.M. End Time: *4:30* P.M.
Objectives: *Develop rapport with client* *Obtain client's orientation to project*	**Reminders:** *Show enthusiasm for project* *Pay for food and beverages*

Agenda:	Approximate Time:
Introduce one another	1 minute
Tour store	10 minutes
Go to restaurant and order food/drinks	10 minutes
Explain course requirements	10 minutes
Describe team member backgrounds/goals	10 minutes
Have client explain her background	10 minutes
Discuss updates to SSR	
Ask for new ideas	10 minutes
Ask for client concerns	10 minutes
Identify business and personal client goals	5 minutes
Ask client about project expectations	5 minutes
Discover if employees know about it	2 minutes
Ask about letter to employees	1 minutes
Request computer system description	1 minutes
Summarize meeting and set follow-up time	5 minutes

General Observations:

Notes on Questions:

BEC FIGURE 4-1
INTERVIEW OUTLINE FOR INITIAL INTERVIEW WITH CARRIE DOUGLASS

by saying that they would provide to Carrie in the next two weeks a detailed schedule for the next phase of the project (the whole analysis phase). Along with this schedule would be a statement of when there would be face-to-face review meetings, the nature of written status reports, and other elements of a communication plan for the project. Finally, Carrie had one new idea about the requirements for the system. She suggested that a useful feature would be a page that would change weekly with comments from a store employee concerning his or her favorite picks for the week in several categories: adventure, mystery, documentary, children's, and so on. Carrie emphasized that she wanted to provide these comments from her own employees, not from outside sources that provide links to outside movie and music review Web sites.

Conducting Requirements Determination

The St. Claire team of students was almost ready to develop a detailed plan for the analysis phase and a general plan for the whole project. Carrie's comments had been helpful; however, they had a few more questions to help them determine how to conduct the requirements determination portion of the analysis phase. First, they asked Carrie what the business goals were for her and her store. Carrie explained that each BEC store had three main goals: (1) to increase dollar income volume by at least 1.5 percent each month, (2) to increase profit by at least 1 percent each month, and (3) to maintain customer overall satisfaction above 95 percent. The store manager has bonus pay incentives to achieve these objectives. Customer satisfaction is measured monthly by random telephone calls placed by an independent market research firm. A sample of customers from each store and in each area where there is a BEC store are contacted and asked to answer a list of 10 questions about their experience at BEC or competing stores. Personally, Carrie wants to perform as an above-average store manager during her first year, which means that she expects to clearly beat the goals given her by BEC. She also wants to be viewed as an innovative store manager, someone with potential for more significant positions later in her career. She likes working for BEC and sees a long-term career within the organization.

Second, the team asked Carrie what expectations she had about how the project would be conducted. Carrie expects the team to act independently without much direction or supervision from her; she does not have the time to work closely with them. She also expects that they will ask any questions of her or her employees and that everyone will cooperate with the project. The team asked if Carrie had told all of the store employees about the project. Carrie said that she had mentioned it to a few employees, but there had been no formal announcement. The team then asked if it could draft a note for Carrie to send to each employee explaining the nature of the project and requesting their assistance. Carrie agreed but reserved the right to edit the note before sending it. Next, Carrie asked about the time line for the project. The team members replied that their project plan would outline this but that their course included 18 more weeks of work over the following 20 calendar weeks. Finally, Carrie expects there to be minimal project expenses. The team said that it would develop and submit a cost estimate as part of the detailed project plan for the analysis phase.

At the end of the meeting the team members asked Carrie for a copy of a description of the computer capabilities available to them through in-store technology. Carrie agreed to send them a copy of an overview of the BEC information systems she obtained in the training program (see the BEC case at the end of Chapter 2). If they need more details, they should ask for specific information, and Carrie will try to get those details from corporate staff.

Case Summary

The St. Claire team of students is off to an enthusiastic start to the BEC customer relationship management system project. The initial meeting with the client seemed to have gone well. Team members liked Carrie, and she seemed to like them and spoke frankly to them. But, there were also signs of risk for the project. First, Carrie might be overly enthusiastic and naïve about the system. As an inexperienced store manager, she may not have well-seasoned or definitive ideas. Second, other store employees, many of whom are part-time help with no long-term commitment to BEC, have not been involved in the development of the project's ideas. The team will have to assess whether the employees or other people are critical stakeholders in the project. Third, Carrie needs the project to be conducted with minimal costs. Until the team can develop a clear understanding of the system requirements and the available technology in the store, additional costs are unclear. Certainly the team can develop the design and proof-of-concept prototype of a system on computers at St. Claire, but if the system is to be used in the store, there may be significant start-up costs. Finally, the team is a little concerned about Carrie's reluctance for corporate IS staff to be involved in the project. Carrie seemed hesitant about the handoff meeting as a way to provide follow-up after the team's work is done, and she wants to be a buffer with corporate about IS details.

The team members agree that the project looks like a great learning experience. There is original analysis and design work to be done. The project schedule and techniques to be used are wide open, within the constraints of their course requirements. There are some interesting stakeholder issues to be handled. The benefits of the system are still vague, as are the costs for a complete implementation. And the team is diverse, with a variety of skills and experience, but with the unknowns of how members will react and work together when critical deadlines must be met.

Case Questions

1. From what you know so far about the customer relationship management system project at BEC, whom do you consider to be the stakeholders in this system? How would you suggest involving each stakeholder in the project in order to gain the greatest insights during requirements determination and to achieve success for the project?

2. Develop a detailed project schedule for the analysis phase and a general project schedule for subsequent phases of this project. This schedule should follow from answers to questions in BEC cases from prior chapters and from any class project guidelines given to you by your instructor. Be prepared to suggest a different overall schedule than the 18 workweeks indicated in the case if your available project time is different from this project length. Also prepare a budget for the project as you would conduct it. What resources do you anticipate the team will need to conduct the project as you outline it? How might these resources be acquired?

3. Probably an early activity in the analysis phase you outlined in your answer to Question 2 is to distribute the project announcement to employees, as discussed in this case. The BEC team offered to write a draft of this note. What are the critical items to communicate in this letter? Draft the letter for Carrie Douglass's approval.

4. Your answer to Question 2 likely included review points with Carrie and other stakeholders. These review points are part of the project's overall communication plan. Explain the overall communication activities you would suggest for this project. How should team members communicate questions, findings, and results to one another? How should the team communicate with stakeholders?

5. In this case the BEC student team asked for details on the in-store computing environment. What other documents or documentation should the team collect during requirements determination? Why?

6. The initial meeting with Carrie Douglass in this BEC case was not intended to be a formal interview as part of requirements determination. Prepare a detailed interview plan for the first interview you would hold with Carrie to explore in depth the needs she sees for a customer relationship management system. As part of this interview plan, consider any opportunities that might exist for reengineering current customer relationship business processes.

7. How would you propose involving BEC store customers (actual and potential) in requirements determination? Do they need to be involved? If not, why not? If so, prepare a plan for questionnaires, interviews, focus group or JAD session, observation, or whatever means you suggest using to elicit their requirements for the system.

8. How would you propose involving BEC store employees in requirements determination? Do they need to be involved? If not, why not? If so, prepare a plan for questionnaires, interviews, focus group or JAD session, observation, or whatever means you suggest using to elicit their requirements for the system.

9. Visit the Web sites of at least three companies that sell or rent merchandise in physical stores. From reviewing these Web sites, what features would you suggest for the BEC customer relationship management system? How would you determine if BEC needs these features for its Web site?

5

Structuring System Requirements: Process Modeling

Objectives

*After studying this chapter,
you should be able to:*

- ➡ Understand the logical modeling of processes through studying examples of data flow diagrams.
- ➡ Draw data flow diagrams following specific rules and guidelines that lead to accurate and well-structured process models.
- ➡ Decompose data flow diagrams into lower-level diagrams.
- ➡ Balance higher-level and lower-level data flow diagrams.
- ➡ Use data flow diagrams as a tool to support the analysis of information systems.
- ➡ Discuss process modeling for Internet applications.
- ➡ Use Structured English and decision tables to represent process logic.

Chapter Preview . . .

In the last chapter, you learned about various methods that systems analysts use to collect the information they need to determine systems requirements. In this chapter, we continue our focus on the systems analysis part of the SDLC, which is highlighted in Figure 5-1. Note that there are three parts to the analysis phase: determining requirements, structuring requirements, and selecting the best alternative design strategy. We focus on a tool analysts use to structure information—data flow diagrams (DFDs). Data flow diagrams allow you to model how data flow through an information system, the relationships among the data flows, and how data come to be stored at specific locations. Data flow diagrams also show the processes that change or transform data. Because data flow diagrams concentrate on the movement of data between processes, these diagrams are called process models.

As the name indicates, a data flow diagram is a graphical tool that allows analysts (and users) to show the flow of data in an information system. The system can be physical or logical, manual or computer based. In this chapter, you learn the mechanics of drawing and revising data flow diagrams as well as the basic symbols and set of rules for drawing them. We also alert you to pitfalls. You learn two important concepts related to data flow diagrams: balancing and decomposition. At the end of the chapter, you learn how to use data flow diagrams as part of the analysis of an information system and as a tool for supporting business process reengineering. You also are briefly introduced to two methods for modeling the logic inside processes: Structured English and decision tables.

FIGURE 5-1
SYSTEMS ANALYSIS
Of the three parts to the analysis phase of the systems development life cycle, we focus on structuring requirements in this chapter.

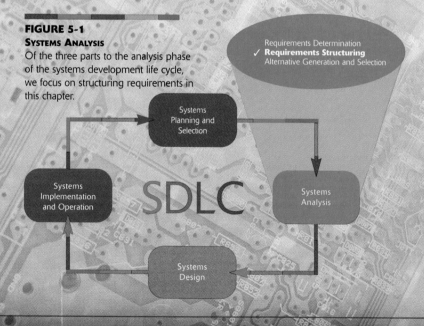

Requirements Determination
✓ **Requirements Structuring**
Alternative Generation and Selection

Process modeling
Graphically representing
the processes that capture,
manipulate, store, and distribute
data between a system and its
environment and among
components within a system.

Data flow diagram (DFD)
A graphic that illustrates the
movement of data between
external entities and the
processes and data stores
within a system.

NET SEARCH
*In addition to data flow
diagrams, there are
numerous ways to
model business
processes. Visit
http://www.prenhall.
com/valacich to
complete an exercise
related to this topic.*

Process Modeling

Process modeling involves graphically representing the processes, or actions, that capture, manipulate, store, and distribute data between a system and its environment and among components within a system. A common form of a process model is a **data flow diagram (DFD).** A data flow diagram is a graphic that illustrates the movement of data between external entities and the processes and data stores within a system. Although several different tools have been developed for process modeling, we focus solely on data flow diagrams because they are commonly used today for process modeling.

Data flow diagramming is one of several structured analysis techniques used to increase software development productivity. Although not all organizations use each structured analysis technique, collectively, these techniques, like data flow diagrams, have had a significant impact on the quality of the systems development process. For example, Raytheon, a defense and commercial electronics developer, reported a savings from 1988 through 1994 of $17.2 million in software costs by applying structured analysis techniques, due mainly to avoiding rework to fix mistakes made during requirements determination. These mistakes include such things as misunderstanding how existing systems will have to work with the new system and incorrect specifications for necessary data, forms, and reports. Raytheon's success represents a doubling of systems developers' productivity and helped it avoid costly system mistakes.

Modeling a System's Process

The analysis team begins the process of structuring requirements with an abundance of information gathered during requirements determination. As part of structuring, you and the other team members must organize the information into a meaningful representation of the information system that exists and of the requirements desired in a replacement system. In addition to modeling the processing elements of an information system and transformation of data in the system, you must also model the structure of data within the system (which we review in Chapter 6). Analysts use both process and data models to establish the specification of an information system. With a supporting tool such as a CASE tool, process and data models can also provide the basis for the automatic generation of an information system.

Deliverables and Outcomes

In structured analysis, the primary deliverables from process modeling are a set of coherent, interrelated data flow diagrams. Table 5-1 lists the progression of deliverables that result from studying and documenting a system's process. First, a context data flow diagram shows the scope of the system, indicating which elements are inside and outside the system. Second, data flow diagrams of the current system specify which people and technologies are used in which processes to move and transform data, accepting inputs and producing outputs. The detail of these diagrams allows analysts to understand the current

TABLE 5-1:   Deliverables for Process Modeling

1. Context DFD

2. DFDs of current physical system

3. DFDs of new logical system

4. Thorough descriptions of each DFD component

system and eventually to determine how to convert the current system into its replacement. Third, technology-independent or logical, data flow diagrams show the data flow, structure, and functional requirements of the new system. Finally, entries for all of the objects included in all diagrams are included in the project dictionary or CASE repository.

This logical progression of deliverables helps you to understand the existing system. You can then reduce this system into its essential elements to show the way in which the new system should meet its information processing requirements, as they were identified during requirements determination. In later steps in the systems development life cycle, you and other project team members make decisions on exactly how the new system will deliver these new requirements in specific manual and automated functions. Because requirements determination and structuring are often parallel steps, data flow diagrams evolve from the more general to the more detailed as current and replacement systems are better understood.

Even though data flow diagrams remain popular tools for process modeling and can significantly increase software development productivity, they are not used in all systems development methodologies. Some organizations, like Electronic Data Systems, have developed their own type of diagrams to model processes. Some methodologies, such as Rapid Application Development (RAD), do not model process separately at all. Instead RAD builds process—the work or actions that transform data so that they can be stored or distributed—into the prototypes created as the core of its development life cycle (see Appendix B for more details on RAD). However, even if you never formally use data flow diagrams in your professional career, they remain a part of systems development's history. DFDs illustrate important concepts about the movement of data between manual and automated steps, and are a way to depict work flow in an organization. DFDs continue to benefit information systems professionals as tools for both analysis and communication. For that reason, we devote this entire chapter to DFDs.

Data Flow Diagramming Mechanics

Data flow diagrams are versatile diagramming tools. With only four symbols, data flow diagrams can represent both physical and logical information systems. The four symbols used in DFDs represent data flows, data stores, processes, and sources/sinks (or external entities). The set of four symbols we use in this book was developed by Gane and Sarson and is illustrated in Figure 5-2.

Phase 2:
Systems
Analysis

NET SEARCH
In this book, we use the DeMarco and Yourdon conventions for data flow diagrams. Visit http://www.prenhall. com/valacich to complete an exercise related to this topic.

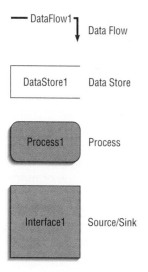

FIGURE 5-2
Gane and Sarson identified four symbols to use in data flow diagrams to represent the flow of data: data flow symbol, data store symbol, process symbol, and source/sink symbol. We use the Gane and Sarson symbols in this book.

A data flow is data that are in motion and moving as a unit from one place in a system to another. A data flow could represent data on a customer order form or a payroll check. It could also represent the results of a query to a database, the contents of a printed report, or data on a data entry computer display form. A data flow can be composed of many individual pieces of data that are generated at the same time and flow together to common destinations.

A **data store** is data at rest. A data store may represent one of many different physical locations for data including a file folder, one or more computer-based file(s), or a notebook. To understand data movement and handling in a system, the physical configuration is not really important. A data store might contain data about customers, students, customer orders, or supplier invoices.

A **process** is the work or actions performed on data so that they are transformed, stored, or distributed. When modeling the data processing of a system, it doesn't matter whether a process is performed manually or by a computer.

Finally, a **source/sink** is the origin and/or destination of the data. Source/sinks are sometimes referred to as *external entities* because they are outside the system. Once processed, data or information leave the system and go to some other place. Because sources and sinks are outside the system we are studying, many of their characteristics are of no interest to us. In particular, we do not consider the following:

- Interactions that occur between sources and sinks
- What a source or sink does with information or how it operates (i.e., a source or sink is a "black box")
- How to control or redesign a source or sink because, from the perspective of the system we are studying, the data a sink receives and often what data a source provides are fixed
- How to provide sources and sinks direct access to stored data because, as external agents, they cannot directly access or manipulate data stored within the system; that is, processes within the system must receive or distribute data between the system and its environment

Definitions and Symbols

DFD symbols are presented in Figure 5-2. A data flow is depicted as an arrow. The arrow is labeled with a meaningful name for the data in motion; for example, customer order, sales receipt, or paycheck. The name represents the aggregation of all the individual elements of data moving as part of one packet; that is, all the data moving together at the same time. A rectangle or square is used for sources/sinks and its name states what the external agent is, such as customer, teller, EPA office, or inventory control system. The symbol for a process is a rectangle with rounded corners. Inside the rectangle are written both the number of the process and a name, which indicates what the process does. For example, the process may generate paychecks, calculate overtime pay, or compute grade point average. The symbol for a data store is a rectangle with the right verticle line missing. Its label includes the number of the data store (e.g., D1 or D2) and a meaningful label, such as student file, transcripts, or roster of classes.

As stated earlier, sources/sinks are always outside the information system and define the system's boundaries. Data must originate outside a system from one or more sources, and the system must produce information to one or more sinks. (These are principles of open systems, and almost every information system is an example of an open system.) If any data processing takes place inside the source/sink, we are not interested in it, as this processing takes place outside of the system we are diagramming. A source/sink might consist of the following:

Data store

Data at rest, which may take the form of many different physical representations.

Process

The work or actions performed on data so that they are transformed, stored, or distributed.

Source/Sink

The origin and/or destination of data; sometimes referred to as *external entities*.

◗ Another organization or organizational unit that sends data to or receives information from the system you are analyzing (e.g., a supplier or an academic department—in either case, this organization is external to the system you are studying)

◗ A person inside or outside the business unit supported by the system you are analyzing and who interacts with the system (e.g., a customer or a loan officer)

◗ Another information system with which the system you are analyzing exchanges information

Many times students learning how to use DFDs become confused about whether a person or activity is a source/sink or a process within a system. This dilemma occurs most often when the data flows in a system cross office or departmental boundaries. In such a case, some processing occurs in one office and the processed data are moved to another office, where additional processing occurs. Students are tempted to identify the second office as a source/sink to emphasize the fact that the data have been moved from one physical location to another. Figure 5-3(A) illustrates an incorrectly drawn DFD showing a process, 3.0 Update Customer Master, as a source/sink, Accounting Department. The reference numbers "1.0" and "2.0" uniquely identify each

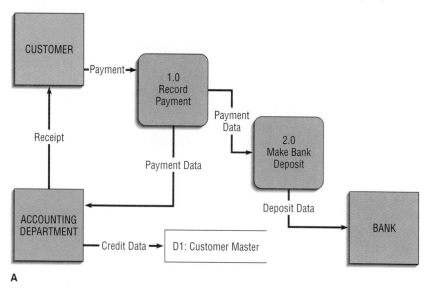

FIGURE 5-3
(A) AN INCORRECTLY DRAWN DFD SHOWING A PROCESS AS A SOURCE/SINK
(B) A DFD SHOWING PROPER USE OF A PROCESS

A

B

process. D1 identifies the first data store in the diagram. However, we are not concerned with where the data are physically located. We are more interested in how they are moving through the system and how they are being processed. If the processing of data in the other office is part of your system, then you should represent the second office as one or more processes on your DFD. Similarly, if the work done in the second office might be redesigned to become part of the system you are analyzing, then that work should be represented as one or more processes on your DFD. However, if the processing that occurs in the other office takes place outside the system you are working on, then it should be a source/sink on your DFD. Figure 5-3(B) is a DFD showing proper use of a process.

Developing DFDs: An Example

Let's work through an example to see how DFDs are used to model the logic of data flows in information systems. Consider Hoosier Burger, a fictional fast-food restaurant in Bloomington, Indiana. Hoosier Burger is owned by Bob and Thelma Mellankamp and is a favorite of students at nearby Indiana University. Hoosier Burger uses an automated food ordering system. The boundary or scope of this system, and the system's relationship to its environment, is represented by a data flow diagram called a **context diagram.** A context diagram is shown in Figure 5-4. Notice that this context diagram contains only one process, no data stores, four data flows, and three sources/sinks. The single process, labeled "0," represents the entire system; all context diagrams have only one process labeled "0." The sources/sinks represent its environmental boundaries. Because the data stores of the system are conceptually inside the one process, no data stores appear on a context diagram.

After drawing the context diagram, the next step for the analyst is to think about which processes are represented by the single process. As you can see in Figure 5-5, we have identified four separate processes, providing more detail of the Hoosier Burger food ordering system. The main processes in the DFD represent the major functions of the system, and these major functions correspond to such actions as the following:

1. Capturing data from different sources (Process 1.0)
2. Maintaining data stores (Processes 2.0 and 3.0)
3. Producing and distributing data to different sinks (Process 4.0)
4. High-level descriptions of data transformation operations (Process 1.0)

Context diagram
A data flow diagram of the scope of an organizational system that shows the system boundaries, external entities that interact with the system, and the major information flows between the entities and the system.

FIGURE 5-4
A CONTEXT DIAGRAM OF HOOSIER BURGER'S FOOD ORDERING SYSTEM
There is one process (Food Ordering System), four data flows (Customer Order, Receipt, Food Order, Management Reports), and three sources/sinks (Customer, Kitchen, and Restaurant Manager).

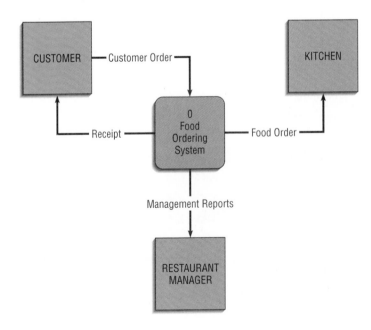

Phase 2:
Systems
Analysis

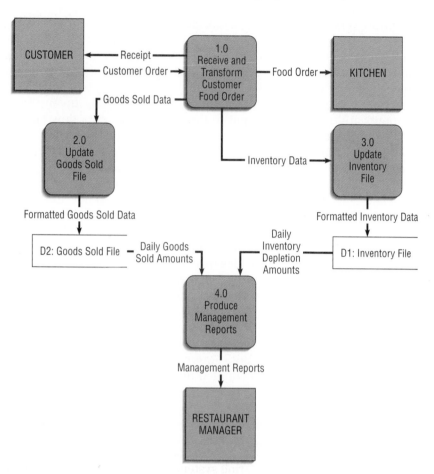

FIGURE 5-5
FOUR SEPARATE PROCESSES OF THE
HOOSIER BURGER FOOD ORDERING
SYSTEM

We see that the system in Figure 5-5 begins with an order from a customer, as was the case with the context diagram. In the first process, labeled "1.0," we see that the customer order is processed. The results are four streams or flows of data: (1) The food order is transmitted to the kitchen, (2) the customer order is transformed into a list of goods sold, (3) the customer order is transformed into inventory data, and (4) the process generates a receipt for the customer.

Notice that the sources/sinks are the same in the context diagram (Figure 5-4) and in this diagram: the customer, the kitchen, and the restaurant's manager. A context diagram is a DFD that provides a general overview of a system. Other DFDs can be used to focus on the details of a context diagram. A **level-0 diagram,** illustrated in Figure 5-5, is an example of such a DFD. Compare the level of detail in Figure 5-5 with that of Figure 5-4. A level-0 diagram represents the primary individual processes in the system at the highest possible level of detail. Each process has a number that ends in .0 (corresponding to the level number of the DFD).

Two of the data flows generated by the first process, Receive and Transform Customer Food Order, go to external entities (Customer and Kitchen), so we no longer have to worry about them. We are not concerned about what happens outside of our system. Let's trace the flow of the data represented in the other two data flows. First, the data labeled Goods Sold go to Process 2.0, Update Goods Sold File. The output for this process is labeled Formatted Goods Sold Data. This output updates a data store labeled Goods Sold File. If the customer order were for two cheeseburgers, one order of fries, and a large soft drink, each of these categories of goods sold in the data store would be incremented appropriately. The Daily Goods Sold Amounts are then used as input to Process 4.0, Produce Management Reports. Similarly, the remaining

Level-0 diagram
A data flow diagram that represents a system's major processes, data flows, and data stores at a high level of detail.

data flow generated by Process 1.0, called Inventory Data, serves as input for Process 3.0, Update Inventory File. This process updates the Inventory File data store, based on the inventory that would have been used to create the customer order. For example, an order of two cheeseburgers would mean that Hoosier Burger now has two fewer hamburger patties, two fewer burger buns, and four fewer slices of American cheese. The Daily Inventory Depletion Amounts are then used as input to Process 4.0. The data flow leaving Process 4.0, Management Reports, goes to the sink Restaurant Manager.

Figure 5-5 illustrates several important concepts about information movement. Consider the data flow Inventory Data moving from Process 1.0 to Process 3.0. We know from this diagram that Process 1.0 produces this data flow and that Process 3.0 receives it. However, we do not know the timing of when this data flow is produced, how frequently it is produced, or what volume of data is sent. Thus, this DFD hides many physical characteristics of the system it describes. We do know, however, that this data flow is needed by Process 3.0, and that Process 1.0 provides this needed data.

Also implied by the Inventory Data data flow is that whenever Process 1.0 produces this flow, Process 3.0 must be ready to accept it. Thus, Processes 1.0 and 3.0 are coupled to each other. In contrast, consider the link between Process 2.0 and Process 4.0. The output from Process 2.0, Formatted Goods Sold Data, is placed in a data store and, later, when Process 4.0 needs such data, it reads Daily Goods Sold Amounts from this data store. In this case, Processes 2.0 and 4.0 are decoupled by placing a buffer, a data store (Goods Sold File), between them. Now, each of these processes can work at its own

NET SEARCH

Fast-food restaurants like Hoosier Burger have a broad range of special-purpose software available to them to support the operation and management of the restaurant. Visit http://www.prenhall. com/valacich to complete an exercise related to this topic.

TABLE 5-2: Rules Governing Data Flow Diagramming

Process

A. No process can have only outputs. It is making data from nothing (a miracle). If an object has only outputs, then it must be a source.

B. No process can have only inputs (a black hole). If an object has only inputs, then it must be a sink.

C. A process has a verb phrase label.

Data Store

D. Data cannot move directly from one data source to another data store. Data must be moved by a process.

E. Data cannot move directly from an outside source to a data store. Data must be moved by a process that receives data from the source and places the data into the data store.

F. Data cannot move directly to an outside sink from a data store. Data must be moved by a process.

G. A data store has a noun phrase label.

Source/Sink

H. Data cannot move directly from a source to a sink. They must be moved by a process if the data are of any concern to our system. Otherwise, the data flow is not shown on the DFD.

I. A source/sink has a noun phrase label.

Data Flow

J. A data flow has only one direction of flow between symbols. It may flow in both directions between a process and a data store to show a read before an update. The latter is usually indicated, however, by two separate arrows because these happen at different times.

K. A fork in a data flow means that exactly the same data go from a common location to two or more different processes, data stores, or sources/sinks (this usually indicates different copies of the same data going to different locations).

L. A join in a data flow means that exactly the same data come from any of two or more different processes, data stores, or sources/sinks to a common location.

M. A data flow cannot go directly back to the same process it leaves. There must be at least one other process that handles the data flow, produces some other data flow, and returns the original data flow to the beginning process.

N. A data flow to a data store means update (delete or change).

O. A data flow from a data store means retrieve or use.

P. A data flow has a noun phrase label. More than one data flow noun phrase can appear on a single arrow as long as all of the flows on the same arrow move together as one package.

Source: Adapted from J. Celko, "I. Data Flow Diagrams," *Computer Language* 4 (January 1987), 41–43.

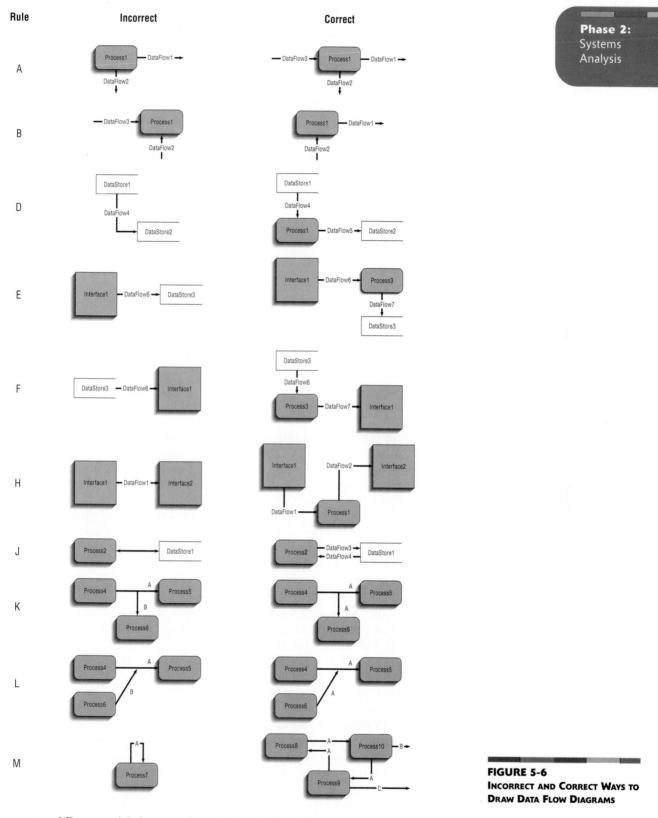

FIGURE 5-6
INCORRECT AND CORRECT WAYS TO DRAW DATA FLOW DIAGRAMS

pace, and Process 4.0 does not have to be vigilant by being able to accept input at any time. Further, the Goods Sold File becomes a data resource that other processes could potentially draw upon for data.

Data Flow Diagramming Rules

You must follow a set of rules when drawing data flow diagrams. These rules, listed in Table 5-2, allow you to evaluate DFDs for correctness. Figure 5-6 illustrates incorrect ways to draw DFDs and the corresponding correct application of the rules. The rules that prescribe naming conventions (rules C, G, I, and P in Table 5-2) and those that explain how to interpret data flows in and out of data stores (rules N and O in Table 5-2) are not illustrated in Figure 5-6.

Besides the rules in Table 5-2, two DFD guidelines apply most of the time:

- ❿ The inputs to a process are different from the outputs of that process: The reason is that processes, to have a purpose, typically transform inputs into outputs, rather than simply passing the data through without some manipulation. The same input may go in and out of a process, but the process also produces other new data flows that are the result of manipulating the inputs.

- ❿ Objects on a DFD have unique names: Every process has a unique name. There is no reason to have two processes with the same name. To keep a DFD uncluttered, however, you may repeat data stores and sources/sinks. When two arrows have the same data flow name, you must be careful that these flows are exactly the same. It is a mistake to reuse the same data flow name when two packets of data are almost the same but not identical. Because a data flow name represents a specific set of data, another data flow that has even one more or one less piece of data must be given a different, unique name.

Decomposition of DFDs

In the Hoosier Burger's food ordering system, we started with a high-level context diagram (see Figure 5-4). After drawing the diagram, we saw that the larger system consisted of four processes. The act of going from a single system to four component processes is called *(functional) decomposition*. Functional decomposition is a repetitive process of breaking the description or perspective of a system down into finer and finer detail. This process creates a set of hierarchically related charts in which one process on a given chart is explained in greater detail on another chart. For the Hoosier Burger system, we broke down or decomposed the larger system into four processes. Each of those processes (or subsystems) is also a candidate for decomposition. Each process may consist of several subprocesses. Each subprocess may also be broken down into smaller units. Decomposition continues until no subprocess can logically be broken down any further. The lowest level of DFDs is called a primitive DFD, which we define later in this chapter.

Let's continue with Hoosier Burger's food ordering system to see how a level-0 DFD can be further decomposed. The first process in Figure 5-5, called Receive and Transform Customer Food Order, transforms a customer's verbal food order (e.g., "Give me two cheeseburgers, one small order of fries, and one large orange soda") into four different outputs. Process 1.0 is a good candidate process for decomposition. Think about all of the different tasks that Process 1.0 has to perform: (1) Receive a customer order, (2) transform the entered order into a printed receipt for the customer, (3) transform the order into a form meaningful for the kitchen's system, (4) transform the order into goods sold data, and (5) transform the order into inventory data. At least these five logically separate functions occur in Process 1.0. We can represent the decomposition of Process 1.0 as another DFD, as shown in Figure 5-7.

Note that each of the five processes in Figure 5-7 are labeled as subprocesses of Process 1.0: Process 1.1, Process 1.2, and so on. Also note that, just as with the other data flow diagrams we have looked at, each of the processes and data flows is named. No sources or sinks are represented. The context and level-0

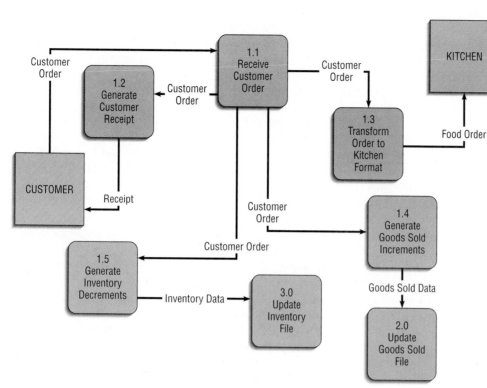

Phase 2:
Systems
Analysis

FIGURE 5-7
**LEVEL-1 DFD SHOWING THE
DECOMPOSITION OF PROCESS 1.0
FROM THE LEVEL-0 DIAGRAM FOR
HOOSIER BURGER'S FOOD
ORDERING SYSTEM**

diagrams show the sources and sinks. The data flow diagram in Figure 5-7 is called a level-1 diagram. If we should decide to decompose Processes 2.0, 3.0, or 4.0 in a similar manner, the DFDs we create would also be called level-1 diagrams. In general, a **level-*n* diagram** is a DFD that is generated from *n* nested decompositions from a level-0 diagram.

Processes 2.0 and 3.0 perform similar functions in that they both use data input to update data stores. Because updating a data store is a singular logical function, neither of these processes needs to be decomposed further. We can, on the other hand, decompose Process 4.0, Produce Management Reports, into at least three subprocesses: Access Goods Sold and Inventory Data, Aggregate Goods Sold and Inventory Data, and Prepare Management Reports. The decomposition of Process 4.0 is shown in the level-1 diagram of Figure 5-8.

Level-*n* diagram
A DFD that is the result of *n* nested decompositions of a series of subprocesses from a process on a level-0 diagram.

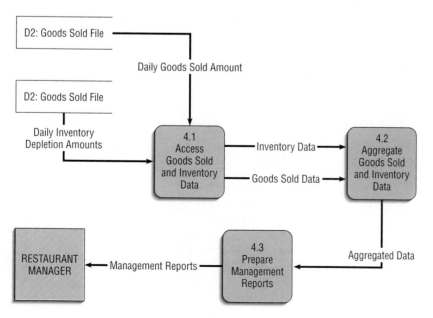

FIGURE 5-8
**LEVEL-1 DIAGRAM SHOWING THE
DECOMPOSITION OF PROCESS 4.0
FROM THE LEVEL-0 DIAGRAM FOR
HOOSIER BURGER'S FOOD
ORDERING SYSTEM**

Each level-1, -2, or -n DFD represents one process on a level-$n+1$ DFD; each DFD should be on a separate page. As a rule of thumb, no DFD should have more than about seven processes in it, because the diagram would be too crowded and difficult to understand.

To continue with the decomposition of Hoosier Burger's food ordering system, we examine each of the subprocesses identified in the two level-1 diagrams, one for Process 1.0 and one for Process 4.0. To further decompose any of these subprocesses, we would create a level-2 diagram showing that decomposition. For example, if we decided to further decompose Process 4.3 in Figure 5-8, we would create a diagram that looks something like Figure 5-9. Again, notice how the subprocesses are labeled.

Just as the labels for processes must follow numbering rules for clear communication, process names should also be clear yet concise. Typically, process names begin with an action verb, such as *receive, calculate, transform, generate,* or *produce.* Often process names are the same as the verbs used in many computer programming languages. Examples include *merge, sort, read, write,* and *print.* Process names should capture the essential action of the process in just a few words, yet be descriptive enough of the process' action so that anyone reading the name gets a good idea of what the process does. Many times, students just learning DFDs will use the names of people who perform the process or the department in which the process is performed as the process name. This practice is not very useful, as we are more interested in the action the process represents than the person performing it or the place where it occurs.

Balancing DFDs

When you decompose a DFD from one level to the next, a conservation principle is at work. You must conserve inputs and outputs to a process at the next level of decomposition. In other words, Process 1.0, which appears in a level-0 diagram, must have the same inputs and outputs when decomposed into a level-1 diagram. This conservation of inputs and outputs is called **balancing.**

Let's look at an example of balancing a set of DFDs. In Figure 5-4, the context diagram for Hoosier Burger's food ordering system, there is one input to the system, the customer order, which originates with the customer. Notice also that there are three outputs: the customer receipt, the food order intended for the kitchen, and management reports. Now look at Figure 5-5, the level-0 diagram for the food ordering system. Remember that all data stores and flows to or from them are internal to the system. Notice that the same single input to the system and the same three outputs represented in the context diagram also appear at level-0. Further, no new inputs to or outputs from the system have been introduced. Therefore, we can say that the context diagram and level-0 DFDs are balanced.

Balancing
The conservation of inputs and outputs to a data flow diagram process when that process is decomposed to a lower level.

FIGURE 5-9
LEVEL-2 DIAGRAM SHOWING THE DECOMPOSITION OF PROCESS 4.3 FROM THE LEVEL-1 DIAGRAM FOR PROCESS 4.0 FOR HOOSIER BURGER'S FOOD ORDERING SYSTEM

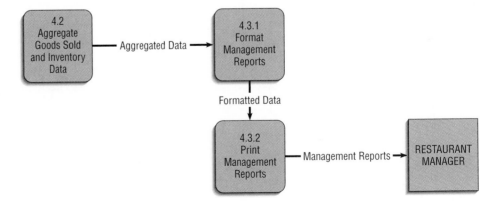

Now look at Figure 5-7, where Process 1.0 from the level-0 DFD has been decomposed. As we have seen before, Process 1.0 has one input and four outputs. The single input and multiple outputs all appear on the level-1 diagram in Figure 5-7. No new inputs or outputs have been added. Compare Process 4.0 in Figure 5-5 to its decomposition in Figure 5-8. You see the same conservation of inputs and outputs.

Figure 5-10(A) shows you one example of what an unbalanced DFD could look like. Here, in the context diagram, there is one input to the system, A, and one output, B. Yet, in the level-0 diagram, Figure 5-10(B), there is an additional input, C, and flows A and C come from different sources. These two DFDs are not balanced. If an input appears on a level-0 diagram, it must also appear on the context diagram. What happened in this example? Perhaps, when drawing the level-0 DFD, the analyst realized that the system also needed C in order to compute B. A and C were both drawn in the level-0 DFD, but the analyst forgot to update the context diagram. In making corrections, the analyst should also include SOURCE ONE and SOURCE TWO on the context diagram. It is very important to keep DFDs balanced, from the context diagram all the way through each level of the diagram you must create.

A data flow consisting of several subflows on a level-n diagram can be split apart on a level-$n+1$ diagram for a process that accepts this composite data flow as input. For example, consider the partial DFDs from Hoosier Burger illustrated in Figure 5-11. In Figure 5-11(A) we see that the payment and coupon always flow together and are input to the process at the same time. In Figure 5-11(B) the process is decomposed (sometimes called exploded or nested) into two subprocesses, and each subprocess receives one of the components of the composite data flow from the higher-level DFD. These diagrams are still balanced because exactly the same data are included in each diagram.

The principle of balancing and the goal of keeping a DFD as simple as possible lead to four additional, advanced rules for drawing DFDs, summarized in Table 5-3. Rule Q covers the situation illustrated in Figure 5-11. Rule R covers a

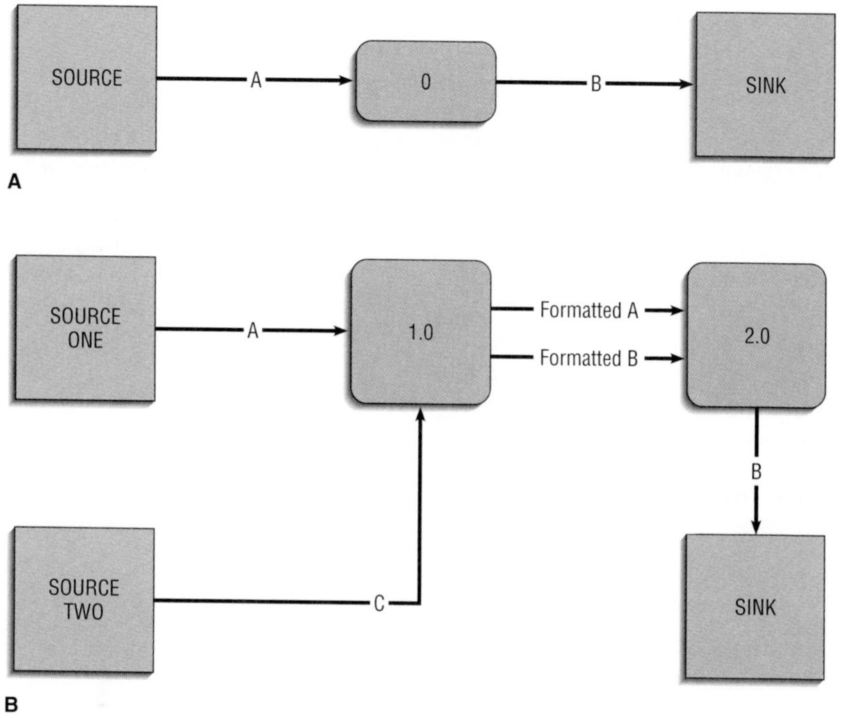

Phase 2:
Systems
Analysis

FIGURE 5-10
AN UNBALANCED SET OF DATA
FLOW DIAGRAMS
(A) A CONTEXT DIAGRAM
(B) A LEVEL-0 DIAGRAM

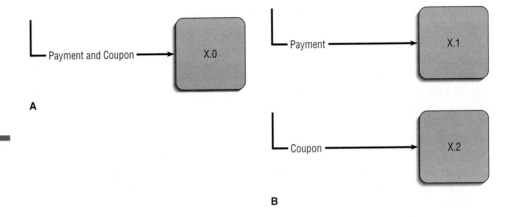

A

B

FIGURE 5-11
EXAMPLE OF A DATA FLOW
SPLITTING
(A) COMPOSITE DATA FLOW
(B) DISAGGREGATED DATA FLOWS

TABLE 5-3:   Advanced Rules Governing Data Flow Diagramming

Q. A composite data flow on one level can be split into component data flows at the next level, but no new data can be added and all data in the composite must be accounted for in one or more subflows.

R. The input to a process must be sufficient to produce the outputs (including data placed in data stores) from the process. Thus, all outputs can be produced, and all data in inputs move somewhere, either to another process or to a data store outside the process or on a more detailed DFD showing a decomposition of that process.

S. At the lowest level of DFDs, new data flows may be added to represent data that are transmitted under exceptional conditions; these data flows typically represent error messages (e.g., "Customer not known; do you want to create a new customer?") or confirmation notices (e.g., "Do you want to delete this record?").

T. To avoid having data flow lines cross each other, you may repeat data store or sources/sinks on a DFD. Use an additional symbol, like a double line on the middle vertical line of a data store symbol, or a diagonal line in a corner of a sink/source square, to indicate a repeated symbol.

Source: Adapted from J. Celko, "I. Data Flow Diagrams," *Computer Language* 4 (January 1987), 41–43.

conservation principle about process inputs and outputs. Rule S addresses one exception to balancing. Rule T tells you how you can minimize clutter on a DFD.

Using Data Flow Diagramming in the Analysis Process

Learning the mechanics of drawing data flow diagrams is important to you because data flow diagrams are essential tools for the structured analysis process. In addition to drawing DFDs that are mechanically correct, you must be concerned about whether the DFDs are complete and consistent across levels. You also need to consider how you can use them as a tool for analysis.

Guidelines for Drawing DFDs

In this section, we consider additional guidelines for drawing DFDs that extend beyond the simple mechanics of drawing diagrams and making sure that the rules listed in Tables 5-2 and 5-3 are followed. These additional guidelines include:

1. Completeness
2. Consistency
3. Timing considerations

4. The iterative nature of drawing DFDs
5. Drawing primitive DFDs

Completeness The concept of **DFD completeness** refers to whether your DFDs include all of the components necessary for the system you are modeling. If your DFD contains data flows that do not lead anywhere, or data stores, processes, or external entities that are not connected to anything else, your DFD is not complete. Most CASE tools have built-in facilities to help find incompleteness in your DFDs. When you draw many DFDs for a system, it is not uncommon to make errors; either CASE tool analysis functions or walk-throughs with other analysts can help you identify such problems.

DFD completeness
The extent to which all necessary components of a data flow diagram have been included and fully described.

Not only must all necessary elements of a DFD be present, each of the components must be fully described in the project dictionary. For most CASE tools, when you define a process, data flow, source/sink, or data store on a DFD, an entry is automatically created in the tool's repository for that element. You must then enter the repository and complete the element's description. Different descriptive information can be kept about each of the four types of elements on a DFD, and each CASE tool has different entry information. A data flow repository entry includes:

- The label or name for the data flow as entered on DFDs
- A short description defining the data flow
- A list of other repository objects grouped into categories by type of object
- The composition or list of data elements contained in the data flow
- Notes supplementing the limited space for the description that go beyond defining the data flow to explaining the context and nature of this repository object
- A list of locations (the names of the DFDs) on which this data flow appears and the names of the sources and destinations for the data flow on each of these DFDs

Consistency The concept of **DFD consistency** refers to whether or not the depiction of the system shown at one level of a DFD is compatible with the depictions of the system shown at other levels. A gross violation of consistency would be a level-1 diagram with no level-0 diagram. Another example of inconsistency would be a data flow that appears on a higher-level DFD but not on lower levels (a violation of balancing). Yet another example is a data flow attached to one object on a lower-level diagram but attached to another object at a higher level. For example, a data flow named Payment, which serves as input to Process 1 on a level-0 DFD, appears as input to Process 2.1 on a level-1 diagram for Process 2.

DFD consistency
The extent to which information contained on one level of a set of nested data flow diagrams is also included on other levels.

You can use the analysis facilities of CASE tools to detect such inconsistencies across nested (or decomposed) data flow diagrams. For example, to avoid making DFD consistency errors when you draw a DFD using a CASE tool, most tools will automatically place the inflows and outflows of a process on the DFD you create when you inform the tool to decompose that process. In manipulating the lower-level diagram, you could accidentally delete or change a data flow, which would cause the diagrams to be out of balance; thus, a consistency check facility with a CASE tool is quite helpful.

Timing You may have noticed in some of the DFD examples we have presented that DFDs do not do a very good job of representing time. On a given DFD, there is no indication of whether a data flow occurs constantly in real time, once per week, or once per year. There is also no indication of when a system would run. For example, many large transaction-based systems may run several large, computing-intensive jobs in batch mode at night, when demands on the computer system are lighter. A DFD has no way of indicating such overnight batch processing. When you draw DFDs, then, draw them as if the system you are modeling has never started and will never stop.

Iterative Development The first DFD you draw will rarely capture perfectly the system you are modeling. You should count on drawing the same diagram over and over again, in an iterative fashion. With each attempt, you will come closer to a good approximation of the system or aspect of the system you are modeling. Iterative DFD development recognizes that requirements determination and requirements structuring are interacting, not sequential, subphases of the analysis phase of the SDLC. One rule of thumb is that it should take you about three revisions for each DFD you draw. Fortunately, CASE tools make revising drawings a lot easier than if you had to draw each revision with pencil and template.

Primitive DFDs One of the more difficult decisions you need to make when drawing DFDs is when to stop decomposing processes. One rule is to stop drawing when you have reached the lowest logical level; however, it is not always easy to know what the lowest logical level is. Other more concrete rules for when to stop decomposing are:

- ❿ When you have reduced each process to a single decision or calculation or to a single database operation, such as retrieve, update, create, delete, or read.
- ❿ When each data store represents data about a single entity, such as a customer, employee, product, or order.
- ❿ When the system user does not care to see any more detail, or when you and other analysts have documented sufficient detail to do subsequent systems development tasks.
- ❿ When every data flow does not need to be split further to show that different data are handled in various ways.
- ❿ When you believe that you have shown each business form or transaction, computer online display, and report as a single data flow (e.g., often means that each system display and report title corresponds to the name of an individual data flow).
- ❿ When you believe there is a separate process for each choice on all lowest-level menu options.

By the time you stop decomposing DFDs, a DFD can become quite detailed. Seemingly simple actions, such as generating an invoice, may pull information from several entities and may also return different results depending on the specific situation. For example, the final form of an invoice may be based on the type of customer (which would determine such things as discount rate), where the customer lives (which would determine such things as sales tax), and how the goods are shipped (which would determine such things as the shipping and handling charges). At the lowest-level DFD, called a **primitive DFD,** all of these conditions would have to be met. Given the amount of detail required in a primitive DFD, perhaps you can see why many experts believe analysts should not spend their time diagramming the current physical information system completely: Much of the detail will be discarded when the current logical DFD is created.

Using these guidelines will help you create DFDs that are more than just mechanically correct. Your data flow diagrams will also be robust and accurate representations of the information system you are modeling. Such primitive

Primitive DFD
The lowest level of decomposition for a data flow diagram.

DFDs also facilitate consistency checks with the documentation produced from other requirements structuring techniques as well as make it easy for you to transition to system design steps. Having mastered the skills of drawing good DFDs, you can now use them to support the analysis process, the subject of the next section.

Using DFDs as Analysis Tools

We have seen that data flow diagrams are versatile tools for process modeling and that they can be used to model both physical and logical systems. Data flow diagrams can also be used for a process called **gap analysis.** In gap analysis, the analyst's role is to discover discrepancies between two or more sets of data flow diagrams or discrepancies within a single DFD.

Gap analysis
The process of discovering discrepancies between two or more sets of data flow diagrams or discrepancies within a single DFD.

Once the DFDs are complete, examine the details of individual DFDs for such problems as redundant data flows, data that are captured but not used by the system, and data that are updated identically in more than one location. These problems may not have been evident to members of the analysis team or to other participants in the analysis process when the DFDs were created. For example, redundant data flows may have been labeled with different names when the DFDs were created. Now that the analysis team knows more about the system it is modeling, analysts can detect such redundancies. Many CASE tools can generate a report listing all the processes that accept a given data element as input (remember, a list of data elements is likely part of the description of each data flow). From the label of these processes, you can determine whether or not it appears as if the data are captured redundantly or if more than one process is maintaining the same data stores. In such cases, the DFDs may accurately mirror the activities occurring in the organization. As the business processes being modeled took many years to develop, with participants in one part of the organization sometimes adapting procedures in isolation from other participants, redundancies and overlapping responsibilities may well have resulted. The careful study of the DFDs created as part of analysis can reveal these procedural redundancies and allow them to be corrected as part of system design.

A wide variety of inefficiencies can also be identified by studying DFDs. Some inefficiencies relate to violations of DFD drawing rules. Consider rule R from Table 5-3: The inputs to a process must be sufficient to produce the outputs from the process. A violation of rule R could occur because obsolete data are captured but never used within a system. Other inefficiencies are due to excessive processing steps. For example, consider the correct DFD in rule M of Figure 5-6: A data flow cannot go directly back to the same process it leaves. Although this flow is mechanically correct, such a loop may indicate potential delays in processing data or unnecessary approval operations.

Similarly, comparing a set of DFDs that models the current logical system to DFDs that model the new logical system can better determine which processes systems developers need to add or revise while building the new system. Processes for which inputs, outputs, and internal steps have not changed can possibly be reused in the construction of the new system. You can compare alternative logical DFDs to identify those few elements that must be discussed in evaluating competing opinions on system requirements. The logical DFDs for the new system can also serve as the basis for developing alternative design strategies for the new physical system. As we saw with the Hoosier Burger example, a process on a new logical DFD can be implemented in several different physical ways.

Using DFDs in Business Process Reengineering

Data flow diagrams also make a useful tool for modeling processes in business process reengineering (BPR), which you read about in Chapter 4. To illustrate their usefulness, let's look at an example from M. Hammer and J. Champy, two

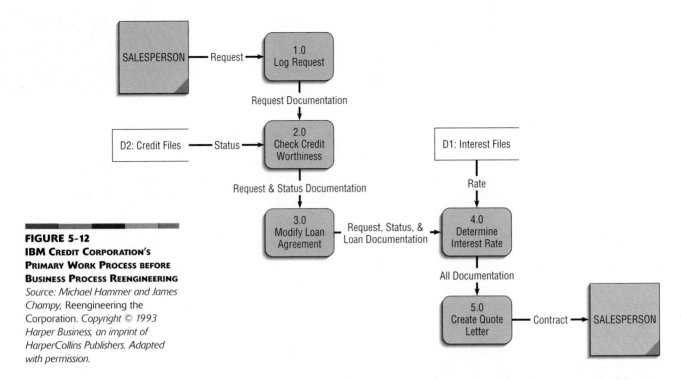

FIGURE 5-12
IBM CREDIT CORPORATION'S
PRIMARY WORK PROCESS BEFORE
BUSINESS PROCESS REENGINEERING
Source: Michael Hammer and James Champy, Reengineering the Corporation. *Copyright © 1993 Harper Business, an imprint of HarperCollins Publishers. Adapted with permission.*

experts of business redesign processes and authors of reengineering books. Hammer and Champy (1993) use IBM Credit Corporation as an example of a firm that successfully reengineered its primary business process. IBM Credit Corporation provides financing for customers making large purchases of IBM computer equipment. Its job is to analyze deals proposed by salespeople and write the final contracts governing those deals.

According to Hammer and Champy, IBM Credit Corporation typically took six business days to process each financing deal. The process worked like this: First, the salesperson called in with a proposed deal. The call was taken by one of a half dozen people sitting around a conference table. Whoever received the call logged it and wrote the details on a piece of paper. A clerk then carried the paper to a second person, who initiated the next step in the process by entering the data into a computer system and checking the client's creditworthiness. This person then wrote the details on a piece of paper and carried the paper, along with the original documentation, to a loan officer. Step 3, the loan officer modified the standard IBM loan agreement for the customer. This involved a separate computer system from the one used in step 2. Details of the modified loan agreement, along with the other documentation, were then sent on to the next station in the process, where a different clerk determined the appropriate interest rate for the loan. Step 4 also involved its own information system. In step 5, the resulting interest rate, and all of the paper generated up to this point, were then carried to the next step, where the quote letter was created. Once complete, the quote letter was sent via overnight mail back to the salesperson.

Only reading about this process makes it seem complicated. We can use data flow diagrams, as illustrated in Figure 5-12, to illustrate how the overall process worked. DFDs help us see that the process is not as complicated as it is tedious and wasteful, especially when you consider that so many different people and computer systems were used to support the work at each step.

According to Hammer and Champy, two IBM managers decided one day to see if they could improve the overall process at IBM Credit Corporation. They took a call from a salesperson and walked him through the system. These managers found the actual work being done on a contract took only 90 minutes. For much of the rest of the six days it took to process the deal, the various bits of documentation were sitting in someone's in basket waiting to be processed.

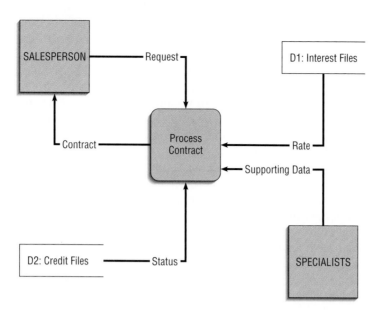

FIGURE 5-13
IBM CREDIT CORPORATION'S
PRIMARY WORK PROCESS AFTER
BUSINESS PROCESS REENGINEERING
Source: Michael Hammer and James Champy, Reengineering the Corporation. *Copyright © 1993 Harper Business, an imprint of HarperCollins Publishers. Adapted with permission.*

IBM Credit Corporation management decided to reengineer its entire process. The five sets of task specialists were replaced with generalists. Now each call from the field comes to a single clerk, who does all the work necessary to process the contract. Instead of having different people check for creditworthiness, modify the basic loan agreement, and determine the appropriate interest rate, now one person does it all. IBM Credit Corporation still has specialists for the few cases that are significantly different from what the firm routinely encounters. There is also a single supporting computer system. The new process is modeled by the DFD in Figure 5-13. The most striking difference between the DFDs in Figures 5-12 and 5-13, other than the number of process boxes in each one, is the lack of documentation flow in Figure 5-13. The resulting process is much simpler and cuts down dramatically on any chance of documentation getting lost between steps. Redesigning the process from beginning to end allowed IBM Credit Corporation to increase the number of contracts it could handle by 100-fold—not 100 percent, which would only be doubling the amount of work. BPR allowed IBM Credit Corporation to handle 100 times more work, in the same amount of time, and with fewer people!

Logic Modeling

Before we move on to logical methods for representing data, we first introduce the topic of logic modeling. Although data flow diagrams are very good for identifying processes, they do not show the logic inside the processes. Even the processes on the primitive-level data flow diagrams do not show the most fundamental processing steps. Just what occurs within a process? How are the input data converted to the output information? Because data flow diagrams are not really designed to show the detailed logic of processes, you must model process logic using other techniques.

Logic modeling involves representing the internal structure and functionality of the processes represented on data flow diagrams. These processes appear on DFDs as little more than black boxes, in that we cannot tell from only their names precisely what they do and how they do it. Yet the structure and functionality of a system's processes are a key element of any information system. Processes must be clearly described before they can be translated into a programming language.

We introduce you to two of the most common methods for modeling system logic. The first is Structured English, a modified version of the English language that is useful for representing the logic in information system processes.

You can use Structured English to represent all three of the fundamental statements necessary for structured programming: choice, repetition, and sequence.

Second, you learn about decision tables. Decision tables allow you to represent a set of conditions and the actions that follow from them in a tabular format. When several conditions and several possible actions can occur, decision tables help you keep track of the possibilities in a clear and concise manner.

Creating diagrams and descriptions of process logic is not an end in itself. Rather, these diagrams and descriptions are created ultimately to serve as part of a clear and thorough explanation of the system's specifications. These specifications are used to explain the system requirements to developers, whether people or automated code generators. Users, analysts, and programmers use logic diagrams and descriptions throughout analysis to incrementally specify a shared understanding of requirements. Logic diagrams do not take into account specific programming languages or development environments. Such diagrams may be discussed during JAD sessions or project review meetings. Alternatively, system prototypes generated from such diagrams may be reviewed, and requested changes to a prototype will be implemented by changing logic diagrams and generating a new prototype from a CASE tool or other code generator.

Modeling Logic with Structured English

You must understand more than just the flow of data into, through, and out of an information system. You also need to understand what each identified process does and how it accomplishes its task. Starting with the processes depicted in the various sets of data flow diagrams you and others on the analysis team have produced, you must now begin to study and document the logic of each process. Structured English is one method used to illustrate process logic.

Structured English

Modified form of the English language used to specify the logic of information system processes. Although there is no single standard, Structured English typically relies on action verbs and noun phrases and contains no adjectives or adverbs.

Structured English is a modified form of English that is used to specify the contents of process boxes in a DFD. It uses a subset of English vocabulary to express information system process procedures. Structured English uses strong verbs, such as *read, write, print, sort, move, merge, add, subtract, multiply,* and *divide;* it also uses noun phrases to describe data structures, such as *patron name* and *patron address*. Unlike regular English, Structured English does not use adjectives or adverbs. The whole point of using Structured English is to represent processes in a shorthand manner that is relatively easy for users and programmers to read and understand. As there is no standard version, each analyst has his or her own particular dialect of Structured English.

It is possible to use Structured English to represent all three processes typical to structured programming: sequence, conditional statements, and repetition. Sequence requires no special structure but can be represented with one sequential statement following another. Conditional statements can be represented with a structure like the following:

```
BEGIN IF
    IF Quantity-in-stock is less than Minimum-order-quantity
    THEN GENERATE new order
    ELSE DO nothing
END IF
```

Another type of conditional statement is a case statement where there are many different actions a program can follow, but only one is chosen. A case statement might be represented as:

```
READ Quantity-in-stock
SELECT CASE
    CASE 1 (Quantity-in-stock greater than Minimum-order-quantity)
        DO nothing
```

```
        CASE 2 (Quantity-in-stock equals Minimum-order-quantity)
            DO nothing
        CASE 3 (Quantity-in-stock is less than Minimum-order-quantity)
            GENERATE new order
        CASE 4 (Stock out)
            INITIATE emergency reorder routine
    END CASE
```

Repetition can take the form of Do-Until loops or Do-While loops. A Do-Until loop might be represented as follows:

```
    DO
        READ Inventory records
        BEGIN IF
            IF Quantity-in-stock is less than Minimum-order-quantity
            THEN GENERATE new order
            ELSE DO nothing
        END IF
    UNTIL End-of-file
```

A Do-While loop might be represented as follows:

```
    READ Inventory records
    WHILE NOT End-of-file DO
        BEGIN IF
            IF Quantity-in-stock is less than Minimum-order-quantity
            THEN GENERATE new order
            ELSE DO nothing
        END IF
    END DO
```

Let's look at an example of how Structured English would represent the logic of some of the processes identified in Hoosier Burger's current logical inventory control system (Figure 5-14).

Four processes are depicted in Figure 5-14: Update Inventory Added, Update Inventory Used, Generate Orders, and Generate Payments. Structured English representations of each process are shown in Figure 5-15. Notice that in this version of Structured English the file names are connected with hyphens and file names and variable names are capitalized. Terms that signify logical comparisons, such as *greater than* and *less than*, are spelled out rather than represented by their arithmetic symbols. Also notice how short the Structured English specifications are, considering that these specifications all describe level-0 processes. The final specifications would model the logic in the lowest-level DFDs only. From reading the process descriptions in Figure 5-15, it should be obvious that much more detail would be required to actually perform the processes described. In fact, creating Structured English representations of processes in higher-level DFDs is one method that can help you decide if a particular DFD needs further decomposition.

Notice how the format of the Structured English process description mimics the format usually used in programming languages, especially the practice of indentation. This is the "structured part" of Structured English. Notice also that the language used is similar to spoken English, using verbs and noun phrases. The language is simple enough for a user who knows nothing about computer programming to understand the steps involved in performing the various processes, yet the structure of the descriptions makes it easy to convert to a programming language eventually. Using Structured English also means not having to worry about initializing variables, opening and closing files, or finding related records in separate files. These more technical details are left for later in the design process.

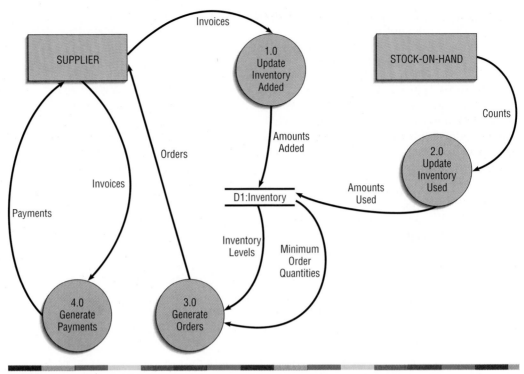

FIGURE 5-14
CURRENT LOGICAL DFD FOR HOOSIER BURGER'S INVENTORY CONTROL SYSTEM

FIGURE 5-15
STRUCTURED ENGLISH REPRESENTATIONS OF THE FOUR PROCESSES DEPICTED IN FIGURE 5-14

```
Process 1.0: Update Inventory Added
DO
    READ next Invoice-item-record
    FIND matching Inventory-record
    ADD Quantity-added from Invoice-item-record to Quantity-in-stock on
        Inventory-record
UNTIL End-of-File
```

```
Process 2.0: Update Inventory Used
DO
    READ next Stock-item-record
    FIND matching Inventory-record
    SUBTRACT Quantity-used on Stock-item-record from Quantity-in-stock
        on Inventory-record
UNTIL End-of-File
```

```
Process 3.0: Generate Orders
DO
    READ next Inventory-record
    BEGIN IF
        IF Quantity-in-stock is less than Minimum-order-quantity
        THEN GENERATE Order
    END IF
UNTIL End-of-File
```

```
Process 4.0: Generate Payments
READ Today's-date
DO
    SORT Invoice-records by Date
    READ next Invoice-record
    BEGIN IF
        IF Date is 30 days or greater than Today's-date
        THEN GENERATE Payments
    END IF
UNTIL End-of-File
```

Modeling Logic with Decision Tables

Structured English can be used to represent the logic contained in an information system process, but sometimes a process's logic can become quite complex. If several different conditions are involved, and combinations of these conditions dictate which of several actions should be taken, then Structured English may not be adequate for representing the logic behind such a complicated choice. Even though Structured English can represent complicated logic, it becomes more difficult to understand and verify. Research has shown, for example, that people become confused in trying to interpret more than three nested IF statements; hence, a diagram may be much clearer than a Structured English statement. A **decision table** is a diagram of process logic where the logic is reasonably complicated. All of the possible choices and the conditions the choices depend on are represented in tabular form, as illustrated in the decision table in Figure 5-16.

The decision table in Figure 5-16 models the logic of a generic payroll system. There are three parts to the table: the **condition stubs,** the **action stubs,** and the **rules.** The condition stubs contain the various conditions that apply in the situation the table is modeling. In Figure 5-16, there are two condition stubs for employee type and hours worked. Employee type has two values: "S," which stands for salaried, and "H," which stands for hourly. Hours worked has three values: less than 40, exactly 40, and more than 40. The action stubs contain all the possible courses of action that result from combining values of the condition stubs. There are four possible courses of action in this table: pay base salary, calculate hourly wage, calculate overtime, and produce Absence Report. You can see that not all actions are triggered by all combinations of conditions. Instead, specific combinations trigger specific actions. The part of the table that links conditions to actions is the section that contains the rules.

To read the rules, start by reading the values of the conditions as specified in the first column: Employee type is "S," or salaried, and hours worked are less than 40. When both of these conditions occur, the payroll system is to pay the base salary. In the next column, the values are "H" and "A40," meaning an hourly worker who worked less than 40 hours. In such a situation, the payroll system calculates the hourly wage and makes an entry in the Absence Report. Rule 3 addresses the situation when a salaried employee works exactly 40 hours. The system pays the base salary, as was the case for rule 1. For an hourly worker who has worked exactly 40 hours, rule 4 calculates the hourly wage. Rule 5 pays the base salary for salaried employees who work more than 40 hours. Rule 5 has the same action as rules 1 and 3, and governs behavior with regard to salaried employees. The number of hours worked does not affect the outcome for rules 1, 3, or 5. For these rules, hours worked is an **indifferent condition** in that its value does not affect the action taken. Rule 6 calculates hourly pay and overtime for an hourly worker who has worked more than 40 hours.

Phase 2: Systems Analysis

Decision table
A matrix representation of the logic of a decision, which specifies the possible conditions for the decision and the resulting actions.

Condition stubs
That part of a decision table that lists the conditions relevant to the decision.

Action stubs
That part of a decision table that lists the actions that result for a given set of conditions.

Rules
That part of a decision table that specifies which actions are to be followed for a given set of conditions.

Indifferent condition
In a decision table, a condition whose value does not affect which actions are taken for two or more rules.

Conditions/ Courses of Action	Rules					
	1	2	3	4	5	6
Employee type	S	H	S	H	S	H
Hours worked	<40	<40	40	40	>40	>40
Pay base salary	X		X		X	
Calculate hourly wage		X		X		X
Calculate overtime						X
Produce Absence Report		X				

Condition Stubs / Action Stubs

FIGURE 5-16
COMPLETE DECISION TABLE FOR PAYROLL SYSTEM EXAMPLE

Because of the indifferent condition for rules 1, 3, and 5, we can reduce the number of rules by condensing rules 1, 3, and 5 into one rule, as shown in Figure 5-17. The indifferent condition is represented with a dash. Whereas we started with a decision table with six rules, we now have a simpler table that conveys the same information with only four rules.

In constructing these decision tables, we have actually followed a set of basic procedures, as follows:

1 *Name the conditions and the values each condition can assume.* Determine all of the conditions that are relevant to your problem, and then determine all of the values each condition can take. For some conditions, the values will be simply "yes" or "no" (called a limited entry). For others, such as the conditions in Figures 5-16 and 5-17, the conditions may have more values (called an extended entry).

2 *Name all possible actions that can occur.* The purpose of creating decision tables is to determine the proper course of action given a particular set of conditions.

3 *List all possible rules.* When you first create a decision table, you have to create an exhaustive set of rules. Every possible combination of conditions must be represented. It may turn out that some of the resulting rules are redundant or make no sense, but these determinations should be made only after you have listed every rule so that no possibility is overlooked. To determine the number of rules, multiply the number of values for each condition by the number of values for every other condition. In Figure 5-16, we have two conditions, one with two values and one with three, so we need 2 X 3, or 6, rules. If we added a third condition with three values, we would need 2 X 3 X 3, or 18, rules.

   When creating the table, alternate the values for the first condition, as we did in Figure 5-16 for type of employee. For the second condition, alternate the values but repeat the first value for all values of the first condition, then repeat the second value for all values of the first condition, and so on. You essentially follow this procedure for all subsequent conditions. Notice how we alternated the values of hours worked in Figure 5-16. We repeated "A40" for both values of type of employee, "S" and "H." Then we repeated "40," and then "B40."

4 *Define the actions for each rule.* Now that all possible rules have been identified, provide an action for each rule. In our example, we were able to figure out what each action should be and whether all of the actions made sense. If an action doesn't make sense, you may want to create an "impossible" row in the action stubs in the table to keep track of impossible actions. If you can't tell what the system ought to do in that situation, place question marks in the action stub spaces for that particular rule.

5 *Simplify the decision table.* Make the decision table as simple as possible by removing any rules with impossible actions. Consult users on the rules where system actions aren't clear, and either decide on an action or remove

FIGURE 5-17
REDUCED DECISION TABLE FOR PAYROLL SYSTEM EXAMPLE

Conditions/ Courses of Action	Rules			
	1	2	3	4
Employee type	S	H	H	H
Hours worked	–	<40	40	>40
Pay base salary	X			
Calculate hourly wage		X	X	X
Calculate overtime				X
Produce Absence Report		X		

the rule. Look for patterns in the rules, especially for indifferent conditions. We were able to reduce the number of rules in the payroll example from six to four, but often greater reductions are possible.

Let's look at an example from Hoosier Burger. The Mellankamps are trying to determine how they reorder food and other items they use in the restaurant. If they are going to automate the inventory control functions at Hoosier Burger, they need to articulate their reordering process. In thinking through the problem, the Mellankamps realize that how they reorder depends on whether the item is perishable. If an item is perishable, such as meat, vegetables, or bread, the Mellankamps have a standing order with a local supplier stating that a pre-specified amount of food is delivered each weekday for that day's use and each Saturday for weekend use. If the item is not perishable, such as straws, cups, and napkins, an order is placed when the stock on hand reaches a certain prede-termined minimum reorder quantity. The Mellankamps also realize the impor-tance of the seasonality of their work. Hoosier Burger's business is not as good during the summer months when the students are off campus as it is during the academic year. They also note that business falls off during Christmas and spring breaks. Their standing orders with all their suppliers are reduced by spe-cific amounts during the summer and holiday breaks. Given this set of condi-tions and actions, the Mellankamps put together an initial decision table (see Figure 5-18).

Three things are distinctive about Figure 5-18. First, the values for the third condition repeat, providing a distinctive pattern for relating the values for all three conditions to each other. Every possible rule is clearly provided in this table. Second, there are 12 rules. Two values for the first condition (type of item) times 2 values for the second condition (time of week) times 3 values for the third condition (season of year) equals 12 possible rules. Third, the action for nonperishable items is the same, regardless of day of week or time of year. For nonperishable goods, both time-related conditions are indifferent. Collapsing the decision table accordingly gives us the decision table in Figure 5-19. Now there are only 7 rules instead of 12.

You have now learned how to draw and simplify decision tables. You can also use decision tables to specify additional decision-related information. For example, if the actions that should be taken for a specific rule are more com-plicated than one or two lines of text can convey, or if some conditions need to be checked only when other conditions are met (nested conditions), you may want to use separate, linked decision tables. In your original decision table,

Conditions/ Courses of Action	Rules											
	1	2	3	4	5	6	7	8	9	10	11	12
Type of item	P	N	P	N	P	N	P	N	P	N	P	N
Time of week	D	D	W	W	D	D	W	W	D	D	W	W
Season of year	A	A	A	A	S	S	S	S	H	H	H	H
Standing daily order	X				X				X			
Standing weekend order			X				X				X	
Minimum order quantity		X		X		X		X		X		X
Holiday reduction										X		X
Summer reduction					X		X					

FIGURE 5-18
COMPLETE DECISION TABLE FOR HOOSIER BURGER'S INVENTORY REORDERING SYSTEM

Type of item:
P = perishable
N = nonperishable

Time of week:
D = weekday
W = weekend

Season of year:
A = academic year
S = summer
H = holiday

Conditions/ Courses of Action	Rules						
	1	2	3	4	5	6	7
Type of item	P	P	P	P	P	P	N
Time of week	D	W	D	W	D	W	–
Season of year	A	A	S	S	H	H	–
Standing daily order	X		X		X		
Standing weekend order		X		X		X	
Minimum order quantity							X
Holiday reduction					X	X	
Summer reduction			X	X			

FIGURE 5-19
REDUCED DECISION TABLE FOR HOOSIER BURGER'S INVENTORY REORDERING SYSTEM

you can specify an action in the action stub that says "Perform Table B." Table B could contain an action stub that returns to the original table, and the return would be the action for one or more rules in Table B. Another way to convey more information in a decision table is to use numbers that indicate sequence rather than Xs where rules and action stubs intersect. For example, for rules 3 and 4 in Figure 5-19, it would be important for the Mellankamps to account for the summer reduction to modify the existing standing order for supplies. "Summer reduction" would be marked with a "1" for rules 3 and 4, whereas "standing daily order" would be marked with a "2" for rule 3, and "standing weekend order" would be marked with a "2" for rule 4.

You have seen how decision tables can model the relatively complicated logic of a process. Decision tables are more useful than Structured English for complicated logic in that they convey information in a tabular rather than a linear, sequential format. As such, decision tables are compact; you can pack a lot of information into a small table. Decision tables also allow you to check for the extent to which your logic is complete, consistent, and not redundant.

ELECTRONIC COMMERCE APPLICATION: PROCESS MODELING

Process modeling for an Internet-based electronic commerce application is no different than the process followed for other applications. In the last chapter, you read how Pine Valley Furniture determined the system requirements for its WebStore project—a project to sell furniture products over the Internet. In this section, we analyze the WebStore's high-level system structure and develop a level-0 DFD for those requirements.

Process Modeling for Pine Valley Furniture's WebStore

After completing the JAD session, senior systems analyst, Jim Woo, went to work on translating the WebStore system structure into a data flow diagram. His first step was to identify the level-0—major system—processes. To begin, he carefully examined the outcomes of the JAD session that focused on defining the system structure of the WebStore. From this analysis, he identified six high-level processes that would become the foundation of the level-0 DFD. These processes, listed in Table 5-4, were the "work" or "action" parts of the Web site; note that these processes correspond to the major processing items listed in the system structure.

Next, Jim determined that it would be most efficient if the WebStore system exchanged information with existing PVF systems rather than capturing and

**TABLE 5-4:   System Structure of the WebStore
and Corresponding Level-0 Processes**

WebStore System	Processes
Main page	Information display (minor/no processes)
Product line (Catalog)	1.0 Browse Catalog
• Desks	2.0 Select Item for Purchase
• Chairs	
• Tables	
• File cabinets	
Shopping cart	3.0 Display Shopping Cart
Checkout	4.0 Check Out Process Order
Account profile	5.0 Add/Modify Account Profile
Order status/history	6.0 Order Status Request
Customer comments	Information display (minor/no processes)
Company information	
Feedback	
Contact information	

storing redundant information. This analysis concluded that the WebStore should exchange information with the Purchasing Fulfillment System—a system for tracking orders (discussed in Chapter 2) and the Customer Tracking System (discussed in Chapter 3). These two existing systems will be "sources" (providers) and "sinks" (receivers) of information for the WebStore system. When a customer opens an account, his or her information will be passed from the WebStore system to the Customer Tracking System. When an order is placed (or when a customer requests status information on a prior order), information will be stored in and retrieved from the Purchasing Fulfillment System.

Finally, Jim found that the system would need to access two additional data sources. First, in order to produce an online product catalog, the system would need to access the inventory database. Second, to store the items a customer wants to purchase in the WebStore's shopping cart, a temporary database would need to be created. Once the transaction was completed, the shopping cart data could be deleted. With this information, Jim was then able to develop the level-0 DFD for the WebStore system shown in Figure 5-20. He understood how information would flow through the WebStore, how a customer would interact with the system, and how the WebStore would share information with existing PVF systems. Each of these high-level processes would eventually need to be further decomposed before system design could proceed. Yet, before doing that, he wanted to get a clear picture of exactly what would occur within each of the major processes.

Logic Modeling for Pine Valley Furniture's WebStore

After defining the level-0) DFD for the WebStore system, the Pine Valley furniture development methodology dictated that Jim needed to represent the logic within each of the unique processes. Since the logic within each process was relatively straightforward, Jim decided to represent each using Structured English. For

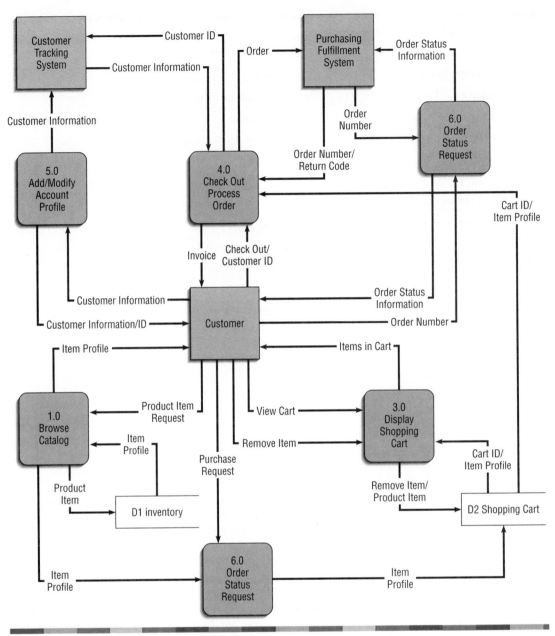

FIGURE 5-20
LEVEL-0 DFD FOR THE WEBSTORE SYSTEM

Process 1.0 and 2.0, the logic was very straightforward (see Table 5-5). However, for Process 3.0, two distinct activities were being performed: (1) displaying the contents of the shopping cart and (2) removing item from the shopping cart. Therefore, Jim concluded that it would be best to first diagram the subprocesses using a DFD (see Figure 5-21), then write the logic for each separate process (see Table 5-6). The logic for the remaining processes is shown in Table 5-7.

Now that the high-level logic of the main processes of the WebStore has been defined, Jim needed to get a clear picture of exactly what data were needed throughout the entire system. We will learn how Jim and the PVF development group address this next analysis activity—conceptual data modeling—in the next chapter.

TABLE 5-5:   Structured English Representations of Processes 1.0 and
2.0 from Figure 5-20

Process 1.0: Browse Catalog	**Process 2.0: Select Item for Purchase**
READ Product-Item Request	READ Purchase-Request
FIND Matching Product-Item from Inventory	READ Inventory-Item Profile
DISPLAY Item-Profile	ADD Item-Profile to Shopping-Cart

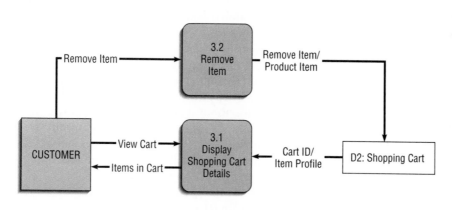

FIGURE 5-21
LEVEL-1 DFD FOR THE WEBSTORE
SYSTEM FOR PROCESS 3.0

TABLE 5-6:   Structured English Representations of Processes 3.1 and
3.2 from Figure 5-21

Process 3.1: Display Shopping Cart Details	**Process 3.2: Remove Item**
READ View-Cart	READ Remove-Item
DO	SUBTRACT Product Item from Shopping Cart
READ Item-Profile	
DISPLAY Item-Profile	
UNTIL Shopping Cart is Empty	

TABLE 5-7:   Structured English Representations of Processes 4.0, 5.0, and 6.0 from Figure 5-20

Process 4.0: Check-Out Process Orders	**Process 5.0 Add/Modify Account Profile**
READ Customer ID	READ Customer-Information
READ Check-Out	ADD Customer-Information to Customer-Tracking-System
FIND Customer-Information from Customer-Tracking-System	DISPLAY Customer-Information
DO	**Process 6.0 Order Status Request**
READ Item-Profile	READ Order-Number
ADD Item-Profile to Order	FIND Order-Status-Information from Purchasing-Fulfillment-System
UNTIL Shopping Cart Is Empty	DISPLAY Order-Status-Information
ADD Order to Purchasing-Fufillment-System	
READ Order-Number from Purchase-Fulfillment-System	
READ Return-Code from Purchasing-Fulfillment-System	
DISPLAY Invoice	

Key Points Review

Data flow diagrams, or DFDs, are very useful for representing the overall data flows into, through, and out of an information system. Data flow diagrams rely on only four symbols to represent the four conceptual components of a process model: data flows, data stores, processes, and sources/sinks.

1. **Understand the logical modeling of processes through studying examples of data flow diagrams.**

   Data flow diagrams are hierarchical in nature, and each level of a DFD can be decomposed into smaller, simpler units on a lower-level diagram. You begin with a context diagram, which shows the entire system as a single process. The next step is to generate a level-0 diagram, which shows the most important high-level processes in the system.

2. **Draw data flow diagrams following specific rules and guidelines that lead to accurate and well-structured process models.**

   Several rules govern the mechanics of drawing DFDs. These are listed in Table 5-2 and illustrated in Figure 5-6. Most of these rules are about the ways in which data can flow from one place to another within a DFD.

3. **Decompose data flow diagrams into lower-level diagrams.**

   Starting with a level-0 diagram, decompose each process, as warranted, until it makes no logical sense to go any further.

4. **Balance higher-level and lower-level data flow diagrams.**

   When decomposing DFDs from one level to the next, it is important that the diagrams be balanced; that is, inputs and outputs on one level must be conserved on the next level.

5. **Use data flow diagrams as a tool to support the analysis of information systems.**

   Data flow diagrams should be mechanically correct, but they should also accurately reflect the information system being modeled. To that end, you need to check DFDs for completeness and consistency and draw them as if the system being modeled were timeless. You should be willing to revise DFDs several times. Complete sets of DFDs should extend to the primitive level where every component reflects certain irreducible properties; for example, a process represents a single database operation and every data store represents data about a single entity. Following these guidelines, you can produce DFDs to aid the analysis process by analyzing the gaps between existing procedures and desired procedures and between current and new systems.

6. **Discuss process modeling for Internet applications.**

   Although the modeling of processes for information systems development is over 20 years old, dating back at least to the beginnings of the philosophy of structured analysis and design, it is just as important for Internet applications as it is for more traditional systems.

7. **Use Structured English and decision tables to represent process logic.**

   Process modeling helps isolate and define the many processes that make up an information system. Once the processes are identified, though, analysts need to begin thinking about what each process does and how to represent that internal logic. Two methods for representing process logic exist: Structured English and decision tables. Both techniques are simple yet powerful.

Key Terms Checkpoint

Here are the key terms from the chapter. The page where each term is first explained is in parentheses after the term.

1. **Action stubs (p. 177)**
2. **Balancing (p. 166)**
3. **Condition stubs (p. 177)**
4. **Context diagram (p. 160)**
5. **Data flow diagram (DFD) (p. 156)**
6. **Data store (p. 158)**
7. **Decision table (p. 177)**
8. **DFD completeness (p. 169)**
9. **DFD consistency (p. 169)**
10. **Gap analysis (p. 171)**
11. **Indifferent condition (p. 177)**
12. **Level-0 diagram (p. 161)**
13. **Level-*n* diagram (p. 165)**
14. **Primitive DFD (p. 170)**
15. **Process (p. 158)**
16. **Process modeling (p. 156)**
17. **Rules (p. 177)**
18. **Source/sink (p. 158)**
19. **Structured English (p. 174)**

Match each of the key terms above with the definition that best fits it.

_____ 1. A picture of the movement of data between external entities and the processes and data stores within a system.

_____ 2. The conservation of inputs and outputs to a data flow diagram process when that process is decomposed to a lower level.

_____ 3. That part of a decision table that lists the conditions relevant to the decision.

_____ 4. A data flow diagram that represents a system's major processes, data flows, and data stores at a high level of detail.

_____ 5. The origin and/or destination of data; sometimes referred to as *external entities*.

_____ 6. In a decision table, a condition whose value does not affect which actions are taken for two or more rules.

_____ 7. A data flow diagram of the scope of an organizational system that shows the system boundaries, external entities that interact with the system, and the major information flows between the entities and the system.

_____ 8. The lowest level of decomposition for a data flow diagram.

_____ 9. The extent to which all necessary components of a data flow diagram have been included and fully described.

_____ 10. A matrix representation of the logic of a decision, which specifies the possible conditions for the decision and the resulting actions.

_____ 11. The extent to which information contained on one level of a set of nested data flow diagrams is also included on other levels.

_____ 12. Modified form of the English language used to specify the logic of information system processes.

_____ 13. A DFD that is the result of n nested decompositions of a series of subprocesses from a process on a level-0 diagram.

_____ 14. The work or actions performed on data so that they are transformed, stored, or distributed.

_____ 15. That part of a decision table that specifies which actions are to be followed for a given set of conditions.

_____ 16. Data at rest, which may take the form of many different physical representations.

_____ 17. Graphically representing the processes that capture, manipulate, store, and distribute data between a system and its environment and among components within a system.

_____ 18. The process of discovering discrepancies between two or more sets of data flow diagrams or discrepancies within a single DFD.

_____ 19. That part of a decision table that lists the actions that result for a given set of conditions.

Review Questions

1. What is a data flow diagram? Why do systems analysts use data flow diagrams?
2. Explain the rules for drawing good data flow diagrams.
3. What is decomposition? What is balancing? How can you determine if DFDs are not balanced?
4. Explain the convention for naming different levels of data flow diagrams.
5. How can data flow diagrams be used as analysis tools?
6. Explain the guidelines for deciding when to stop decomposing DFDs.
7. How do you decide if a system component should be represented as a source/sink or as a process?
8. What unique rules apply to drawing context diagrams?
9. Explain what the term *DFD consistency* means and provide an example.

10. Explain what the term *DFD completeness* means and provide an example.
11. How well do DFDs illustrate timing considerations for systems? Explain your answer.
12. How can data flow diagrams be used in business process redesign?
13. What is the purpose of logic modeling? What techniques are used to model decision logic and what techniques are used to model temporal logic?
14. What is Structured English? How can Structured English be used to represent sequence, conditional statements, and repetition in an information systems process?
15. What are the steps in creating a decision table? How do you reduce the size and complexity of a decision table?
16. What verbs are used in Structured English? What type of words are not used in Structured English?
17. What formula is used to calculate the number of rules a decision table must cover?

Problems and Exercises

1. Using the example of a retail clothing store in a mall, list relevant data flows, data stores, processes, and sources/sinks. Observe several sales transactions. Draw a context diagram and a level-0 diagram that represent the selling system at the store. Explain why you chose certain elements as processes versus sources/sinks.
2. Choose a transaction that you are likely to encounter, perhaps ordering a cap and gown for graduation, and develop a high-level DFD, or context diagram. Decompose this to a level-0 diagram.
3. Evaluate your level-0 DFD from Problem and Exercise 2 using the rules for drawing DFDs in this chapter. Edit your DFD so that it does not break any of these rules.
4. Choose an example like that in Problem and Exercise 2, and draw a context diagram.

Decompose this diagram until it doesn't make sense to continue. Be sure that your diagrams are balanced, as discussed in this chapter.
5. Refer to Figure 5-22, which contains drafts of a context and level-0 DFD for a university class registration system. Identify and explain potential violations of rules and guidelines on these diagrams.
6. Why should you develop both logical and physical DFDs for systems? What advantage is there for drawing a logical DFD before a physical DFD for a new information system?
7. This chapter has shown you how to model, or structure, just one aspect, or view, of an information system, namely the process view. Why do you think analysts have different types of diagrams and other documentation to depict different views (e.g., process, logic, and data) of an information system?

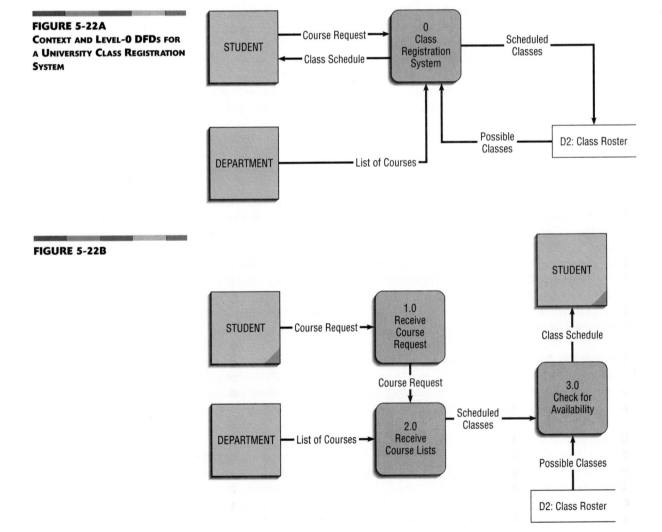

FIGURE 5-22A
CONTEXT AND LEVEL-0 DFDS FOR A UNIVERSITY CLASS REGISTRATION SYSTEM

FIGURE 5-22B

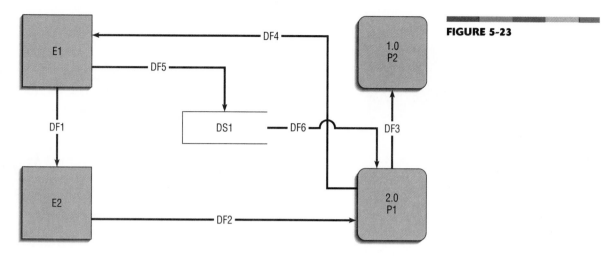

FIGURE 5-23

8. Consider the DFD in Figure 5-23. List three errors (rule violations) on this DFD.
9. Consider the three DFDs in Figure 5-24. List three errors (rule violations) on these DFDs.
10. Starting with a context diagram, draw as many nested DFDs as you consider necessary to represent all the details of the employee hiring system described in the following narrative. You must draw at least a context diagram and a level-0 diagram. In drawing these diagrams, if you discover that the narrative is incomplete, make up reasonable explanations to complete the story. Supply these extra explanations along with the diagrams.

   Projects, Inc., is an engineering firm with approximately 500 engineers of different types. The company keeps records on all employees, their skills, projects assigned, and departments worked in. New employees are hired by the personnel manager based on data in an application form and evaluations collected from other managers who interview the job candidates. Prospective employees may apply at any time. Engineering managers notify the personnel manager when a job opens and list the characteristics necessary to be eligible for the job. The personnel manager compares the qualifications of the available pool of applicants with the characteristics of an open job, then schedules interviews between the manager in charge of the open position and the three best candidates from the pool. After receiving evaluations on each interview from the manager, the personnel manager makes the hiring decision based upon the evaluations and applications of the candidates and the characteristics of the job, and then notifies the interviewees and the manager about the decision. Applications of rejected applicants are retained for one year, after which time the application is purged. When hired, a new engineer completes a nondisclosure agreement, which is filed with other information about the employee.

11. a. Starting with a context diagram, draw as many nested DFDs as you consider necessary to represent all the details of the system described in the following narrative. In drawing these diagrams, if you discover that the narrative is incomplete, make up reasonable explanations to complete the story. Supply these extra explanations along with the diagrams.

   Maximum Software is a developer and supplier of software products to individuals and businesses. As part of its operations, Maximum provides an 800 telephone number help desk for clients with questions about software purchased from Maximum. When a call comes in, an operator inquires about the nature of the call. For calls that are not truly help desk functions, the operator redirects the call to another unit of the company (such as Order Processing or Billing). Because many customer questions require in-depth knowledge of a product, help desk consultants are organized by product. The operator directs the call to a consultant skilled on the software that the caller needs help with. Because a consultant is not always immediately available, some calls must be put into a queue for the next available consultant. Once a consultant answers the call, he determines if this is the first call from this customer about this problem. If so, he creates a new call report to keep track of all information about the problem. If not, he asks the customer for a call report number, and retrieves the open call report to determine the status of the inquiry. If the caller does not know the call report number, the consultant collects other identifying information such as the caller's name, the software involved, or the name of the consultant

FIGURE 5-24

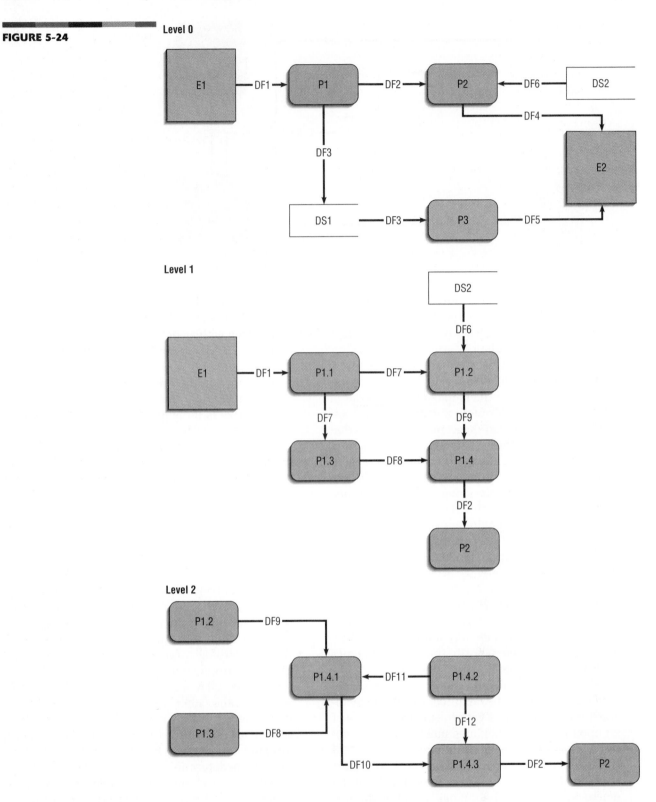

who has handled the previous calls on the problem in order to conduct a search for the appropriate call report. If a resolution of the customer's problem has been found, the consultant informs the client what that resolution is, indicates on the report that the customer has been notified, and closes out the report. If resolution has not been discovered, the consultant finds out if the consultant handling this problem is on duty. If so, he transfers the call to the other consultant (or puts the call into the queue of calls waiting to be handled by that consultant). Once the proper consultant receives the call, he records any new details the customer may have. For continuing problems and for new call reports, the consultant tries to discover an answer to the problem by using the relevant software and looking up information in reference manuals. If he can resolve the problem, he tells the customer how to deal with the problem, and closes the call report. Otherwise, the consultant files the report for continued research and tells the customer that someone at Maximum will get back to him, or if the customer discovers new information about the problem, to call back identifying the problem with a specified call report number.

    b. Analyze the DFDs you created in Part a. What recommendations for improvements in the help desk system at Maximum can you make based upon this analysis? Draw new logical DFDs that represent the requirements you would suggest for an improved help desk system. Remember, these are to be logical DFDs, so consider improvements independent of technology that can be used to support the help desk.

12. Represent the decision logic in the decision table of Figure 5-16 in Structured English.

13. In one company the rules for buying personal computers are that any purchase over $15,000 has to go out for bid and the Request for Proposals must be approved by the Purchasing department. If the purchase is under $15,000, the personal computers can simply be bought from any approved vendor; however, the purchase order must still be approved by the Purchasing department. If the purchase goes out for bid, there must be at least three proposals received for the bid. If not, the RFP must go out again. If there still are not enough proposals, then the process can continue with the one or two vendors that have submitted proposals. The winner of the bid must be on an approved list of vendors for the company and, in addition, must not have any violations against it for affirmative action or environmental matters. At this point, if the proposal is complete, the Purchasing department can issue a purchase order. Use Structured English to represent the logic in this process. Notice the similarities between the text in this question and format of your answer.

14. In a relatively small company that sells thin, electronic keypads and switches, the rules for selling products are such that sales representatives are assigned to unique regions of the country. Sales come either from cold calling, referrals, or current customers with new orders. A sizable portion of its business comes from referrals from larger competitors, who send their excess and/or "difficult" projects to this company. The company tracks these and, similarly, returns the favors to these competitors by sending business their way. The sales reps receive a 10 percent commission

on purchases, not on orders, in their region. They can collaborate on a sale with reps in other regions and share the commissions, with 8 percent going to the "home" rep, and 2 percent to the "visiting" rep. For any sales beyond the rep's previously stated and approved individual annual sales goals, he or she receives an additional 5 percent commission, an additional end-of-the-year bonus determined by management, and a special vacation for his or her family. Customers receive a 10 percent discount for any purchases over $100,000 per year, which are factored into the rep's commissions. In addition, the company focuses on customer satisfaction with the product and service, so there is an annual survey of customers in which they rate the sales rep. These ratings are factored into the bonuses such that a high rating increases the bonus amount, a moderate rating does nothing, and a low rating can lower the bonus amount. The company also wants to ensure that the reps close all sales. Any differences between the amount of orders and actual purchases are also factored into the rep's bonus amount. As best you can, present the logic of this business process first using Structured English, then using a decision table. Write down any assumptions you have to make. Which of these techniques is most helpful for this problem? Why?

15. The following example demonstrates the rules of the tenure process for faculty at many universities. Present the logic of this business process first using Structured English, then using a decision table, and then using a decision tree. Write down any assumptions you have to make. Which of these techniques is most helpful for this problem? Why?

    A faculty member applies for tenure in his or her sixth year by submitting a portfolio summarizing his or her work. In rare circumstances he or she can come up for tenure earlier than the sixth year, but only if the faculty member has permission of the department chair and college dean. New professors, who have worked at other universities before taking their current jobs, rarely, if ever, come in with tenure. They are usually asked to undergo one "probationary" year during which they are evaluated and only then can be granted tenure. Top administrators coming in to a new university job, however, can often negotiate for retreat rights that enable them to become a tenured faculty member should their administrative post end. These retreat arrangements generally have to be approved by faculty. The tenure review process begins with an evaluation of the candidate's portfolio by a committee of faculty within the candidate's department. The committee then writes a recommendation on tenure and sends it to the department's chairperson, who then makes a recommendation and passes the portfolio and recommendation on to the next level, a college-wide faculty committee. This committee does the same as the department committee and passes its recommendation, the department's recommendation, and the portfolio on to the next level, a university-wide faculty committee. This committee does the same as the other two committees and passes everything on to the provost (or sometimes the academic vice president). The provost then writes his or her own recommendation and passes everything to the president, the final decision maker. This process, from the time the candidate creates his or her portfolio until the time the president makes a decision, can take an entire academic year. The focus of the evaluation is on research, which could be grants, presentations, and publications, although preference is given for empirical research that has been published in top-ranked, refereed journals and where the publication makes a contribution to the field. The candidate must also do well in teaching and service (i.e., to the university, to the community, or to the discipline) but the primary emphasis is on research.

16. An organization is in the process of upgrading microcomputer hardware and software for all employees. Hardware will be allocated to each employee in one of three packages. The first hardware package includes a standard microcomputer with a color monitor of moderate resolution and moderate storage capabilities. The second package includes a high-end microcomputer with a high-resolution color monitor and a great deal of RAM and ROM. The third package is a high-end notebook-size microcomputer. Each computer comes with a network interface card so that it can be connected to the network for printing and e-mail. The notebook computer comes with a modem for the same purpose. All new and existing employees will be evaluated in terms of their computing needs (e.g., the types of tasks they perform, how much and in what ways they can use the computer). Light users receive the first hardware package. Heavy users receive the second package. Some moderate users will receive the first package and some will receive the second package, depending on their needs. Any employee who is deemed to be primarily mobile (e.g., most of the sales force) will receive the third package. Each employee will also be considered for additional hardware. For example, those who need scanners will receive them and those needing their own printers will receive them. A determination will be made regarding whether or not the user receives a color or black-and-white scanner, and a slow or fast, or color or black-and-white printer. In addition, each employee will receive a suite of software, including a word processor, spreadsheet, and presentation maker. All employees will be evaluated for their additional software needs. Depending on their needs, some will receive a desktop publishing package, some will receive a database management system (and some will also receive a developer's kit for the DBMS), and some will receive a programming language. Every 18 months those employees with the high-end systems will receive new hardware and then their old systems will be passed on to those who previously had the standard systems. All those employees with the portable systems will receive new notebook computers. Present the logic of this business process first using Structured English and then using a decision table. Write down any assumptions you have to make. Which of these techniques is most helpful for this problem? Why?

Discussion Questions

1. Discuss the importance of diagramming tools for process modeling. Without such tools, what would an analyst do to model diagrams?
2. Think and write about how data flow diagrams might be modified to allow for time considerations to be adequately incorporated.
3. How would you answer someone who told you that data flow diagrams were too simple and took too long to draw to be of much use? What if they also said that keeping data flow diagrams up-to-date took too much effort, compared to the potential benefits?
4. Find another example of where data flow diagrams were successfully used to support business process reengineering. Write a report, complete with DFDs, about what you found.

Case Problems

1. **Pine Valley Furniture**

As a Pine Valley Furniture intern, you have gained valuable insights into the systems development process. Jim Woo has made it a point to discuss with you both the WebStore and the Customer Tracking System projects. The data requirements for both projects have been collected and are ready to be organized into data flow diagrams. Jim has prepared the data flow diagrams for the WebStore; however, he has requested your help in preparing the data flow diagrams for the Customer Tracking System.

You recall that Pine Valley Furniture distributes its products to retail stores, sells directly to customers, and is in the process of developing its WebStore, which will support online sales in the areas of corporate furniture buying, home office furniture purchasing, and student furniture purchasing. You also know that the Customer Tracking System's primary objective is to track and forecast customer buying patterns.

Information collected during the requirements determination activity suggests that the Customer Tracking System should collect customer purchasing activity data. Customers will be tracked based on a variety of factors, including customer type, geographic location, type of sale, and promotional item purchases. The Customer Tracking System should support trend analysis, facilitate sales information reporting, enable managers to generate ad hoc queries, and interface with the WebStore.

a. Construct a context data flow diagram, illustrating the Customer Tracking System's scope.
b. Construct a level-0 diagram for the Customer Tracking System.
c. Using the level-0 diagram that you constructed above, select one of the level-0 processes and prepare a level-1 diagram.
d. Exchange your diagrams with another class member. Ask your class member to review your diagrams for completeness and consistency. What errors did he or she find? Correct these errors.

2. **Hoosier Burger**

As one of Build a Better System's lead analysts on the Hoosier Burger project, you have spent significant time discussing the current and future needs of the restaurant with Bob and Thelma Mellankamp. In one of these conversations, Bob

and Thelma mentioned that they were in the process of purchasing the empty lot next to Hoosier Burger. In the future, they would like to expand Hoosier Burger to include a drive-through, build a larger seating area in the restaurant, include more items on the Hoosier Burger menu, and provide delivery service to Hoosier Burger customers. After several discussions and much thought, the decision was made to implement the drive-through and delivery service and wait on the activities requiring physical expansion. Implementing the drive-through service will require only minor physical alterations to the west side of the Hoosier Burger building. Many of Hoosier Burger's customers work in the downtown area, so Bob and Thelma think a noon delivery service will offer an additional convenience to their customers.

One day while having lunch at Hoosier Burger with Bob and Thelma, you discuss how the new delivery and drive-through services will work. Customer order taking via the drive-through window will mirror in-house dining operations. Therefore, drive-through window operations will not require information system modifications. Until a new system is implemented, the delivery service will be operated manually; each night Bob will enter necessary inventory data into the current system.

Bob envisions the delivery system operating as follows. When a customer calls and places a delivery order, a Hoosier Burger employee records the order on a multiform order ticket. The employee captures such details as customer name, business or home address, phone number, order placement time, items ordered, and amount of sale. The multiform document is sent to the kitchen where it is separated when the order is ready for delivery. Two copies accompany the order; a third copy is placed in a reconciliation box. When the order is prepared, the delivery person delivers the order to the customer, removes one order ticket from the food bag, collects payment for the order, and returns to Hoosier Burger. Upon arriving at Hoosier Burger, the delivery person gives the order ticket and the payment to Bob. Each evening Bob reconciles the order tickets stored in the reconciliation box with the delivery payments and matching order tickets returned by the delivery person. At the close of business each evening, Bob uses the data from the order tickets to update the goods sold and inventory files.

a. Modify the Hoosier Burger context-level data flow diagram (Figure 5-4) to reflect the changes mentioned in the case.

b. Modify Hoosier Burger's level-0 diagram (Figure 5-5) to reflect the changes mentioned in the case.

c. Prepare level-1 diagrams to reflect the changes mentioned in the case.

d. Exchange your diagrams with those of another class member. Ask your classmate to review your diagrams for completeness and consistency. What errors did he or she find? Correct these errors.

3. Evergreen Nurseries

Evergreen Nurseries offers a wide range of lawn and garden products to its customers. Evergreen Nurseries conducts both wholesale and retail operations. Although the company serves as a wholesaler to nurseries all over the United States, the company's founder and president has restricted its retail operations to California, the company's home state. The company is situated on 150 acres and wholesales its bulbs, perennials, roses, trees, shrubs, and Evergreen Accessory products. Evergreen Accessory products include a variety of fertilizers, plant foods, pesticides, and gardening supplies.

In the past five years, the company has seen a phenomenal sales growth. Unfortunately, its information systems have been left behind. Although many of Evergreen Nurseries' processing activities are computerized, these activities require reengineering. You are part of the project team hired by Seymour Davis, the com-

pany's president, to renovate its wholesale division. Your project team was hired to renovate the billing, order taking, and inventory control systems.

From requirements determination, you discovered the following. An Evergreen Nurseries customer places a call to the nursery. A sales representative takes the order, verifies the customer's credit standing, determines if the items are in stock, notifies the customer of the product's status, informs the customer if any special discounts are in effect, and communicates the total payment due. Once an order is entered into the system, the customer's account is updated, product inventory is adjusted, and ordered items are pulled from stock. Ordered items are then packed and shipped to the customer. Once each month, a billing statement is generated and sent to the customer. The customer has 30 days to remit payment in full; otherwise a 15 percent penalty is applied to the customer's account.

a. Construct a context data flow diagram, illustrating Evergreen Nurseries' wholesale system.

b. Construct a level-0 diagram for Evergreen Nurseries' wholesale system.

c. Using the level-0 diagram that you constructed in part b, select one of the level-0 processes, and prepare a level-1 diagram.

d. Exchange your diagrams with those of another class member. Ask your classmate to review your diagrams for completeness and consistency. What errors did he or she find? Correct these errors.

CASE: BROADWAY ENTERTAINMENT COMPANY, INC.

Process Modeling for the Web-Based Customer Relationship Management System

Case Introduction

The BEC student team of Tracey Wesley, John Whitman, Missi Davies, and Aaron Sharp left the first meeting with Carrie Douglass, manager of the BEC store in Centerville, Ohio, eager to begin investigating the requirements for the customer relationship management system. Before they began requirements determination, they structured what they had already learned. Based on the System Service Request and the initial meeting with Carrie, the team developed a context diagram for the system (see BEC Figure 5-1).

The context diagram using Microsoft Visio shows the system in the middle, the major external entities (Customer, Employee, and the Entertainment Tracker BEC in-store information system) that interact with the system on the outside, and the data flows between the system and the external entities.

Not too surprisingly, most of the data flows are between the system and customers. For this reason, the team decided to repeat the Customer, using one copy of Customer as a source and another copy as a sink of data flows.

The context diagram helped the team to organize for requirements determination. This data collection part of the analysis phase would be used to verify this overview model of the customer relationship management system and to gather details for each

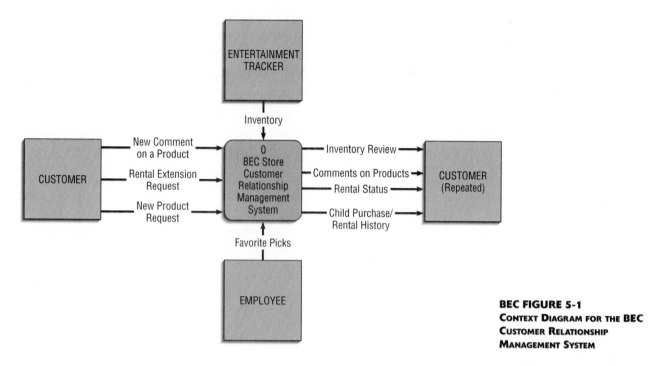

BEC FIGURE 5-1
CONTEXT DIAGRAM FOR THE BEC
CUSTOMER RELATIONSHIP
MANAGEMENT SYSTEM

data flow, processing activity, and data storage component inside the system.

The team needed one more result before beginning the detailed work of analysis and design—a catchy name for the system it was designing. "The BEC Customer Relationship Management System" was too long and dull. With the cooperation of Carrie Douglass, team members ran a contest among the other teams in their class to give each member of the team with the best name suggestion (as selected by Carrie) a free movie rental at the Centerville BEC store. Some teams tried to create acronyms using the words and acronyms *BEC, Broadway Entertainment Company,* and *customer relationship management,* but most of these were not pronounceable nor very meaningful. Other teams created phrases that conveyed the Web technology to be used to build the system (e.g., one team suggested VideosByBEC, similar to AutoByTel for automobile sales and information on the Web). But, Carrie wanted a name that conveyed the personal relationship the system created with the customer. Thus, one suggested name stood out from the rest. The winner was MyBroadway.

Structuring the High-Level Process Findings from Requirements Determination

The BEC student team used various methods to understand the requirements for MyBroadway. The following sections explain how they approached studying each data flow on the context diagram and what they discovered from their analysis.

Inventory and Rental Extension Request

The team studied documentation of the Entertainment Tracker system provided to store employees and the manager. From this documentation the team understood the data about products and product sales and rentals maintained in store records. This was a necessary step to determine what data could be in the Inventory data flow. It was clear that MyBroadway would not be the system of record to operate the store; Entertainment Tracker was this official record. For example, the official record of when a rented product was due to be returned would be recorded in the Entertainment Tracker database. Thus, product inventory, sales, and rental data needed by MyBroadway would be periodically extracted from Entertainment Tracker to be stored in MyBroadway for faster access and to keep the two systems as decoupled as possible. Because of the role of Entertainment Tracker, any activity in MyBroadway that changed data in Entertainment Tracker would have to submit a transaction to Entertainment Tracker that Entertainment Tracker understood. The only instance of this the team discovered related to the Rental Extension Request data flow. The process handling this inflow would find the due date in the MyBroadway database and then interact with Entertainment Tracker to request the extension and to inform the customer whether the extension was accepted. Entertainment Tracker, however, would make the decision, based on its own rules, whether to accept the extension. Fortunately, requesting an extension is a transaction in Entertainment Tracker handled from a point-of-sale terminal in the store, so MyBroadway would simply need to simulate this transaction.

Favorite Picks

The team also surveyed employees and customers to understand what would be useful related to the employee Favorite Picks data flow. Both employees and customers agreed that there were only two broad groups of items for favorite picks: new releases and classics. Each week a different store employee would select one or two new release or classic products in a given product category. For example, each week one employee would select one or two new release children's videos, another employee would select one or two new release jazz and new age CDs, and yet another employee would select one or two classic romance DVDs. It is not possible to cover every category of videotape, DVD, and CD each week, but over time most categories would be selected. Selections would be retained for two years. Each week, five store employees would make selections each in a different product category. An employee would be given a list of those ten product categories for which favorite picks had not been made for the longest time. Each employee would be matched with the category with which he or she is most familiar and given a list of those new release and classic products in that category. A classic product is one that continues to be rented or sold at least 10 years after its initial release. An employee selects one or two products on this list and provides a quality grade for each (A, A-, B, . . . F), a description of its contents relevant to language and sexually explicit references, and a few sentences of personal comments about the product that a parent might want to know. The date of the entry would be recorded with the rest of the data.

New Product Request

The team used interviews with customers, Carrie Douglass, and the assistant store manager as well as observation of people using similar Web-based systems from major online bookstores and other shopping enterprises to determine the nature of the other six data flows on the context diagram. For New Product Request, MyBroadway will collect all the requests and at the request of Carrie print a list, in decreasing order of frequency of request, of each requested product. Carrie will then use this report to send a letter to the BEC purchasing department requesting the acquisition of these items. New Product Requests will be kept for two months and then purged.

New Comment on a Product

For New Comment on a Product, MyBroadway will show a parent or child basic information about the product (such as title, publisher, artist, and date released) and then allow him or her to enter an unstructured comment about that product. There is no limit on the length of the comment. Each comment is stored separately, and the same person may comment on the same item many times. The date and time of the comment are stored with the comment. An issue that required some discussion with Carrie was whether the person had to identify himself or herself for the comment to be recorded. Carrie was unsure what to do. So, the team convened a focus group of a few parents to explore this issue. The team discovered that the parents would not consider a comment valid unless it were attributed and that the parents thought they and their children would enter a more helpful comment if it were attributed. Carrie, however, saw no need to retain data about customers in MyBroadway, but how would bogus customer names be identified? Entertainment Tracker maintains data about each customer, including each child with a membership card. Thus, it was decided that the customer would have to enter his or her membership number along with the comment. This number would be sent to Entertainment Tracker for matching with its record of customer numbers after the comment was entered but before it was available to be reported to other customers. Whether the comment is entered by a parent or a child is also recorded with the comment. If the number does not match a membership number for a BEC customer, the comment will be dropped. When the comment is displayed, the name of the member entering the comment as well as whether that person is a parent or child will be shown with the comment.

Inventory Review

The Inventory Review data flow consolidates several data flows. A customer can ask to see product data by specific title, or to see data for all the products by artist, category (e.g., new age or jazz CD), publisher, release month, or any combination of these factors. In each case, for each product identified by the search criteria, the product title, artist, publisher, release date, media, description, and sale and rental price are shown.

Comments on Products

Comments on Products is produced when a customer enters the name of the product (and possibly searching through a set of products with approximately that name until the exact product is found). Once the exact product the customer is interested in is identified, then all the comments previously entered by customers are available for display. For the purpose of this data flow, Favorite Picks records are also considered comments. The customer may ask to see only those comments entered since some date they specify and may ask to see comments only by parents, only by children, only by employees (i.e., only Favorite Picks), or all comments. Comments are shown in reverse chronology entry order.

Rental Status

For this data flow, the customer enters her or his membership number and then MyBroadway displays a list of all the product titles and return due dates for all outstanding rented items. Often, customers obtain a Rental Status before they submit a Rental Extension Request, but the team decided to consider these separate data flows.

Child Purchase/Rental History

The team discovered that this is arguably the most complex of the data flows on the context diagram. The team decided to model this data flow at a fairly high level first and then decompose the process producing Child Purchase/Rental History later. At a high level, to produce this data flow MyBroadway needs access to sales and rental history data, including what products have been bought and rented by whom. Customers indicated that a simple history would not be sufficient. They also wanted to see the customer comments and favorite picks ratings for each item. So, an instance of the Child Purchase/Rental History data flow is a report that shows for a given child the title of each item he or she has bought or rented in the past six months, and for each item the rating entered by each employee who has rated that product, and the five most recent parent comments recorded about that item.

Case Summary

Accurately and thoroughly documenting business processes can be tedious and time-consuming, but very insightful. The student team working on the analysis and design of MyBroadway quickly discovered how extensive a system Carrie and the store employees and customers wanted for this customer relationship management system. The team was unsure whether it could do a thorough analysis and design for all the desired features. Starting with a context diagram and successively decomposing processes, however, allowed the team to show the total scope of the system as desired by the project sponsor and system users and yet focus attention on one piece of the system at a time. If only parts of the system could be built during the course project, at least the team would be able to show how those pieces fit into the complete system. The team members also recognized that structuring processes and data flows were only part of the systems analysis. They would also need to identify all the data stored inside MyBroadway (in data stores) and then struc-

ture these data into a database specification. Each primitive process on the lowest-level DFDs would have to be specified in sufficient detail for a programmer to build that functionality into the information system.

The BEC student team had made the decision to use automated tools to draw DFDs (and other system diagrams) and to record project dictionary data about system objects, such as external entities, data flows, data stores, and processes. (Because you will use whatever tools your instructor recommends, we do not refer to any specific tools by name in this or subsequent cases.) These automated tools are critical for making it easy to change diagrams, to produce clean documentation about the system requirements, and to make all aspects of the documentation consistent with each other. Drawing the initial diagrams and recording all the dictionary entries is very time-consuming. This automated data, however, can be changed by any team member, and team members can prepare new diagrams and dictionary reports at any time with minimal effort.

Case Questions

1. Does the context diagram in BEC Figure 5-1 represent an accurate and complete overview of the system as described in this case for requirements collected during the analysis phase? If not, what is wrong or missing? If necessary, draw a new context diagram in light of what is explained in this case. Why might a context diagram initially drawn at the end of project initiation and planning need to be redrawn during the analysis phase?
2. In the context diagram of BEC Figure 5-1, why is the Rental Extension Request data flow shown as an inflow to the system? Why is the Rental Status data flow shown as an outflow from the system? Do you agree with these designations of the two data flows? Why or why not?
3. The store manager is not shown in the context diagram in BEC Figure 5-1, except implicitly as an Employee who enters Favorite Picks. Based on the descriptions in this case, does it make sense that store manager does not appear on the context diagram? If not on the context diagram, where might store manager appear? As an external entity on a lower-level diagram? As a process or data store on a lower-level diagram? Based on the description in this case, are there any external entities missing on the context diagram of BEC Figure 5-1?
4. Based on the descriptions in this case of each data flow from the context diagram, draw a level-0 data flow diagram for MyBroadway using Microsoft Visio (begin by drawing the context diagram, then explode to level-0). Be sure it is

balanced with the context diagram you might have drawn in answer to Question 1.

5. Write project dictionary entries (using standards given to you by your instructor) for all the data stores shown in the level-0 diagram in your answer for Question 4. Are there other data stores hidden inside processes for your level-0 diagram? If so, what kinds of data do you anticipate are retained in these hidden data stores? Why are these data stores hidden inside processes rather than appearing on the level-0 diagram?

6. Write project dictionary entries (using standards given to you by your instructor) for all the data flows shown in the level-0 diagram in your answer to Question 4. How detailed are these entries at this point? How detailed must these entries be for primitive DFDs?

7. Explain how you modeled in your answer to Question 4 the process that receives the New Product Request data flow. Was this a difficult process to model on the DFD? Did you consider several alternative ways to show this process? If so, explain the alternatives and why you chose the representation you drew in the level-0 diagram for question 4.

8. Look at your answer to Question 4 and focus attention on the process for the Rental Extension Request data flow. Using Microsoft Visio, draw a level-1 diagram for this process based on the description of this data flow in the case and in the following explanation. A customer provides his or her customer number or name and a prod-

uct number or title and then MyBroadway finds in its records the rental information for this customer's outstanding rental of this product, including the due date. Then the customer may decide that he or she can return the item by the due date, in which case no request for extension is made. If the customer decides to extend the due date, the customer can request a one-day or two-day extension, each with a different fee, which will be due when the product is returned. MyBroadway will then send a Rental Extension Request transaction to Entertainment Tracker as if it were a point-of-sale terminal from which the same request was being made. Entertainment Tracker may reject the request if the customer has delinquent fees. Once Entertainment Tracker makes its decision, it returns a code to MyBroadway indicating a yes or the reason for a no to the request. If the decision is no, the customer is given a message to explain rejection. If yes, MyBroadway rental data are updated to reflect the extension, and the user is given a confirmation message.

9. Does your answer to Question 7 necessitate any changes to your answer to Question 4? If so, what are these changes? Prepare a new level-0 diagram for MyBroadway.

10. Investigate the capabilities of Microsoft Visio to store and report project dictionary entries for objects on dataflow diagrams. What capabilities of CASE tools does Visio not provide?

6

Structuring System Requirements: Conceptual Data Modeling

Objectives

*After studying this chapter,
you should be able to:*

- Concisely define each of the following key data-modeling terms: *conceptual data model, entity-relationship diagram, entity type, entity instance, attribute, candidate key, multivalued attribute, relationship, degree, cardinality,* and *associative entity*.

- Ask the right kinds of questions to determine data requirements for an information system.

- Draw an entity-relationship (E-R) diagram to represent common business situations.

- Explain the role of conceptual data modeling in the overall analysis and design of an information system.

- Distinguish between unary, binary, and ternary relationships, and give an example of each.

- Distinguish between a relationship and an associative entity, and use associative entities in a data model when appropriate.

- Relate data modeling to process and logic modeling as different ways of describing an information system.

Chapter Preview . . .

In Chapter 5 you learned how to model and analyze the flow of data (data in motion) between manual or automated steps and how to show data stores (data at rest) in a data flow diagram. Data flow diagrams show how, where, and when data are used or changed in an information system, but they do not show the definition, structure, and relationships within the data. Data modeling, the subject of this chapter, develops this missing, and crucial, piece of the description of an information system.

Systems analysts perform data modeling during the systems analysis phase, as highlighted in Figure 6-1. Data modeling is typically done at the same time as other requirements structuring steps. Many systems developers believe that a data model is the most important part of the information system requirements statement for four reasons. First, the characteristics of data captured during data modeling are crucial in the design of databases, programs, computer screens, and printed reports. For example, facts such as these—a data element is numeric, a product can be in only one product line at a time, a line item on a customer order can never be moved to another customer order—are all essential in ensuring an information system's data integrity.

Second, data rather than processes are the most complex aspects of many modern information systems. For example, transaction processing systems can have considerable complexity in validating data, reconciling errors,

FIGURE 6-1
Systems analysts perform data modeling during the systems analysis phase. Data modeling typically occurs in parallel with other requirements structuring steps.

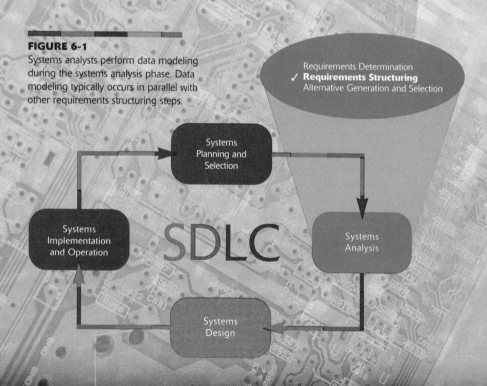

and coordinating the movement of data to various databases. Management information systems (such as sales tracking), decision support systems (such as short-term cash investment), and executive support systems (such as product planning) are data intensive and require extracting data from various data sources. Third, the characteristics about data (such as format and relationships with other data) are permanent. In contrast, who receives which data, the format of reports, and what reports are used change constantly over time. A data model explains the inherent nature of the organization, not its transient form. So, an information system design based on data, rather than processes or logic, should have a longer useful life. Finally, structural information about data is essential to generate programs automatically. For example, the fact that a customer order has many line items as opposed to just one affects the automatic design of a computer form in Microsoft Access for entry of customer orders.

In this chapter, we discuss the key concepts of data modeling including the most common format used for data modeling, entity-relationship (E-R) diagramming. During the systems analysis phase of the SDLC, you use data flow diagrams to show data in motion and E-R diagrams to show the relationships among data objects. We also illustrate E-R diagrams drawn using Microsoft's Visio tool, highlighting this tool's capabilities and limitations.

Conceptual Data Modeling

Conceptual data model
A detailed model that shows the overall structure of organizational data while being independent of any database management system or other implementation considerations.

A **conceptual data model** is a representation of organizational data. The purpose of a conceptual data model is to show as many rules about the meaning and interrelationships among data as possible.

Entity-relationship (E-R) data models are commonly used diagrams that show how data are organized in an information system. The main goal of conceptual data modeling is to create accurate E-R diagrams. As a systems analyst, you typically do conceptual data modeling at the same time as other requirements analysis and structuring steps during systems analysis. You can use methods such as interviewing, questionnaires, and JAD sessions to collect information for conceptual data modeling. On larger systems development teams, a subset of the project team concentrates on data modeling while other team members focus attention on process or logic modeling. You develop (or use from prior systems development) a conceptual data model for the current system and build a conceptual data model that supports the scope and requirements for the proposed or enhanced system.

The work of all team members is coordinated and shared through the project dictionary or repository. As discussed in Chapter 2, this repository and associated diagrams may be maintained by a CASE tool or a specialized tool such as Microsoft's Visio. Whether automated or manual, the process flow, decision logic, and data model descriptions of a system must be consistent and complete because each describes different but complementary views of the same information system. For example, the names of data stores on primitive-level DFDs often correspond to the names of data entities in entity-relationship diagrams, and the data elements in data flows on DFDs must be attributes of entities and relationships in entity-relationship diagrams.

The Process of Conceptual Data Modeling

You typically begin conceptual data modeling by developing a data model for the system being replaced, if a system exists. This is essential for planning the conversion of the current files or database into the database of the new system. Further, this is a good, but not a perfect, starting point for your understanding of the new system's data requirements. Then, you build a new conceptual data model that includes all of the data requirements for the new system. You discovered these requirements from the fact-finding methods used during requirements determination. Today, given the popularity of prototyping and other rapid development methodologies, these requirements often evolve through various iterations of a prototype, so the data model is constantly changing.

Conceptual data modeling is only one kind of data modeling and database design activity done throughout the systems development process. Figure 6-2 shows the different kinds of data modeling and database design that occur during the systems development life cycle. The conceptual data modeling methods we discuss in this chapter are suitable for various tasks in the planning and analysis phases. These phases of the SDLC address issues of system scope, general requirements, and content. An E-R data model evolves from project identification and selection through analysis as it becomes more specific and is validated by more detailed analysis of system needs.

In the design phase, the final E-R model developed in analysis is matched with designs for systems inputs and outputs and is translated into a format that

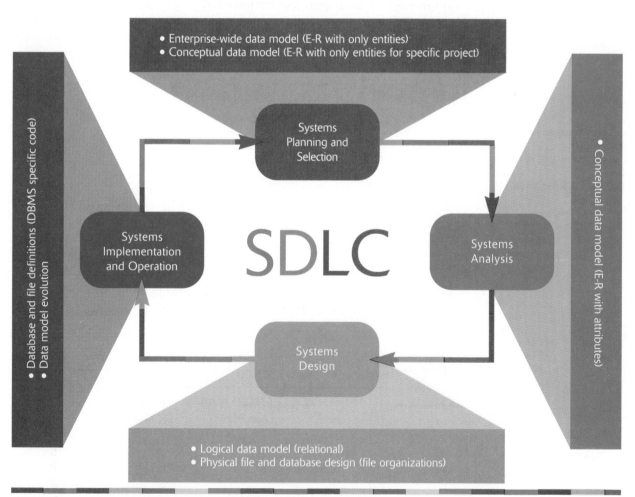

FIGURE 6-2
RELATIONSHIP BETWEEN DATA MODELING AND THE SYSTEMS DEVELOPMENT LIFE CYCLE

enables physical data storage decisions. During physical design, specific data storage architectures are selected, and then, in implementation, files and databases are defined as the system is coded. Through the use of the project repository, a field in a physical data record can, for example, be traced back to the conceptual data attribute that represents it on an E-R diagram. Thus, the data modeling and design steps in each of the SDLC phases are linked through the project repository.

Deliverables and Outcomes

Most organizations today do conceptual data modeling using entity-relationship modeling, which uses a special notation of rectangles, diamonds, and lines to represent as much meaning about data as possible. Thus, the primary deliverable from the conceptual data modeling step within the analysis phase is an entity-relationship (E-R) diagram. A sample E-R diagram appears in Figure 6-3(A). This figure shows the major categories of data (rectangles in the diagram) and the business relationships between them (lines connecting rectangles). For example, Figure 6-3(A) describes that, for the business represented, a SUPPLIER sometimes supplies ITEMs to the company, and an ITEM is always supplied by one to four SUPPLIERS. The fact that a supplier only sometimes supplies items implies that the business wants to keep track of some suppliers without designating what they can supply. This diagram includes two names on each line giving you explicit language to read a relationship in each direction. For simplicity, we will not typically include two names on lines in E-R diagrams in this book; however, many organizations use this standard.

It is very common that E-R diagrams are developed using CASE tools or other smart drawing packages. These tools provide functions to facilitate consistency of data models across different systems development phases, reverse engineering an existing database definition into an E-R diagram, and provide documentation of objects on a diagram. One popular tool is Microsoft Visio®. Figure 6-3(B)

FIGURE 6-3
SAMPLE CONCEPTUAL DATA MODEL DIAGRAMS
(A) STANDARD E-R NOTATION

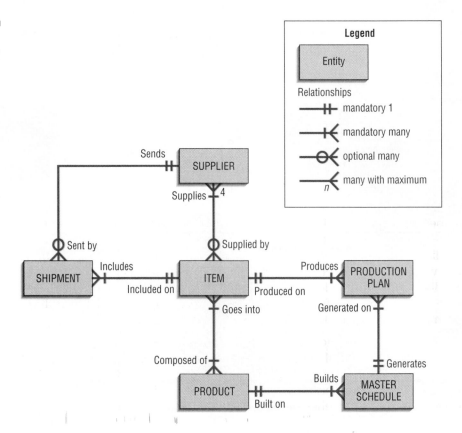

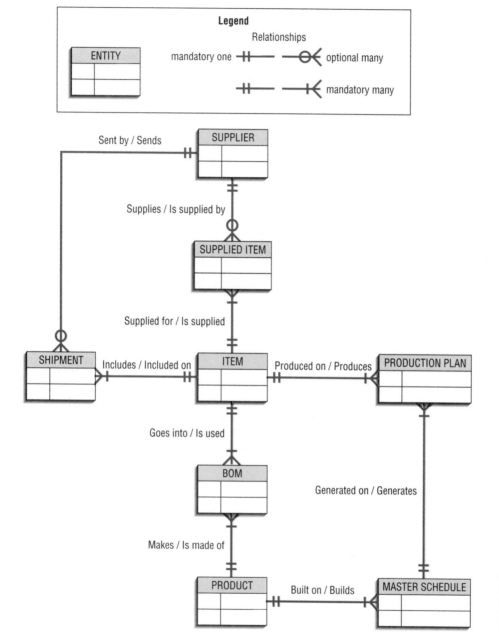

FIGURE 6-3
SAMPLE CONCEPTUAL DATA MODEL
DIAGRAMS
(B) VISIO E-R NOTATION

shows the equivalent of Figure 6-3(A) using Visio. This diagram is developed using the Database Model Diagram tool. The Database|Options|Document settings are specified as relational symbol set, conceptual names on the diagram, optionality is shown, and relationships are shown using the crow's foot notation with forward and inverse relationship names. These settings cause Visio to draw an E-R diagram that most closely resembles the standards used in this text.

There are some key differences between the standard E-R notation illustrated in Figure 6-3(A) and the notation used in Visio, including:

⦿ Relationships such as Supplies/Supplied by between SUPPLIER and ITEM in Figure 6-3(A) require an intermediate category of data (called SUPPLIED ITEM in Figure 6-3(B)) because Visio does not support representing these so-called many-to-many relationships.

◗ Relationships may be named in both directions, but these names appear on a text box on the relationship line, separated by a forward slash.

◗ Limitations such as an ITEM is always supplied by at most four SUPPLIERS is not shown on the diagram, but rather documented in the Miscellaneous set of Database Properties of the relationship, which are part of Visio's version of a CASE repository.

◗ The symbol for each category of data (e.g., SHIPMENT) includes space for listing other properties of each data category (such as all the attributes or columns of data we know about that data category); we will illustrate these components later in this chapter.

We concentrate on the traditional E-R diagramming notation in this chapter; however, we will include the equivalent Visio version on several occasions so you can see how to show data modeling concepts in this popular database design tool.

There may be as many as four E-R diagrams produced and analyzed during conceptual data modeling:

1. An E-R diagram that covers just the data needed in the project's application. (This allows you to concentrate on the data requirements without being constrained or confused by unnecessary details.)
2. An E-R diagram for the application system being replaced. (Differences between this diagram and the first show what changes you have to make to convert databases to the new application.) This is, of course, not produced if the proposed system supports a completely new business function.
3. An E-R diagram for the whole database from which the new application's data are extracted. (Because many applications share the same database or even several databases, this and the first diagram show how the new application shares the contents of more widely used databases.)
4. An E-R diagram for the whole database from which data for the application system being replaced is drawn. (Again, differences between this diagram and the third show what global database changes you have to make to implement the new application.) Even if no system is being replaced, an understanding of the existing data systems is necessary to see where the new data will fit in or if existing data structures must change to accommodate new data.

The other deliverable from conceptual data modeling is a set of entries about data objects to be stored in the project dictionary or repository. The repository is the mechanism to link data, process, and logic models of an information system. For example, there are explicit links between a data model and a data flow diagram. Some important links are briefly explained here.

◗ Data elements included in data flows also appear in the data model and vice versa. You must include in the data model any raw data captured and retained in a data store. The data model can include only data that have been captured or are computed from captured data. Because a data model is a general business picture of data, both manual and automated data stores will be included.

◗ Each data store in a process model must relate to business objects (what we call data entities) represented in the data model. For example, in Figure 5-5, the Inventory File data store must correspond to one or several data objects on a data model.

Gathering Information for Conceptual Data Modeling

Requirements determination methods must include questions and investigations that take a data focus rather than only a process and logic focus. For example, during interviews with potential system users, you must ask specific questions to gain the perspective on data needed to develop a data model. In later sections of this chapter, we introduce some specific terminology and constructs used in data modeling. Even without this specific data modeling language, you can begin to understand the kinds of questions that must be answered during requirements determination. These questions relate to understanding the rules and policies by which the area supported by the new information system operates. That is, a data model explains what the organization does and what rules govern how work is performed in the organization. You do not, however, need to know how or when data are processed or used to do data modeling.

You typically do data modeling from a combination of perspectives. The first perspective is called the *top-down approach*. It derives the data model from an intimate understanding of the nature of the business, rather than from any specific information requirements in computer displays, reports, or business forms. Table 6-1 summarizes key questions to ask system users and business managers so that you can develop an accurate and complete data model. The questions are purposely posed in business terms. Of course, technical terms do not mean much to a business manager, so you must learn how to frame your questions in business terms.

TABLE 6-1:   Questions to Ask to Develop Accurate and Complete Data Models

Category of Questions	Questions to Ask System Users and Business Managers
1. Data entities and their descriptions	What are the subjects/objects of the business? What types of people, places, things, and materials are used or interact in this business, about which data must be maintained? How many instances of each object might exist?
2. Candidate key	What unique characteristic(s) distinguishes each object from other objects of the same type? Could this distinguishing feature change over time or is it permanent? Could this characteristic of an object be missing even though we know the object exists?
3. Attributes and secondary keys	What characteristic describes each object? On what basis are objects referenced, selected, qualified, sorted, and categorized? What must we know about each object in order to run the business?
4. Security controls and understanding who really knows the meaning of data	How do you use these data? That is, are you the source of the data for the organization, do you refer to the data, do you modify them, and do you destroy them? Who is not permitted to use these data? Who is responsible for establishing legitimate values for these data?
5. Cardinality and time dimensions of data	Over what period of time are you interested in these data? Do you need historical trends, current "snapshot" values, and /or estimates or projections? If a characteristic of an object changes over time, must you know the obsolete values?
6. Relationships and their cardinality and degrees	What events occur that imply associations between various objects? What natural activities or transactions of the business involve handling data about several objects of the same or different type?
7. Integrity rules, minimum and maximum cardinality, time dimensions of data	Is each activity or event always handled the same way or are there special circumstances? Can an event occur with only some of the associated objects, or must all objects be involved? Can the associations between objects change over time (e.g., employees change departments)? Are values for data characteristics limited in any way?

FIGURE 6-4
CUSTOMER ORDER FORM USED AT
PINE VALLEY FURNITURE

PVF CUSTOMER ORDER

ORDER NO: 61384   CUSTOMER NO: 1273

NAME: Contemporary Designs
ADDRESS: 123 Oak St.
CITY-STATE-ZIP: Austin, TX 28384

ORDER DATE: 11/04/2003   PROMISED DATE: 11/21/2003

PRODUCT NO	DESCRIPTION	QUANTITY ORDERED	UNIT PRICE
M128	Bookcase	4	200.00
B381	Cabinet	2	150.00
R210	Table	1	500.00

Alternatively, you can gather the information for data modeling by reviewing specific business documents—computer displays, reports, and business forms—handled within the system. This second perspective of gaining an understanding of data is often called a *bottom-up approach*. These business documents will appear as data flows on DFDs and will show the data processed by the system, which probably are the data that must be maintained in the system's database. Consider, for example, Figure 6-4, which shows a customer order form used at Pine Valley Furniture.

From the form in Figure 6-4, we determine that the following data must be kept in the database:

ORDER NO   CUSTOMER NO
ORDER DATE   NAME
PROMISED DATE   ADDRESS
PRODUCT NO   CITY-STATE-ZIP
DESCRIPTION
QUANTITY ORDERED
UNIT PRICE

We also see that each order is from one customer, and an order can have multiple line items, each for one product. We use this kind of understanding of an organization's operation to develop data models.

Introduction to Entity-Relationship Modeling

The basic entity-relationship modeling notation uses three main constructs: data entities, relationships, and their associated attributes. Several different E-R notations exist, and many CASE tools support multiple notations. For simplicity, we have adopted one common notation for this book, the so-called crow's foot notation. If you use another notation in courses or work, you should be able to easily translate between notations.

An **entity-relationship diagram** (or **E-R diagram**) is a detailed, logical, and graphical representation of the data for an organization or business area.

Entity-relationship diagram (E-R diagram)
A detailed, logical, and graphical representation of the entities, associations, and data elements for an organization or business area.

The E-R diagram is a model of entities in the business environment, the relationships or associations among those entities, and the attributes or properties of both the entities and their relationships. A rectangle is used to represent an entity and a diamond is used to represent the relationship between two or more entities. The notation for E-R diagrams appears in Figure 6-5.

Entities

An **entity** is a person, place, object, event, or concept in the user environment about which the organization wishes to maintain data. As noted in Table 6-1, the first requirements determination question an analyst should ask concerns data entities. An entity has its own identity, which distinguishes it from every other entity. Some examples of entities follow:

- Person: EMPLOYEE, STUDENT, PATIENT
- Place: STATE, REGION, COUNTRY, BRANCH
- Object: MACHINE, BUILDING, AUTOMOBILE, PRODUCT
- Event: SALE, REGISTRATION, RENEWAL
- Concept: ACCOUNT, COURSE, WORK CENTER

There is an important distinction between entity *types* and entity *instances*. An **entity type** is a collection of entities that share common properties or characteristics. Each entity type in an E-R model is given a name. Because the name represents a set of entities, it is singular. Also, because an entity is an object, we use a simple noun to name an entity type. We use capital letters in

Phase 2:
Systems
Analysis

NET SEARCH
Investigate the origins and variations of the entity-relationship notation. Visit http:// www.prenhall.com/ valacich to complete an exercise related to this topic.

Entity
A person, place, object, event, or concept in the user environment about which the organization wishes to maintain data.

Entity type
A collection of entities that share common properties or characteristics.

Basic symbols

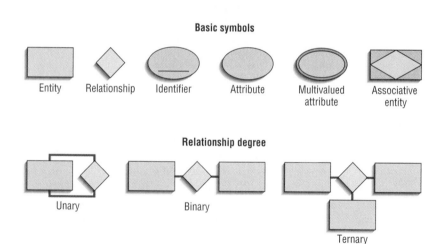

Entity   Relationship   Identifier   Attribute   Multivalued attribute   Associative entity

Relationship degree

Unary   Binary   Ternary

Relationship cardinality

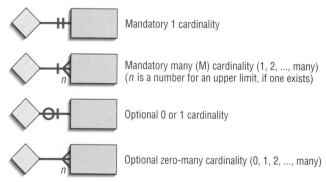

Mandatory 1 cardinality

Mandatory many (M) cardinality (1, 2, ..., many) (*n* is a number for an upper limit, if one exists)

Optional 0 or 1 cardinality

Optional zero-many cardinality (0, 1, 2, ..., many)

FIGURE 6-5
ENTITY-RELATIONSHIP DIAGRAM NOTATIONS: BASIC SYMBOLS, RELATIONSHIP DEGREES, AND RELATIONSHIP CARDINALITY
A rectangle represents an entity, and a diamond represents the relationship between two or more entities.

naming an entity type, and in an E-R diagram, the name is placed inside a rectangle representing the entity, for example:

An **entity instance** (or **instance**) is a single occurrence of an entity type. An entity type is described just once in a data model, whereas many instances of that entity type may be represented by data stored in the database. For example, there is one EMPLOYEE entity type in most organizations, but there may be hundreds (or even thousands) of instances of this entity type stored in the database.

A common mistake made in learning to draw E-R diagrams, especially if you already know how to do data flow diagramming, is to confuse data entities with sources/sinks, system outputs, or system users, and to confuse relationships with data flows. A simple rule to avoid such confusion is that a true data entity will have many possible instances, each with a distinguishing characteristic, as well as one or more other descriptive pieces of data. Consider the entity types below that might be associated with a church expense system:

In this situation, the church treasurer manages accounts and records expense transactions against each account. However, do we need to keep track of data about the treasurer and her supervision of accounts as part of this accounting system? The treasurer is the person entering data about accounts and expenses and making inquiries about account balances and expense transactions by category. Because there is only one treasurer, TREASURER data do not need to be kept. On the other hand, if each account has an account manager (e.g., a church committee chair) who is responsible for assigned accounts, then we may wish to have an ACCOUNT MANAGER entity type, with pertinent attributes as well as relationships to other entity types.

In this same situation, is an expense report an entity type? Because an expense report is computed from expense transactions and account balances, it is a data flow, not an entity type. Even though there will be multiple instances of expense reports over time, the report contents are already represented by the ACCOUNT and EXPENSE entity types.

Often when we refer to entity types in subsequent sections we simply say *entity*. This is common among data modelers. We will clarify that we mean an entity by using the term *entity instance*.

Attributes

Each entity type has a set of attributes associated with it. An **attribute** is a property or characteristic of an entity that is of interest to the organization (relationships may also have attributes, as we see in the section on relationships). Asking about attributes is the third question noted in Table 6-1. Following are some typical entity types and associated attributes:

STUDENT: Student_ID,Student_Name,Address,Phone_Number,Major
AUTOMOBILE: Vehicle_ID,Color,Weight,Horsepower
EMPLOYEE: Employee_ID,Employee_Name,Address,Skill

We use nouns with an initial capital letter followed by lowercase letters in naming an attribute. In E-R diagrams, we represent an attribute by placing its name in an ellipse with a line connecting it to the associated entity. Sometimes attributes are listed within the entity rectangle under the entity name. In many E-R drawing tools, such as Microsoft Visio, attributes are listed within the entity retangle under the entity name.

Candidate Keys and Identifiers

Every entity type must have an attribute or set of attributes that distinguishes one instance from other instances of the same type. A **candidate key** is an attribute (or combination of attributes) that uniquely identifies each instance of an entity type. A candidate key for a STUDENT entity type might be Student_ID.

Sometimes more than one attribute is required to identify a unique entity. For example, consider the entity type GAME for a basketball league. The attribute Team_Name is clearly not a candidate key, because each team plays several games. If each team plays exactly one home game against every other team, then the combination of the attributes Home_Team and Visiting_Team is a candidate key for GAME.

Some entities may have more than one candidate key. One candidate key for EMPLOYEE is Employee_ID; a second is the combination of Employee_Name and Address (assuming that no two employees with the same name live at the same address). If there is more than one candidate key, the designer must choose one of the candidate keys as the identifier. An **identifier** is a candidate key that has been selected to be used as the unique characteristic for an entity type.

Identifiers should be selected carefully because they are critical for the integrity of data. You should apply the following identifier selection rules:

1. Choose a candidate key that will not change its value over the life of each instance of the entity type. For example, the combination of Employee_Name and Address would probably be a poor choice as a primary key for EMPLOYEE because the values of Employee_Name and Address could easily change during an employee's term of employment.

2. Choose a candidate key such that, for each instance of the entity, the attribute is guaranteed to have valid values and not be null. To ensure valid values, you may have to include special controls in data entry and maintenance routines to eliminate the possibility of errors. If the candidate key is a combination of two or more attributes, make sure that all parts of the key have valid values.

3. Avoid the use of so-called intelligent keys, whose structure indicates classifications, locations, and other entity properties. For example, the first two digits of a key for a PART entity may indicate the warehouse location. Such codes are often modified as conditions change, which renders the primary key values invalid.

4. Consider substituting single-attribute surrogate keys for large composite keys. For example, an attribute called Game_ID could be used for the entity GAME instead of the combination of Home_Team and Visiting_Team.

Candidate key
An attribute (or combination of attributes) that uniquely identifies each instance of an entity type.

Identifier
A candidate key that has been selected as the unique, identifying characteristic for an entity type.

For each entity, the name of the identifier is underlined on an E-R diagram. The following diagram shows the representation for a STUDENT entity type using E-R notation:

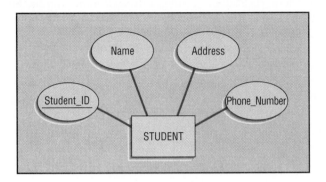

The equivalent representation using Microsoft Visio is the following:

In the Visio notation, the primary key is listed immediately below the entity name with the notation PK and the primary key is underlined. All required attributes (that is, an instance of STUDENT must have values for Student_ID and Name) are in bold.

Multivalued Attributes

Multivalued attribute
An attribute that may take on more than one value for each entity instance.

A **multivalued attribute** may take on more than one value for each entity instance. Suppose that Dep_Name (dependent name) is one of the attributes of EMPLOYEE. If each employee can have more than one dependent, Dep_Name is a multivalued attribute. During conceptual design, there are two common special symbols or notations to highlight multivalued attributes. The first is to use a double-line ellipse, so that the EMPLOYEE entity with its attributes is diagrammed as follows:

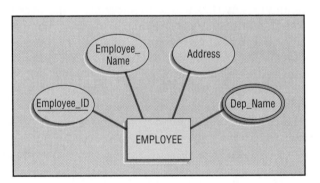

Many E-R drawing tools, such as Microsoft Visio, do not support multivalued attributes within an entity. Thus, a second approach is to separate the repeating data into another entity, called a *weak* (or *attributive*) entity, and then using a relationship (relationships are discussed in the next section), link the weak entity to its associated regular entity. The approach also easily handles several attributes that repeat together, called a **repeating group.** For example, dependent name, age, and relation to employee (spouse, child, parent, etc.) are multivalued attributes about an employee, and these attributes repeat together. We

Repeating group
A set of two or more multivalued attributes that are logically related.

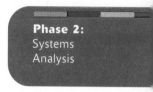

can show this using an attributive entity, DEPENDENT, and a relationship, shown here simply by a line between DEPENDENT and EMPLOYEE. The crow's foot next to DEPENDENT means that there may be many DEPENDENTs for the same EMPLOYEE. Some E-R notations and CASE tools use a special symbol to signify a weak entity. Common notations are a double-line border on the entity box or a mark on the relationship line.

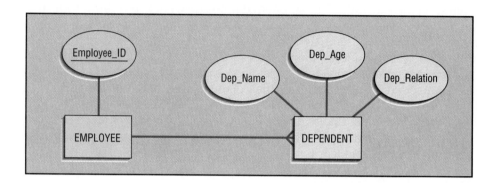

Relationships

Relationships are the glue that hold together the various components of an E-R model. In Table 6-1 questions 5, 6, and 7 deal with relationships. A **relationship** is an association between the instances of one or more entity types that is of interest to the organization. An association usually means that an event has occurred or that some natural linkage exists between entity instances. For this reason, relationships are labeled with verb phrases. For example, a training department in a company is interested in tracking which training courses each of its employees has completed. This leads to a relationship (called Completes) between the EMPLOYEE and COURSE entity types that we diagram as follows:

Relationship
An association between the instances of one or more entity types that is of interest to the organization.

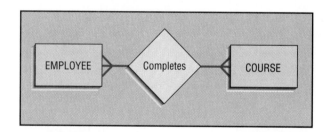

As indicated by the lines, this is a many-to-many relationship: Each employee may complete more than one course, and each course may be completed by more than one employee. More significantly, we can use the Completes relationship to determine the specific courses that a given employee has completed. Conversely, we can determine the identity of each employee who has completed a particular course.

To avoid cluttering an E-R diagram with excess symbols, many organizations and tools (for example, Microsoft Visio) will not include the relationship diamond and simply place the verb phrase for the relationship name near the line. Figures 6-3(A) and (B), for example, do not include relationship diamonds. Other tools, like the E-R diagrammer in Microsoft Access, do not even include the relationship names. We always use the relationship name in this book, and we sometimes use two verb phrases so that there is an explicit name for the relationship in each direction. Your organization will determine the standards you follow.

Conceptual Data Modeling and the E-R Model

The last section introduced the fundamentals of the E-R data modeling notation—entities, attributes, and relationships. The goal of conceptual data modeling is to capture as much of the meaning of data as possible. The more details (or what some systems analysts call *business rules*) about data that we can model, the better the system we can design and build. Further, if we can include all these details in an automated repository, such as a CASE tool, and if a CASE tool can generate code for data definitions and programs, then the more we know about data, the more code can be generated automatically. This will make system building more accurate and faster. More importantly, if we can keep a thorough repository of data descriptions, we can regenerate the system as needed as the business rules change. Because maintenance is the largest expense with any information system, the efficiencies gained by maintaining systems at the rule, rather than code, level drastically reduce the cost.

In this section, we explore more advanced concepts needed to more thoroughly model data and learn how the E-R notation represents these concepts.

Degree of a Relationship

The **degree** of a relationship, question 6 in Table 6-1, is the number of entity types that participate in that relationship. Thus, the relationship Completes illustrated previously is of degree two because there are two entity types: EMPLOYEE and COURSE. The three most common relationships in E-R diagrams are unary (degree one), binary (degree two), and ternary (degree three). Higher-degree relationships are possible, but they are rarely encountered in practice, so we restrict our discussion to these three cases. Examples of unary, binary, and ternary relationships appear in Figure 6-6.

Degree
The number of entity types that participate in a relationship.

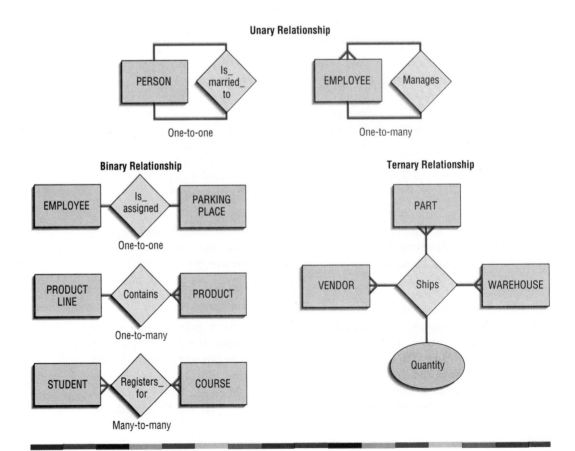

FIGURE 6-6
EXAMPLES OF THE THREE MOST COMMON RELATIONSHIPS IN E-R DIAGRAMS: UNARY, BINARY, AND TERNARY

Unary Relationship Also called a recursive relationship, a **unary relationship** is a relationship between the instances of one entity type. Two examples are shown in Figure 6-6. In the first example, Is_married_to is shown as a one-to-one relationship between instances of the PERSON entity type. That is, each person may be currently married to one other person. In the second example, Manages is shown as a one-to-many relationship between instances of the EMPLOYEE entity type. Using this relationship, we could identify (for example) the employees who report to a particular manager or, reading the Manages relationship in the opposite direction, who the manager is for a given employee.

Unary relationship
   (recursive
   relationship)
A relationship between the
instances of one entity type.

Binary Relationship A **binary relationship** is a relationship between instances of two entity types and is the most common type of relationship encountered in data modeling. Figure 6-6 shows three examples. The first (one-to-one) indicates that an employee is assigned one parking place, and each parking place is assigned to one employee. The second (one-to-many) indicates that a product line may contain several products, and each product belongs to only one product line. The third (many-to-many) shows that a student may register for more than one course, and that each course may have many student registrants.

Binary relationship
A relationship between instances
of two entity types.

Ternary Relationship A **ternary relationship** is a simultaneous relationship among instances of three entity types. In the example shown in Figure 6-6, the relationship Ships tracks the quantity of a given part that is shipped by a particular vendor to a selected warehouse. Each entity may be a one or a many participant in a ternary relationship (in Figure 6-6, all three entities are many participants).

Ternary relationship
A simultaneous relationship
among instances of three entity
types.

Note that a ternary relationship is not the same as three binary relationships. For example, Quantity is an attribute of the Ships relationship in Figure 6-6. Quantity cannot be properly associated with any of the three possible binary relationships among the three entity types (such as that between PART and VENDOR) because Quantity is the amount of a particular PART shipped from a particular VENDOR to a particular WAREHOUSE.

Cardinalities in Relationships

Suppose that two entity types, A and B, are connected by a relationship. The **cardinality** of a relationship (see the fifth, sixth, and seventh questions in Table 6-1) is the number of instances of entity B that can (or must) be associated with each instance of entity A. For example, consider the following relationship for videotapes and movies:

Cardinality
The number of instances of
entity B that can (or must) be
associated with each instance of
entity A.

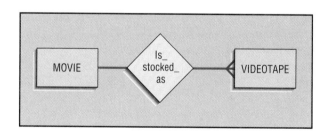

Clearly, a video store may stock more than one videotape of a given movie. In the terminology we have used so far, this example is intuitively a "many" relationship. Yet it is also true that the store may not have a single tape of a particular movie in stock. We need a more precise notation to indicate the range of cardinalities for a relationship. This notation of relationship cardinality was introduced in Figure 6-5, which you may want to review at this point.

NET SEARCH
Investigate the concept of business rules (cardinality is one type of business rule). Visit http://www.prenhall. com/valacich to complete an exercise related to this topic.

Minimum and Maximum Cardinalities The minimum cardinality of a relationship is the minimum number of instances of entity B that may be associated with each instance of entity A. In the preceding example, the minimum number of videotapes available for a movie is zero, in which case we say that VIDEOTAPE is an optional participant in the Is_stocked_as relationship. When the minimum cardinality of a relationship is one, then we say entity B is a mandatory participant in the relationship. The maximum cardinality is the maximum number of instances. For our example, this maximum is "many" (an unspecified number greater than one). Using the notation from Figure 6-5, we diagram this relationship as follows:

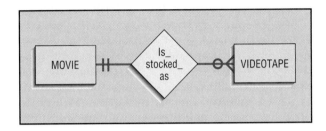

The zero through the line near the VIDEOTAPE entity means a minimum cardinality of zero, whereas the crow's foot notation means a "many" maximum cardinality. It is possible for the maximum cardinality to be a fixed number, not an arbitrary "many" value. For example, see the Supplies relationship in Figure 6-3, which indicates that there are at most four suppliers for each item.

Associative Entities As seen in the examples of the Ships ternary relationship in Figure 6-6, attributes may be associated with a many-to-many relationship as well as with an entity. For example, suppose that the organization wishes to record the date (month and year) when an employee completes each course. Some sample data follow:

Employee_ID	Course_Name	Date_Completed
549–23–1948	Basic Algebra	March 2003
629–16–8407	Software Quality	June 2003
816–30–0458	Software Quality	Feb 2003
549–23–1948	C Programming	May 2003

From this limited data you can conclude that the attribute Date_Completed is not a property of the entity EMPLOYEE (because a given employee such as 549–23–1948 has completed courses on different dates). Nor is Date_Completed a property of COURSE because a particular course (such as Software Quality) may be completed on different dates. Instead, Date_Completed is a property of the relationship between EMPLOYEE and COURSE. The attribute is associated with the relationship and diagrammed as follows:

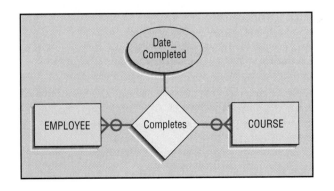

Phase 2:
Systems
Analysis

FIGURE 6-7
EXAMPLE OF AN ASSOCIATIVE
ENTITY

Associative entity
An entity type that associates
the instances of one or more
entity types and contains
attributes that are peculiar to
the relationship between those
entity instances.

Because many-to-many and one-to-one relationships may have associated attributes, the E-R diagram poses an interesting dilemma: Is a many-to-many relationship actually an entity in disguise? Often the distinction between entity and relationship is simply a matter of how you view the data. An **associative entity** is a relationship that the data modeler chooses to model as an entity type. Figure 6-7 shows the E-R notation for representing the Completes relationship as an associative entity. The diamond symbol is included within the entity rectangle as a reminder that the entity was derived from a relationship (some organizations and tools do not include the diamond but rather use the standard entity symbol, a rectangle, for an associative entity). The lines from CERTIFICATE to the two entities are not two separate binary relationships, so they do not have labels. Note that EMPLOYEE and COURSE have mandatory one cardinality because an instance of Completes must have an associated EMPLOYEE and COURSE. The implicit identifier of Completes is the combination of the identifiers of EMPLOYEE and COURSE, Employee_ID, and Course_Name, respectively.

E_R drawing tools that do not support many-to-many relationships require that any such relationship be converted into an associative entity, whether it has attributes or not. You have already seen an example of this in Figure 6-3 for Microsoft Visio, in which the Supplies/Supplied by relationship from Figure 6-3(A) was converted in Figure 6-3(B) into the SUPPLIED ITEM entity (actually, associative entity) and two mandatory one-to-many relationships.

One situation in which a relationship must be turned into an associative entity is when the associative entity has other relationships with entities besides the relationship that caused its creation. For example, consider the following E-R model, which represents price quotes from different vendors for purchased parts stocked by Pine Valley Furniture:

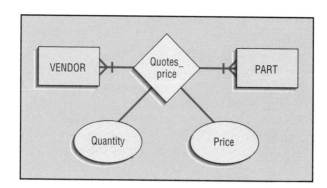

Now, suppose that we also need to know which price quote is in effect for each part shipment received. This additional data requirement necessitates that the Quotes_price relationship be transformed into an associative entity. This new relationship is represented in Figure 6-8.

In this case, PRICE QUOTE is not a ternary relationship. Rather, PRICE QUOTE is a binary many-to-many relationship (associative entity) between

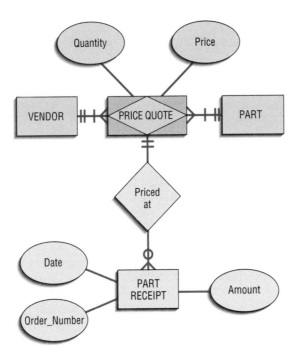

FIGURE 6-8
AN E-R MODEL THAT REPRESENTS
EACH PRICE QUOTE FOR EACH
PART SHIPMENT RECEIVED BY PINE
VALLEY FURNITURE

VENDOR and PART. In addition, each PART RECEIPT, based on Amount, has an applicable, negotiated Price. Each PART RECEIPT is for a given PART from a specific VENDOR, and the Amount of the receipt dictates the purchase price in effect by matching with the Quantity attribute. Because the PRICE QUOTE pertains to a given PART and given VENDOR, PART RECEIPT does not need direct relationships with these entities.

An Example of Conceptual Data Modeling at Hoosier Burger

Chapter 5 structured the process and data flow requirements for a food ordering system for Hoosier Burger. Figure 6-9 describes requirements for a new system using Microsoft Visio. The purpose of this system is to monitor and report changes in raw material inventory levels and to issue material orders and payments to suppliers. Thus, the central data entity for this system will be an INVENTORY ITEM, shown in Figure 6-10, corresponding to data store D1 in Figure 6-9.

Changes in inventory levels are due to two types of transactions: receipt of new items from suppliers and consumption of items from sales of products. Inventory is added upon receipt of new raw materials, for which Hoosier Burger receives a supplier INVOICE (see Process 1.0 in Figure 6-9). Figure 6-10 shows that each INVOICE indicates that the supplier has sent a specific quantity of one or more INVOICE ITEMs, which correspond to Hoosier's INVENTORY ITEMs. Inventory is used when customers order and pay for PRODUCTs. That is, Hoosier makes a SALE for one or more ITEM SALEs, each of which corresponds to a food PRODUCT. Because the real-time customer order processing system is separate from the inventory control system, a source, STOCK-ON-HAND in Figure 6-9, represents how data flow from the order processing to the inventory control system. Finally, because food PRODUCTs are made up of various INVENTORY ITEMs (and vice versa), Hoosier maintains a RECIPE to indicate how much of each INVENTORY ITEM goes into making one PRODUCT. From this discussion, we have identified the data entities required in a data model for the new Hoosier Burger inventory control system:

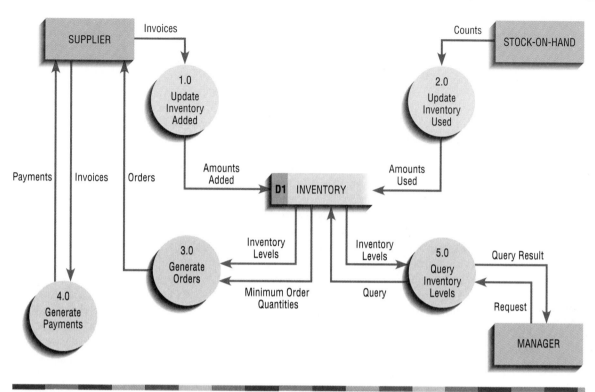

FIGURE 6-9
LEVEL-0 DATA FLOW DIAGRAM FOR HOOSIER BURGER'S NEW LOGICAL INVENTORY CONTROL SYSTEM

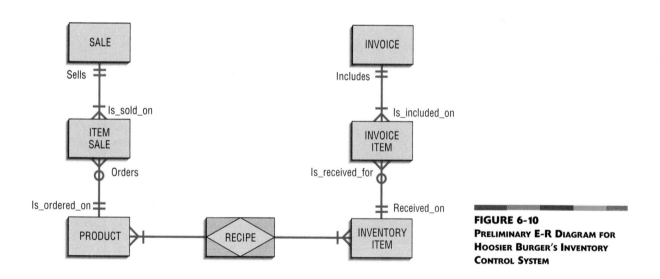

FIGURE 6-10
PRELIMINARY E-R DIAGRAM FOR HOOSIER BURGER'S INVENTORY CONTROL SYSTEM

INVENTORY ITEM, INVOICE, INVOICE ITEM, PRODUCT, SALE, ITEM SALE, and RECIPE. To complete the E-R diagram, we must determine necessary relationships between these entities as well as attributes for each entity.

The wording in the previous description tells us much of what we need to know to determine relationships:

- An INVOICE includes one or more INVOICE ITEMs, each of which corresponds to an INVENTORY ITEM. Obviously, an INVOICE ITEM cannot exist without an associated INVOICE, and over time there will be zero-to-many receipts, or INVOICE ITEMs, for an INVENTORY ITEM.

◐ Each PRODUCT has a RECIPE of INVENTORY ITEMs. Thus, RECIPE is an associative entity between PRODUCT and INVENTORY ITEM.

◐ A SALE indicates that Hoosier sells one or more ITEM SALEs, each of which corresponds to a PRODUCT. An ITEM SALE cannot exist without an associated SALE, and over time there will be zero-to-many ITEM SALEs for a PRODUCT.

Figure 6-10 shows an E-R diagram with the entities and relationships described above. We include on this diagram two labels for each relationship, one to be read in either relationship direction (e.g., an INVOICE Includes one-to-many INVOICE ITEMs, and an INVOICE ITEM Is_included_on exactly one INVOICE). RECIPE, because it is an associative entity, also serves as the label for the many-to-many relationship between PRODUCT and INVENTORY ITEM. Now that we understand the entities and relationships, we must decide which data elements are associated with the entities and associative entities in this diagram.

You may wonder at this point why only the INVENTORY data store is shown in Figure 6-9 when seven entities and associative entities are on the E-R diagram. The INVENTORY data store corresponds to the INVENTORY ITEM entity in Figure 6-10. The other entities are hidden inside other processes for which we have not shown lower-level diagrams. In actual requirements structuring steps, you would have to match all entities with data stores: Each data store represents some subset of an E-R diagram, and each entity is included in one or more data stores. Ideally, each data store on a primitive DFD will be an individual entity.

To determine data elements for an entity, we investigate data flows in and out of data stores that correspond to the data entity, and supplement this with a study of decision logic that uses or changes data about the entity. Six data flows are associated with the INVENTORY data store in Figure 6-9. The description of each data flow in the project dictionary or repository would include the data flow's composition, which then tells us what data are flowing in or out of the data store. For example, the Amounts Used data flow coming from Process 2.0 indicates how much to decrease an attribute Quantity_in_Stock due to use of the INVENTORY ITEM to fulfill a customer sale. Thus, the Amounts Used data flow implies that Process 2.0 will first read the relevant INVENTORY ITEM record, then update its Quantity_in_Stock attribute, and finally store the updated value in the record. Structured English for Process 2.0 would depict this logic. Each data flow would be analyzed similarly (space does not permit us to show the analysis for each data flow).

After having considered all data flows in and out of data stores related to data entities, plus all decision logic related to inventory control, we derive the full E-R diagram, with attributes, shown in Figure 6-11(A). A Microsoft Visio version of the Hoosier Burger inventory control system database appears in Figure 6-11(B). In Visio, the ITEM SALE, RECIPE, and INVOICE ITEM entities participate in what are called identifying relationships. Thus, Visio treats all of them as associative entities, not just the RECIPE entity. Visio automatically includes the primary keys of the identifying entities as primary keys in the identified (associative) entities. In fact, ITEM SALE and INVOICE ITEM could have been represented as associative entities in Figure 6-11(A) because they arise from many-to-many relationships with their identifying entities, they have mandatory relationships with their identifying entities, and they do not have primary keys of their own. Also note that in Visio, because it cannot represent many-to-many relationships, there are two mandatory relationships on either side of RECIPE. We have not included relationship names in the case to emphasize that RECIPE was represented as an associative entity in Figure 6-11(A).

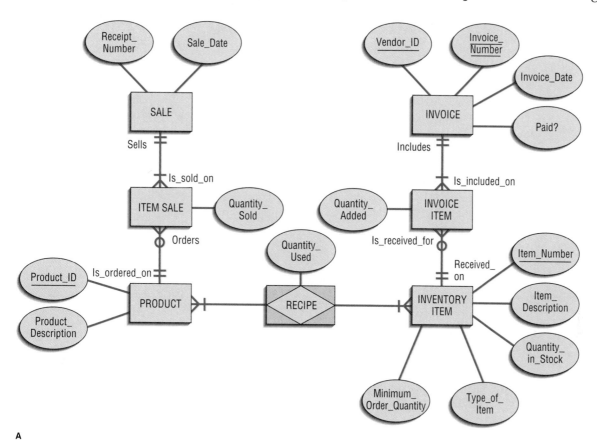

A

B

FIGURE 6-11
FINAL E-R DIAGRAM FOR HOOSIER BURGER'S INVENTORY
CONTROL SYSTEM
(A) STANDARD E-R NOTATION
(B) VISIO NOTATION

ELECTRONIC COMMERCE APPLICATION: CONCEPTUAL DATA MODELING

Conceptual data modeling for an Internet-based electronic commerce application is no different than the process followed when analyzing the data needs for other types of applications. In the last chapter, you read how Jim Woo analyzed the flow of information within the WebStore and developed a data flow diagram. In this section, we examine the process he followed when developing the WebStore's conceptual data model.

Conceptual Data Modeling for Pine Valley Furniture's WebStore

To better understand what data would be needed within the WebStore, Jim Woo carefully reviewed the information from the JAD session and his previously developed data flow diagram. Table 6-2 summarizes the customer and inventory information identified during the JAD session. Jim wasn't sure if this information was complete but knew that it was a good starting place for identifying what information the WebStore needed to capture, store, and process. To identify additional information, he carefully studied the level-0 DFD shown in Figure 6-12. In this diagram, two data stores—Inventory and Shopping Cart—are clearly identified; both were strong candidates to become entities within the conceptual data model. Finally, Jim examined the data flows from the DFD as additional possible sources for entities. Hence, he identified five general categories of information to consider:

- Customer
- Inventory
- Order
- Shopping Cart
- Temporary User/System Messages

After identifying these multiple categories of data, his next step was to define each item carefully. To do this, he again examined all data flows within the DFD and recorded each one's source and destination. By carefully listing these flows, he could move more easily through the DFD and understand more thoroughly what information needed to move from point to point. This activity resulted in the creation of two tables that documented Jim's growing understanding of the WebStore's requirements. The first, Table 6-3, lists each of the data flows within each data category and its corresponding description. The second, Table 6-4, lists each of the unique data flows within each data category. Jim then felt ready to construct an entity-relationship diagram for the WebStore.

He concluded that Customer, Inventory, and Order were all unique entities and would be part of his E-R diagram. Recall that an entity is a person, place, or object; all three of these items meet this criteria. Because the Temporary User/System Messages data were not permanently stored items—nor were they a person, place, or object—he concluded that this should not be an

TABLE 6-2: Customer and Inventory Information for WebStore

Corporate Customer	Home Office Customer	Student Customer	Inventory Information
Company name	Name	Name	SKU
Company address	Doing business as (company's name)	School	Name
Company phone		Address	Description
Company fax	Address	Phone	Finished product size
Company preferred shipping method	Phone	E-mail	Finished product weight
Buyer name	Fax		Available materials
Buyer phone	E-mail		Available colors
Buyer e-mail			Price
			Lead time

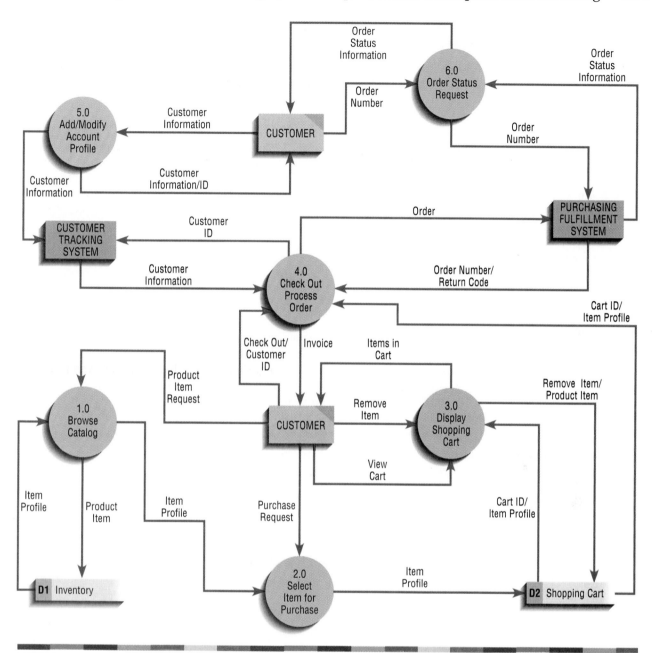

FIGURE 6-12
LEVEL-0 DATA FLOW DIAGRAM FOR THE WEBSTORE

entity in the conceptual data model. Alternatively, although the shopping cart was also a temporarily stored item, its contents needed to be stored for at least the duration of a customer's visit to the WebStore and should be considered an object. As shown in Figure 6-12, Process 4, Check Out Process Order, moves the Shopping Cart contents to the Purchasing Fulfillment System, where the order details are stored. Thus, he concluded that Shopping Cart—along with Customer, Inventory, and Order—would be entities in his E-R diagram.

The final step was to identify the interrelationships between these four entities. After carefully studying all the related information, he concluded the following:

1. Each Customer <u>owns</u> *zero-to-many* Shopping Cart Instances; each Shopping Cart Instance <u>is-owned-by</u> *one-and-only-one* Customer.

TABLE 6-3: Data Category, Data Flow, and Data Flow Descriptions for the WebStore DFD

Data Category

Data Flow	Description
Customer Related	
Customer ID	Unique identifier for each customer (generated by Customer Tracking System)
Customer Information	Detailed customer information (stored in Customer Tracking System)
Inventory Related	
Product Item	Unique identifier for each product item (stored in Inventory Database)
Item Profile	Detailed product information (stored in Inventory Database)
Order Related	
Order Number	Unique identifier for an order (generated by Purchasing Fulfillment System)
Order	Detailed order information (stored in Purchasing Fulfillment System)
Return Code	Unique code for processing customer returns (generated by/stored in Purchasing Fulfillment System)
Invoice	Detailed order summary statement (generated from order information stored in Purchasing Fulfillment System)
Order Status Information	Detailed summary information on order status (stored/generated by Purchasing Fulfillment System)
Shopping Cart	
Cart ID	Unique identifier for shopping cart
Temporary User/System Messages	
Product Item Request	Request to view information on a catalog item
Purchase Request	Request to move an item into the shopping cart
View Cart	Request to view the contents of the shopping cart
Items in Cart	Summary report of all shopping cart items
Remove Item	Request to remove item from shopping cart
Check Out	Request to check out and process order

2. Each Shopping Cart Instance <u>contains</u> *one-and-only-one* Inventory item; each Inventory item <u>is-contained-in</u> *zero-to-many* Shopping Cart Instances.

3. Each Customer <u>places</u> *zero-to-many* Orders; each Order <u>is-placed-by</u> *one-and-only-one* Customer.

4. Each Order <u>contains</u> *one-to-many* Shopping Cart Instances; each Shopping Cart Instance <u>is-contained-in</u> *one-and-only-one* Order.

With these relationships defined, Jim drew the E-R diagram shown in Figure 6-13. He had a very good understanding of the requirements, the flow of information within the WebStore, the flow of information between the WebStore and existing PVF systems, and now the conceptual data model. Over the next few hours, Jim planned to refine his understanding further by listing the specific attributes for each entity and then compare these lists with the existing inventory, customer, and order database tables. He had to make sure that all attributes were accounted for before beginning the process of selecting a final design strategy.

TABLE 6-4:   Data Category, Data Flow, and the Source/Destination
of Data Flows within the WebStore DFD

Data Category

Data Flow	**From/To**
Customer Related	
Customer ID	From Customer to Process 4.0
	From Process 4.0 to Customer Tracking System
	From Process 5.0 to Customer
Customer Information	From Customer to Process 5.0
	From Process 5.0 to Customer
	From Process 5.0 to Customer Tracking System
	From Customer Tracking System to Process 4.0
Inventory Related	
Product Item	From Process 1.0 to Data Store D1
	From Process 3.0 to Data Store D2
Item Profile	From Data Store D1 to Process 1.0
	From Process 1.0 to Process 2.0
	From Process 2.0 to Data Store D2
	From Data Store D2 to Process 3.0
	From Data Store D2 to Process 4.0
Order Related	
Order Number	From Purchasing Fulfillment System to Process 4.0
	From Customer to Process 6.0
	From Process 6.0 to Purchasing Fulfillment System
Order	From Process 4.0 to Purchasing Fulfillment System
Return Code	From Purchasing Fulfillment System to Process 4.0
Invoice	From Process 4.0 to Customer
Order Status Information	From Process 6.0 to Customer
	From Purchasing Fulfillment System to Process 6.0
Shopping Cart	
Cart ID	From Data Store D2 to Process 3.0
	From Data Store D2 to Process 4.0
Temporary User/System Messages	
Product Item Request	From Customer to Process 1.0
Purchase Request	From Customer to Process 2.0
View Cart	From Customer to Process 3.0
Items in Cart	From Process 3.0 to Customer
Remove Item	From Customer to Process 3.0
	From Process 3.0 to Data Store D2
Check Out	From Customer to Process 4.0

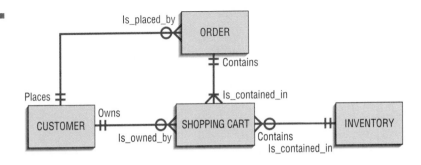

FIGURE 6-13
ENTITY-RELATIONSHIP DIAGRAM
FOR THE WEBSTORE SYSTEM

Key Points Review

1. **Concisely define each of the following key data modeling terms:** *conceptual data model, entity-relationship diagram, entity type, entity instance, attribute, candidate key, multivalued attribute, relationship, degree, cardinality,* **and** *associative entity.*

    A conceptual data model represents the overall structure of organizational data, independent of any database technology. An E-R diagram is a detailed representation of the entities, associations, and attributes for an organization or business area. An entity type is a collection of entity instances that share common properties or characteristics. An attribute is a named property or characteristic of an entity. One or a combination of attributes of each entity instance that uniquely identifies that instance is called a candidate key. A multivalued attribute may take on more than one value for an entity instance. A relationship is an association between the instances of one or more entity types, and the number of entity types participating in a relationship is the degree of the relationship. Cardinality is the number of entity instances associated with a given entity instance. Data that are simultaneously associated with several entity instances are stored in an associative entity.

2. **Ask the right kinds of questions to determine data requirements for an information system.**

    Information is gathered for conceptual data modeling as part of each phase of the systems development life cycle. You must ask questions in business, rather than data modeling, terms so that business managers can explain the nature of the business; the systems analyst represents the objects and events of the business through a data model. Questions include what are the objects of the business, what uniquely characterizes each object, what characteristics describe each object, how are data used, what history of data must be retained, what events occur that relate different kinds of data, and are there special data handling procedures? (See Table 6-1 for details.)

3. **Draw an entity-relationship (E-R) diagram to represent common business situations.**

    An E-R diagram uses symbols for entity, relationship, identifier, attribute, multivalued attribute, and associative entity and shows the degree and cardinality of relationships (see Figure 6-5 for all the symbols discussed in this chapter and Figures 6-3 and 6-11 for example diagrams). Exercises at the end of this chapter give you practice at drawing E-R diagrams.

4. **Explain the role of conceptual data modeling in the overall analysis and design of an information system.**

    Conceptual data modeling occurs in parallel with other requirements analysis and structuring steps during systems development. Information for conceptual data modeling is collected during interviews, from questionnaires, and in JAD sessions. Conceptual data models may be developed for a new information system and the system it is replacing, as well as the whole database for current and new systems. A conceptual data model is useful input to subsequent data-oriented steps in the analysis, design, and implementation phases of systems development where logical data models, physical file designs, and database file coding are done.

5. **Distinguish between unary, binary, and ternary relationships, and give an example of each.**

    A unary relationship is between instances of the same entity type (e.g., Is_married_to relates different instances of a PERSON entity type). A binary relationship is between instances of two entity types (e,g., Registers_for relates instances of STUDENT and COURSE entity types). A ternary relationship is a simultaneous association between instances of three entity types (e.g., Ships relates instances of PART, VENDOR, and WAREHOUSE entity types).

6. **Distinguish between a relationship and an associative entity, and use associative entities in a data model when appropriate.**

   Sometimes many-to-many and one-to-one relationships have associated attributes. When this occurs it is best to change the relationship into an associative entity. For example, if we needed to know the date an employee completed a course, Date_Completed is neither an attribute of EMPLOYEE nor COURSE but of the relationship between these entities. In this case, we would create a CERTIFICATE associative entity (see Figure 6-7), associate Date_Completed with CERTIFICATE, and draw mandatory one-to-many relationships from CERTIFICATE to each of EMPLOYEE and COURSE. An associative entity, like any entity, then may be related to other entities, as shown in Figure 6-8.

7. **Relate data modeling to process and logic modeling as different ways of describing an information system.**

   Process and logic modeling represent the movement and use of data, whereas data modeling represents the meaning and structure of data. A data model is usually a more permanent representation of the data requirements of an organization than are models of data flow and use. There must, however, be consistency between these models of different views of an information system. For example, all the data in an E-R diagram for an information system must be in data stores on associated data flow diagrams.

Key Terms Checkpoint

Here are the key terms from the chapter. The page where each term is first explained is in parentheses after the term.

1. **Associative entity (p. 215)**
2. **Attribute (p. 208)**
3. **Binary relationship (p. 213)**
4. **Candidate key (p. 209)**
5. **Cardinality (p. 213)**
6. **Conceptual data model (p. 200)**
7. **Degree (p. 212)**
8. **Entity (p. 207)**
9. **Entity instance (instance) (p. 208)**
10. **Entity-relationship diagram (E-R diagram) (p. 206)**
11. **Entity type (p. 207)**
12. **Identifier (p. 209)**
13. **Multivalued attribute (p. 210)**
14. **Relationship (p. 211)**
15. **Repeating group (p. 210)**
16. **Ternary relationship (p. 213)**
17. **Unary relationship (recursive relationship) (p. 213)**

Match each of the key terms above with the definition that best fits it.

_____ 1. A detailed, logical, and graphical representation of the entities, associations, and data elements for an organization or business area.

_____ 2. A single occurrence of an entity type.

_____ 3. An attribute that may take on more than one value for each entity instance.

_____ 4. A simultaneous relationship among instances of three entity types.

_____ 5. A collection of entities that share common properties or characteristics.

_____ 6. A relationship between instances of two entity types.

_____ 7. An entity type that associates the instances of one or more entity types and contains attributes that are peculiar to the relationship between those entity instances.

_____ 8. A named property or characteristic of an entity that is of interest to the organization.

_____ 9. The number of instances of entity B that can (or must) be associated with each instance of entity A.

_____ 10. A candidate key that has been selected as the unique, identifying characteristic for an entity type.

_____ 11. An association between the instances of one or more entity types that is of interest to the organization.

_____ 12. An attribute (or combination of attributes) that uniquely identifies each instance of an entity type.

_____ 13. The number of entity types that participate in a relationship.

_____ 14. A relationship between the instances of one entity type.

_____ 15. A detailed model that shows the overall structure of organizational data while being independent of any database management system or other implementation considerations.

_____ 16. A set of two or more multivalued attributes that are logically related.

_____ 17. A person, place, object, event, or concept in the user environment about which the organization wishes to maintain data.

Review Questions

1. What characteristics of data are represented in an E-R diagram?
2. What elements of a data flow diagram should be analyzed as part of data modeling?
3. Explain why a ternary relationship is not the same as three binary relationships.
4. When must a many-to-many relationship be modeled as an associative entity?
5. Which of the following types of relationships can have attributes associated with them: one-to-one, one-to-many, many-to-many?
6. What is the degree of a relationship? Give an example of each of the relationship degrees illustrated in this chapter.
7. Give an example of a ternary relationship (different from any example in this chapter).
8. List the deliverables from the conceptual data modeling.
9. Explain the relationship between minimum cardinality and optional and mandatory participation.
10. List the ideal characteristics of an entity identifier attribute.
11. List the four types of E-R diagrams produced and analyzed during conceptual data modeling.
12. What notation is used on an E-R diagram to show a lower-bound or upper-bound limit on the "many" side of a one-to-many relationship?
13. Explain the difference between a candidate key and the identifier of an entity type.
14. What distinguishes a repeating group from a simple multivalued attribute?

Problems and Exercises

1. Assume that at Pine Valley Furniture each product (described by Product No., Description, and Cost) is comprised of at least three components (described by Component No., Description, and Unit of Measure) and components are used to make one or many products (i.e., must be used in at least one product). In addition, assume that components are used to make other components and that raw materials are also considered to be components. In both cases of components being used to make products and components being used to make other components, we need to keep track of how many components go into making something else. Draw an E-R diagram for this situation and place minimum and maximum cardinalities on the diagram.
2. A software training program is divided into training modules, and each module is described by module name and the approximate practice time. Each module sometimes has prerequisite modules. Model this situation of training programs and modules with an E-R diagram.
3. Each semester, each student must be assigned an adviser who counsels students about degree requirements and helps students register for classes. Students must register for classes with the help of an adviser, but if their assigned adviser is not available, they may register with any adviser. We must keep track of students, their assigned adviser, and with whom the student registered for the current term. Represent this situation of students and advisers with an E-R diagram.

4. Consider the E-R diagram in Figure 6-7.

   a. What is the identifier for the CERTIFICATE associative entity?
   b. Now, assume that the same employee may take the same course multiple times, on different dates. Does this change your answer to part a? Why or why not?
   c. Now, assume we do know the instructor who issues each certificate to each employee for each course. Include this new entity in Figure 6-7 and relate it to the other entities. How did you choose to relate INSTRUCTOR to CERTIFICATE and why?

5. Study the E-R diagram of Figure 6-14. Based on this E-R diagram, answer the following questions:

   a. How many PROJECTs can an employee work on?
   b. What is the degree of the Includes relationship?
   c. Are there any associative entities on this diagram? If so, name them.
   d. How else could the attribute Skill be modeled?
   e. Is it possible to attach any attributes to the Includes relationship?
   f. Could TASK be modeled as an associative entity?

6. An airline reservation is an association between a passenger, a flight, and a seat. Select a few pertinent attributes for each of these entity types and represent a reservation in an E-R diagram.
7. Consider the E-R diagram in Figure 6-15. Are all three relationships—Holds, Goes_on, and

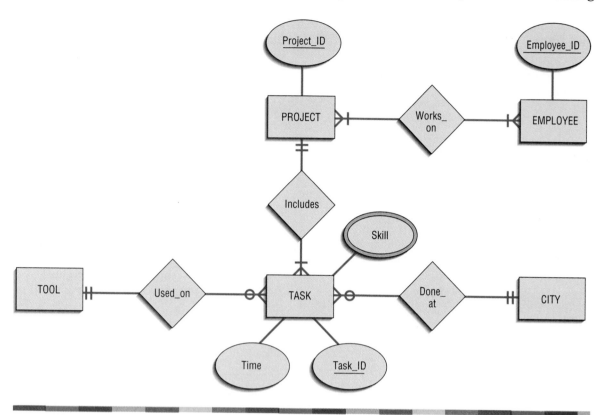

FIGURE 6-14
E-R Diagram for Problem and Exercise 5

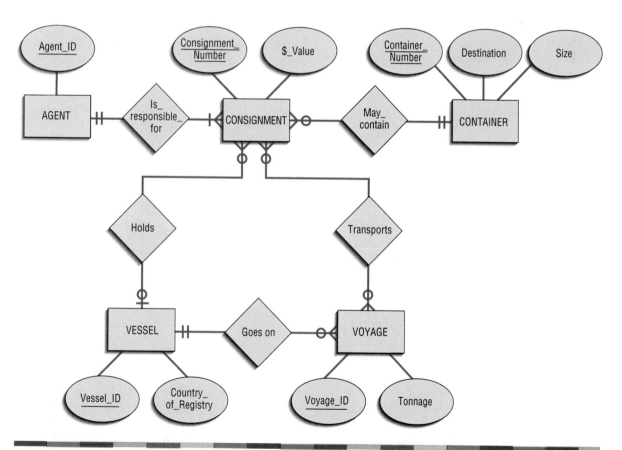

FIGURE 6-15
E-R Diagram for Problem and Exercise 7

Transports—necessary (i.e., can one of these be deduced from the other two)? Are there reasonable assumptions that make all three relationships necessary? If so, what are these assumptions?

8. Draw an E-R diagram to represent the sample customer order in Figure 6-4.

9. In a real estate database, there is an entity called PROPERTY, which is a property for sale by the agency. Each time a potential property buyer makes a purchase offer on a property, the agency records the date, offering price, and name of the person making the offer.

a. Represent the PROPERTY entity and its purchase offer attributes using the notation for multivalued attributes.

b. Represent the PROPERTY entity and its purchase offer attributes using two entity types.

c. Finally, assume the agency decides to also keep data about buyers and potential buyers, including their name, address, and phone number. Augment your answer to part b to accommodate this new entity type.

10. Consider the Is_married_to unary relationship in Figure 6-6.

a. Draw minimum and maximum cardinalities for each end of this relationship.

b. Assume we wanted to know the date on which a marriage occurred. Augment this E-R diagram to include a Date_married attribute.

c. Because persons sometimes remarry after the death of a spouse or divorce, redraw this E-R diagram to show the whole history of marriages (not just the current marriage) for PERSONs. Show the Date_married attribute on this diagram.

Discussion Questions

1. Discuss why some systems developers believe that a data model is one of the most important parts of the statement of information system requirements.

2. Using Table 6-1 as a guide, develop a script of at least 10 questions you would ask during an interview of the customer order processing department manager at Pine Valley Furniture. Assume the focus is on analyzing the requirements for a new order entry system. The purpose of the interview is to develop a preliminary E-R diagram for this system.

3. If possible, contact a systems analyst in a local organization. Discuss with this systems analyst the role of conceptual data modeling in the overall systems analysis and design of information systems at his or her company. How, and by whom, is conceptual data modeling performed?

What training in this technique is given? At what point(s) is this done in the development process? Why?

4. Talk to MIS professionals at a variety of organizations and determine the extent to which CASE tools are used in the creation and editing of entity-relationship diagrams. Try to determine whether or not they use CASE tools for this purpose; which CASE tools are used; and why, when, and how they are used. In companies that do not use CASE tools for this purpose, determine why not and what would have to change in order to use them.

5. Ask a systems analyst to give you a copy of the standard notation he or she uses to draw E-R diagrams. In what ways is this notation different from notation in this text? Which notation do you prefer and why? What is the meaning of any additional notation?

Case Problems

1. Pine Valley Furniture

In order to determine the requirements for the new Customer Tracking System, several JAD sessions, interviews, and observations were conducted. Resulting information from these requirements determination methods was very useful in the preparation of the Customer Tracking System's data flow diagrams.

One afternoon while you are working on the Customer Tracking System's data flow diagrams,

Jim Woo stops by your desk and assigns you the task of preparing a conceptual entity-relationship diagram for the Customer Tracking System. Later that afternoon you review the requirements determination phase deliverables, including the data flow diagrams you have just finished preparing.

Your review of these deliverables suggests that the Customer Tracking System's primary objective is to track and forecast customer buying patterns. Additionally, in order to track a customer's

buying habits, an order history must be established, satisfaction levels assessed, and a variety
of demographic data collected. The demographic
data will categorize the customer according to
type, geographic location, and type of purchase.
Customer Tracking System information will
enable Pine Valley Furniture to better forecast its
product demand, control its inventory, and solicit
customers. Also, the Customer Tracking System's
ability to interface with the WebStore is very
important to the project.

a. What entities are identified in the above scenario? Can you think of additional entities?
   What interrelationships exist between the
   entities?
b. For each entity, identify its set of associated
   attributes. Specify identifiers for each entity.
c. Based on the case scenario and your answers
   to parts a and b, prepare an entity relationship
   diagram. Be sure to specify the cardinalities
   for each relationship.
d. How does this conceptual model differ from
   the WebStore's conceptual model?

2. Hoosier Burger

   Although Hoosier Burger is well recognized for
   its fast foods, especially the Hoosier Burger
   Special, plate lunches are also offered. These
   include such main menu items as barbecue ribs,
   grilled steak, meatloaf, and grilled chicken
   breast. The customer can choose from a variety
   of side items, including roasted garlic mashed
   potatoes, twice-baked potatoes, coleslaw, corn,
   baked beans, and Caesar salad.

   Many downtown businesses often call and place
   orders for Hoosier Mighty Meals. These are combination meals, consisting of a selection of main
   menu items and three side orders. The customer
   can request Hoosier Mighty Meals to feed 5, 10, 15,
   or 20 individuals. As a convenience to its business
   customers, Bob and Thelma allow business customers to charge their order. Once each month, a
   bill is generated and sent to those business customers who have charged their orders. Bob and
   Thelma have found that many of their business
   customers are repeat customers and often place
   orders for the same Hoosier Mighty Meals. Bob

asks you if it is possible to track a customer's order
history, and you indicate that it is indeed possible.

a. Based on the information provided in the case
   scenario, what entities will Hoosier Burger
   need to store information about?
b. For the entities identified in part a, identify a
   set of attributes for each entity.
c. Specify an identifier for each entity. What rules
   did you apply when selecting the identifier?
d. Modify Figure 6-10 to reflect the addition of
   these new entities. Be sure to specify the cardinalities for each relationship.

3. Corporate Technology Center

   Five years ago, Megan Thomas was a busy
   executive seeking to keep herself and her
   employees current with new technology. She
   realized that many small companies were facing
   the same dilemma. Using her life savings and
   money from investors, Megan founded Corporate
   Technology Center. Corporate Technology
   Center's primary objective is to offer technology
   update seminars to local business executives and
   their employees. A wide variety of seminars is
   offered, including ones covering operating systems, spreadsheets, word processing, database
   management, Internet, Web page design, and
   telecommunications.

   Although Corporate Technology Center offers
   seminars at its own campus, it also provides on-
   site training for local companies. One-day, two-
   day, or four-day seminars are offered. Courses
   are open to a minimum of 20 students and a maximum of 40 students. Although several staff
   members are capable of teaching any given
   course, generally only one staff member teaches
   a given course on a given date.

a. What entities are identified in the above scenario? Can you identify additional entities?
b. For each entity identified in part a, specify a
   set of associated attributes.
c. Select an identifier for each entity. What rules
   did you apply when selecting the identifier?
d. Based on the case scenario and your answers
   to a, b, and c, prepare an entity relationship
   diagram. Be sure to specify the cardinalities
   for each relationship.

CASE: BROADWAY ENTERTAINMENT COMPANY, INC.

Conceptual Data Modeling for the Web-Based Customer Relationship Management System

Case Introduction

Requirements determination activities for the MyBroadway project yielded what at times seemed to the student team to be an overwhelming amount of data. The team of students from St. Claire Community College has several hundred pages of notes from various data collection activities including twelve interviews with employees and customers, six hours of observation of employees using online shopping services, a one-hour focus group session with customers, and investigations of Broadway Entertainment documents. Structuring these requirements for the analysis of the MyBroadway information system is a much bigger effort than any class exercise the team members had ever encountered.

Also adding to the complexity of requirements structuring activities in the analysis phase of the project is that work is not easily compartmentalized. It seems to the team members that while they are documenting data movement and processing requirements, they also have to find ways to understand the meaning of data the system will handle. Conceptual data modeling techniques, primarily entity-relationship diagramming, help but changes are frequent. The steps are very repetitive. As the team decomposes a business process, members need to redesign the E-R diagram for MyBroadway. When they change the E-R model, they gain new insights about the data and suggest issues of data handling processes for validation, special cases, and capturing relationships.

Structuring the High-Level Data Modeling Findings from Requirements Determination

The various BEC student team members have taken responsibility for different requirements collection activities and for developing the explosions of each process on the level-0 DFD. So, no one team member has a complete picture of all the data needs. This is not uncommon on real development projects. The team has yet to appoint someone to be the data administrator for the project. To gain a shared understanding of the database needs for MyBroadway, the team members read all of their notes carefully in preparation for a team meeting.

At the team meeting each member suggests the data entities he or she thinks are needed in his or her part of the system. After some discussion the team

concludes that six entity types are referenced repeatedly in data flows and data stores across all business processes. See BEC Figure 6-1 for the initial entity-relationship diagram the team draws. The entity types identified by the team follow.

- PRODUCT: An item made available for sale or rent by BEC to customers. Each product is a CD, DVD, or videocassette title. For example, a product is the movie *Star Wars Episode I: The Phantom Menace* on videocassette. Although the Entertainment Tracker operational system must keep track of each copy of a movie available for rent at a store, MyBroadway simply needs to track titles, not individual copies. For items for sale, a product is the generic title, not an individual copy, of the product for sale.

- REQUEST: An inquiry by a customer asking BEC to stock a product. BEC may choose never to stock that product. Yet if enough requests for the same product are submitted, it is likely that the item will eventually appear on the store shelves.

- SALE: A record that a particular product (by title) was sold to a specified customer on what date. Entertainment Tracker keeps the official record of each sale transaction, including _which items are sold in the same transaction. MyBroadway, however, does not need this information but rather needs only

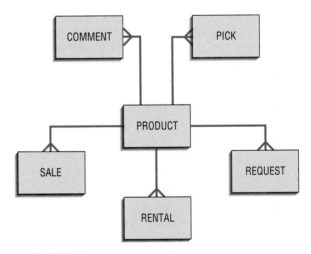

BEC FIGURE 6-1
INITIAL E-R DIAGRAM FOR MYBROADWAY

when and to whom each product was ever sold.

◗ RENTAL: A record that a particular product (by title) was rented by a specified customer on what date. As with the SALE entity type, MyBroadway does not need to track the rental transaction in detail.

◗ COMMENT: An unstructured statement by a customer about a specified product (by title).

◗ PICK: An unstructured comment and rating _by an employee about a specified product (by title).

◗ The MyBroadway team finds it is very interesting that although both customers and employees are very prominent actors in the system and data about each are needed in data flows and are retained in data stores, neither of these appear to be useful entities themselves. The team concluded this because data about these objects of the business seem to have no usefulness on their own (in MyBroadway), but only when associated with other data. For example, MyBroadway needs to know what members bought which products when, but customer data such as name, address, standard credit card number, and so forth—prominent in Entertainment Tracker—never appear in any data flow.

The team also concluded that attributes about PRODUCT, SALE, and RENTAL would not be captured within MyBroadway; rather, data for these entities would come from the Entertainment Tracker database. The team's initial thought was that MyBroadway probably needs only a minor subset of the data from Entertainment Tracker on these data entities. For example, the team had not identified any data flow that needed product price, cost, location in store, or a host of other product attributes in the Entertainment Tracker database useful for transaction processing and management reporting. In MyBroadway, the transactions of interest are entries of comments, favorite picks, and product requests. Each comment, pick, or request is considered an independent data item, whereas product sales and rentals frequently appear together all in one transaction (e.g., someone rents three movies and buys one CD all in the same point-of-sale transaction). These observations suggest to the team that the structure of the MyBroadway database may be simpler than for most operational databases.

Case Summary

Of course, whether these six entities are all the team needs still remains to be finalized. The team must carefully match this list of data entities with the data stores and data flows from data flow diagrams they are developing. For example, every attribute of a data flow going into a data store must be an attribute of some entity type. Also, there must be an attribute either in a data store or directly passed through the system to generate all the attributes of each data flow leaving MyBroadway to some external entity. The team has many more questions to answer before it can produce an E-R diagram for the MyBroadway system.

Case Questions

1. Review the data flow diagrams you developed for questions in the BEC case at the end of Chapter 5 (or diagrams given to you by your instructor). Study the data flows and data stores on these diagrams and decide if you agree with the team's conclusion that there are only the six entity types listed in this case and in BEC Figure 6-1. If you disagree, define additional entity types, explain why they are necessary, and modify BEC Figure 6-1.

2. Again, review the DFDs you developed for the MyBroadway system (or those given to you by your instructor). Use these DFDs to identify the attributes of each of the six entities listed in this case plus any additional entities identified in your answer to Question 1. Write an unambiguous definition for each attribute. Then, redraw BEC Figure 6-1 by placing the six (and additional) entities in this case on the diagram along with their associated attributes.

3. Using your answer to Question 2, designate which attribute or attributes form the identifier for each entity type. Explain why you chose each identifier.

4. Using your answer to Question 3, draw the relationships between entity types needed by the system. Remember, a relationship is needed only if the system wants data about associated entity instances. Give a meaningful name to each relationship. Specify cardinalities for each relationship and explain how you decided on each minimum and maximum cardinality on each end of each relationship. State any assumptions you made if the BEC cases you have read so far and the answers to questions in these cases do not provide the evidence to justify the cardinalities you choose. Redraw your final E-R diagram in Microsoft Visio.

5. Now that you have developed in your answer to Question 4 a complete E-R diagram for the MyBroadway database, what are the consequences of not having customer or employee

entity types on this diagram? Assuming only the attributes you show on the E-R diagram, would any attribute be moved from the entity it is currently associated with to a customer or employee entity type if such entity types were on the diagram? Why or why not?

6. Write project dictionary entries (using standards given to you by your instructor) for all the entities, attributes, and relationships shown in the E-R diagram in your answer to Question 4. How detailed are these entries at this point? What other details still must be filled in? Are any of the entities on the E-R diagram in your answer to Question 4 weak entities? Why? In particular, is the REQUEST entity type a weak entity? If so, why? If not, why not?

7. What date-related attributes did you identify in each of the entity types in your answer to Question 4? Why are each of these needed? Can you make some general observations about why date attributes must be kept in a database based on your analysis of this database?

7

Selecting the Best Alternative Design Strategy

> **Objectives**

After studying this chapter, you should be able to:

- Describe the different sources of software.
- Assemble the various pieces of an alternative design strategy.
- Generate at least three alternative design strategies for an information system.
- Select the best design strategy using both qualitative and quantitative methods.
- Update a Baseline Project Plan based on the results of the analysis phase.
- Understand how selecting the best design strategy applies to development for Internet applications.

Chapter Preview . . .

You have now reached the point in the analysis phase where you are ready to transform all of the information you have gathered and structured into some concrete ideas about the design for the new or replacement information system. This is called the design strategy. From requirements determination, you know what the current system does. You also know what the users would like the replacement system to do. From requirements structuring, you know what forms the replacement system's process flow and data should take, at a logical level independent of any physical implementation. However, there still may be some uncertainty about the capabilities of a new system. There are two sources of this uncertainty: (1) competing ideas from different users on what they would like the system to do, and (2) the number of existing alternatives for an implementation environment for any new system. As Figure 7-1 shows, there are three parts to systems analysis: determining requirements, structuring requirements, and selecting the best alternative design strategy. To bring analysis to a conclusion, your job is to take these structured requirements and transform them into several alternative design strategies. One of these strategies will be pursued in the design phase of the life cycle.

FIGURE 7-1
There are three parts to systems analysis: determining requirements, structuring requirements, and selecting the best alternative design strategy.

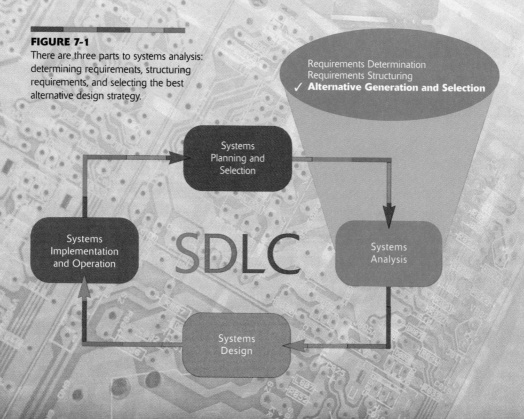

Requirements Determination
Requirements Structuring
✓ **Alternative Generation and Selection**

SDLC

Systems Planning and Selection

Systems Implementation and Operation

Systems Analysis

Systems Design

Part of generating a design strategy involves tapping into sources inside and outside an organization to determine the best way that organization can acquire the replacement system. If you advise the organization to proceed with development inside the organization, you and your team will have to answer general questions about software: Will software be built in-house, purchased off-the-shelf, or contracted to software development companies? You will also have to answer general questions about hardware and system software: Will the new system run on a mainframe platform, with stand-alone personal computers, or on a client/server platform? Can the system run on existing hardware?

It is also not too early to begin thinking about data conversion issues and how much training users will need. Determine whether you can build and implement the system given the funding and management support you can count on. You have to address all these concerns for each alternative generated so that you can update the Baseline Project Plan with detailed activities and resource requirements for the next life cycle phase—systems design. In this step of the analysis phase, you bring the current phase to a close, prepare a report and presentation to management concerning continuation of the project, and get ready to move the project into design.

In this chapter, you learn why you need to come up with alternative design strategies and about guidelines for generating alternatives. You then learn the different issues that must be addressed for each alternative. Once you have generated your alternatives, you will have to choose the best design strategy to pursue. We include a discussion of one technique that analysts and users often use to help them agree on the best approach for the new information system.

Throughout this chapter we emphasize the need for sound project management. Now that you have seen the various techniques and steps of the analysis phase, we outline what a typical analysis phase project schedule might look like. We also discuss the execution of the analysis phase and the transition from analysis to design.

Selecting the Best Alternative Design Strategy

Selecting the best alternative system involves at least two basic steps: (1) generating a comprehensive set of alternative design strategies and (2) selecting the one that is most likely to result in the desired information system, given all of the organizational, economic, and technical constraints that limit what can be done. A system **design strategy** represents a particular approach to developing the system. Selecting a strategy requires you to answer questions about the system's functionality, hardware and system software platform, and method for acquisition. We use the term *design strategy* in this chapter rather than *alternative system* because, at the end of analysis, we are still quite a long way from specifying an actual system. This delay is purposeful because we do not want to invest in design efforts until there is agreement on which direction to take the project and the new system. The best we can do at this point is to outline rather broadly the approach we can take in moving from logical system specifications to a working physical system. The overall process of selecting the best system strategy and the deliverables from this step in the analysis process are discussed next.

Design strategy
A particular approach to developing an information system. It includes statements on the system's functionality, hardware and system software platform, and method for acquisition.

The Process of Selecting the Best Alternative Design Strategy

There are three parts to systems analysis: determining requirements, structuring requirements, and selecting the best alternative design strategy. After the system requirements have been structured in terms of process flow and data, analysts again work with users to package the requirements into different system configurations. Shaping alternative system design strategies involves the following processes:

> ❿ Dividing requirements into different sets of capabilities, ranging from the bare minimum that users would accept (the required features) to the most elaborate and advanced system the company could afford to develop (which includes all the features desired across all users). Alternatively, different sets of capabilities may represent the position of different organizational units with conflicting notions about what the system should do.

> ❿ Enumerating different potential implementation environments (hardware, system software, and network platforms) that could be used to deliver the different sets of capabilities. (Choices on the implementation environment may place technical limitations on the subsequent design phase activities.)

> ❿ Proposing different ways to source or acquire the various sets of capabilities for the different implementation environments.

In theory, if there are three sets of requirements, two implementation environments, and four sources of application software, there would be 24 possible design strategies. In practice, some combinations are usually infeasible, and only a small number—typically three—can be easily considered. Selecting the best alternative is usually done with the help of a quantitative procedure, an example of which comes later in the chapter. Analysts will recommend what they believe to be the best alternative, but management (a combination of the steering committee and those who will fund the rest of the project) will make the ultimate decision about which system design strategy to follow. At this point in the life cycle, it is also certainly possible for management to end a project before the more expensive phases of system design or system implementation and operation are begun. Reasons for ending a project might include the costs or risks outweighing the benefits, the needs of the organization having changed since the project began, or other competing projects having become more important while development resources remain limited.

Deliverables and Outcomes

The primary deliverables from generating alternative design strategies and selecting the best one are outlined in Table 7-1. The primary deliverable that is carried forward into design is an updated Baseline Project Plan detailing the work necessary to turn the selected design strategy into the desired replacement information system. Of course, that plan cannot be assembled until a strategy has been selected, and no strategy can be selected until alternative strategies have been generated and compared. Therefore, all three objects—the alternatives, the selected alternative, and the plan—are listed as deliverables in Table 7-1. Further, these three deliverables plus the supporting deliverables from requirements determination and structuring steps are necessary to conduct systems design. All of this information is stored in the project dictionary or CASE repository for reference in later phases.

TABLE 7-1:   Deliverables for Generating
Alternatives and Selecting the Best One

1. At least three substantively different system design strategies for building the replacement information system

2. A design strategy judged most likely to lead to the most desirable information system

3. A Baseline Project Plan for turning the most likely design strategy into a working information system

Generating Alternative Design Strategies

The solution to an organizational problem may seem obvious to an analyst. Typically, the analyst is very familiar with the problem, having conducted an extensive analysis of it and how it has been solved in the past. On the other hand, the analyst may be very familiar with a particular solution that he or she attempts to apply to all organizational problems encountered. For example, if an analyst is an expert at using advanced database technology to solve problems, then he or she tends to recommend advanced database technology as a solution to every possible problem. Or if the analyst designed a similar system for another customer or business unit, the "natural" design strategy would be the one used before. Given the role of experience in the solutions analysts suggest, analysis teams typically generate at least two alternative solutions for every problem they work on.

A good number of alternatives for analysts to generate is three. Why three? Three alternatives can neatly represent low, middle, and high ranges of potential solutions. One alternative represents the low end of the range. Low-end alternatives are the most conservative in terms of the effort, cost, and technology involved in developing a new system. Some low-end solutions may not involve computer technology at all, focusing instead on making paper flows more efficient or reducing redundancies in current processes. A low-end strategy provides all the required functionality users demand with a system that is minimally different from the current system.

Another alternative represents the high end of the range. High-end alternatives go beyond simply solving the problem in question and focus instead on systems that contain many extra features users may desire. Functionality, not cost, is the primary focus of high-end alternatives. A high-end alternative will provide all desired features using advanced technologies that often allow the system to expand to meet future requirements. Finally, the third alternative lies between the extremes of the low-end and high-end systems. Such alternatives combine the frugality of low-end alternatives with the focus on functionality of high-end alternatives. Midrange alternatives represent compromise solutions. There are certainly other possible solutions that exist outside of these three alternatives. Defining the low, middle, and high possibilities allows the analyst to draw bounds around what can be reasonably done.

How do you know where to draw bounds around the potential solution space? The analysis team has already gathered the information it needs to identify the solution, but first that information must be systematically organized. There are two major considerations. The first is determining the minimum requirements for the new system. These are the mandatory features, and if any of them are missing, the design strategy is useless. Mandatory features are those that everyone agrees are necessary to solve the problem or meet the opportunity. Which features are mandatory can be determined from a survey of users and others who have been involved in requirements determination. You would conduct this survey near the end of the analysis phase after all requirements have been structured and analyzed. In this survey, users rate features discovered during requirements determination or categorize features on some scale, and an arbitrary breakpoint is used to divide mandatory from desired features. Some organizations will break the features into three categories: mandatory, essential, and desired. Whereas mandatory features screen out possible solutions, essential features are the important capabilities of a system that serve as the primary basis for comparison of different design strategies. Desired features are those that users could live without but that are used to select between design strategies that are of almost equal value in terms of essential features. Features can take many different forms, as illustrated in Figure 7-2, and might include:

> *Data kept in system files:* For example, multiple customer addresses so that bills can be sent to addresses different from where we ship goods.

> *System outputs:* Printed reports, online displays, transaction documents, for example, a paycheck or sales summary graph.

> *Analyses to generate the information in system outputs:* For example, a sales forecasting module or an installment billing routine.

> *Expectations on accessibility, response time, or turnaround time for system functions:* For example, online, real-time updating of inventory files.

The second consideration in drawing bounds around alternative design strategies is determining the constraints on system development. Constraints, some of which also appear in Figure 7-2, may include:

> *A date when the replacement system is needed.*

> *Available financial and human resources.*

> *Elements of the current system that cannot change.*

> *Legal and contractual restrictions:* For example, a software package bought off-the-shelf cannot be legally modified, or a license to use a particular software package may limit the number of concurrent users to 25.

> *The importance or dynamics of the problem that may limit how the system can be acquired:* For example, a strategically important system that uses highly proprietary data probably cannot be outsourced or purchased.

Phase 2: Systems Analysis

Features

Database

Data   Output   Analyses

Constraints

Time   Financial   Legal

FIGURE 7-2

Essential features to consider during systems development include data (such as customer addresses), output (such as a printed report like a sales summary graph), and analyses (such as a sales forecast). Constraints on systems developing may include time, finances, and legal issues.

Remember, be impertinent and question whether stated constraints are firm. You may want to consider some design alternatives that violate constraints you consider to be flexible.

Both requirements and constraints must be identified and ranked in order of importance. The reason behind such a ranking should be clear. Whereas you can design a high-end alternative to fulfill every wish users have for a new system, you design low-end alternatives to fulfill only the most important wishes. The same is true of constraints. Low-end alternatives will meet every constraint; high-end alternatives will ignore all but the most daunting constraints.

Issues to Consider in Generating Alternatives

The requirements and constraints of the replacement system raise many issues that analysts must consider when they develop alternative design strategies. Most of the substantive debate about alternative design strategies hinges on the relative importance of system features. Issues of functionality help determine software and hardware selection, implementation, organizational limitations such as available funding levels, and whether the system should be developed and run in-house. This list is not complete, but it does remind you that an information system is more than just software. We now discuss each issue, beginning with outsourcing.

Outsourcing

Outsourcing
The practice of turning over responsibility of some or all of an organization's information systems applications and operations to an outside firm.

If another organization develops or runs a computer application for your organization, that practice is called outsourcing. **Outsourcing** includes a spectrum of working arrangements. At one extreme is having a firm develop and run your application on its computers—you only supply input and take output. A common example is a company that runs payroll applications for clients so that clients don't have to develop an independent in-house payroll system. Instead they simply provide employee payroll information to the company and, for a fee, the company returns completed paychecks, payroll accounting reports, and tax and other statements for employees. For many organizations, payroll is a very cost-effective operation when outsourced in this way. In another example of outsourcing arrangements, you hire a company to run your applications at your site on your computers. In some cases, an organization employing such an arrangement will dissolve some or all of its information systems unit and fire all of its information systems employees. Many times the company brought in to run the organization's computing will hire many of the information systems unit's employees.

Why would an organization outsource its information systems operations? As we saw in the payroll example, outsourcing may be cost-effective. If a company specializes in running payroll for other companies, it can leverage the economies of scale it achieves from running one very stable computer application for many organizations into very low prices. But why would an organization dissolve its entire information processing unit and bring in an outside firm to manage its computer applications? One reason may be to overcome operating problems the organization faces in its information systems unit. For example, the city of Grand Rapids, Michigan, hired an outside firm to run its computing 30 years ago in order to manage its computing center employees better. Union contracts and civil service constraints then in force made it difficult to fire people, so the city brought in a facilities management organization to run its computing operations, and it was able to get rid of problem employees at the same time. Another reason for total outsourcing is that an organization's management may feel its core mission does not involve managing an information systems unit and that it might achieve more effective computing by turning over all of its operations to a more experienced, computer-oriented company.

Phase 2:
Systems
Analysis

Kodak decided in the late 1980s that it was not in the computer applications business and turned over management of its mainframes to IBM and management of its personal computers to Businessland.

Outsourcing is an alternative analysts need to be aware of. When generating alternative system development strategies for a system, you as an analyst should consult organizations in your area that provide outsourcing services. It may well be that at least one such organization has already developed and is running an application very close to what your users are asking for. Perhaps outsourcing the replacement system should be one of your alternatives. Knowing what your system requirements are before you consider outsourcing means that you can carefully assess how well the suppliers of outsourcing services can respond to your needs. However, should you decide not to consider outsourcing, you need to consider whether some software components of your replacement system should be purchased and not built.

NET SEARCH
Outsourcing has become very popular and many believe that it will continue to grow as a way for organizations to develop and operate their information systems. Visit http:// www.prenhall.com/ valacich to complete an exercise related to this topic.

Sources of Software

We can group organizations that produce software into five major categories: hardware manufacturers, packaged software producers, custom software producers, enterprise-wide solutions, and in-house developers.

Hardware Manufacturers   At first it may seem counterintuitive that hardware manufacturers would develop information systems or software. Yet hardware manufacturers are among the largest producers of software; for example, IBM is a leader in software development. Table 7-2 ranks the top 10 global software companies and their revenues in 2001. However, IBM actually develops relatively little application software (roughly 15 percent of its software revenue is from application software). Rather, IBM's leadership comes from its operating systems and utilities (like sort routines or database management systems) for the hardware it manufactures, as well as its middleware, the software that links one set of software services to another.

Packaged Software Producers   The growth of the software industry has been phenomenal since its beginnings in the mid-1960s. Now, some of the largest computer companies in the world, as measured by *Software Magazine*,

TABLE 7-2:   The Top 10 Global Software Companies

Rank among Software Companies	Company	2001 Software Revenues (in millions)	Software Specialization
1	IBM	$47,895	Middleware
2	Microsoft	24,666	Operating systems
3	EDS	21,543	Custom systems
4	Accenture	13,348	Custom systems
5	Oracle	10,859	Databases
6	Computer Sciences Corporation	10,524	Custom systems
7	Compaq	7,746	Custom systems
8	PriceWaterhouse	7,481	Custom systems
9	Cap Gemini Ernst & Young	7,454	Custom systems
10	NTT Data	6,460	Custom systems

Source: Adapted from the *Software Magazine* Web site, www.softwaremag.com.

NET SEARCH

Lists ranking the top IS firms and software companies are readily available. Visit http://www.prenhall. com/valacich to complete an exercise related to this topic.

are companies that produce software exclusively (see Table 7-2). Consulting firms, such as Accenture (number 10), also rank in the top 50 packaged software producers.

Software companies develop what are sometimes called prepackaged or off-the-shelf systems. Microsoft's Project and Intuit's Quicken, QuickPay, and QuickBooks are popular examples of such software. The packaged software development industry serves many market segments. Its software offerings range from general, broad-based packages, such as general ledger, to very narrow, niche packages, such as software to help manage a day-care center. Software companies develop software to run on many different computer platforms, from microcomputers to large mainframes. The companies range in size from just a few people to thousands of employees. Software companies consult with system users after the initial software design has been completed and an early version of the system has been built. The systems are then tested in actual organizations to determine whether there are any problems or if any improvements can be made. Until testing is completed, the system is not offered for sale to the public.

Some off-the-shelf software systems cannot be modified to meet the specific, individual needs of a particular organization. Such application systems are sometimes called turnkey systems. The producer of a turnkey system will make changes to the software only when a substantial number of users ask for a specific change. Other off-the-shelf application software can be modified or extended, however, by the producer or the user to fit the needs of the organization more closely. Even though many organizations perform similar functions, no two organizations do the same thing in quite the same way. A turnkey system may be good enough for a certain level of performance, but it will never perfectly match the way a given organization does business. A reasonable estimate is that off-the-shelf software can at best meet 70 percent of an organization's needs. Thus, even in the best case, 30 percent of the software system doesn't match the organization's specifications.

Custom Software Producers If a company needs an information system but does not have the expertise or the personnel to develop the system in-house and a suitable off-the-shelf system is not available, the company will likely consult a custom software company. Consulting firms, such as PriceWaterhouseCoopers or Electronic Data Systems (EDS), will help a firm develop custom information systems for internal use. These firms employ people with expertise in the development of information systems. Their consultants may also have expertise in a given business area. For example, consultants who work with banks understand financial institutions as well as information systems. Consultants use many of the same methodologies, techniques, and tools that companies use to develop systems in-house.

Enterprise Solutions Software As mentioned in Chapter 1, more and more organizations are choosing complete software solutions, called **enterprise solutions** or **enterprise resource planning (ERP) systems,** to support their operations and business processes. These ERP software solutions consist of a series of integrated modules. Each module supports an individual traditional business function, such as accounting, distribution, manufacturing, and human resources. The difference between the modules and traditional approaches is that the modules are integrated to focus on business processes rather than on business functional areas. For example, a series of modules will support the entire order entry process, from receiving an order to adjusting inventory to shipping to billing to after-the-sale service. The traditional approach would use different systems in different functional areas of the business, such as a billing system in accounting and an inventory system in the warehouse. Using enterprise software solutions, a firm can integrate all parts of a business process in a unified information system. All aspects of a single

Enterprise solutions software or enterprise resource planning (ERP) systems

A system that integrates individual traditional business functions into a series of modules so that a single transaction occurs seamlessly within a single information system rather than several separate systems.

transaction occur seamlessly within a single information system, rather than in a series of disjointed, separate systems focused on business functional areas.

The benefits of the enterprise solutions approach include a single repository of data for all aspects of a business process and the flexibility of the modules. A single repository ensures more consistent and accurate data, as well as less maintenance. The modules are very flexible because additional modules can be added as needed once the basic system is in place. Added modules are immediately integrated into the existing system.

There are also disadvantages to enterprise solutions software. The systems are very complex, so implementation can take a long time to complete. Organizations typically do not have the necessary expertise in-house to implement the systems, so they must rely on consultants or employees of the software vendor, which can be very expensive. In some cases, organizations must change how they do business in order to benefit from a shift toward enterprise solutions.

There are several major vendors of enterprise solution software. The best known is probably SAP AG, a German firm, known for its flagship product R/3. *SAP* stands for *Systems, Applications, and Products in Data Processing.* SAP AG was founded in 1972, but most of its growth has occurred since 1992. In 2001, SAP AG was the 11th largest supplier of software in the world (see Table 7-2). Another supplier of enterprise solution software is PeopleSoft, Inc., a U.S. firm founded in 1987. PeopleSoft began with enterprise solutions that focused on human resources management but has since expanded to cover financials, materials management, distribution, and manufacturing. Two other major vendors of enterprise solutions, both U.S.-based firms, are Oracle Corp. and J. D. Edwards & Company.

In-House Development We have talked about four different types of external organizations that serve as sources of software, but in-house development remains an option. Of course, in-house development need not entail development of all of the software that will comprise the total system. Hybrid solutions involving some purchased and in-house software components are common. Table 7-3 compares the five different software sources.

If you choose to acquire software from outside sources, this choice is made at the end of the analysis phase. Choosing between a package or external supplier

Phase 2: Systems Analysis

NET SEARCH
There is much interest in and information available on enterprise resource planning (ERP) systems. Visit http://www.prenhall. com/valacich to complete an exercise related to this topic.

TABLE 7-3:   Comparison of Five Different Sources of Software Components

Producers	Source of Application Software?	When to Go to This Type of Organization for Software	Internal Staffing Requirements
Hardware manufacturers	Generally not	For system software and utilities	Varies
Packaged software producers	Yes	When supported task is generic	Some IS and user staff to define requirements and evaluate packages
Custom software producers	Yes	When task requires custom support and system can't be built internally	Internal staff may be needed, depending on application
Enterprise-wide solutions	Yes	For complete systems that cross functional boundaries	Some internal staff necessary but mostly need consultants
In-house developers	Yes	When resources and staff are available and system must be built from scratch	Internal staff necessary though staff size may vary

will be determined by your needs, not by what the supplier has to sell. As we discuss, the results of your analysis study will define the type of product you want to buy and will make working with an external supplier much easier, more productive and worthwhile.

Choosing Off-the-Shelf Software

Once you have decided to purchase off-the-shelf software rather than write some or all of the software for your new system, how do you decide what to buy? There are several criteria to consider, and special ones may arise with each potential software purchase. For each standard, an explicit comparison should be made between the software package and the process of developing the same application in-house. The most common criteria, highlighted in Figure 7-3, are as follows:

- Cost
- Functionality
- Vendor support
- Viability of vendor
- Flexibility
- Documentation
- Response time
- Ease of installation

The relative importance of these standards will vary from project to project and from organization to organization. If you had to choose two criteria that would always be among the most important, those two would probably be vendor support and vendor viability. You don't want to license software from a vendor with a reputation for poor support. Similarly, you don't want to get involved with a vendor that might not be in business tomorrow. How you rank the importance of the remaining criteria depends very much on your specific situation.

Cost involves comparing the cost of developing the same system in-house to the cost of purchasing or licensing the software package. Be sure to include a comparison of the cost of purchasing vendor upgrades or annual license fees with the costs you would incur to maintain your own software. Costs for purchasing and developing in-house can be compared based on the economic fea-

FIGURE 7-3
COMMON CRITERIA FOR CHOOSING OFF-THE-SHELF SOFTWARE

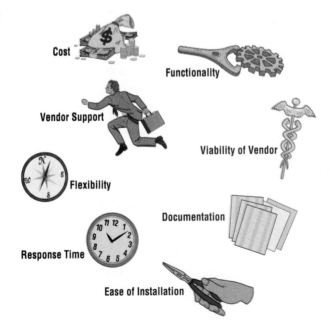

Cost

Functionality

Vendor Support

Viability of Vendor

Flexibility

Documentation

Response Time

Ease of Installation

sibility measures. Functionality refers to the tasks the software can perform and the mandatory, essential, and desired system features. Can the software package perform all or just some of the tasks your users need? If some, can it perform the necessary core tasks? Note that meeting user requirements occurs at the end of the analysis phase because you cannot evaluate packaged software until user requirements have been gathered and structured. Purchasing application software is not a substitute for conducting the systems analysis phase.

As we said earlier, vendor support refers to whether and how much support the vendor can provide. Support includes assistance to install the software, to train user and systems staff on the software, and to provide help as problems arise after installation. Recently, many software companies have significantly reduced the amount of free support they provide customers, so the cost to use telephone, on-site, fax, or computer bulletin board support facilities should be considered. Related to support is the vendor's viability. This latter point should not be minimized. The software industry is quite dynamic, and innovative application software is created by entrepreneurs working from home offices—the classic cottage industry. Such organizations, even with outstanding software, often do not have the resources or business management ability to stay in business very long. Further, competitive moves by major software firms can render the products of smaller firms outdated or incompatible with operating systems. One software firm we talked to while developing this book was struggling to survive just trying to make its software work on any supposedly IBM-compatible PC (given the infinite combination of video boards, monitors, BIOS chips, and other components). Keeping up with hardware and system software change may be more than a small firm can handle, and good off-the-shelf application software is lost.

Flexibility refers to how easy it is for you, or the vendor, to customize the software. If the software is not very flexible, your users may have to adapt the way they work to fit the software. Are they likely to adapt in this manner? Purchased software can be modified in several ways. Sometimes, the vendor will make custom changes for you if you are willing to pay for the redesign and programming. Some vendors design the software for customization. For example, the software may include several different ways of processing data and, at installation time, the customer chooses which to initiate. Also, displays and reports may be easily redesigned if these modules are written in a fourth-generation language. Reports, forms, and displays may be easily customized using a process whereby your company name and chosen titles for reports, displays, forms, and column headings are selected from a table of parameters you provide. You may want to employ some of these same customization techniques for in-house developed systems so that the software can be easily adapted for different business units, product lines, or departments.

Documentation includes the user's manual as well as technical documentation. How understandable and up-to-date is the documentation? What is the cost for multiple copies, if required? Response time refers to how long it takes the software package to respond to the user's requests in an interactive session. Another measure of time would be how long it takes the software to complete running a job. Finally, ease of installation is a measure of the difficulty of loading the software and making it operational.

Validating Purchased Software Information   One way to get all of the information you want about a software package is to collect it from the vendor. Some of this information may be contained in the software documentation and technical marketing literature. Other information can be provided upon request. For example, you can send prospective vendors a questionnaire asking specific questions about their packages. This may be part of a request for proposal (RFP) or request for quote (RFQ) process your organization requires when major purchases are made (see next section for more).

NET SEARCH
Requests for Proposals and Requests for Quotes are common ways for organizations to purchase software. Visit http://www. prenhall.com/valacich to complete an exercise related to this topic.

Of course, there is no replacement for actually using the software yourself and running it through a series of tests based on the criteria for selecting software. Remember to test not only the software but also the documentation, the training materials, and even the technical support facilities. One requirement you can place on prospective software vendors as part of the bidding process is that they install (free or at an agreed-upon cost) their software for a limited amount of time on your computers. This way you can determine how their software works in your environment, not in some optimized environment they have.

One of the most reliable and insightful sources of feedback is other users of the software. Vendors will usually provide a list of customers (remember, they will naturally tell you about satisfied customers, so you may have to probe for a cross-section of customers) and people who are willing to be contacted by prospective customers. Here is where your personal network of contacts, developed through professional groups, college friends, trade associations, or local business clubs, can be a resource; do not hesitate to find some contacts on your own. Such current or former customers can provide a depth of insight on use of a package at their organizations.

To gain a range of opinion about possible packages, you can use independent software testing services that periodically evaluate software and collect user opinions. Such surveys are available for a fee either as subscription services or on demand. Occasionally unbiased surveys appear in trade publications. Often, however, articles in trade publications, even software reviews, are actually seeded by the software manufacturer and are not unbiased.

If you are comparing several software packages, you can assign scores for each package on each criterion and compare the scores using the quantitative method we demonstrate at the end of the chapter for comparing alternative system design strategies.

Hardware and System Software Issues

The first question to ask about hardware and system software is whether the new system that follows a particular design strategy can be run on your firm's existing hardware and software platform. System software refers to such key components as operating systems, database management systems, programming languages, code generators, and network software. To determine if current hardware and system software are sufficient, you should consider such factors as the age and capacity of the current hardware and system software, the fit between the hardware and software and your new application's goals and proposed functionality, and, if some of your system components are off-the-shelf software, whether the software can run on the existing hardware and system software. The advantages to running your new system on the existing platform are persuasive:

1. Lower costs as little, if any, new hardware and system software have to be purchased and installed.
2. Your information systems staff is quite familiar with the existing platform and how to operate and maintain it.
3. The odds of integrating your new application system with existing applications are enhanced.
4. No added costs of converting old systems to a new platform, if necessary, or of translating existing data between current technology and the new hardware and system software you have to acquire for your system.

On the other hand, there are also very persuasive reasons for acquiring new hardware or system software:

1. Some software components of your new system will run only on particular platforms with particular operating systems.

2. Developing your system for a new platform gives your organization the opportunity to upgrade or expand its current technology holdings.
3. New platform requirements may allow your organization to change its computing operations radically, as in moving from mainframe-centered processing to a database machine or a client/server architecture.

As the determination of whether to acquire new hardware and system software is so context dependent, it is essential to provide platform options as part of your design strategy alternatives.

If you decide that new hardware or system software is a strong possibility, you may want to issue a **request for proposal (RFP)** to vendors. The RFP will ask the vendors to propose hardware and system software that will meet the requirements of your new system. Issuing an RFP gives you the opportunity to have vendors conduct the research you need in order to decide among various options. You can request that each bid submitted by a vendor contain certain information essential for you to decide on what best fits your needs. For example, you can ask for performance information related to speed and number of operations per second. You can ask about machine reliability and service availability and whether there is an installation nearby that you can visit for more information. You can ask to take part in a demonstration of the hardware. Of course the bid will also include information on cost. You can then use the information you have collected in generating your alternative design strategies.

Request for proposal (RFP)
A document provided to vendors to ask them to propose hardware and system software that will meet the requirements of your new system.

Implementation Issues

As you will see in Chapter 10, implementing a new information system is just as much an organizational change process as it is a technical process. Implementation involves more than installing a piece of software, turning it on, and moving on to the next software project. New systems often entail new ways of performing the same work, new working relationships, and new skills. Users have to be trained. Disruptions in work procedures have to be found and addressed. In addition, system implementation may be phased in over many weeks or even months. You must address the technical and social aspects of implementation as part of any alternative design strategy. Management and users will want to know how long the implementation will take, how much training will be required, and how disruptive the process will be.

Developing Design Strategies for Hoosier Burger's New Inventory Control System

As an example of alternative generation and selection, let's look at an inventory control system that Hoosier Burger wants developed. Figure 7-4 lists ranked requirements and constraints for the enhanced information system

SYSTEM REQUIREMENTS (in descending priority)	SYSTEM CONSTRAINTS (in descending order)
1. Must be able to easily enter shipments into system as soon as they are received.	1. System development can cost no more than $50,000.
2. System must automatically determine whether and when a new order should be placed.	2. New hardware can cost no more than $50,000.
3. Management should be able to determine at any time approximately what inventory levels are for any given item in stock.	3. The new system must be operational in no more than six months from the start of the contract.
	4. Training needs must be minimal; i.e., the new system must be very easy to use.

FIGURE 7-4
RANKED SYSTEM REQUIREMENTS AND CONSTRAINTS FOR HOOSIER BURGER'S INVENTORY SYSTEM

being considered by Hoosier Burger. The requirements represent a sample of those developed from the requirements determination and structuring carried out in prior analysis steps. The system in question is an upgrade to the company's existing inventory system. Before deciding to get a new inventory system, Bob Mellankamp, one of the owners of Hoosier Burger, had to follow several steps in his largely manual inventory control system, as identified in Figure 7-5.

Using the current manual system, Bob first receives invoices from suppliers, and he records their receipt on an invoice log sheet. He puts the actual invoices in his accordion file. Using the invoices, Bob records the amount of stock delivered on the stock logs, paper forms posted near the point of storage for each inventory item. The stock logs include minimum order quantities as well as spaces for posting the starting amount, amount delivered, and the amount used for each item. Amounts delivered are entered on the sheet when Bob logs stock deliveries; amounts used are entered after Bob has compared the amounts of stock used, according to a physical count, and according to the numbers on the inventory report generated by the food ordering system. Some Hoosier Burger items, especially perishable goods, have standing orders for daily delivery.

The Mellankamps want to improve their inventory system so that new orders are immediately accounted for, so that the system can determine when new orders should be placed, and so that management can obtain accurate inventory levels at any time of the day. All three of these system requirements have been ranked in order of descending priority in Figure 7-4. A logical data flow diagram showing the key processes in the desired inventory system is shown in Figure 7-6. The goal of having new orders automatically accounted for is reflected in Process 1.0. The goal of having the system determine when new orders should be placed is realized in Process 3.0. The third goal for the new system, of allowing managers to obtain accurate inventory levels at any time, is captured by Process 5.0. The two other processes in Figure 7-6, generating payments (4.0) and updating inventory levels due to usage (2.0), are part of the existing manual system.

The constraints on developing an enhanced inventory system at Hoosier Burger are also listed in Figure 7-4, again in order of descending priority. The first two constraints cover costs for systems development and for new computer hardware. Development can cost no more than $50,000. New hardware can cost no more than $50,000. The third constraint involves time for development—Hoosier Burger wants the system to be installed and in operation in no more than six months from the beginning of the development project. Finally, Hoosier Burger would prefer that training for the system be simple; the new system must be designed so that it is easy to use. However, as this is the fourth most important constraint, the demands it makes are more flexible than those contained in the other three.

FIGURE 7-5
THE STEPS IN HOOSIER BURGER'S
INVENTORY CONTROL SYSTEM

1. Meet delivery trucks before opening restaurant.
2. Unload and store deliveries.
3. Log invoices and file in accordion file.
4. Manually add amounts received to stock logs.
5. After closing, print inventory report.
6. Count physical inventory amounts.
7. Compare inventory reports totals to physical count totals.
8. Compare physical count totals to minimum order quantities; if the amount is less, make order; if not, do nothing.
9. Pay bills that are due and record them as paid.

Any set of alternative solutions to Hoosier Burger's inventory system problems must be developed with the company's prioritized requirements and constraints in mind. Figure 7-7 illustrates how each of three possible alternatives meets (or exceeds) the criteria implied in Hoosier Burger's requirements and constraints. Alternative A is a low-end solution. It meets only the first requirement completely and partially satisfies the second requirement, but it does not meet the final one. However, Alternative A is relatively inexpensive to develop and requires hardware that is much less expensive than the largest amount Hoosier Burger is willing to pay. Alternative A also meets the requirements for the other two constraints: It will take only three months to become operational and users will require only one week of training. Alternative C is the high-end solution. Alternative C meets all of the requirements criteria. On the other hand, Alternative C violates two of the four constraints: Development costs are high at $65,000 and time to operation is nine months. If Hoosier Burger really wants to satisfy all three of its requirements for its new inventory system, the

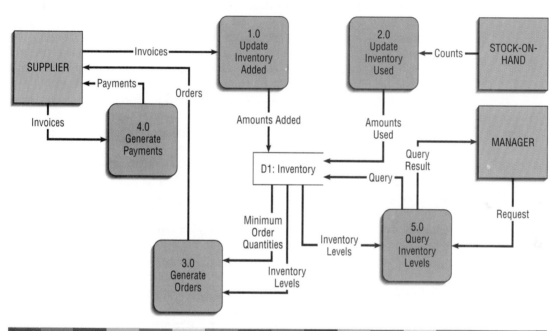

FIGURE 7-6
A Logical Data Flow Diagram Showing the Key Processes in Hoosier Burger's Desired Inventory System

CRITERIA	ALTERNATIVE A	ALTERNATIVE B	ALTERNATIVE C
Requirements			
1. Easy real-time entry of new shipment data	Yes	Yes	Yes
2. Automatic reorder decisions	For some items	For all items	For all items
3. Real-time data on inventory levels	Not available	Available for some items only	Fully available
Constraints			
1. Cost to develop	$25,000	$50,000	$65,000
2. Cost of hardware	$25,000	$50,000	$50,000
3. Time to operation	Three months	Six months	Nine months
4. Ease of training	One week of training	Two weeks of training	One week of training

FIGURE 7-7
Description of Three Alternative Systems That Could Be Developed for Hoosier Burger's Inventory System

company will have to pay more than it wants and wait longer for development. Once operational, however, Alternative C will take just as much time to train people to use as Alternative A. Alternative B is in the middle. This alternative solution meets the first two requirements, partially satisfies the third, and does not violate any of the constraints.

Now that three plausible alternative solutions have been generated for Hoosier Burger, the analyst hired to study the problem has to decide which one to recommend to management for development. Management will then decide whether to continue with the development project (incremental commitment) and whether the system recommended by the analyst should be developed.

Selecting the Most Likely Alternative

One method we can use to decide among the alternative solutions to Hoosier Burger's inventory system problem is illustrated in Figure 7-8. On the left, you see that we have listed all three system requirements and all four constraints from Figure 7-4. These are our decision criteria. We have weighted requirements as a group and constraints as a group equally; that is, we believe that requirements are just as important as constraints. We do not have to weight requirements and constraints equally; it is certainly possible to make requirements more or less important than constraints. Weights are arrived at in discussions among the analysis team, users, and sometimes managers. Weights tend to be fairly subjective and, for that reason, should be determined through a process of open discussion to reveal underlying assumptions, followed by an attempt to reach consensus among stakeholders. We have also assigned weights to each individual requirement and constraint. Notice that the total of the weights for both requirements and constraints is 50. Our weights correspond with our prioritization of the requirements and constraints.

Our next step is to rate each requirement and constraint for each alternative, on a scale of 1 to 5. A rating of 1 indicates that the alternative does not meet the requirement very well or that the alternative violates the constraint. A rating of 5 indicates that the alternative meets or exceeds the requirement or clearly abides by the constraint. Ratings are even more subjective than weights and

Criteria	Weight	Alternative A		Alternative B		Alternative C	
		Rating	Score	Rating	Score	Rating	Score
Requirements							
Real-time data entry	18	5	90	5	90	5	90
Auto reorder	18	3	54	5	90	5	90
Real-time data query	14	1	14	3	42	5	70
	50		158		222		250
Constraints							
Development costs	20	5	100	4	80	3	60
Hardware costs	15	5	75	4	60	4	60
Time to operation	10	5	50	4	40	3	30
Ease of training	5	5	25	3	15	5	25
	50		250		195		175
Total	100		408		417		425

FIGURE 7-8
WEIGHTED APPROACH FOR COMPARING THE THREE ALTERNATIVE SYSTEMS FOR HOOSIER BURGER'S INVENTORY SYSTEM

should also be determined through open discussion among users, analysts, and managers. The next step is to multiply the rating for each requirement and each constraint by its weight, and follow this procedure for each alternative. The final step is to add up the weighted scores for each alternative. Notice that we have included three sets of totals: for requirements, for constraints, and overall totals. If you look at the totals for requirements, Alternative C is the best choice (score of 250), as it meets or exceeds all requirements. However, if you look only at constraints, Alternative A is the best choice (score of 250), as it does not violate any constraints. When we combine the totals for requirements and constraints, we see that the best choice is Alternative C (score of 425), even though it had the lowest score for constraints, as it has the highest overall score.

Alternative C, then, appears to be the best choice for Hoosier Burger. Whether Alternative C is actually chosen for development is another issue. The Mellankamps may be concerned that Alternative C violates two constraints, including the most important one, development costs. On the other hand, the owners (and chief users) at Hoosier Burger may so want the full functionality Alternative C offers that they are willing to ignore the constraints violations. Or Hoosier Burger's management may be so interested in cutting costs that it prefers Alternative A, even though its functionality is severely limited. What may appear to be the best choice for a systems development project may not always be the one that ends up being developed.

Updating the Baseline Project Plan

You will recall that the Baseline Project Plan was developed during systems planning and selection (see Chapter 3) to explain the nature of the requested system and the project to develop it. We presented the plan originally in Figure 3-13 and reproduce it here as Figure 7-9. The plan includes a preliminary description of the system as requested, an assessment of the feasibility or justification for the system (the business case), and an overview of management issues for the system and project. This plan was presented to a steering committee or other body who approved the commitment of funds to conduct the analysis phase just completed. Thus, it is time to report back (in written and oral form) to this group on the project's progress and to update the group on the findings from analysis. This group will make the final decision on the design strategy to be followed and approve the commitment of resources outlined from the logical (and possibly physical) design steps. Of course, this group could determine that the business case has not developed as originally thought and either stop or drastically redirect the project.

The outline of the Baseline Project Plan can still be used for the analysis phase status report. The updated plan will typically be longer as more is known on each topic. Further, the various process, logic, and data models are often included to make the system description more specific. Usually only high-level versions of the diagrams are included within section 2.0, and more detailed versions are provided as appendices.

Every section of the Baseline Project Plan Report is updated at this point. For example, section 1.0.B will now contain the recommendation for the design strategy chosen by the analysis team. Section 2.0.A provides the descriptions of the competing strategies studied during alternative generation and selection, often including the types of comparison charts shown earlier in this chapter. Section 3.0 is typically significantly changed because you now know much better than you did during project initiation and planning what the needs of the organization are. For example, economic benefits that were intangible before may now be tangible. Risks, especially operational ones, are likely better understood.

BASELINE PROJECT PLAN REPORT

1.0 *Introduction*
   A. Project Overview—Provides an executive summary that specifies the project's scope, feasibility, justification, resource requirements, and schedules. Additionally, a brief statement of the problem, the environment in which the system is to be implemented, and constraints that affect the project are provided.
   B. Recommendation—Provides a summary of important findings from the planning process and recommendations for subsequent activities.

2.0 *System Description*
   A. Alternatives—Provides a brief presentation of alternative system configurations.
   B. System Description—Provides a description of the selected configuration and a narrative of input information, tasks performed, and resultant information.

3.0 *Feasibility Assessment*
   A. Economic Analysis—Provides an economic justification for the system using cost-benefit analysis.
   B. Technical Analysis—Provides a discussion of relevant technical risk factors and an overall risk rating of the project.
   C. Operational Analysis—Provides an analysis of how the proposed system solves business problems or takes advantage of business opportunities in addition to an assessment of how current day-to-day activities will be changed by the system.
   D. Legal and Contractual Analysis—Provides a description of any legal or contractual risks related to the project (e.g., copyright or nondisclosure issues, data capture or transferring, and so on).
   E. Political Analysis—Provides a description of how key stakeholders within the organization view the proposed system.
   F. Schedules, Time Line, and Resource Analysis—Provides a description of potential time frame and completion date scenarios using various resource allocation schemes.

4.0 *Management Issues*
   A. Team Configuration and Management—Provides a description of the team member roles and reporting relationships.
   B. Communication Plan—Provides a description of the communication procedures to be followed by management, team members, and the customer.
   C. Project Standards and Procedures—Provides a description of how deliverables will be evaluated and accepted by the customer.
   D. Other Project-Specific Topics—Provides a description of any other relevant issues related to the project uncovered during planning.

FIGURE 7-9
OUTLINE OF BASELINE PROJECT PLAN

Section 3.0.F will now show the actual activities and their durations during the analysis phase, as well as include a detailed schedule for the activities in the design phases and whatever additional details can be anticipated for later phases. Many Gantt charting packages can show actual progress versus planned activities. It is important to show in this section how well the actual conduct of the analysis phase matched the planned activities. This helps you and management understand how well the project is understood and how likely it is that the stated future schedule will occur. Knowing those activities whose actual durations differed significantly from planned durations may be very useful to you in estimating future activity durations. For example, a longer than expected task to analyze a certain process on a DFD may suggest that the design of system features to support this process may take longer than originally anticipated.

Often the design phase activities will be driven by the capabilities chosen for the recommended design strategy. For example, you place specific design activities on the schedule for such design deliverables as the following:

- Layout of each report and data input and display screens (DFDs include data flows for each of these)
- Structuring of data into logical tables or files (E-R diagrams identify what data entities are involved in this)
- Programs and program modules that need to be described
- Training on new technologies to be used in implementing the system

Many design phase activities result in developing design specifications for one or more examples of the types of design deliverables listed previously.

Section 4.0 is also updated. It is likely that the project team needs to change as new skills are needed in the next and subsequent project phases. Also, because project team members are often evaluated after each phase, the project leader may request the reassignment of a team member who has not performed as required. Reassessing the communication plan shows whether other communication methods are necessary. New standards and procedures will be needed as the team discovers that some current procedures are inadequate for the new tasks. Section 4.0.D is often used to outline issues for management that have been discovered during analysis. Recall, for example, that we discussed in Chapter 4 how you might find redundancies and inconsistencies in job descriptions and the way people actually do their jobs. Because these issues must be resolved by management and must be addressed before you can progress into detailed system design, now is the last time to call them to the attention of management.

As the project leader, you and other analysts also must ensure that the project workbook and CASE repository are completely up-to-date as you finalize the analysis phase. Because the project team composition will likely change and, as time passes, you forget facts learned in earlier stages, the workbook and repository are necessary to transfer information between phases. This is also a good time for the project leader to do a final check that all elements of project execution have been properly handled.

Besides the written Baseline Project Plan Report update, an oral presentation is typically made, and it may be at this meeting that a decision to approve your recommendations, redirect your recommendations, or kill the project is made. It is not uncommon for the analysis team to follow this project review meeting with a suitable celebration for reaching an important project milestone.

Before and After Baseline Project Plans for Hoosier Burger

Even though its inventory control system was relatively small, Hoosier Burger developed a Baseline Project Plan for the project. The plan included information for each area listed in Figure 7-9. Now that the analysis phase of the life cycle has ended, the plan must be updated. Those updated sections are reproduced below. The first item we consider is the cost-benefit analysis, Section 3.0.A (economic analysis) of the Baseline Project Plan. Hoosier Burger's initial cost-benefit analysis for the inventory project is shown in Figure 7-10. The spreadsheet format is the same as that used to summarize the cost and benefits of the Pine Valley Customer Tracking System in Chapter 3 (see Figure 3-11).

The numbers in the spreadsheet are based in part on the constraints listed in Figure 7-4 (budget, time frame, and training needs). The Mellankamps used the

Hoosier Burger							
Economic Feasibility Analysis							
Inventory Control System							
				Year of Product			
	Year 0	Year 1	Year 2	Year 3	Year 4	Year 5	TOTALS
Net economic benefit	$0	$30,000	$30,000	$30,000	$30,000	$30,000	
Discount rate (12%)	1	0.8928571	0.7971939	0.7117802	0.6355181	0.5674269	
PV of benefits	$0	$26,786	$23,916	$21,353	$19,066	$17,023	
NPV of all BENEFITS	$0	$26,786	$50,702	$72,055	$91,120	$108,143	$108,143
One-time COSTS	($100,000)						
Recurring Costs	$0	($2,000)	($2,000)	($2,000)	($2,000)	($2,000)	
Discount rate (12%)	1	0.8928571	0.7971939	0.7117802	0.6355181	0.5674269	
PV of Recurring Costs	$0	($1,786)	($1,594)	($1,424)	($1,271)	($1,135)	
NPV of all COSTS	($100,000)	($101,786)	($103,380)	($104,804)	($106,075)	($107,210)	($107,210)
Overall NPV							$934
Overall ROI - (Overall NPV / NPV of all COSTS)							0.01

FIGURE 7-10
HOOSIER BURGER'S INITIAL COST-BENEFIT ANALYSIS FOR ITS INVENTORY CONTROL SYSTEM PROJECT SUMMARIZED IN A SPREADSHEET

figures in Table 7-4 to determine the values in the spreadsheet in Figure 7-10. They estimated that the benefits of the new system could be quantified in two ways: First, instant entry of shipment data would lead to more accurate inventory data; second, Hoosier Burger would be less likely to run out of stock with an automatic order determination as part of the system. Savings from more accurate data amount to $1,500 per month or $18,000 per year; savings from fewer stock-outs amount to $1,000 per month or $12,000 per year. As you can see from Figure 7-10, although a new inventory control system for Hoosier Burger would break even, it is not a very good investment, with only a 1 percent return, given a 12 percent discount rate.

Figure 7-11 shows the cost-benefit analysis after the analysis phase has ended. This information appears in the updated Baseline Project Plan. Notice that developing the new system, represented by Alternative C in Figure 7-7, is now a better investment, with a 15 percent return. Yet the overall costs of Alternative C exceed the costs of the original estimation of the system. What happened?

Much of Figure 7-11 is the same as Figure 7-10. Recurring costs are the same; the discount rate is the same. What has changed, in addition to larger one-time costs, is that net benefits are now estimated to be larger than in Figure 7-10. Details on these new estimates are shown in Table 7-5. The Mellankamps reestimated the savings from the new system, using more accurate data, and found they had been too optimistic. The expected savings from instantly logging new shipments of supplies were reduced from $18,000 per year to $15,000 per year. But the Mellankamps also found a new benefit: They realized that better management information, and its ready availability through a new query capability, could be quantified at about $1,000 per month. The Mellankamps did not just dream up the

TABLE 7-4:   Hoosier Burger's Initial Economic Analysis Worksheet

One-time costs: Development	$50,000
One-time costs: Hardware	$50,000
Recurring costs: Maintenance	$2,000 per year
Savings: Fewer stock-outs due to automatic reordering	$12,000 per year
Savings: More accurate data from shipment logging	$18,000 per year
Intangible benefit: Better management information	

Hoosier Burger
Economic Feasibility Analysis
Inventory Control System

	Year 0	Year 1	Year 2	Year 3	Year 4	Year 5	TOTALS
				Year of Product			
Net economic benefit	$0	$39,000	$39,000	$39,000	$39,000	$39,000	
Discount rate (12%)	1	0.8928571	0.7971939	0.7117802	0.6355181	0.5674269	
PV of benefits	$0	$34,821	$31,091	$27,759	$24,785	$22,130	
NPV of all BENEFITS	$0	$34,821	$65,912	$93,671	$118,457	$140,586	$140,586
One-time COSTS	($115,000)						
Recurring Costs	$0	($2,000)	($2,000)	($2,000)	($2,000)	($2,000)	
Discount rate (12%)	1	0.8928571	0.7971939	0.7117802	0.6355181	0.5674269	
PV of Recurring Costs	$0	($1,786)	($1,594)	($1,424)	($1,271)	($1,135)	
NPV of all COSTS	($115,000)	($116,786)	($118,380)	($119,804)	($121,075)	($122,210)	($122,210)
Overall NPV							$18,377
Overall ROI - (Overall NPV / NPV of all COSTS)							0.15

FIGURE 7-11
Hoosier Burger's Revised Cost-Benefit Analysis for Its Inventory Control Project

$1,000 per month savings estimate. They developed it from thinking about how timely, accurate management information would affect their ability to prepare reports as well as their ability to improve their operations through better inventory control. So even though Alternative C was more expensive to develop than the other alternatives, it actually resulted in a higher level of tangible benefits.

Figure 7-12 shows the project schedule from the initial version of the Baseline Project Plan. Notice that the schedule covers only the design and implementation phases of the life cycle and that the schedule is very general. The physical aspects of design are not broken down into constituent parts. The task times in the schedule are also driven by two of the constraints listed in Figure 7-4. The entire schedule spans exactly six months of activity and training takes only one week. The estimates of how long each task should take to complete are all very rough.

TABLE 7·5: Hoosier Burger's Updated Economic Analysis Worksheet

One-time costs: Development	$65,000
One-time costs: Hardware	$50,000
Recurring costs: Maintenance	$2,000 per year
Savings: Fewer stock-outs due to automatic reordering	$12,000 per year
Savings: More accurate data from shipment logging	$15,000 per year
Savings: Better management information and availability	$12,000 per year

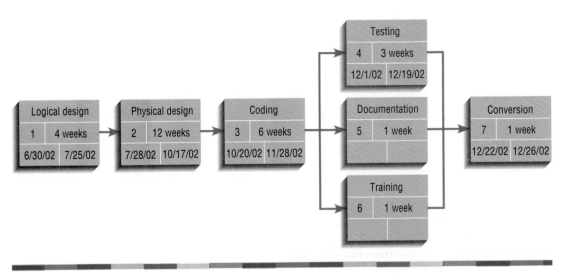

FIGURE 7-12
HOOSIER BURGER'S INITIAL SCHEDULE FOR ITS INVENTORY CONTROL SYSTEM PROJECT

Compare Figure 7-12 to Figure 7-13, the revised schedule that goes in the updated Baseline Project Plan. The schedule in Figure 7-13 is more detailed, and it more closely reflects the development time necessary for Alternative C. Training still takes only one week, but now that estimate is based on a clear understanding of the requirements of a particular system rather than primarily on positive thinking. Also, the entire schedule now spans nine months, the time necessary to fully develop and implement Alternative C. Some design tasks in Figure 7-13 have been decomposed into four different subtasks, many of which can be worked on concurrently: interface design, designing the internals, physical database design, and data conversion. Note that even this schedule presents the project at a very high level. It would be typical in actual projects to show not only the major steps but also the individual activities needed to complete each step. For example, interface design might be broken down into many steps for each display or report and for different activities, such as meetings with users to walk through the tentative designs.

We have shown you only two parts to the Hoosier Burger Baseline Project Plan for its Inventory Control System. Even for a project this small, a complete Baseline Project Plan would be too much to include in this book. From these examples, though, you should get a good general idea of both what an initial Baseline Project Plan contains and how it changes when a major life cycle phase, like analysis, ends.

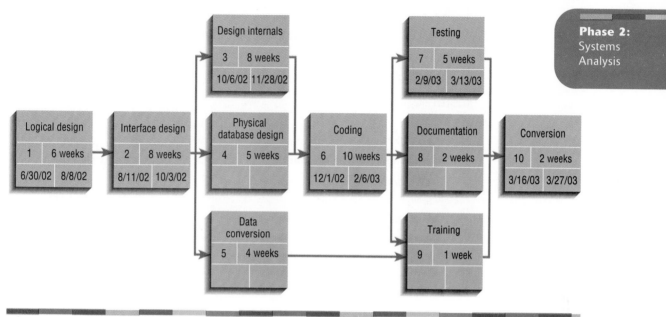

FIGURE 7-13
HOOSIER BURGER'S REVISED SCHEDULE FOR ITS INVENTORY CONTROL SYSTEM PROJECT

ELECTRONIC COMMERCE APPLICATION: SELECTING THE BEST ALTERNATIVE DESIGN STRATEGY

Selecting the best design strategy for an Internet-based electronic commerce application is no different than the process followed when selecting the optimal design strategy for other types of applications. In the last chapter, you read how Jim Woo modeled the data requirements for the WebStore system. In this section, we examine the process he followed when assessing and selecting a design strategy for the WebStore.

Selecting the Best Alternative Design Strategy for Pine Valley Furniture's WebStore

As Jim Woo began to evaluate the possible design options for the WebStore, he quickly realized that he and PVF's technical group had limited understanding of Internet application development. Consequently, he recommended to PVF management that a consulting firm be hired to assist in setting the WebStore design options. Management quickly approved this recommendation, and Jim retained a small consulting organization that had a strong reputation for designing and developing very high-quality electronic commerce solutions. Once under contract, Jim worked with the consulting firm to solidify the system requirements and constraints. During this process, they organized the requirements into three categories: minimum system requirements, essential system requirements, and desired system requirements, summarized in Table 7-6. In addition to the system requirements, they also identified four significant constraints that any design must address, also summarized in Table 7-6.

Next, Jim and the consultants defined three alternative system designs, with advantages and disadvantages for each. PVF management requested that three alternative designs be defined so that clear comparisons could be made between low-end (low cost and limited features), high-end (high cost and extensive features), and midlevel designs (moderate cost and features). Table 7-7 summarizes the results of this analysis. Now that both the system requirements

TABLE 7-6:   WebStore System Requirements and Constraints

Requirements	Constraints
Minimum System Requirements	Christmas season rollout
Full integration with current inventory, sales, and customer tracking systems	Small development/support staff
99.9% uptime and availability	Transaction-style interaction with current systems
Essential System Requirements	Limited external consultation budget
Flexibility and scalability for future systems integration	
Efficient and cost-effective system management	
Desired System Requirements	
Available support and/or emergency response	
Documentation	

TABLE 7-7:   Three Alternative Systems and Their Advantages and Disadvantages

1. Outsource Application Service Provider (Low-end)

Advantages	*Disadvantages*
All hardware located off-site	Inflexible
Application developed and professionally managed off-site	Difficult to integrate with current systems
Excellent emergency response	Shared resources with other clients

2. Enterprise Resource Planning System (High-end)

Advantages	*Disadvantages*
Stability	Requires skilled internal staff
Available documentation	Expensive hardware and software
	Big learning curve

3. Application Server/Object Framework (Moderate)

Advantages	*Disadvantages*
Excellent integration with current system	Requires internal development (and/or a professional consultation)
Scalability	Proprietary
Flexible	Documentation crucial during planning and development

and constraints, as well as the alternative system designs, had been defined, a meeting was held with PVF management to select a design strategy for the WebStore. At this meeting, it was unanimously agreed upon that option 3 in Table 7-7, Application Server/Object Framework, the moderate alternative, best suited both PVF's current needs and future growth initiatives.

The proposed system would incorporate a scalable three-tier architecture to integrate the WebStore with the current systems. A **scalable** system has the

Scalable
The ability to seamlessly upgrade the capabilities of the system through either hardware upgrades, software upgrades, or both.

ability to seamlessly upgrade the capabilities of the system through hardware upgrades, software upgrades, or both. As Figure 7-14 shows, Tier 1, the Web server layer, processes incoming Internet requests. For example, a *scalable* electronic commerce system would be one that could effectively handle six requests per second with one server, and by adding a second server, twelve requests per second could be effectively handled. The **Web server** is a computer that is connected to the Internet and stores files written in HTML—hypertext markup language—which are publicly available through an Internet connection. As shown in Figure 7-14, the Web server layer communicates with Tier 2, the application server layer. The **application server** is a "middle tier" software and hardware combination that lies between the Web server and the corporate network and systems such as the Customer Tracking System, Inventory Control System, and Order Fulfillment System. In other words, the Web server manages the client interaction and broker requests to the middle-tier application server. The application server manages the data specific to running the WebStore (shopping carts, promotions, site logs, etc.) and also manages all interactions with existing PVF systems for managing customers, inventory, and orders. A third server, the fail-over server, is an emergency backup system on standby, ready to take the place of either server should one fail. Each of these separate components—Web server, application server, fail-over server—can be thought of as an *object* (see Appendix A), each with a well-defined role that can be easily defined, designed, implemented, and modified. For this reason, option 3 is referred to as an Application Server/Object Framework architecture.

Phase 2:
Systems
Analysis

Web server
A computer that is connected to the Internet and stores files written in HTML—hypertext markup language—which are publicly available through an Internet connection.

Application server
A "middle tier" software and hardware combination that lies between the Web server and the corporate network and systems.

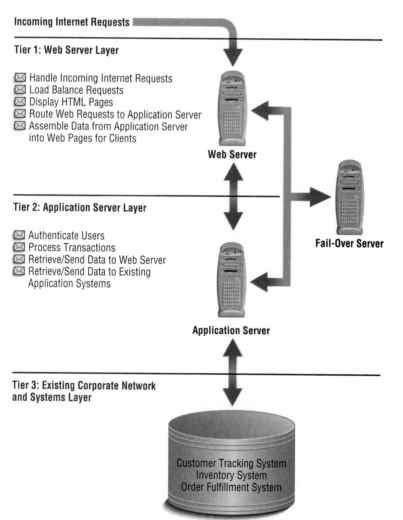

FIGURE 7-14
WEBSTORE MULTITIER SYSTEM
ARCHITECTURE

Incoming Internet Requests

Tier 1: Web Server Layer

- Handle Incoming Internet Requests
- Load Balance Requests
- Display HTML Pages
- Route Web Requests to Application Server
- Assemble Data from Application Server into Web Pages for Clients

Web Server

Fail-Over Server

Tier 2: Application Server Layer

- Authenticate Users
- Process Transactions
- Retrieve/Send Data to Web Server
- Retrieve/Send Data to Existing Application Systems

Application Server

Tier 3: Existing Corporate Network and Systems Layer

Customer Tracking System
Inventory System
Order Fulfillment System

Now that the basic architecture of the WebStore had been defined, Jim worked with the consultants to further refine the specifications of the system. The PVF development staff will use these specifications as a blueprint in their development efforts and will implement all six system requirements and comply with each of the four constraints listed in Table 7-6. In the next chapter, we learn how Jim and the PVF development group designed the human interface for the WebStore.

Key Points Review

In selecting the best alternative design strategy, you develop alternative solutions to the organization's information system problem. A design strategy is a combination of system features, hardware and system software platform, and acquisition method that characterizes the nature of the system and how it will be developed. A good number of alternative design strategies to develop is three, as three alternatives can represent the high end, middle, and low end of the spectrum of possible systems that can be built.

1. **Describe the different sources of software.**

   You can obtain application (and system) software from hardware vendors, packaged software vendors, custom software developers, and enterprise-wide solution vendors as well as from internal systems development resources. You can even hire an organization to handle all of your systems development needs, called outsourcing. You must also know which criteria to use to be able to choose among off-the-shelf software products. These criteria include cost, functionality, vendor support, vendor viability, flexibility, documentation, response time, and ease of installation. You must also determine whether new hardware and system software are needed. Requests for proposals are one way to collect more information about hardware and system software, their performance, and costs. In addition to hardware and software issues, you must consider implementation issues and broader organizational concerns, such as the availability of funding and management support.

2. **Assemble the various pieces of an alternative design strategy.**

   Alternative design strategies are developed after a system's requirements and constraints have been identified and prioritized. You can start initially with a simple list of features and obstacles for each alternative you are considering. From there, you can assign numerical weights to each of your system requirements and constraints. Finally, you can rate each alternative on each attribute and use these numbers to determine numerical rankings of your alternatives.

3. **Generate at least three alternative design strategies for an information system.**

   Generating different alternatives is something you would do in actual systems analysis or as part of a class project. Three is not a magic number. It represents instead the endpoints and midpoints of a series of alternatives, such as the most expensive, the least expensive, and an alternative somewhere in the middle.

4. **Select the best design strategy using both qualitative and quantitative methods.**

   Once developed, alternatives can be compared to each other through quantitative methods, but the actual decision may depend on other criteria, such as organizational politics. In this chapter, you were introduced to one way to compare alternative design strategies quantitatively.

5. **Update a Baseline Project Plan based on the results of the analysis phase.**

   Because generating and selecting alternative design strategies completes the analysis phase of the SDLC, the systems development project has reached a major milestone. Once an analysis of alternative design strategies is completed, you and other members of the analysis team present your findings to a management steering committee and/or the client requesting the system change. In this presentation (both written and oral), you summarize the requirements discovered, evaluate alternative design strategies, and justify the recommended alternative as well as present an updated Baseline Project Plan for the project to follow if the committee decides to fund the next life cycle phase.

6. **Understand how selecting the best design strategy applies to development for Internet applications.**

   Most of the same techniques used for selecting the best design strategy for traditional systems can also be fruitfully applied to the development of Internet applications. In the PVF example, you saw how analysts generated three alternative designs for the WebStore Internet application, based on system requirements and constraints, just as was done in the Hoosier Burger example.

Although Internet applications are very different in some ways from traditional systems, many of the same analysis tools and techniques can be used profitably in each.

Key Terms Checkpoint

Here are the key terms from the chapter. The page where each term is first explained is in parentheses after the term.

1. **Application server (p. 259)**
2. **Design strategy (p. 236)**
3. **Enterprise solutions or enterprise resource planning (ERP) systems (p. 242)**
4. **Outsourcing (p. 240)**
5. **Request for proposal (RFP) (p. 247)**
6. **Scalable (p. 258)**
7. **Web server (p. 259)**

Match each of the key terms above with the definition that best fits it.

_____ 1. A computer that is connected to the Internet and stores files written in HTML—hypertext markup language—which are publicly available through an Internet connection.

_____ 2. A particular approach to developing an information system. It includes statements on the system's functionality, hardware and system software platform, and method for acquisition.

_____ 3. The ability to seamlessly upgrade the capabilities of the system through either hardware upgrades, software upgrades, or both.

_____ 4. A system that integrates individual traditional business functions into a series of modules so that a single transaction occurs seamlessly within a single information system rather than several separate systems.

_____ 5. The practice of turning over responsibility of some or all of an organization's information systems applications and operations to an outside firm.

_____ 6. A document provided to vendors to ask them to propose hardware and system software that will meet the requirements of your new system.

_____ 7. A "middle tier" software and hardware combination that lies between the Web Server and the corporate network and systems.

Review Questions

1. What are the deliverables from selecting the best alternative design strategy?
2. Why generate at least three alternatives?
3. Describe the five sources of software.
4. How do you decide among various off-the-shelf software options? What criteria do you use?
5. What issues are considered when analysts try to determine whether new hardware or system software is necessary?
6. What is an RFP and how do analysts use one to gather information on hardware and system software?
7. What issues other than hardware and software must analysts consider in preparing alternative system design strategies?
8. How do analysts generate alternative solutions to information systems problems?
9. How do managers decide which alternative design strategy to develop?
10. Which elements of a Baseline Project Plan might be updated during the alternative generation and selection step of the analysis phase of the SDLC?
11. What methods can a systems analyst employ to verify vendor claims about a software package?
12. What are enterprise resource planning systems? What are the benefits and disadvantages of such systems as a design strategy?

Problems and Exercises

1. Find the most current Software 500 list from *Software Magazine*. How much has the rank order of the top software companies changed compared to the list in Table 7-2? Try to determine why your list is different from that of Table 7-2. What changes are occurring in the computer industry that might affect this list?

2. Research how to prepare a request for proposal. Prepare an outline of an RFP for Hoosier Burger to use in collecting information on its new inventory system hardware.

3. Re-create the spreadsheet in Figure 7-8 in your spreadsheet package. Change the weights and compare the outcome to Figure 7-8. Change the rankings. Add criteria. What additional information does this "what if" analysis provide for you as a decision maker? What insight do you gain into the decision-making process involved in choosing the best alternative system design?

4. Prepare a list for evaluating computer hardware and system software that is comparable to the list of criteria for selecting off-the-shelf application software presented earlier.

5. The method for evaluating alternatives used in Figure 7-8 is called weighting and scoring. This method implies that the total utility of an alternative is the product of the weights of each criterion times the weight of the criterion for the alternative. What assumptions are characteristic of this method for evaluating alternatives? That is, what conditions must be true for this to be a valid method of evaluation alternatives?

6. Weighting and scoring (see Problem and Exercise 5) is only one method for comparing alternative solutions to a problem. Go to the library, find a book or articles on qualitative and quantitative decision making and voting methods, and outline two other methods for evaluating alternative solutions to a problem. What are the pros and cons of these methods compared to the weighting and scoring method? Under weighting and scoring and the other alternatives you find, how would you incorporate the opinions of multiple decision makers?

7. Prepare an agenda for a meeting at which you would present the findings of the analysis phase of the SDLC to Bob Mellankamp concerning his request for a new inventory control system. Use information provided in Chapters 4 through 7 as background in preparing this agenda. Concentrate on which topics to cover, not the content of each topic.

8. Review the criteria for selecting off-the-shelf software presented in this chapter. Use your experience and imagination and describe other criteria that are or might be used to select off-the-shelf software in the real world. For each new criterion, explain how its use might be functional (i.e., it is useful to use this criterion), dysfunctional, or both.

9. The owner of two pizza parlors located in adjacent towns wants to computerize and integrate sales transactions and inventory management within and between both stores. The point-of-sale component must be very easy to use and flexible enough to accommodate a variety of pricing strategies and coupons. The inventory management, which will be linked to the point-of-sale component, must also be easy to use and fast. The systems at each store need to be linked so that sales and inventory levels can be determined instantly for each store and for both stores combined. The owner can allocate $40,000 for hardware and $20,000 for software and must have the new system operational in three months. Training must be very short and easy. Briefly describe three alternative systems for this situation and explain how each would meet the requirements and constraints. Are the requirements and constraints realistic? Why or why not?

10. Compare the alternative systems from Problem and Exercise 9 using the weighted approach demonstrated in Figure 7-8. Which system would you recommend? Why? Was the approach taken in this and Problem and Exercise 9 useful even for this relatively small system? Why or why not?

11. Suppose that an analysis team did not generate alternative design strategies for consideration by a project steering committee or client. What might the consequences be of having only one design strategy? What might happen during the oral presentation of project progress if only one design strategy is offered?

12. In the section on choosing off-the-shelf software, eight criteria are proposed for evaluating alternative packages. Suppose the choice was between alternative custom software developers rather than prewritten packages. What criteria would be appropriate to select and compare among competing bidders for custom development of an application? Define each of these criteria.

13. How might the project team recommending an enterprise resource planning design strategy justify its recommendation as compared to other types of design strategies?

Discussion Questions

1. Consider the purchase of a new PC to be used by you at your work (or by you at a job that you would like to have). Describe in detail three alternatives for this new PC that represent the low, middle, and high points of a continuum of potential solutions. Be sure that the low-end PC meets at least your minimum requirements and the high-end PC is at least within a reasonable budget. At this point, without quantitative analysis, which alternative would you choose?

2. For the new PC described above, develop ranked lists of your requirements and constraints as displayed in Figure 7-8. Display the requirements and constraints, along with the three alternatives, as done in Figure 7-8, and note how each alternative is rated on each requirement and constraint. Calculate scores for each alternative on each criterion and compute total scores. Which alternative has the highest score? Why? Does this choice fit with your selection in the previous question? Why or why not?

3. One of the most competitive software markets today is electronic spreadsheets. Pick three packages (e.g., Microsoft Excel, Lotus 1-2-3, and Quattro Pro—but any three spreadsheet packages would do). Study how to use spreadsheet packages for school, work, and personal financial management. Develop a list of criteria important to you on which to compare alternative packages. Then contact each vendor and ask for the information you need to evaluate its package and company. Request a demonstration copy or trial use of its software. If the company cannot provide a sample copy, then try to find a computer software dealer or club where you can test the software and documentation. Based on the information you receive and the software you use, rate each package using your chosen criteria. Which package is best for you? Why? Talk to other students and find out which package they rated as best. Why are there differences between what different students determined as best?

4. Obtain copies of actual requests for proposals used for information systems developments and/or purchases. If possible, obtain RFPs from public and private organizations. Find out how they are used. What are the major components of these proposals? Do these proposals seem to be useful? Why or why not? How and why are the RFPs different for the public versus the private organizations?

Case Problems

1. Pine Valley Furniture

During your time as a Pine Valley Furniture intern, you have learned much about the systems analysis and design process. You have been able to observe Jim Woo as he serves as the lead analyst on the WebStore project, and you have also received hands-on experience with the Customer Tracking System project. The requirements determination and requirements structuring activities for the Customer Tracking System are now complete, and it is time to begin generating alternative design strategies.

On Monday afternoon, Jim Woo stops by your desk and requests that you attend a meeting scheduled for tomorrow morning. He mentions that during tomorrow's meeting, the Customer Tracking System's requirements and constraints, weighting criteria, and alternative design strategy ratings will be discussed. He also mentions that during the previously conducted systems planning and selection phase, Jackie Judson and he prepared a Baseline Project Plan. At the time the initial Baseline Project Plan was prepared, the in-house development option was the preferred design strategy. The marketing group's unique information needs seemed to indicate that in-house development was the best option. However, other alternative design strategies have since been investigated.

During Tuesday's meeting, several end users, managers, and systems development team members meet, discuss, and rank the requirements and constraints for the new Customer Tracking System. Also, weights and rankings are assigned to the three alternative design strategies. At the end of the meeting, Jim Woo assigns you the task of arranging this information into a table and calculating the overall scores for each alternative. He would like to review this information later in the afternoon. Tables 7-8 and 7-9 summarize the information obtained from Tuesday's meeting.

TABLE 7-8

Criteria	Alternative A	Alternative B	Alternative C
New Requirements			
Ease of use	Acceptable	Fair	Good
Easy real-time updating of customer profiles	Yes	Yes	Yes
Tracks customer purchasing activity	No	Yes	Yes
Supports sales forecasting	Some forecasting models are supported	Some forecasting models are supported	Provides support for all necessary forecasting models
Ad hoc report generation	No	Yes	Yes
Constraints			
Must interface with existing systems	Requires significant modifications	Minor modifications	Minor modifications
Costs to develop	$150,000	$200,000	$350,000
Cost of hardware	$80,000	$80,000	$100,000
Time to operation	6 months	7 months	9 months
Must interface with existing systems	Requires significant modifications	Minor modifications	Minor modifications
Ease of training	3 weeks of training	3 weeks of training	2 weeks of training
Legal restrictions	Cannot be modified	Allows for customization	None

TABLE 7-9

Criteria	Weight	Alternative A		Alternative B		Alternative C	
Requirements		Rating	Score	Rating	Score	Rating	Score
Ease of use	15	2		3		5	
Real-time customer profile updating	12	3		3		4	
Tracks customer purchasing activity	12	1		3		3	
Sales forecasting	8	2		2		3	
Ad hoc report generation	3	1		2		3	
Total	50						
Constraints							
Interfaces with existing systems	15	3		4		2	
Development costs	10	5		4		2	
Hardware costs	10	5		4		2	
Time to operation	5	4		1		2	
Ease of training	5	2		2		4	
Legal restrictions	5	1		2		5	
Total	50						

a. Generally speaking, what alternative design strategies were available to Pine Valley Furniture?
b. Of the alternative design strategies available to Pine Valley Furniture, which were the most viable? Why?
c. Using the information provided in Table 7-9, calculate the scores for each alternative.
d. Based on the information provided in Tables 7-8 and 7-9, which alternative do you recommend?

2. Hoosier Burger

As the lead analyst on the Hoosier Burger project, you have been busy collecting, structuring, and evaluating the new system's requirements. During a Monday morning meeting with Bob and Thelma, the three of you review the system requirements, system constraints, and alternative design strategies. The proposed alternative design strategies address low-end, midrange, and high-end solutions. Additionally, weights are assigned to the evaluation criteria, and the alternatives are ranked according to the criteria.

Bob has stated repeatedly that his main priority is to implement an inventory control system. However, you are aware that, if at all possible, Bob would like to also implement a delivery system. You inform Bob that two of the alternative design strategies support a delivery system, but will increase the system's development cost by at least $20,000 and add $10,000 in recurring costs

to the new system. Bob feels that the addition of the new delivery system will result in $25,000 in yearly benefits over the life of the new system.

The inclusion of a delivery system necessitates the addition of several new requirements and the modification of system constraints. Table 7-10 outlines these changes. The weights, ratings, and scores also require adjustments. Table 7-11 contains information about these adjustments.

a. Generally speaking, what alternative design strategies are available to Hoosier Burger?
b. Is an enterprise resource planning system a viable option for Hoosier Burger? Why or why not?
c. Modify Figure 7-8 to incorporate the criteria mandated by the new delivery system. Which alternative should be chosen?
d. Assuming that Alternative C is still chosen, update Hoosier Burger's economic feasibility analysis to reflect the changes mentioned in this scenario.

3. Whistler Car Wash

Whistler Car Wash was founded by Bradley James almost 30 years ago, and has its headquarters in Atlanta, Georgia. The company is a leading national manufacturer of car wash equipment, parts, and supplies. Whistler's mission is to provide its customers with excellent service both before and after the sale. To fulfill its mission, the

TABLE 7-10

Criteria	Alternative A	Alternative B	Alternative C
New Requirements			
Easy real-time entry of new shipment data	Yes	Yes	Yes
Automatic reorder decisions	For some items	For all items	For all items
Real-time data on inventory levels	Not available	Available for some items only	Fully available
Facilitates forecasting	Not available	Available	Available
Track delivery sales	Available	Available	Available
Customer billing	Not available	Not available	Available
Constraints			
Costs to develop	$45,000	$70,000	$85,000
Cost of hardware	$25,000	$50,000	$50,000
Time to operation	4 months	7 months	10 months
Ease of training	1 week of training	3 weeks of training	3 weeks of training

TABLE 7-11

Criteria	Weight	Alternative A		Alternative B		Alternative C	
Requirements		Rating	Score	Rating	Score	Rating	Score
Real-time data entry	12	5		5		5	
Auto reorder	12	3		5		5	
Real-time data query	10	1		3		5	
Facilitates forecasting	8	1		2		3	
Track delivery sales	5	3		3		3	
Customer billing	3	1		1		3	
Total	50						
Constraints							
Development costs	20	5		4		2	
Hardware costs	15	5		4		3	
Time to operation	10	5		4		3	
Ease of training	5	2		1		5	
Total	50						

company offers a variety of self-service,-automatic, and tunnel car wash alternatives. Additionally, the company helps its customers select car wash sites, determine car wash layouts, and build and install car washes. Whistler has an excellent technical support team available 24-hours a day, 7 days a week. Self-service operations, convenience stores, car dealerships, and rental car agencies all across the United States use Whistler Car Wash equipment, products, and supplies.

While Whistler executives believe the company is achieving its mission, they have watched over the years as the competition has slowly eroded Whistler's once dominant market share. In recent years, Whistler's Information Systems Department has been inundated with systems development requests. The manufacturing department has submitted a request to renovate its inventory tracking system. While the Sales Department has requested an updated order-entry system, the Accounting Department has recently submitted a request to reengineer its

accounts receivable and accounts payable systems. In addition, information exchange between the functional areas is cumbersome at best.

As a senior systems analyst in Whistler's Information Systems Department, you have recently been appointed to the company's steering committee. The steering committee is in charge of evaluating all information systems requests, and is responsible for approving or rejecting these requests.

a. Based on the systems development requests mentioned above, what alternative design strategies are available?
b. Would enterprise resource solution software be viable? If so, for which systems development requests?
c. It is likely that a request for proposal will be sent to several vendors. Locate a Web site that provides information about requests for proposals. Briefly summarize your findings.
d. When shaping alternative design strategies, what processes are involved?

CASE: BROADWAY ENTERTAINMENT COMPANY, INC.

Formulating a Design Strategy for the Web-Based Customer Relationship Management System

Case Introduction

Defining a design strategy for a systems development project is typically a crucial step for a project team. The team must select the scope of functionality, the implementation platform, and the method of acquisition for the system. There may be many alternatives, and different sets of users may support different design strategies. The design strategy, chosen near the end of the analysis phase, is also crucial because it sets the direction for the rest of the project. In contrast, a development approach that uses a prototype to prove the concept for the value of a system need not be perfect. Further, the prototype does not need to take the exact form of the final system. For these reasons, some choices can be made to meet shorter-term rather than longer-term objectives. This is the case with MyBroadway, the customer relationship information system being developed by a team of students from St. Claire Community College for Carrie Douglass, manager of the Broadway Entertainment Company store in Centerville, Ohio.

Determining the System Functionality

The activities of systems analysis for the MyBroadway project have identified possible system requirements. Usually a prototyping methodology follows a repeti-tive development process of building, testing, and rebuilding a system until the user agrees that the system functions as desired. In the case of MyBroadway, the prototype will be used more during the design (rather than analysis) phase to refine system requirements, screens, and report layouts, and to test customer acceptance (see BEC Figure 7-1). So far, a structured approach has been taken to determine initial requirements. The team took this combined approach of initially a structured and then a prototyping methodology because data flow and storage issues needed to be addressed. For example, structured techniques allowed the team to study connections between MyBroadway and the Entertainment Tracker store management system. Prototyping works best once the data requirements are in place, but the use of the data still needs to be decided. Thus, the team fully expects changes to be made to the functional requirements during design.

Because prototyping will be used in subsequent steps, the student team decided not to propose several alternative (or extreme) functional scopes to Carrie Douglass. The team members suggest that alternatives should be considered in an evolutionary fashion during design. The only functional scope requirement the team raised with Carrie was a big one and needed to be clarified before the prototyping began. The issue related to one finding from the observation of customers using other, non-competing online shopping services. Several customers voiced their interest in being able to preview products before buying or renting. Their

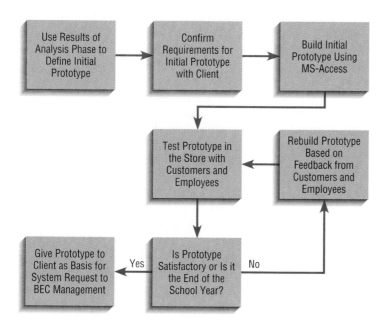

BEC FIGURE 7-1
DESIGN PROCESS FOR MYBROADWAY SYSTEM

suggestion was to include a video or sound clip with the other data about products and then be able to play these as streaming video or audio via the Web.

The team explained to Carrie that although this capability is definitely technically feasible, it would present several problems. First, the data storage requirements for MyBroadway would greatly increase. Second, some Web browsers might have difficulty playing the video or audio streams, which would cause user frustration. Third, although client tools for playing multimedia streams are inexpensive, the server-side software is rather costly (in terms of Carrie's budget), so platform cost would increase. Finally, there would be considerable effort to create the video or audio streams. The team suggested that once the system being developed for Carrie is proven without this feature and BEC management has a chance to see the system, they could decide to include this capability in a company-wide rollout of the product. Carrie also pointed out that for a company-wide rollout, there would be multiple-language capabilities required of streaming multimedia, because BEC operates stores in both Spanish- and French-speaking locations. Carrie agreed to keep this feature out of the prototype. She added, however, that it had been her experience in using Amazon.com and other services that once she actually used the system, her concept of what features she wanted available to her changed. Thus, she wanted the team to know that although its analysis so far had been good, the functional capabilities should be able to change as easily as possible during the design prototyping.

Determining the System Platform

The platform chosen by the students was somewhat of a surprise to Carrie—the team suggested using Microsoft Access. Carrie thought the team would choose a specialized Web development tool, such as Microsoft FrontPage. The St. Claire student team explained that it felt that Access was a better prototyping platform than FrontPage. The final system would likely be built using a system like FrontPage, using active server pages accessing a backend database. However, the team members suggested that the prototype system be tested in the store on a stand-alone PC, and not be placed on the Web. They felt that a Web site, although it would allow use from home, was too risky for the prototype. The prototype would likely change frequently, and the changing user interface could be confusing to customers. Testing could be more controlled and monitored by allowing customers to test the prototype only in the store. Also, and very importantly, Carrie would not have to rent space from an Internet service provider (ISP) for the prototype.

Instead of contracting with an ISP, the prototype could be placed on a Web server at St. Claire Community College, although customers might be confused by the St. Claire URL being involved in the site name.

Microsoft Access screens could be designed to look very similar to browser screens. Both customers and employees could use the system. The students are quite familiar with Access, so there would be minimal, if any, training time. Access should provide fast response during prototype testing. Also, the Computer Information Systems department at St. Claire has a laptop computer and extra 17-inch monitor the team had arranged to use for the project. The monitor could be left at the store, and the team could develop the prototype on the laptop and bring it to the store for customer and employee use. The laptop would have sufficient power for one user at a time. The laptop already had the latest version of Access loaded on it, part of the campus computing requirement instituted two years ago, through which each student received a license for all Microsoft Office and Visual Studio tools.

Carrie was pleased with the prototype platform recommendation, primarily because it meant no additional money to be spent on the project. She had already bought more than enough drinks, pizza, and wings at a nearby restaurant for the various project status report sessions.

Determining the System Acquisition Methodology

The St. Claire student team felt that it had sufficient skills and time to build the prototype. Thus, no third party would be needed during design and implementation. Also, because prototyping would be used to fine-tune requirements and to develop experience with usage of the system, it did not seem feasible to use a package. Actually, the team never considered a package, and Carrie never asked. Carrie was already committed to a custom-developed system, one that she could say would be hers.

The team did raise one acquisition question with Carrie. The question was whether Carrie knew if any other BEC store had tried to build such a system, or if the BEC corporate IT organization was developing a prototype of a Web-based system. Carrie did not know the answer to either question and was still reluctant to let other BEC stores or corporate employees know about the project she was sponsoring. Carrie and the student team spent about half an hour using Web search engines to try to find a similar service from any video and audio media store but were unable to find such a system. Not that Carrie would have supported buying a system from another store, but it would have at least been interesting to see what others might be doing.

Case Summary

The St. Claire student team had accomplished a lot since the first meeting with Carrie Douglass. Carrie is pleased with the progress. Carrie gave the team the go-ahead to start to build a prototype, using the design strategy the team suggested. But, Carrie was still concerned about the rest of the project. What would happen once customers and employees saw a real working system? Would they start asking for many additional features? How could their expectations be tempered? How would the team know when the project was done, if people kept suggesting new features? Even though the student team felt confident that a laptop computer would work well to simulate both a client and server for a Web-based customer relationship management system, would this assessment be right? When should Carrie expose BEC corporate IS people to the project? So far, the team members had been able to find out everything they needed to know about Entertainment Tracker from the user documentation provided to the store, but would this continue as more technical design issues arose?

Case Questions

1. Was the St. Claire student team wise to not suggest several alternative sets of functional capabilities for MyBroadway? What are the risks of considering only one starting point for prototyping?
2. Although the St. Claire student team and Carrie could find no other entertainment media store that had a Web site, certainly companies in competition with BEC have Web sites. Visit the Web sites for Blockbuster and at least one other BEC competitor. From reviewing those sites, what other features would you anticipate might be suggested during customer and employee use of the initial MyBroadway prototype? Given your answers to questions in previous BEC cases in this text, will the architecture of the prototype (including the database model and process flow models for acquiring and reporting data) be changed much by some of the Web site features you find? Why or why not? Is scalability an issue when using prototyping as the systems development methodology?
3. What is your assessment of the recommendation to use Microsoft Access as the prototyping platform? Would you suggest an alternative? Why? Under what circumstances would an alternative platform be better?
4. Carrie is concerned with user expectation management as the project moves forward. Given the additional information you know now about the system being developed, MyBroadway, what would you recommend the team do to limit user expectations or to handle frustrations if users are underwhelmed by what they see in the prototypes?
5. This case states that the student team thought that including video and audio clips would be a problem, at least for the prototype. Do you agree? Why or why not? One might also argue that because a BEC store carries over 4,000 different products, there is already a database problem even without including multimedia product clips data. What prototype implementation issues does this large product inventory have on the development of the prototypes? What suggestions would you make for changes (or refinements) to the design strategy to handle this extensive set of products? Justify your suggestions.
6. The acceptance of a design strategy concludes the analysis phase of a project. Hence, this is a natural time to reassess how to conduct the rest of the project. The statement of how to progress is done by updating the Baseline Project Plan. In several questions for the BEC case at the end of Chapter 3, you developed components of the BPP. Given your answers to those questions and what you now know about the MyBroadway system, answer the following questions related to the components of a BPP.

   a. List tangible and intangible benefits and costs for this project. Be sure to quantify tangible costs and benefits.
   b. What are the remaining risks of the project? How would you suggest the project team deal with these risks?
   c. How would you continue to use the concept of incremental commitment for the rest of the project?
   d. What would be the next steps in the project? If possible, develop a PERT or Gantt chart for the remaining steps of the project. Is such a project schedule chart possible for the design strategy recommended by the student team?
   e. What contact, if any, do you think the student team should have during the design phase with BEC corporate IT staff, especially those responsible for the Entertainment Tracker system? Under what circumstance would such contact not be necessary? Can the team organize design activities so that it does not need to know anything about the technical architecture of Entertainment Tracker?

8

Designing the Human Interface

→ Objectives

After studying this chapter, you should be able to:

→ Explain the process of designing forms and reports and the deliverables for their creation.

→ Apply the general guidelines for formatting forms and reports.

→ Format text, tables, and lists effectively.

→ Explain the process of designing interfaces and dialogues and the deliverables for their creation.

→ Describe and apply the general guidelines for interface design, including guidelines for layout design, structuring data entry fields, providing feedback, and system help.

→ Design human-computer dialogues, including the use of dialogue diagramming.

→ Discuss interface design guidelines unique to the design of Internet-based electronic commerce systems.

Chapter Preview . . .

Analysts must complete two important activities in the systems design phase, as illustrated in Figure 8-1: designing the human interface and designing databases. In this chapter, you learn guidelines to follow when designing the human-computer interface. In the next section, we describe the process of designing forms and reports and provide guidance on the deliverables produced during this process. Properly formatted segments of information are the building blocks for designing all forms and reports. We present guidelines for formatting information and for designing interfaces and dialogues. Next, we show you a method for representing human-computer dialogues called dialogue diagramming. Finally, we close the chapter by examining various human-computer interface design issues for Internet-based applications, specifically as they apply to Pine Valley Furniture's WebStore.

FIGURE 8-1
The systems design phase consists of two important activities: designing the human interface and designing databases.

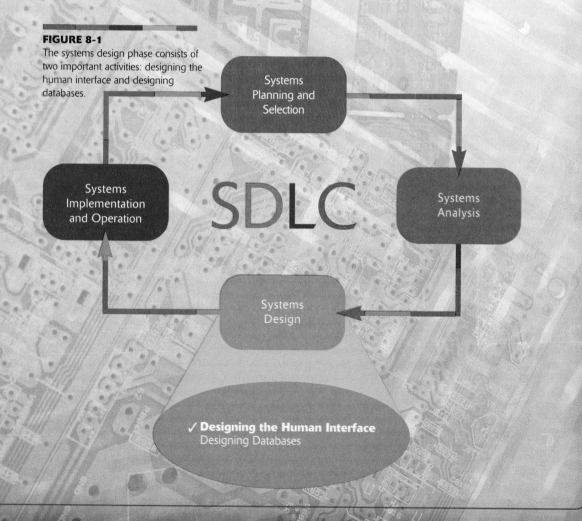

Systems Planning and Selection

Systems Analysis

Systems Design

Systems Implementation and Operation

SDLC

✓ **Designing the Human Interface**
Designing Databases

Designing Forms and Reports

System inputs and outputs—forms and reports—are produced at the end of the systems analysis phase of the SDLC. During systems analysis, however, you may not have been concerned with the precise appearance of forms and reports. Instead, you focused on which forms and reports needed to exist and the content they needed to contain. You may have distributed to users the prototypes of forms and reports that emerged during analysis as a way to confirm requirements. Forms and reports are integrally related to the DFD and E-R diagrams developed during requirements structuring. For example, every input form is associated with a data flow entering a process on a DFD, and every output form or report is a data flow produced by a process on a DFD. This means that the contents of a form or report correspond to the data elements contained in the associated data flow. Further, the data on all forms and reports must consist of data elements in data stores and on the E-R data model for the application or else be computed from these data elements. (In rare instances, data simply go from system input to system output without being stored within the system.) It is common to discover flaws in DFDs and E-R diagrams as you design forms and reports; these diagrams should be updated as designs evolve.

If you are unfamiliar with computer-based information systems, it will be helpful to clarify exactly what we mean by a form or report. A **form** is a business document containing some predefined data and often includes some areas where additional data are to be filled in. Most forms have a stylized format and are usually not in simple rows and columns. Examples of business forms are product order forms, employment applications, and class registration sheets. Traditionally, forms have been displayed on a paper medium, but today, video display technology allows us to duplicate the layout of almost any printed form, including an organizational logo or any graphic, on a video display terminal. Forms on a video display may be used for data display or data entry. Additional examples of forms are an electronic spreadsheet, computer sign-on or menu, and an automated teller machine (ATM) transaction layout. On the Internet, form interaction is the standard method of gathering and displaying information when consumers order products, request product information, or query account status.

A **report** is a business document containing only predefined data; it is a passive document used solely for reading or viewing. Examples of reports are invoices, weekly sales summaries by region and salesperson, and a pie chart of population by age categories. We usually think of a report as printed on paper, but it may be printed to a computer file, a visual display screen, or some other medium such as microfilm. Often a report has rows and columns of data, but a report may consist of any format—for example, mailing labels. Frequently, the differences between a form and a report are subtle. A report is only for reading and often contains data about multiple unrelated records in a computer file. On the other hand, a form typically contains data from only one record or is, at least, based on one record, such as data about one customer, one order, or one student. The guidelines for the design of forms and reports are very similar.

The Process of Designing Forms and Reports

Designing forms and reports is a user-focused activity that typically follows a prototyping approach (see Figure 1-17 to review the prototyping method). First, you must gain an understanding of the intended user and task objectives during the requirements determination process. During this process, the intended user must answer several questions that attempt to answer the who, what, when, where, and how related to the creation of all forms or reports, as listed in Table 8-1. Gaining an understanding of these questions is a required first step in the creation of any form or report.

Form
A business document that contains some predefined data and may include some areas where additional data are to be filled in. An instance of a form is typically based on one database record.

Report
A business document that contains only predefined data; it is a passive document used only for reading or viewing. A report typically contains data from many unrelated records or transactions.

**TABLE 8-1:   Fundamental Questions
When Designing Forms and Reports**

1. Who will use the form or report?

2. What is the purpose of the form or report?

3. When is the form or report needed and used?

4. Where does the form or report need to be delivered and used?

5. How many people need to use or view the form or report?

Understanding the skills and abilities of the users helps you create an effective design. Are your users experienced computer users or novices? What is their educational level, business background, and task-relevant knowledge? Answers to these questions provide guidance for both the format and the content of your designs. Also, what is the purpose of the form or report? What task will users be performing, and what information is needed to complete this task? Other questions are also important to consider. Where will the users be when performing this task? Will users have access to online systems or will they be in the field? How many people will need to use this form or report? If, for example, a report is being produced for a single user, the design requirements and usability assessment will be relatively simple. A design for a larger audience, however, may need to go through a more extensive requirements collection and usability assessment process.

After collecting the initial requirements, you structure and refine this information into an initial prototype. Structuring and refining the requirements are completed without assistance from the users, although you may occasionally need to contact users to clarify some issue overlooked during analysis. Finally, you ask users to review and evaluate the prototype; then they may accept the design or request that changes be made. If changes are needed, repeat the construction-evaluate-refinement cycle until the design is accepted. Usually, several repetitions of this cycle occur during the design of a single form or report. As with any prototyping process, you should make sure that these iterations occur rapidly in order to gain the greatest benefit from this design approach.

The initial prototype may be constructed in numerous environments, including DOS, UNIX, Windows, Linux, or Apple. The obvious choice is to employ standard development tools used within your organization. Often, initial prototypes are simply mock screens that are not working modules or systems. Mock screens can also be produced from a word processor, computer graphics design package, or electronic spreadsheet. It is important to remember that the focus of this phase within the SDLC is on the design—content and layout. How specific forms or reports are implemented (e.g., the programming language or screen painter code) is left for a later stage. Nonetheless, tools for designing forms and reports are rapidly evolving. In the past, inputs and outputs of all types were typically designed by hand on a coding or layout sheet. For example, Figure 8-2 shows the layout of a data input form using a coding sheet.

Although coding sheets are still used, their importance has diminished due to significant changes in system operating environments and the evolution of automated design tools. Prior to the creation of graphical operating environments, for example, analysts designed many inputs and outputs that were 80 columns (characters) by 25 rows, the standard dimensions for most video displays. These limits in screen dimensions are radically different in graphical operating environments such as Microsoft's Windows, where font sizes and screen dimensions can often be changed from user to user. Consequently, the

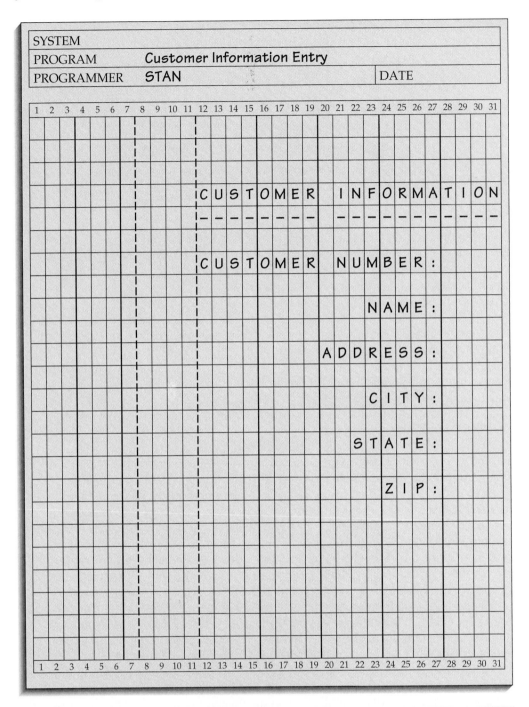

FIGURE 8-2
THE LAYOUT OF A DATA INPUT FORM USING A CODING SHEET

creation of new tools and development environments was needed to help analysts and programmers develop these graphical and flexible designs. Figure 8-3 shows an example of the same data input form as designed in Microsoft's Visual Basic. Note the variety of fonts, sizes, and highlighting that was used. Online graphical tools for designing forms and reports are rapidly becoming the standard in most professional development organizations.

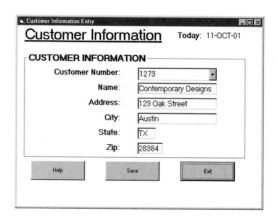

FIGURE 8-3
A DATA INPUT SCREEN DESIGNED
IN MICROSOFT'S VISUAL BASIC

Deliverables and Outcomes

Each SDLC activity helps you to construct a system. In order to move from phase to phase, each activity produces some type of deliverable that is used in a later activity. For example, within the systems planning and selection phase of the SDLC, the Baseline Project Plan serves as input to many subsequent SDLC activities. In the case of designing forms and reports, design specifications are the major deliverables and are inputs to the system implementation and operation phase. Design specifications have three sections:

1. Narrative overview
2. Sample design
3. Testing and usability assessment

The narrative overview provides a general overview of the characteristics of the target users, tasks, system, and environmental factors in which the form or report will be used. Its purpose is to explain to those who will actually develop the final form why this form exists and how it will be used so that they can make the appropriate implementation decisions. In this section, you list general information and the assumptions that helped shape the design. For example, Figure 8-4 shows an excerpt of a design specification for a Customer Account Status form for Pine Valley Furniture. The first section of the specification, Figure 8-4(A), provides a narrative overview containing the information relevant to developing and using the form within PVF. The overview explains the tasks supported by the form, where and when the form is used, characteristics of the people using the form, the technology delivering the form, and other pertinent information. For example, if the form is delivered on a visual display terminal, this section would describe the capabilities of this device, such as whether it has a touch screen and whether color and a mouse are available.

In the second section of the specification, Figure 8-4(B), a sample design of the form is shown. This design may be hand-drawn using a coding sheet although, in most instances, it is developed using CASE or standard development tools. Using actual development tools allows the design to be more thoroughly tested and assessed. The final section of the specification, Figure 8-4(C), provides all testing and usability assessment information. Some specification information may be irrelevant when designing certain forms and reports. For example, the design of a simple yes/no selection form may be so straightforward that no usability assessment is needed. Also, much of the narrative overview may be unnecessary unless intended to highlight some exception that must be considered during implementation.

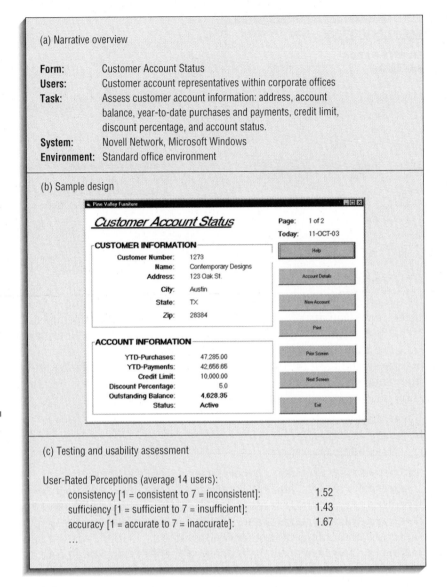

(a) Narrative overview

Form:	Customer Account Status
Users:	Customer account representatives within corporate offices
Task:	Assess customer account information: address, account balance, year-to-date purchases and payments, credit limit, discount percentage, and account status.
System:	Novell Network, Microsoft Windows
Environment:	Standard office environment

(b) Sample design

Pine Valley Furniture

Customer Account Status

Page: 1 of 2
Today: 11-OCT-03

CUSTOMER INFORMATION

Customer Number:	1273
Name:	Contemporary Designs
Address:	123 Oak St.
City:	Austin
State:	TX
Zip:	28384

ACCOUNT INFORMATION

YTD-Purchases:	47,285.00
YTD-Payments:	42,656.66
Credit Limit:	10,000.00
Discount Percentage:	5.0
Outstanding Balance:	4,628.35
Status:	Active

Help
Account Details
New Account
Print
Prior Screen
Next Screen
Exit

(c) Testing and usability assessment

User-Rated Perceptions (average 14 users):
consistency [1 = consistent to 7 = inconsistent]:	1.52
sufficiency [1 = sufficient to 7 = insufficient]:	1.43
accuracy [1 = accurate to 7 = inaccurate]:	1.67

...

FIGURE 8-4
A DESIGN SPECIFICATION FOR A CUSTOMER ACCOUNT STATUS FORM FOR PINE VALLEY FURNITURE
(A) THE NARRATIVE OVERVIEW CONTAINING THE INFORMATION RELEVANT TO DEVELOPING AND USING THE FORM WITHIN PVF
(B) A SAMPLE DESIGN OF THE PVF FORM
(C) TESTING AND USABILITY ASSESSMENT INFORMATION

Formatting Forms and Reports

A wide variety of information can be provided to users of information systems ranging from text to video to audio. As technology continues to evolve, a greater variety of data types will be used. A definitive set of rules for delivering every type of information to users has yet to be defined because these rules are continuously evolving along with the rapid changes in technology. Research conducted by computer scientists on human-computer interaction has provided numerous general guidelines for formatting information. Many of these guidelines undoubtedly will apply to the formatting of all evolving information types on yet-to-be-determined devices. Keep in mind that designing usable forms and reports requires your active interaction with users. If this single and fundamental activity occurs, you will likely create effective designs.

For example, personal digital assistants (PDAs) like the Palm Pilot and pocket PCs based on Microsoft's Windows CE operating system are becoming increasingly popular. PDAs are used to manage personal schedules, send and receive electronic mail, and browse the Web. One of the greatest challenges of palm computing is in the design of the human-computer interface because the video display is significantly smaller than full-size displays and many devices do not use a color display. These two characteristics represent significant challenges for application designers. For example, surfing the Web on a PDA is very difficult because most Internet sites still assume that users will have a full-size, color display. To address this problem, the Web browser in Windows CE is "smart" and automatically shrinks images so that the user's viewing experience is good. Alternatively, a growing number of Web sites are designed with the PDA user specifically in mind. For example, these sites provide a vast array of information preformatted for smaller screens. As these and other computing devices evolve and gain popularity, standard guidelines will emerge to make the process of designing interfaces for them much less challenging.

General Formatting Guidelines

Over the past several years, industry and academic researchers have investigated how information formatting influences individual task performance and perceptions of usability. Through this work, several guidelines for formatting information have emerged, as highlighted in Table 8-2. These guidelines reflect some of the general truths of formatting most types of information. The differences between a well-designed form or report and a poorly designed one often will be obvious. For example, Figure 8-5(A) shows a poorly designed form for viewing a current account balance for a PVF customer. Figure 8-5(B) (page 2 of 2) is a better design, incorporating several general guidelines from Table 8-2.

The first major difference between the two forms has to do with the title. The title in Figure 8-5(A) (Customer Information) is ambiguous, whereas the title in Figure 8-5(B) (Detail Customer Account Information) clearly and specifically describes the contents of the form. The form in Figure 8-5(B) also includes the date (October 11, 2003) the form was generated so that, if printed, it will be

TABLE 8-2:   Guidelines for Designing Forms and Reports

Guideline	Description
Use meaningful titles	Clear and specific titles describing content and use of form or report
	Revision date or code to distinguish a form or report from prior versions
	Current date that identifies when the form or report was generated
	Valid date that identifies on what date (or time) the data in the form or report were accurate
Include meaningful information	Only needed information displayed
	Information provided in a usable manner without modification
Balance the layout	Information balanced on the screen or page
	Adequate spacing and margins used
	All data and entry fields clearly labeled
Design an easy navigation system	Clearly show how to move forward and backward
	Clearly show where you are (e.g., page 1 of 3)
	Notify user of the last page of a multipage sequence

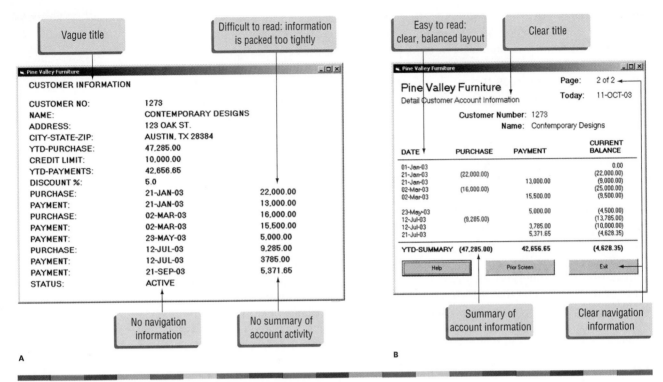

FIGURE 8-5
CONTRAST OF A POORLY DESIGNED AND A WELL-DESIGNED FORM
(A) A POORLY DESIGNED FORM FOR VIEWING A CURRENT ACCOUNT BALANCE FOR A PVF CUSTOMER
(B) A BETTER DESIGN, WHICH INCORPORATES SEVERAL GENERAL GUIDELINES FROM TABLE 8-2

clear to the reader when this occurred. Figure 8-5(A) displays information extraneous to the intent of the form—viewing the current account balance—and provides information that is not in the most useful format for the user. For example, Figure 8-5(A) provides all customer data as well as account transactions and a summary of year-to-date purchases and payments. The form does not, however, provide the current outstanding balance of the account, leaving the reader to perform a manual calculation. The layout of information between the two forms also varies in balance and information density. Gaining an understanding of the skills of the intended system users and the tasks they will be performing is invaluable when constructing a form or report. By following these general guidelines, your chances of creating effective forms and reports will be enhanced. In the next sections we discuss specific guidelines for highlighting information, displaying text, and presenting numeric tables and lists.

Highlighting Information

As display technologies continue to improve, a greater variety of methods will be available to highlight information. Table 8-3 lists the most commonly used methods for highlighting information. Given this vast array of options, it is important to consider how highlighting can be used to enhance an output without being a distraction. In general, highlighting should be used sparingly to draw the user to or away from certain information and to group together related information. In several situations highlighting can be a valuable technique for conveying special information:

- Notifying users of errors in data entry or processing
- Providing warnings to users regarding possible problems such as unusual data values or an unavailable device

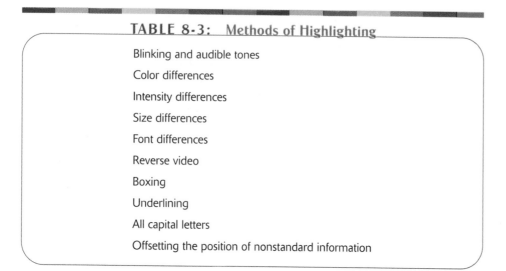

TABLE 8-3: Methods of Highlighting

Blinking and audible tones

Color differences

Intensity differences

Size differences

Font differences

Reverse video

Boxing

Underlining

All capital letters

Offsetting the position of nonstandard information

Phase 3:
Systems
Design

- Drawing attention to keywords, commands, high-priority messages, and data that have changed or gone outside normal operating ranges

Highlighting techniques can be used singularly or in tandem, depending upon the level of emphasis desired by the designer. Figure 8-6 shows a form where several types of highlighting are used. In this example, boxes clarify different categories of data; capital letters and different fonts distinguish labels from actual data; and bolding is used to draw attention to important data.

Highlighting should be used conservatively. For example, blinking and audible tones should be used only to highlight critical information requiring the user's immediate response. Once a response is made, these highlights should be turned off. Additionally, highlighting methods should be consistently selected and used based upon the level of importance of the emphasized information. It is also important to examine how a particular highlighting method

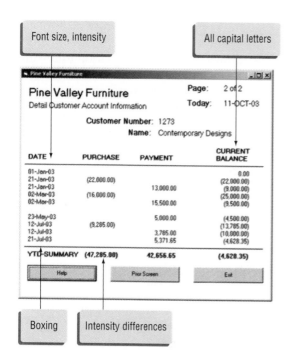

FIGURE 8-6
A FORM IN WHICH SEVERAL TYPES OF HIGHLIGHTING ARE USED

NET SEARCH
There are millions of commercial Web sites that have emerged over the past few years, some having a highly usable interface, whereas others continue to have a relatively weak interface. Visit http://www.prenhall.com/valacich to complete an exercise related to this topic.

appears on all possible output devices that could be used with the system. For example, some color combinations may convey appropriate information on one display configuration but wash out and reduce legibility on another.

Recent advances in the development of graphical operating environments such as Windows, Macintosh, or UNIX have provided designers with some standard highlighting guidelines. However, because these guidelines are continuously evolving, they are often quite vague and leave a great deal of control in the hands of the systems developer. To realize the benefits of using standard graphical operating environments—such as reduced user training time and interoperability among systems—you must be disciplined in how you use highlighting.

Displaying Text

In business-related systems, textual output is becoming increasingly important as text-based applications such as electronic mail, bulletin boards, and information services (e.g., Dow Jones Industrial Average stock index) are more widely used. The display and formatting of system help screens, which often contain lengthy textual descriptions and examples, is one example of textual data that can benefit from following the simple guidelines that have emerged from systems design research. These guidelines appear in Table 8-4. The first one is simple: You should display text using common writing conventions such as mixed upper- and lowercase and appropriate punctuation. For large blocks of text, and if space permits, text should be double-spaced. However, if the text is short, or rarely used, it may make sense to use single spacing and place a blank line between each paragraph. You should also left-justify text with a ragged right margin—research shows that a ragged right margin makes it easier to find the next line of text when reading than when text is both left- and right-justified.

When displaying textual information, you should also be careful not to hyphenate words between lines or use obscure abbreviations and acronyms. Users may not know whether the hyphen is a significant character if it is used to continue words across lines. Information and terminology that are not widely understood by the intended users may significantly influence the usability of the system. Thus, you should use abbreviations and acronyms only if they are significantly shorter than the full text and are commonly known by the intended system users. Figure 8-7 shows two versions of a help screen from an application system at PVF. Figure 8-7(A) shows many violations of the general guidelines for displaying text, whereas 8-7(B) shows the same information following the general guidelines. Formatting guidelines for the entry of text and alphanumeric data are also very important and will be discussed later in the chapter.

TABLE 8-4:   Guidelines for Displaying Text

Case	Display text in mixed upper- and lowercase and use conventional punctuation.
Spacing	Use double spacing if space permits. If not, place a blank line between paragraphs.
Justification	Left-justify text and leave a ragged right margin.
Hyphenation	Do not hyphenate words between lines.
Abbreviations	Use abbreviations and acronyms only when they are widely understood by users and are significantly shorter than the full text.

Designing Tables and Lists

Unlike textual information, where context and meaning are derived through reading, the context and meaning of tables and lists are derived from the format of the information. Consequently, the usability of information displayed in tables and alphanumeric lists is likely to be much more influenced by effective layout than most other types of information display. As with the display of textual information, tables and lists can also be greatly enhanced by following a few simple guidelines. These are summarized in Table 8-5.

Figure 8-8 displays two versions of a form design from a Pine Valley Furniture application system that displays customer year-to-date transaction information in a table format. Figure 8-8(A) displays the information without consideration of the guidelines presented in Table 8-5, and Figure 8-8(B) (only page 2 of 2 is shown) displays this information after consideration of these guidelines.

One key distinction between these two display forms relates to labeling. The information reported in Figure 8-8(B) has meaningful labels that stand out more clearly compared to the display in Figure 8-8(A). Transactions are sorted by date, and numeric data are right-justified and aligned by decimal point in Figure 8-8(B), which helps to facilitate scanning. Adequate space is left between columns, and blank lines are inserted after every five rows in Figure 8-8(B) to help ease the finding and reading of information. Such spacing also provides room for users to annotate data that catch their attention. Using the guidelines presented in Table 8-5 helped create an easy-to-read layout of the information for the user.

Most of the guidelines in Table 8-5 are rather obvious, but this and other tables serve as a quick reference to validate that your form and report designs will be usable. It is beyond our scope here to discuss each of these guidelines, but you should read each carefully and think about why it is appropriate. For example, why should labels be repeated on subsequent screens and pages (the

Phase 3:
Systems
Design

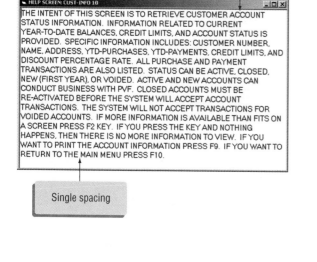

A

B

FIGURE 8-7

CONTRASTING TWO HELP SCREENS FROM AN APPLICATION SYSTEM AT PVF

(A) A POORLY DESIGNED HELP SCREEN WITH MANY VIOLATIONS OF THE GENERAL GUIDELINES FOR DISPLAYING TEXT

(B) AN IMPROVED DESIGN FOR A HELP SCREEN

TABLE 8-5:   General Guidelines for Displaying Tables and Lists

Guideline	Description
Use meaningful labels	All columns and rows should have meaningful labels.
	Labels should be separated from other information by using highlighting.
	Redisplay labels when the data extend beyond a single screen or page.
Formatting columns, rows, and text	Sort in a meaningful order (e.g., ascending, descending, or alphabetic).
	Place a blank line between every five rows in long columns.
	Similar information displayed in multiple columns should be sorted vertically (i.e., read from top to bottom, not left to right).
	Columns should have at least two spaces between them.
	Allow white space on printed reports for user to write notes.
	Use a single typeface, except for emphasis.
	Use same family of typefaces within and across displays and reports.
	Avoid overly fancy fonts.
Formatting numeric, textual, and alphanumeric data	Right-justify *numeric data* and align columns by decimal points or other delimiter.
	Left-justify *textual data.* Use short line length, usually 30 to 40 characters per line (this is what newspapers use, and it is easy to speed-read).
	Break long sequences of *alphanumeric data* into small groups of three to four characters each.

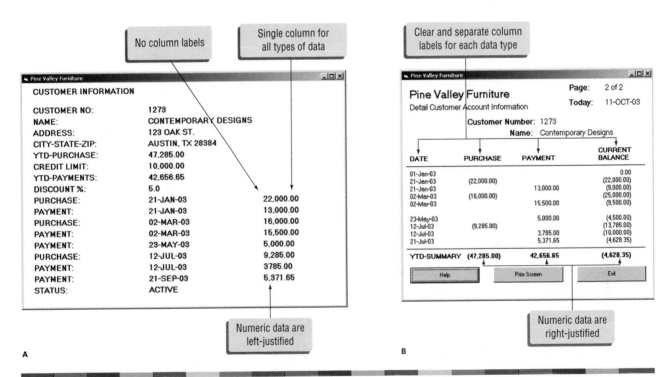

FIGURE 8-8
CONTRASTING TWO PINE VALLEY FURNITURE FORMS
(A) A POORLY DESIGNED FORM
(B) AN IMPROVED DESIGN FORM

third guideline in Table 8-5)? One explanation is that pages may be separated or copied and the original labels will no longer be readily accessible to the reader of the data. Why should long alphanumeric data (see the last guideline) be broken into small groups? (If you have a credit card or bank check, look at how your account number is displayed.) Two reasons are that the characters will be easier to remember as you read and type them and there will be a natural and consistent place to pause when you speak them over the phone, for example, when you are placing a phone order for products in a catalog.

When you design the display of numeric information, you must determine whether a table or a graph should be used. In general, tables are best when the user's task involves finding an individual data value from a larger data set, whereas line and bar graphs are more appropriate for analyzing data changes over time. For example, if the marketing manager for Pine Valley Furniture needed to review the actual sales of a particular salesperson for a particular quarter, a tabular report like the one shown in Figure 8-9 would be most useful. This report has been annotated to emphasize good report design practices. The report has both a printed date as well as a clear indication, as part of the report title, of the period over which the data apply. There is also sufficient white space to provide some room for users to add personal comments

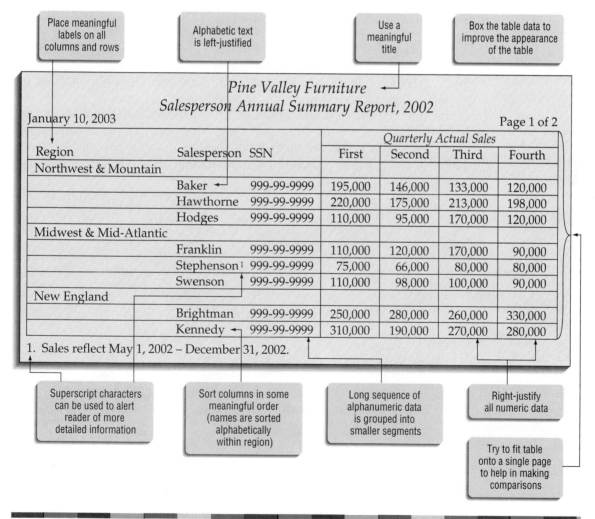

FIGURE 8-9
TABULAR REPORT ILLUSTRATING GOOD REPORT DESIGN GUIDELINES

and observations. Often, to provide such white space, a report must be printed in landscape, rather than portrait, orientation. Alternatively, if the marketing manager wished to compare the overall sales performance of each sales region, a line or bar graph would be more appropriate, as illustrated in Figure 8-10.

Paper versus Electronic Reports

When a report is produced on paper rather than on a computer display, you need to consider some additional things. For example, laser printers (especially color laser printers) and ink-jet printers allow you to produce a report that looks exactly as it does on the display screen. Thus, when using these types of printers, you can follow our general design guidelines to create a report with high usability. However, other types of printers cannot closely reproduce the display screen image onto paper. For example, many business reports are produced using high-speed impact printers that produce characters and a limited range of graphics by printing a fine pattern of dots. The advantages of impact printers are that they are very fast, very reliable, and relatively inexpensive. Their drawbacks are that they have a limited ability to produce graphics and have a somewhat lower print quality. In other words, they are good at rapidly producing reports that contain primarily alphanumeric information but cannot exactly replicate a screen report onto paper. Because of this, impact printers are mostly used for producing large batches of reports, like a batch of phone bills for your telephone company, on a wide range of paper widths and types. When designing reports for impact printers, you use a coding sheet like that displayed in Figure 8-2, although coding sheets for designing printer reports typically can have up to 132 columns. Like the process for designing all forms and reports, you follow a prototyping process and carefully control the spacing of characters in order to produce a high-quality report. However, unlike other form and report designs, you may be limited in the range of formatting, text types, and highlighting options. Nonetheless, you can easily produce a highly usable report of any type if you carefully and creatively use the available formatting options.

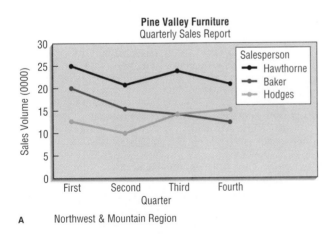
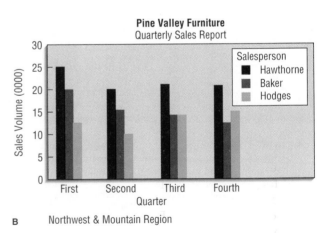

FIGURE 8-10
GRAPHS SHOWING QUARTERLY SALES AT PINE VALLEY FURNITURE
(A) LINE GRAPH
(B) BAR GRAPH

Designing Interfaces and Dialogues

Interface and dialogue design focuses on how information is provided to and captured from users. Dialogues are analogous to a conversation between two people. The grammatical rules followed by each person during a conversation are analogous to the human-computer interface. The design of interfaces and dialogues involves defining the manner in which humans and computers exchange information. A good human-computer interface provides a uniform structure for finding, viewing, and invoking the different components of a system. In this section we describe how to design interfaces and dialogues.

The Process of Designing Interfaces and Dialogues

Similar to designing forms and reports, the process of designing interfaces and dialogues is a user-focused activity. You follow a prototyping methodology of iteratively collecting information, constructing a prototype, assessing usability, and making refinements. To design usable interfaces and dialogues, you must answer the same who, what, when, where, and how questions used to guide the design of forms and reports (see Table 8-1). Thus, this process parallels that of designing forms and reports.

Deliverables and Outcomes

The deliverable and outcome from system interface and dialogue design is the creation of a design specification. This specification is similar to the specification produced for form and report designs—with one exception. Recall that the design specification for forms and reports had three sections (see Figure 8-4):

1. Narrative overview
2. Sample design
3. Testing and usability assessment

For interface and dialogue designs, one additional subsection is included: a section outlining the dialogue sequence—the ways a user can move from one display to another. Later in the chapter you learn how to design a dialogue sequence by using dialogue diagramming. An outline for a design specification for interfaces and dialogues is shown in Figure 8-11.

Designing Interfaces

In this section we discuss the design of interface layouts. This discussion provides guidelines for structuring and controlling data entry fields, providing feedback, and designing online help. Effective interface design requires you to gain a thorough understanding of each of these concepts.

Designing Layouts

To ease user training and data recording, use standard formats for computer-based forms and reports similar to paper-based forms and reports for recording or reporting information. A typical paper-based form for reporting customer sales activity is shown in Figure 8-12. This form has several general areas common to most forms:

- Header information
- Sequence and time-related information
- Instruction or formatting information

Design Specification

1. Narrative Overview
   a. Interface/Dialogue Name
   b. User Characteristics
   c. Task Characteristics
   d. System Characteristics
   e. Environmental Characteristics

2. Interface/Dialogue Designs
   a. Form/Report Designs
   b. Dialogue Sequence Diagram(s) and Narrative Description

3. Testing and Usability Assessment
   a. Testing Objectives
   b. Testing Procedures
   c. Testing Results
      i) Time to Learn
      ii) Speed of Performance
      iii) Rate of Errors
      iv) Retention over Time
      v) User Satisfaction and Other Perceptions

FIGURE 8-11
AN OUTLINE FOR A DESIGN SPECIFICATION FOR INTERFACES AND DIALOGUES

- Body or data details
- Totals or data summary
- Authorization or signatures
- Comments

In many organizations, data are often first recorded on paper-based forms and then later recorded within application systems. When designing layouts to record or display information on paper-based forms, try to make both as similar as possible. Additionally, data entry displays should be consistently formatted across applications to speed data entry and reduce errors. Figure 8-13 shows an equivalent computer-based form to the paper-based form shown in Figure 8-12.

The design of between-field navigation is another item to consider when designing the layout of computer-based forms. Because you can control the sequence for users to move between fields, standard screen navigation should flow from left-to-right and top-to-bottom just as when you work on paper-based forms. For example, Figure 8-14 contrasts the flow between fields on a form used to record business contacts. Figure 8-14(A) uses a consistent left-to-right, top-to-bottom flow. Figure 8-14(B) uses a flow that is nonintuitive. When appropriate, you should also group data fields into logical categories with labels describing the contents of the category. Areas of the screen not used for data entry or commands should be inaccessible to the user.

When designing the navigation procedures within your system, flexibility and consistency are primary concerns. Users should be able to move freely forward and backward or to any desired data entry fields. Users should be able to navigate each form in the same way or in as similar a manner as possible. Additionally, data should not usually be permanently saved by the system until

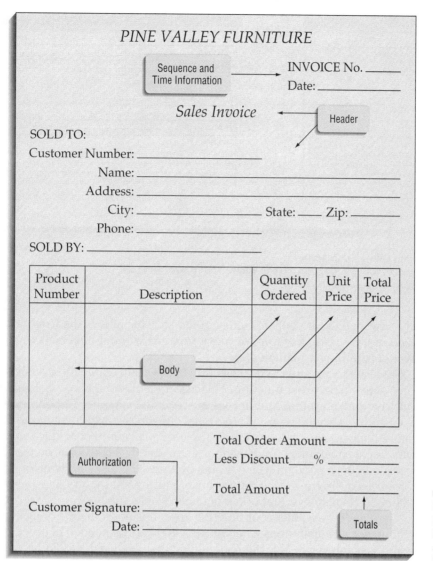

FIGURE 8-12
PAPER-BASED FORM FOR
REPORTING CUSTOMER SALES
ACTIVITY AT PINE VALLEY
FURNITURE

FIGURE 8-13
COMPUTER-BASED FORM FOR
REPORTING CUSTOMER SALES
ACTIVITY AT PINE VALLEY
FURNITURE

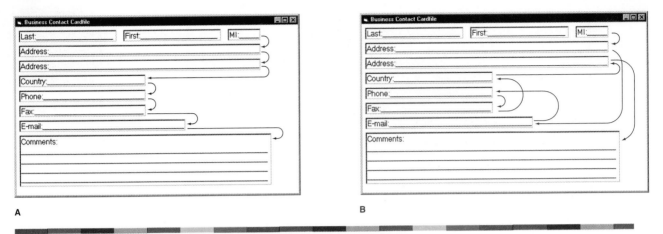

A

B

FIGURE 8-14
CONTRASTING THE NAVIGATION FLOW WITHIN A DATA ENTRY FORM
(A) PROPER FLOW BETWEEN DATA ENTRY FIELDS WITH A CONSISTENT LEFT-TO-RIGHT, TOP-TO-BOTTOM FLOW
(B) POOR FLOW BETWEEN DATA ENTRY FIELDS WITH INCONSISTENT FLOW

the user makes an explicit request to do so. This allows the user to abandon a data entry screen, back up, or move forward without adversely impacting the contents of the permanent data.

Consistency extends to the selection of keys and commands. Assign each key or command only one function. This assignment should be consistent throughout the entire system and across systems, if possible. Depending upon the application, various types of functional capabilities will be required to provide smooth navigation and data entry. Table 8-6 provides a checklist for testing the functional capabilities for providing smooth and easy navigation within a form. For example, a good interface design provides a consistent way for moving the cursor to different places on the form, editing characters and fields, moving among form displays, and obtaining help. These functions may be provided by keystrokes, mouse, menu, or function keys. It is possible that, for a single application, not all capabilities listed in Table 8-6 may be needed in order to create a good user interface. Yet, the capabilities that are used should be consistently applied to provide an optimal user environment. Table 8-6 provides you with a checklist for validating the usability of user interface designs.

Structuring Data Entry

You should consider several guidelines when structuring data entry fields on a form. These guidelines are listed in Table 8-7. The first is simple yet is often violated by designers. To minimize data entry errors and user frustration, never require the user to enter information that is already available within the system or information that can be easily computed by the system. For example, never require the user to enter the current date and time, because each of these values can be easily retrieved from the computer system's internal calendar and clock. By allowing the system to do this, the user simply confirms that the calendar and clock are working properly.

Other guidelines are equally important. For example, suppose that a bank customer is repaying a loan on a fixed schedule with equal monthly payments. Each month when a payment is sent to the bank, a clerk needs to record that the payment has been received into a loan processing system. Within such a system, default values for fields should be provided whenever appropriate. This means that the clerk has to enter data into the system only when the customer pays more or less than the scheduled amount. In all other cases, the clerk simply verifies that the check is for the default amount provided by the system and presses a single key to confirm the receipt of payment.

**TABLE 8-6:   Checklist for Validating the
Usability of User Interface**

Cursor-Control Capabilities

Move the cursor forward to the next data field.

Move the cursor backward to the previous data field.

Move the cursor to the first, last, or some other designated data field.

Move the cursor forward one character in a field.

Move the cursor backward one character in a field.

Editing Capabilities

Delete the character to the left of the cursor.

Delete the character under the cursor.

Delete the whole field.

Delete data from the whole form (empty the form).

Exit Capabilities

Transmit the screen to the application program.

Move to another screen/form.

Confirm the saving of edits or go to another screen/form.

Help Capabilities

Get help on a data field.

Get help on a full screen/form.

Source: Adapted from J. S. Dumas (1988). *Designing User Interfaces for Software.* Upper Saddle River, NJ: Prentice Hall.

When entering data, do not require the user to specify the dimensional units of a particular value; for example, whether an amount is in dollars or a weight is in tons. Use field formatting and the data entry prompt to make clear the type of data being requested. In other words, place a caption describing the data to be entered adjacent to each data field so that the user knows what type of data is being requested. As with the display of information, all data entered onto a form should automatically justify in a standard format (e.g., date, time, money). Table 8-8 illustrates display design options for printed forms. For data entry on video display terminals, highlight the area in which text is entered so that the exact number of characters per line and number of lines are clearly shown. You can also use check-off boxes or radio buttons to allow users to choose standard textual responses. Use data entry controls to ensure that the proper type of data (alphabetic or numeric, as required) is entered. Data entry controls are discussed next.

Controlling Data Input

One objective of interface design is to reduce data entry errors. As data are entered into an information system, steps must be taken to ensure that the input is valid. As a systems analyst, you must anticipate the types of errors users may make and design features into the system's interfaces to avoid, detect, and correct data entry mistakes. Several types of data errors are summarized in Table 8-9. Data errors can occur from appending extra data onto a field, truncating characters off a field, transcripting the wrong characters into a field, or transposing one or more characters within a field. Systems designers

TABLE 8-7: Guidelines for Structuring Data Entry Fields

Entry	Never request data that are already online or that can be computed; for example, do not request customer data on an order form if that data can be retrieved from the database, and do not request extended prices that can be computed from quantity sold and unit prices.
Defaults	Always provide default values when appropriate; for example, assume today's date for a new sales invoice, or use the standard product price unless overridden.
Units	Make clear the type of data units requested for entry; for example, indicate quantity in tons, dozens, pounds, etc.
Replacement	Use character replacement when appropriate; for example, allow the user to look up the value in a table or automatically fill in the value once the user enters enough significant characters.
Captioning	Always place a caption adjacent to fields; see Table 8-8 for caption options.
Format	Provide formatting examples when appropriate; for example, automatically show standard embedded symbols, decimal points, credit symbol, or dollar sign.
Justify	Automatically justify data entries; numbers should be right-justified and aligned on decimal points, and text should be left-justified.
Help	Provide context-sensitive help when appropriate; for example, provide a hot key, such as the F1 key, that opens the help system on an entry that is most closely related to where the cursor is on the display.

TABLE 8-8: Display Design Options for Entering Text

Options	**Example**
Line caption	Phone Number ( ) -
Drop caption	( ) - Phone Number
Boxed caption	Phone Number
Delimited characters	‖(| | |)| | | |-| | | |‖ Phone Number
Check-off boxes	Method of payment (check one) ❏ Check ❏ Cash ❏ Credit card: Type

have developed numerous tests and techniques for detecting invalid data before saving or transmission, thus improving the likelihood that data will be valid. Table 8-10 summarizes these techniques. These tests and techniques are often incorporated into both data entry screens and when data are transferred from one computer to another.

Correcting erroneous data is much easier to accomplish before it is permanently stored in a system. Online systems can notify a user of input problems as

TABLE 8-9: Types of Data Errors

Data Error	Description
Appending	Adding additional characters to a field
Truncating	Losing characters from a field
Transcripting	Entering invalid data into a field
Transposing	Reversing the sequence of one or more characters in a field

TABLE 8-10: Techniques Used by Systems Designers to Detect Data Errors before Saving or Transmission

Validation Test	Description
Class or composition	Test to assure that data are of proper type (e.g., all numeric, all alphabetic, alphanumeric)
Combinations	Test to see if the value combinations of two or more data fields are appropriate or make sense (e.g., does the quantity sold make sense given the type of product?)
Expected values	Test to see if data are what is expected (e.g., match with existing customer names, payment amount, etc.)
Missing data	Test for existence of data items in all fields of a record (e.g., is there a quantity field on each line item of a customer order?)
Pictures/templates	Test to assure that data conform to a standard format (e.g., are hyphens in the right places for a student ID number?)
Range	Test to assure data are within a proper range of values (e.g., is a student's grade point average between 0 and 4.0?)
Reasonableness	Test to assure data are reasonable for situation (e.g., pay rate for a specific type of employee)
Self-checking digits	Test where an extra digit is added to a numeric field in which its value is derived using a standard formula (see Figure 8-15)
Size	Test for too few or too many characters (e.g., is social security number exactly nine digits?)
Values	Test to make sure values come from a set of standard values (e.g., two-letter state codes)

data are being entered. When data are processed online as events occur, it is much less likely that data validity errors will occur and not be caught. In an online system, most problems can be easily identified and resolved before permanently saving data to a storage device using many of the techniques described in Table 8-10. However, in systems where data inputs are stored and entered (or transferred) in batches, the identification and notification of errors are more difficult. Batch processing systems can, however, reject invalid inputs and store them in a log file for later resolution.

Most of the straightforward tests and techniques shown in Table 8-10 are widely used. Some can be handled by data management technologies, such as a database management system (DBMS), to ensure that they are applied for all data maintenance operations. If a DBMS cannot perform these tests, then you must design the tests into program modules. Self-checking digits, shown in Figure 8-15, is an example of a sophisticated program. The figure provides a description and an outline of how to apply the technique. A short example then shows how a check digit is added to a field before data entry or transfer. Once entered or transferred, the check digit algorithm is again applied to the field to "check" whether the check digit received obeys the calculation. If it does, it is likely (but not guaranteed, because two different values could yield the same check digit) that no data transmission or entry error occurred. If not equal, then some type of error occurred.

In addition to validating the data values entered into a system, controls must be established to verify that all input records are correctly entered and processed only once. A common method used to enhance the validity of entering batches of data records is to create an **audit trail** of the entire sequence of data entry, processing, and storage. In such an audit trail, the actual sequence, count, time, source location, and human operator are recorded in a separate transaction log in the event of a data input or processing error. If an error occurs, corrections can be made by reviewing the contents of the log. Detailed logs of data inputs not only are useful for resolving batch data entry errors and system audits but also serve as a powerful method for performing backup and recovery operations in the case of a catastrophic system failure.

Audit trail

A record of the sequence of data entries and the date of those entries.

Providing Feedback

When you talk with friends, you expect them to give you feedback by nodding and replying to your questions and comments. Without feedback, you would be concerned that they were not listening. Similarly, when designing system interfaces, providing appropriate feedback is an easy way to make a user's interaction

---

FIGURE 8-15
HOW A CHECK DIGIT IS CALCULATED

Description	Techniques where extra digits are added to a field to assist in verifying its accuracy
Method	1. Multiply each digit of a numeric field by weighting factor (e.g., 1,2,1,2, . . .). 2. Sum the results of weighted digits. 3. Divide sum by modulus number (e.g., 10). 4. Subtract remainder of division from modulus number to determine check digit. 5. Append check digits to field.

Example	Assume a numeric part number of: 12473
	  1–2. Multiply each digit of part number by weighting factor from right to left and sum the results of weighted digits:
	
	$$\begin{array}{ccccc} 1 & 2 & 4 & 7 & 3 \\ \times 1 & \times 2 & \times 1 & \times 2 & \times 1 \\ \hline 1 + & 4 + & 4 + & 14 + & 3 = 26 \end{array}$$
	  3. Divide sum by modulus number.
	     26/10 = 2 remainder 6
	  4. Subtract remainder from modulus number to determine check digit.
	     check digit = 10 − 6 = 4
	  5. Append check digits to field.
	     Field value with appended check digit = 124734

more enjoyable; not providing feedback is a sure way to frustrate and confuse. System feedback can consist of three types:

1. Status information
2. Prompting cues
3. Error and warning messages

Status Information Providing status information is a simple technique for keeping users informed of what is going on within a system. For example, relevant status information such as displaying the current customer name or time, placing appropriate titles on a menu or screen, and identifying the number of screens following the current one (e.g., Screen 1 of 3) all provide needed feedback to the user. Providing status information during processing operations is especially important if the operation takes longer than a second or two. For example, when opening a file you might display, "Please wait while I open the file," or when performing a large calculation, flash the message, "Working . . . " to the user. Further, it is important to tell the user that besides working, the system has accepted the user's input and the input was in the correct form. Sometimes it is important to give the user a chance to obtain more feedback. For example, a function key could toggle between showing a "Working . . . " message and giving more specific information as each intermediate step is accomplished. Providing status information reassures users that nothing is wrong and makes them feel in command of the system, not vice versa.

Prompting Cues A second feedback method is to display prompting cues. When prompting the user for information or action, it is useful to be specific in your request. For example, suppose a system prompted users with the following request:

READY FOR INPUT:_____

With such a prompt, the designer assumes that the user knows exactly what to enter. A better design would be specific in its request, possibly providing an example, default values, or formatting information. An improved prompting request might be as follows:

Enter the customer account number (123-456-7):_____-_____-__

Errors and Warning Messages A final method available to you for providing system feedback is using error and warning messages. Following a few simple guidelines can greatly improve the usefulness of these messages. First, make messages specific and free of error codes and jargon. Additionally, messages should never scold the user but attempt to guide the user toward a resolution. For example, a message might say, "No customer record found for that customer ID. Please verify that digits were not transposed." Messages should be in user, not computer, terms. Terms such as end of file, disk I/O error, or write protected may be too technical and not helpful for many users. Multiple messages can be useful so that a user can get more detailed explanations if wanted or needed. Also, make sure error messages appear in roughly the same format and placement each time so that they are recognized as error messages and not as some other information. Examples of bad and good messages are provided in Table 8-11. Use these guidelines to provide useful feedback in your designs. A special type of feedback is answering help requests from users. This important topic is described next.

NET SEARCH
Standard error messages have emerged for Internet-related computing that some believe are very cryptic and difficult to understand. Visit http://www.prenhall.com/valacich to complete an exercise related to this topic.

Providing Help

Designing a help system is one of the most important interface design issues you will face. When designing help, you need to put yourself in the user's place. When accessing help, the user likely does not know what to do next, does not

TABLE 8-11: Examples of Poor and Improved Error Messages

Poor Error Messages	Improved Error Messages
ERROR 56 OPENING FILE	The file name you typed was not found. Press F2 to list valid file names.
WRONG CHOICE	Please enter an option from the menu.
DATA ENTRY ERROR	The prior entry contains a value outside the range of acceptable values. Press F9 for list of acceptable values.
FILE CREATION ERROR	The file name you entered already exists. Press F10 if you want to overwrite it. Press F2 if you want to save it with a new name.

TABLE 8-12: Guidelines for Designing System Help

Guideline	Explanation
Simplify	Use short, simple wording, common spelling, and complete sentences. Give users only what they need to know, with ability to find additional information.
Organize	Use lists to break information into manageable pieces.
Show	Provide examples of proper use and the outcomes of such use.

understand what is being requested, or does not know how the requested information needs to be formatted. A user requesting help is much like a ship in distress, sending an SOS. In Table 8-12, we provide our SOS guidelines for the design of system help: Simplify, Organize, and Show. Our first guideline, *simplify*, suggests that help messages should be short, to the point, and use words that users can understand. This leads to our second guideline, *organize*, which means the information in help messages should be easily absorbed by users. Long paragraphs of text are often difficult for people to understand. A better design organizes lengthy information in a manner easier for users to digest through the use of bulleted and ordered lists. Finally, it is often useful to explicitly *show* users how to perform an operation and the outcome of procedural steps. Figure 8-16 contrasts the designs of two help screens, one that employs our guidelines and one that does not.

Many commercially available systems provide extensive system help. For example, Table 8-13 lists the range of help available in a popular electronic spreadsheet. Many systems are also designed so that users can vary the level of detail provided. Help may be provided at the system level, screen or form level, and individual field level. The ability to provide field-level help is often referred to as "context-sensitive" help. For some applications, providing context-sensitive help for all system options is a tremendous undertaking that is virtually a project in itself. If you do decide to design an extensive help system with many levels of detail, you must be sure that you know exactly what the user needs help with, or your efforts may confuse users more than help them. After leaving a help screen, users should always return back to where they were prior to requesting help. If you follow these simple guidelines, you will likely design a highly usable help system.

As with the construction of menus, many programming environments provide powerful tools for designing system help. For example, Microsoft's Help

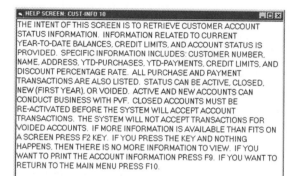

A

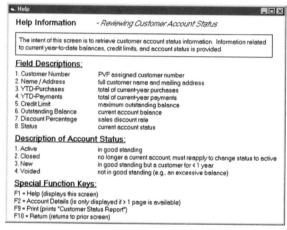

B

FIGURE 8-16
CONTRASTING HELP SCREENS
(A) A POORLY DESIGNED HELP
SCREEN
(B) AN IMPROVED DESIGN FOR A
HELP SCREEN

TABLE 8-13: Types of Help

Type of Help	Example of Question
Help on help	How do I get help?
Help on concepts	What is a customer record?
Help on procedures	How do I update a record?
Help on messages	What does "Invalid File Name" mean?
Help on menus	What does "Graphics" mean?
Help on function keys	What does each function key do?
Help on commands	How do I use the "Cut" and "Paste" commands?
Help on words	What do "merge" and "sort" mean?

Compiler allows you to construct hypertext-based help systems quickly. In this environment, you use a text editor to construct help pages that can be easily linked to other pages containing related or more specific information. Linkages are created by embedding special characters into the text document that make words hypertext buttons—that is, direct linkages—to additional information. The Help Compiler transforms the text document into a hypertext document. For example, Figure 8-17 shows a hypertext-based help screen from Microsoft. Hypertext-based help systems have become the standard environment for most

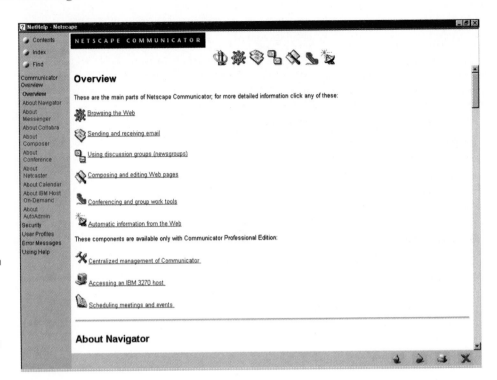

FIGURE 8-17
HYPERTEXT-BASED HELP SYSTEM FROM MICROSOFT
Source: Copyright © 2002 Microsoft Corporation. All rights reserved. Protected by the copyright laws of the United States and international treaties.

commercial operating environments. This has occurred for two primary reasons. First, standardizing system help across applications eases user training. Second, hypertext allows users to selectively access the level of help they need, making it easier to provide effective help for both novice and experienced users within the same system.

Designing Dialogues

Dialogue
The sequence of interaction between a user and a system.

The process of designing the overall sequences that users follow to interact with an information system is called dialogue design. A **dialogue** is the sequence in which information is displayed to and obtained from a user. As with other design processes, designing dialogues is a three-step process:

1. Designing the dialogue sequence
2. Building a prototype
3. Assessing usability

The primary design guideline for designing dialogues is consistency; dialogues need to be consistent in sequence of actions, keystrokes, and terminology. In other words, use the same labels for the same operations on all screens and the same location of the same information on all displays.

One example of these guidelines concerns removing data from a database or file (see the Reversal entry in Table 8-14). It is good practice to display the information that will be deleted before making a permanent change to the file. For example, if the customer service representative wanted to remove a customer from the database, the system should ask only for the customer ID in order to retrieve the correct customer account. Once found, and before allowing the confirmation of the deletion, the system should display the account information. For actions making permanent changes to system data files and when the action is not commonly performed, many system designers use the double-confirmation technique where the users must confirm their intention twice before being allowed to proceed.

Designing the Dialogue Sequence

Your first step in dialogue design is to define the sequence. In other words, you must have a clear understanding of the user, task, technological, and environmental characteristics when designing dialogues. Suppose that the marketing manager at Pine Valley Furniture (PVF) wants sales and marketing personnel to be able to review the year-to-date transaction activity for any PVF customer. After talking with the manager, you both agree that a typical dialogue between a user and the Customer Information System for obtaining this information might proceed as follows:

1. Request to view individual customer information.
2. Specify the customer of interest.
3. Select the year-to-date transaction summary display.
4. Review customer information.
5. Leave system.

As a designer, once you understand how a user wishes to use a system, you can then transform these activities into a formal dialogue specification.

A method for designing and representing dialogues is **dialogue diagramming.** Dialogue diagrams, illustrated in Figure 8-18, have only one symbol, a box with three sections; each box represents one display (which might be a full screen or a specific form or window) within a dialogue. The three sections of the box are used as follows:

1. Top: Contains a unique display reference number used by other displays for referencing it
2. Middle: Contains the name or description of the display

Phase 3:
Systems
Design

Dialogue diagramming
A formal method for designing and representing human-computer dialogues using box and line diagrams.

TABLE 8-14:   Guidelines for the Design of Human-Computer Dialogues

Guideline	Explanation
Consistency	Dialogues should be consistent in sequence of actions, keystrokes, and terminology (e.g., use the same labels for the same operations on all screens and the same location of the same information on all displays).
Shortcuts and sequence	Allow advanced users to take shortcuts using special keys (e.g., CTRL-C to copy highlighted text). A natural sequence of steps should be followed (e.g., enter first name before last name, if appropriate).
Feedback	Feedback should be provided for every user action (e.g., confirm that a record has been added, rather than simply putting another blank form on the screen).
Closure	Dialogues should be logically grouped and have a beginning, middle, and end (e.g., the last in the sequence of screens should indicate that there are no more screens).
Error handling	All errors should be detected and reported; suggestions on how to proceed should be made (e.g., suggest why such errors occur and what the user can do to correct the error). Synonyms for certain responses should be accepted (e.g., accept either "t," "T," or "TRUE").
Reversal	Dialogues should, when possible, allow the user to reverse actions (e.g., undo a deletion); data should not be deleted without confirmation (e.g., display all the data for a record the user has indicated is to be deleted).
Control	Dialogues should make the user (especially an experienced user) feel in control of the system (e.g., provide a consistent response time at a pace acceptable to the user).
Ease	Dialogues should provide simple means for users to enter information and navigate between screens (e.g., provide means to move forward, backward, and to specific screens, such as first and last).

Source: Adapted from B. Shneiderman (2002). *Designing the User Interface: Strategies for Effective Human-Computer Interaction, Third Edition.* Reading, MA: Addison-Wesley.

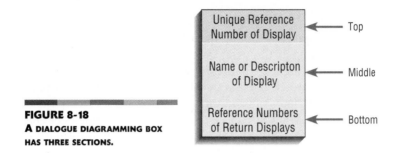

FIGURE 8-18
A DIALOGUE DIAGRAMMING BOX
HAS THREE SECTIONS.

3. Bottom: Contains display reference numbers that can be accessed from the current display

All lines connecting the boxes within dialogue diagrams are assumed to be bidirectional and thus do not need arrowheads to indicate direction. This means that users are allowed to always move forward and backward between adjacent displays. If you desire only unidirectional flows within a dialogue, arrowheads should be placed on one end of the line. Within a dialogue diagram, you can easily represent the sequencing of displays, the selection of one display over another, or the repeated use of a single display (e.g., a data entry display). These three concepts—sequence, selection, and iteration—are illustrated in Figure 8-19.

Continuing with our PVF example, Figure 8-20 shows a partial dialogue diagram for processing the marketing manager's request. In this diagram, the analyst placed the request to view year-to-date customer information within the context of the overall Customer Information System. The user must first gain access to the system through a log-on procedure (item 0). If log-on is successful, a main menu is displayed that has four items (item 1). Once the user selects

FIGURE 8-19
DIALOGUE DIAGRAM ILLUSTRATING
SEQUENCE, SELECTION, AND
ITERATION

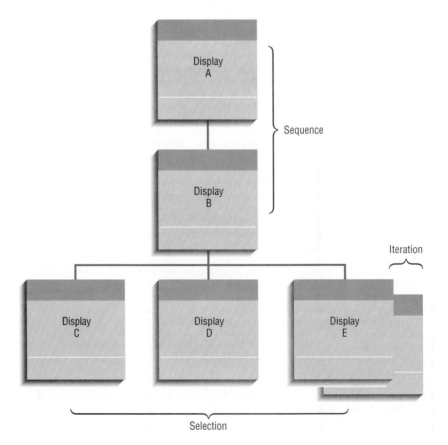

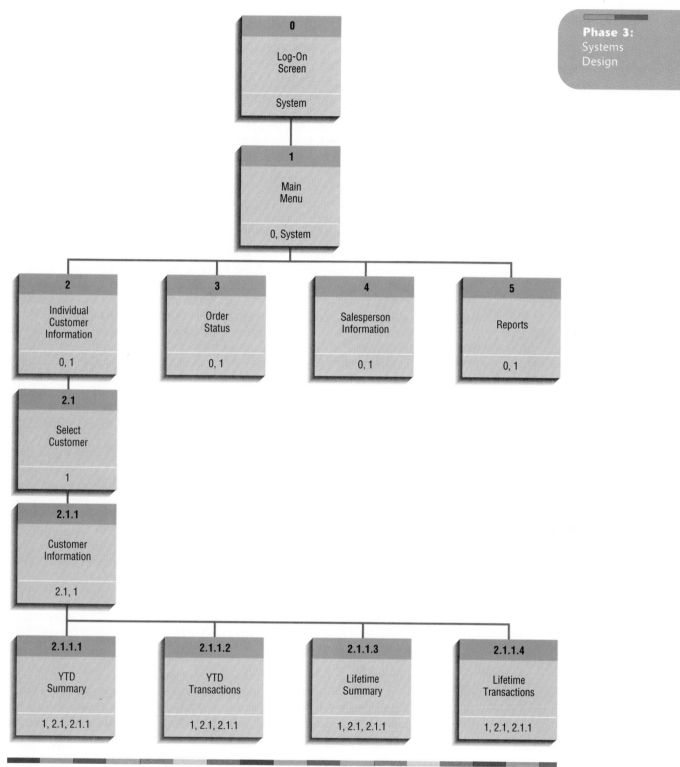

FIGURE 8-20
Dialogue Diagram for the Customer Information System at Pine Valley Furniture

the Individual Customer Information (item 2), control is transferred to the Select Customer display (item 2.1). After a customer is selected, the user is presented with an option to view customer information four different ways (item 2.1.1). Once the user views the customer's year-to-date transaction activity (item 2.1.1.2), the system will allow the user to back up to select a different customer, or back up to the main menu (see bottom of item 2.1.1.2).

Building Prototypes and Assessing Usability

Building dialogue prototypes and assessing usability are often optional activities. Some systems may be very simple and straightforward. Others may be more complex but are extensions to existing systems where dialogue and display standards have already been established. In either case, you may not be required to build prototypes and do a formal assessment. However, for many other systems, it is critical that you build prototype displays and then assess the dialogue; this can pay numerous dividends later in the systems development life cycle (e.g., it may be easier to implement a system or train users on a system they have already seen and used).

Building prototype displays is often a relatively easy activity if you use graphical development environments such as Microsoft's Visual Basic. Some systems development environments include easy-to-use input and output (form, report, or window) design utilities. Also several tools called "Prototypers" or "Demo Builders" allow you to design displays quickly and show how an interface will work within a full system. These demo systems allow users to enter data and move through displays as if they were using the actual system. Such activities are useful not only for showing how an interface will look and feel but also for assessing usability and performing user training long before actual systems are completed.

ELECTRONIC COMMERCE APPLICATION: DESIGNING THE HUMAN INTERFACE

Designing the human interface for an Internet-based electronic commerce application is a central and critical design activity. Because this is where a customer will interact with a company, much care must be put into its design. Like the process followed when designing the interface for other types of systems, a prototyping design process is most appropriate when designing the human interface for an Internet electronic commerce system. Although the techniques and technology for building the human interface for Internet sites are rapidly evolving, several general design guidelines have emerged. In this section, we examine some of these as they apply to the design of Pine Valley Furniture's WebStore.

General Guidelines for Designing Web Interfaces

Over the years, interaction standards have emerged for virtually all of the commonly used desktop computing environments such as Windows or the Machintosh. However, some interface design experts believe that the growth of the Web has resulted in a big step backwards for interface design. One problem is that countless nonprofessional developers are designing commercial Web applications. In addition to this, there are four other important contributing factors (Johnson, 2000):

- Web's single "click-to-act" method of loading static hypertext documents (i.e., most buttons on the Web do not provide click feedback)
- Limited capabilities of most Web-browsers to support finely grained user interactivity
- Limited agreed-upon standards for encoding Web content and control mechanisms
- Lack of maturity of Web scripting and programming languages as well as limitations in commonly used Web GUI component libraries

In addition to these contributing factors, designers of Web interfaces and dialogues are often guilty of many design errors. Although not inclusive of all possible errors, Table 8-15 summarizes those errors that are particularly troublesome.

General Guidelines for Web Layouts

As mentioned above, the rapid deployment of Internet Web sites has resulted in having countless people design sites who, arguably, have limited ability to do so. To put this into perspective consider the following quote from Web design guru, Jakob Nielsen (199; pp. 65–66):

> If the [Web's] growth rate does not slow down, the Web will reach 200 million sites sometime during 2003 . . . . The world has about 20,000 user interface professionals. If all sites were to be professionally designed by a single UI professional, we can conclude that every UI professional in the world would need to design one Web site every working hour from now on to meet demand. This is obviously not going to happen. . . .

There are three possible solutions to the problem:

❿ Make it possible to design reasonably useable sites without having UI expertise

❿ Train more people in good Web design

❿ Live with poorly designed sites that are hard to use

TABLE 8-15: Common Errors when Designing the Interface and Dialogues of Web Sites

Error	Description
Opening New Browser Window	Avoid opening a new browser window when a user clicks on a link unless it is clearly marked that a new window will be opened; users may not see that a new window has been opened, which will complicate navigation, especially when moving backwards.
Breaking or Slowing Down the Back Button	Make sure users can use the back button to return to prior pages. Avoid opening new browser windows, using immediate redirect where, when a user clicks the back button, they are pushed forward to an undesired location, or prevent caching such that each click of the back button requires a new trip to the server.
Complex URLs	Avoid overly long and complex URLs since it makes it more difficult for users to understand where they are and can cause problems if users want to e-mail page locations to colleagues.
Orphan Pages	Avoid having pages with no "parent" that can be reached by using a back button; requires users to "hack" the end of the URL to get back to a prior page.
Scrolling Navigation Pages	Avoid placing navigational links below where a page opens, since many users may miss these important options that are below the opening window.
Lack of Navigation Support	Make sure your pages conform to users expectations by providing commonly used icon links such as a site logo at the top or other major elements. Also place these elements on pages in a consistent manner.
Hidden Links	Make sure you leave a border around images that are links, don't change link colors from normal defaults, and avoid embedding links within long blocks of text.
Links that Don't Provide Enough Information	Avoid not turning off link marking borders so that links clearly show which links users have clicked and which they have not. Make sure users know which links are internal anchor points versus external links and indicate if a link brings up a separate browser window from those that do not. Finally, make sure link images and text provide enough information to the user so that they understand the meaning of the link.
Buttons that Provide No Click Feedback	Avoid using image buttons that don't clearly change when being clicked; use Web GUI toolkit button, HTML from-submit buttons, or simple textual links.

When designing forms and reports there are several errors that are specific to Web site design. It is unfortunately beyond the scope of this book to critically examine all possible design problems with contemporary Web sites. Here, we will simply summarize those errors that commonly occur and are particularly detrimental to the user's experience (see Table 8-16). Fortunately, there are numerous excellent sources for learning more about designing useful Web sites (Johnson, 2000; Flanders and Willis, 1998; Nielson, 1999; Nielson, 2000; www.useit.com; www.webpagesthatsuck.com).

Designing the Human Interface at Pine Valley Furniture

The first design activity that Jim Woo and the PVF development team focused on was the human-computer interface. To begin, they reviewed many popular electronic commerce Web sites and established the following design guidelines:

- Menu-driven navigation with cookie crumbs
- Lightweight graphics
- Form and data integrity rules
- Template-based HTML

In order to assure that all team members understood what was meant by each guideline, Jim organized a design briefing to explain how each would be incorporated into the WebStore interface design.

TABLE 8-16: Common Errors when Designing the Layout of Web Pages

Error	Recommendation
Nonstandard Use of GUI Widgets	Make sure that when using standard design items, that they behave in accordance to major interface design standards. For example, the rules for radio buttons state that they are used to select one item along a set of items, that is, not confirmed until "OK'ed" by a user. In many Web sites, selecting radio buttons are used as both *selection* and *action*.
Anything that Looks Like Advertising	Since research on Web traffic has shown that many users have learned to stop paying attention to Web advertisement, make sure that you avoid designing any legitimate information in a manner that resembles advertising (eg., banners, animations, pop-ups).
Bleeding-Edge Technology	Make sure that users don't need the latest browsers or plug-ins to view your site.
Scrolling Text and Looping Animators	Avoid scrolling text and animations since they are both hard to read and users often equate such content with advertising.
Nonstandard Link Colors	Avoid using nonstandard colors to show links and for showing links that users have already used; nonstandard colors will confuse the user and reduce ease of use.
Outdated Information	Make sure that your site is continuously updated so that users "feel" that the site is regularly maintained and updated. Outdated content is a sure way to lose credibility.
Slow Download Times	Avoid using large images, lots of images, unnecessary animations, or other time-consuming content that will slow the downloading time of a page.
Fixed-Formatted Text	Avoid fixed-formatted text that requires users to scroll horizontally to view contents or links.
Displaying Long Lists as Long Pages	Avoid requiring users to scroll down a page to view information, especially navigational controls. Manage information by showing only *N* items at a time, using multiple pages, or by using a scrolling container within the window.

Menu-Driven Navigation with Cookie Crumbs

After reviewing several sites, the team concluded that menus should stay in the exact same place throughout the entire site. They concluded that placing a menu in the same location on every page will help customers to become familiar with the site more quickly and therefore to navigate through the site more rapidly. Experienced Web developers know that the quicker customers can reach a specific destination at a site, the quicker they can purchase the product they are looking for or get the information they set out to find. Jim emphasized this point by stating, "These details may seem silly, but the second users find themselves 'lost' in our site, they're gone. One mouse click and they're no longer shopping at Pine Valley but at one of our competitor's sites."

A second design feature, and one that is being used on many electronic commerce sites, is cookie crumbs. **Cookie crumbs** are a technique for showing users where they are in the site by placing "tabs" on a Web page that remind users where they are and where they have been. These tabs are hypertext links that can allow users to move backward quickly in the site. For example, suppose that a site is four levels deep, with the top level called "Entrance," the second "Products," the third "Options," and the fourth "Order." As the user moves deeper into the site, a tab is displayed across the top of the page showing the user where she is and giving her the ability to jump backward quickly one or more levels. In other words, when first entering the store, a tab is displayed at the top (or some other standard place) of the screen with the word "Entrance." After moving down a level, two tabs are displayed, "Entrance" and "Products." After selecting a product on the second level, a third level is displayed where a user can choose product options. When this level is displayed, a third tab is produced with the label "Options." Finally, if the customer decides to place an order and selects this option, a fourth-level screen is displayed and a fourth tab displayed with the label "Order." In summary:

Level 1: Entrance
Level 2: Entrance ➔ Products
Level 3: Entrance ➔ Products ➔ Options
Level 4: Entrance ➔ Products ➔ Options ➔ Order

By using cookie crumbs, users know exactly how far they have wandered from 'home.' If each tab is a link, users can quickly jump back to a broader part of the store should they not find exactly what they are looking for. Cookie crumbs serve two important purposes. First, they allow users to navigate to a point previously visited and will ensure that they are not lost. Second, it clearly shows users where they have been and how far they have gone from home.

Lightweight Graphics

In addition to easy menu and page navigation, the PVF development team wants a system where Web pages load quickly. A technique to assist in making pages load quickly is through **lightweight graphics.** Lightweight graphics is the use of small simple images that allow a page to load as quickly as possible. "Using lightweight graphics allows pages to load quickly and helps users to reach their final location in the site—hopefully the point of purchase area—as quickly as possible. Large color images will only be used for displaying detailed product pictures that customers explicitly request to view," Jim explained. Experienced Web designers have found that customers are not willing to wait at each hop of navigation for a page to load, just so they have to click and wait again. The quick feedback that a Web site with lightweight graphics can provide will help to keep customers at the WebStore longer.

Cookie crumbs
A technique for showing users where they are in a Web site by placing a series of "tabs" on a Web page that shows users where they are and where they have been.

Lightweight graphics
The use of small simple images to allow a Web page to be displayed more quickly.

NET SEARCH
A well-designed corporate Web site can help attract potential customers. Visit http://www.prenhall.com/valacich to complete an exercise related to this topic.

Forms and Data Integrity

Because the goal of the WebStore is to have users place orders for products, all forms that request information should be clearly labeled and provide adequate room for input. If a specific field requires a specific input format such as a date of birth or phone number, it must provide a clear example for the user so that data errors can be reduced. Additionally, the site must clearly designate which fields are optional, which are required, and which have a range of values.

Jim emphasized, "All of this to me seems a bit like overkill, but it makes processing the data much simpler. Our site checks all data before submitting it to the server for processing. This allows us to provide quicker feedback to the user on any data entry error and eliminate the possibility of writing erroneous data into the permanent database. Additionally, we want to provide a disclaimer to reassure our customers that the data will be used only for processing orders, will never be sold to marketers, and will be kept strictly confidential."

Template-based HTML
Templates to display and process common attributes of higher-level, more abstract items.

Template-Based HTML

When Jim talked with the consultants about the WebStore during the analysis phase, they emphasized the advantages of using **template-based HTML.** He was told that when displaying individual products, it would be very advantageous to try and have a few "templates" that could be used to display the entire product line. In other words, not every product needs its own page; the development time for that would be far too great. Jim explained, "We need to look for ways to write a module once and reuse it. This way, a change requires modifying one page, not 700. Using HTML templates will help us create an interface that is very easy to maintain. For example, a desk and a filing cabinet are two completely different products. Yet, both have an array of finishes to choose from. Logically, each item requires the same function—namely: 'display all finishes.' If designed correctly, this function can be applied to all products in the store. On the other hand, if we write a separate module for each product, it would require us to change each and every module every time we make a product change, like adding a new finish. But, a function such as 'display all finishes,' written once and associated with all appropriate products, will require the modification of one generic or 'abstract' function, not hundreds."

Key Points Review

1. **Explain the process of designing forms and reports and the deliverables for their creation.**

   Forms and reports are created through a prototyping process. Once created, designs may be stand-alone or integrated into actual working systems. The purpose of the prototyping process, however, is to show users what a form or report will look like when the system is implemented. The outcome of this activity is the creation of a specification document where characteristics of the users, tasks, system, and environment are outlined along with each form and report design. Performance testing and usability assessments may also be included in the design specification.

2. **Apply the general guidelines for formatting forms and reports.**

   Guidelines should be followed when designing forms and reports. These guidelines, proven over years of experience with human-computer interaction, help to create professional, usable systems. There are guidelines for the use of titles, layout of fields, navigation between pages or screens, highlighting information, format of text, and the appropriate use and layout of tables and lists.

3. **Format text, tables, and lists effectively.**

   Textual output is becoming increasingly important as text-based applications such as electronic mail, bulletin boards, and information services become more popular. Text should be displayed using common writing conventions such as mixed upper- and lowercase, appropriate punctuation, left-justified, and a minimal amount of obscure abbreviations. Words should not be hyphenated between lines, and blocks of text should be double-spaced or, minimally, a blank line should be placed between each paragraph. Tables and lists should have meaningful labels that clearly stand out. Information should be sorted and arranged in a meaningful way. Numeric data should be right-justified.

4. **Explain the process of designing interfaces and dialogues and the deliverables for their creation.**

    Designing interfaces and dialogues is a user-focused activity that follows a prototyping methodology of iteratively collecting information, constructing a prototype, assessing usability, and making refinements. The deliverable and outcome from interface and dialogue design are the creation of a specification that can be used to implement the design.

5. **Describe and apply the general guidelines for interface design, including guidelines for layout design, structuring data entry fields, providing feedback, and system help.**

    To have a usable interface, users must be able to move the cursor position, edit data, exit with different consequences, and obtain help. Numerous techniques for structuring and controlling data entry as well as providing feedback, prompting, error messages, and a well-organized help function can be used to enhance usability.

6. **Design human-computer dialogues, including the use of dialogue diagramming.**

    Human-computer dialogues should be consistent in design, allowing for shortcuts, providing feedback and closure on tasks, handling errors, allowing for operations reversal, and giving the user a sense of control and ease of navigation. Dialogue diagramming is a technique for representing human-computer dialogues. The technique uses boxes to represent screens, forms, or reports and lines to show the flow between each.

7. **Discuss interface design guidelines unique to the design of Internet-based electronic commerce systems.**

    The human-computer interface is a central and critical aspect of any Internet-based electronic commerce system. Using menu-driven navigation with cookie crumbs ensures that users can easily understand and navigate a system. Using lightweight graphics ensures that Web pages load quickly. Ensuring data integrity means that customer information is processed quickly, accurately, and securely. Using common templates ensures a consistent interface that is easy to maintain.

Key Terms Checkpoint

Here are the key terms from the chapter. The page where each term is first explained is in parentheses after the term.

1. **Audit trail (p. 292)**
2. **Cookie crumbs (p. 303)**
3. **Dialogue (p. 296)**
4. **Dialogue diagramming (p. 297)**
5. **Form (p. 272)**
6. **Lightweight graphics (p. 303)**
7. **Report (p. 272)**
8. **Template-based HTML (p. 304)**

Match each of the key terms above with the definition that best fits it.

_____ 1. Templates to display and process common attributes of higher-level, more abstract items.

_____ 2. A formal method for designing and representing human-computer dialogues using box and line diagrams.

_____ 3. A business document that contains only predefined data; it is a passive document used only for reading or viewing. A form typically contains data from many unrelated records or transactions.

_____ 4. A technique for showing users where they are in a Web site by placing a series of "tabs" on a Web page that shows users where they are and where they have been.

_____ 5. The sequence of interaction between a user and a system.

_____ 6. A business document that contains some predefined data and may include some areas where additional data are to be filled in. An instance of a form is typically based on one database record.

_____ 7. A record of the sequence of data entries and the date of those entries.

_____ 8. The use of small simple images to allow a Web page to be displayed more quickly.

Review Questions

1. Describe the prototyping process of designing forms and reports. What deliverables are produced from this process? Are these deliverables the same for all types of system projects? Why or why not?
2. To which initial questions must the analyst gain answers in order to build an initial prototype of a system output?
3. How should textual information be formatted on a help screen?
4. What type of labeling can you use in a table or list to improve its usability?
5. What column, row, and text formatting issues are important when designing tables and lists?
6. Describe how numeric, textual, and alphanumeric data should be formatted in a table or list.
7. Provide some examples where variations in user, task, system, and environmental characteristics might impact the design of system forms and reports.
8. Describe the process of designing interfaces and dialogues. What deliverables are produced from this process? Are these deliverables the same for all types of system projects? Why or why not?
9. List and describe the functional capabilities needed in an interface for effective entry and navigation. Which capabilities are most impor-

tant? Why? Will this be the same for all systems? Why or why not?
10. Describe the general guidelines for structuring data entry fields. Can you think of any instances when it would be appropriate to violate these guidelines?
11. Describe four types of data errors.
12. Describe the types of system feedback. Is any form of feedback more important than the others? Why or why not?
13. Describe the general guidelines for designing usable help. Can you think of any instances when it would be appropriate to violate these guidelines?
14. What steps do you need to follow when designing a dialogue? Of the guidelines for designing a dialogue, which is most important? Why?
15. Describe what is meant by a cookie crumb. How do these help prevent users from getting lost?
16. Describe why you might want to use lightweight graphics on some Web pages and large detailed graphics on others.
17. Why is it especially important to eliminate data entry errors on an electronic commerce Web site?
18. How can template-based HTML help to make a large electronic commerce site more maintainable?

Problems and Exercises

1. Imagine that you are to design a budget report for a colleague at work using a spreadsheet package. Following the prototyping discussed in the chapter (see also Figure 1–17), describe the steps you would take to design a prototype of this report.
2. Consider a system that produces budget reports for your department at work. Alternatively, consider a registration system that produces enrollment reports for a department at a university. For whichever system you choose, answer the following design questions: Who will use the output? What is the purpose of the output? When is the output needed and when is the information that will be used within the output available? Where does the output need to be delivered? How many people need to view the output?
3. Imagine the worst possible reports from a system. What is wrong with them? List as many problems as you can. What are the consequences of such reports? What could go wrong as a result? How does the prototyping process help guard against each problem?

4. Given the guidelines presented in this chapter, identify flaws in the design of the Report of Customers shown below. What assumptions about users and tasks did you make in order to assess this design? Redesign this report to correct these flaws.

Report of Customers—26-Oct-01	
Cust-ID	**Organization**
AC-4	A.C. Nielson Co.
ADTRA-20799	Adran
ALEXA-15812	Alexander & Alexander, Inc.
AMERI-1277	American Family Insurance
AMERI-28157	American Residential Mortgage
ANTAL-28215	Antalys
ATT-234	AT&T Residential Services
ATT-534	AT&T Consumer Services
. . .	
DOLE-89453	Dole United, Inc.
DOME-5621	
DO-67	Doodle Dandies
. . .	
ZNDS-22267	Zenith Data System

5. Consider the design of a registration system for a hotel. Following design specification items in Figure 8-11, briefly describe the relevant users, tasks, and displays involved in such a system.

6. Examine the help systems for some software applications that you use. Evaluate each using the general guidelines provided in Table 8-12.

7. Design one sample data entry screen for a hotel registration system using the data entry guidelines provided in this chapter (see Table 8-7). Support your design with arguments for each of the design choices you made.

8. Describe some typical dialogue scenarios between users and a hotel registration system. For hints, reread the section in this chapter that provides sample dialogue between users and the Customer Information System at Pine Valley Furniture.

9. Represent the dialogues from the previous question through the use of dialogue diagrams.

10. When developing an Internet-based electronic commerce system, why is the design of the human-computer interface one of the most critical elements? What makes the interface for an Internet-based electronic commerce system good? What makes it bad?

Discussion Questions

1. Discuss the differences between a form and a report. What characteristics make a form or report good (bad) and effective (ineffective)?

2. Discuss the various ways that information can be highlighted on a computer display. Which methods are most effective? Are some methods better than others? If so, why and when?

3. What problems can occur if a system fails to provide clear feedback and error messages to users?

4. How would you assess a system's usability? How would you know when a system was "usable"?

Case Problems

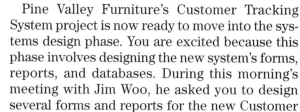

1. Pine Valley Furniture

Pine Valley Furniture's Customer Tracking System project is now ready to move into the systems design phase. You are excited because this phase involves designing the new system's forms, reports, and databases. During this morning's meeting with Jim Woo, he asked you to design several forms and reports for the new Customer Tracking System.

During the requirements determination phase, Jackie Judson requested that a customer profile be created for each customer. The customer profile is established when new customers place their first order. Customers will have the option of not completing a profile; however, to encourage customer participation, a 10 percent discount on the customer's total order will be given to each customer who completes a profile. In the beginning, existing customers will also be given the opportunity to participate in the customer profiling process. Customer profile information will be collected via a Customer Profile Form.

Gracie Breshers, a marketing executive, has requested that the Customer Tracking System generate a Products by Demographics Summary Report. This summary report should identify Pine Valley Furniture's major furniture categories, such as business furniture, living room, dining room, home office, and kitchen. Within each furniture category, she would like the total sales by region and customer age reported. She has also requested that several detailed reports be prepared; these reports will associate customer demographics with specific furniture category items.

Thi Hwang, a Pine Valley Furniture sales executive, would like to know, in terms of percentages, how many of Pine Valley Furniture's customers are repeat customers and how often they make purchases. Additionally, he would like to have this information categorized by customer type. For each customer type, he would like to know the frequency of the purchases. For instance, does this type of customer place an order at least once a month, at least every six months, at least once a year, or longer than one year? To be considered a repeat customer, the customer must have made two separate purchases within a two-year period.

a. What data will the Customer Profile Form need to collect? Using the guidelines presented in the chapter, design the Customer Profile Form.

b. Using the guidelines presented in the chapter, design the Products by Demographics Summary Report.

c. Using the guidelines presented in the chapter, design the Customer Purchasing Frequency Report.

d. Modify the dialogue diagram presented in Figure 8-20 to reflect the addition of the Customer Profile Form, Products by Demographics Summary Report, and the Customer Purchasing Frequency Report.

2. Hoosier Burger

As the lead analyst for the Hoosier Burger project, you have worked closely with Bob and Thelma Mellankamp. Having completed the systems analysis phase, you are now ready to begin designing the new Hoosier Burger information system. As the lead analyst on this project, you are responsible for overseeing the development of the forms, reports, and databases required by the new system. Because the inventory system is being automated and a new delivery system is being implemented, the Hoosier Burger system requires the development of several forms and reports.

Using your data flow diagrams and entity-relationship diagrams, you begin the task of identifying all the necessary forms and reports. You readily identify the need for a Delivery Customer Order Form, a Customer Account Balance Form, a Low-in-Stock Report, and a Daily Delivery Summary Report. The Delivery Customer Order Form will capture order details for those customers placing delivery orders. Bob will use the Customer Account Balance Form to look up a customer's current account balance. The Low-in-Stock Report will be generated daily to identify all food items or supplies that are low in stock. The Daily Delivery Summary Report will summarize each day's delivery sales by menu item sold.

a. What data will the Delivery Customer Order Form need to collect? Using the design guidelines presented in the chapter, design the Delivery Customer Order Form.

b. What data will the Customer Account Balance Form need to show? Using the design guidelines presented in the chapter, design the Customer Account Balance Form.

c. Using the design guidelines presented in the chapter, design the Daily Delivery Summary Report.

d. Using the design guidelines presented in the chapter, design the Low-in-Stock Report.

3. Pet Nanny

Pet owners often have difficulty locating pet-sitters for their pets, boarding their pets, or just getting the pets to the veterinarian. Recognizing these needs, Gladys Murphy decided to open Pet Nanny, a business providing specialized pet-care services to busy pet owners. The company provides a multitude of services, including pet grooming, massage, day care, home care, aroma therapy, boarding, and pickup and delivery. The company has been experiencing a steady increase in demand for its services.

Initially, when the company was founded, all pet-care records were kept manually. However, Gladys recognized the need to renovate Pet Nanny's existing systems and hired your consulting firm to perform the renovation. Your analyst team has just completed the requirements structuring phase and has selected an alternative design strategy. You are now ready to begin the systems design phase.

During the analysis phase, you determined that several forms and reports were necessary, including a Pet Enrollment Form, Pet Service Form, Pickup and Delivery Schedule Report, and Daily Boarding Report. When a customer wishes to use Pet Nanny's services for a new pet, the customer must provide basic information about the pet. For instance, the customer is asked to provide his or her name, address, phone number, the pet's name, birthdate (if known), and special care instructions. When a customer requests a special service for the pet, such as grooming or a massage, a service record is created. Because the pickup and delivery service is one of the most popular services offered by Pet Nanny, Gladys wants to make sure that no pets are forgotten. Each morning a report listing the pet pickups and deliveries is created. She also needs a report listing the pets being boarded, their special needs, and their length of stay.

a. What data should the Pet Enrollment Form collect? Using the guidelines provided in the chapter, design the Pet Enrollment Form.

b. What data should the Pet Service Form collect? Using the guidelines provided in the chapter, design the Pet Service Form.

c. Using the guidelines provided in the chapter, design the Pickup and Delivery Schedule Report.

d. Using the guidelines provided in the chapter, design the Daily Boarding Report.

CASE: BROADWAY ENTERTAINMENT COMPANY, INC.

Designing the Human Interface for the Customer Relationship Management System

Case Introduction

The students from St. Claire Community College are eager to begin building a prototype of MyBroadway, the Web-based customer relationship management system for Carrie Douglass, manager of the Centerville, Ohio, Broadway Entertainment Company (BEC) store. Prototyping seems like an ideal design approach for this system, because the final project product is not intended to be a production system. Rather, the student team is producing a proof-of-concept, initial system version, to be used to justify full development by BEC. Before building the prototype in Microsoft Access, the team is ready to plan the structure for the human interface of the system. For a Web-based system, the human interface is, to the customer, the system. Although the MyBroadway prototype system is not meant to be extensive, the prototype will be effective only if the human interface delights BEC customers. The students first decide to do a pencil-and-paper prototype before development in Access. This initial prototype will be used primarily for discussion among the team and for sharing with other teams in their information systems projects class at St. Claire. Professor Tann, their instructor, encourages collaborative learning, and the members of other teams will be valued, impartial evaluators of the usability of the system's human interface design.

Identifying the Human Interfaces

The human interfaces for MyBroadway are clearly visible from the system's context diagram (see BEC Figure 5-1 at the end of Chapter 5). The main human interfaces are data flows from and to each human external entity—customers and employees. The team decides to concentrate on the customer interfaces for the purpose of the pencil-and-paper prototype. Table BEC 8-1 summarizes the seven customer-related and one employee-related data flows and categorizes them as either system input or system output oriented.

The team quickly realizes that these data flows are not the only human interfaces. Any Web page that is needed in a navigation path leading to these data

flows is also a human interface. For example, to produce the Inventory Review output page, the customer must enter criteria for selecting which inventory items to display.

The student team decides that a customer should start using MyBroadway with a home or welcome page, with a catchy graphic and menu selections for accessing different parts of the system. One way to categorize system functions would logically be to group all inputs, or data entry pages, together and all system outputs, or form and report display pages, together into a second group. The team decides, however, that this is a system-centric view, not a user-centric view of the system's functionality. After some brainstorming, the team decides that it would be more logical for users to understand and use the system if pages were grouped by the type of data the users want to use. There appear to be two natural data groupings: product and purchase/rental data.

Designing the Dialogue between MyBroadway and Users

BEC Figure 8-1 is a dialogue diagram that represents the relationships between system Web pages developed by the team using a data orientation for human interfaces. Page 0 is the welcome page. Besides information to introduce MyBroadway to customers, this page provides menu options or buttons for the user to indicate which data group he or she wants to use. If the user wants to work with product data, then page 1 provides the user a way to enter the request for a new product (page 1.3, which is input 2 from BEC Table 8-1) or to work with existing product data (pages 1.1 and 1.2). Page 1.1 guides the user either to enter a new comment on a product (page 1.1.1 for input 1) or to view existing product comments (page 1.1.2 for output 2). Page 1.1, thus, must provide a way for the user to select or enter data to identify the product for use in subordinate pages.

In general, each system input or output is a terminal (or leaf) node of the dialogue diagram. Each superior node above a leaf is a step for guiding the user to a system input or output. Sometimes a system output can be the basis for a customer to create a system input. For example, consider pages 2.1, 2.1.1, and 2.1.1.1. Page 2.1 is the Rental Status report, output 3. The team decides that users will want to see this report before requesting an extension to a particular rental, which is done on page 2.1.1, representing input 3.

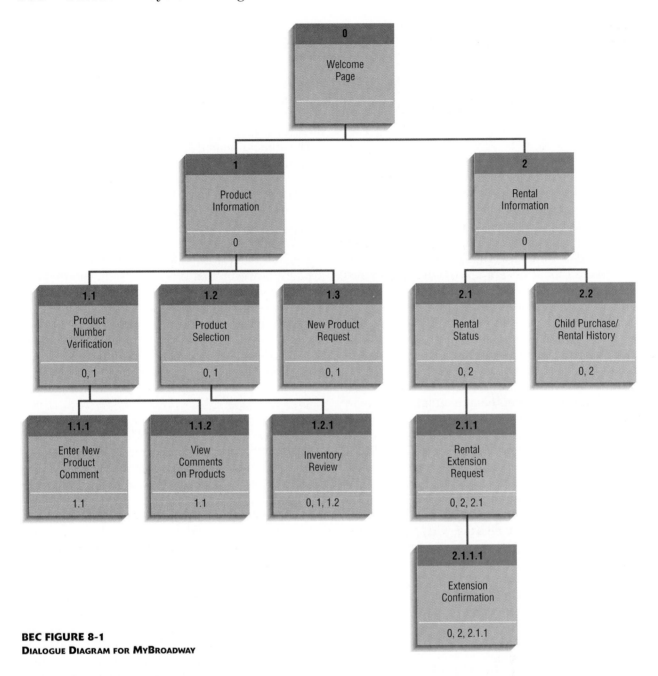

BEC FIGURE 8-1
DIALOGUE DIAGRAM FOR MYBROADWAY

BEC TABLE 8-1:   Human Interfaces
for MyBroadway

Inputs from Users	Outputs to Users
I1 New Comment on a Product	O1 Inventory Review
I2 New Product Request	O2 Comments on Products
I3 Rental Extension Request	O3 Rental Status
I4 Favorite Picks	O4 Child Purchase/Rental History

Thus, page 2.1 not only displays the Rental Status report but also provides a way for a user to select a particular outstanding rental for which to request an extension in page 2.1.1. Page 2.1.1.1 is a message page (possibly not a totally separate page but rather a message window to overlay on top of page 2.1.1) that will say whether the extension request is accepted.

Designing Forms and Reports for MyBroadway

Each of the 13 pages identified in BEC Figure 8-1 needs to be designed for customer usability. The team realizes that usability means that the page is easy to understand, helps the customer perform a given task, and is efficient for the customer to use. From their courses at St. Claire Community College, the students are familiar with many usability guide-

BEC FIGURE 8-2
INTERFACE DESIGN FOR PAGE 1.2

PRODUCT SELECTION

You can see what products we carry in our BEC store by entering values below. For each value you want to search on, click in the check box next to that criterion. If you know the name of the video, music, or game, enter the title—we'll scan our inventory to find the products with a title that comes closest to what you enter. You can also look for products by category (for example, R&B, hard rock, jazz), publisher, or release date. For these criteria, select a value from the available options from the drop-down menu. If you enter or select values from more than one criterion, we'll look for products that satisfy all of the selections you make.

☐ Title: [_____]

☐ Category: [_____▼]

☐ Publisher: [_____▼]

                    Year   Month
☐ Release Date: [____▼]   [____▼]

[Submit]   [Reset]   [Return to Welcome Page]   [Back]

lines for human-computer interfaces. Also from their education the students know that usability is improved if there are frequent reviews of the proposed human interfaces. This is another reason why prototyping will be an effective development strategy for MyBroadway. Initial designs for each page will be reviewed by the team's classmates from St. Claire, and then working prototype iterations will be evaluated by actual customers in the Centerville store.

Two of the pencil-and-paper page prototypes appear in BEC Figures 8-2 and 8-3. Page 1.2 (BEC Figure 8-2) is an intermediate page that helps the user formulate the criteria to specify which items from inventory to include in the Inventory Review system output, output 1 from BEC Table 8-1. Page 1.2 has a title and explanation of its purpose. Because the page is not too full of data, the team decides to include an explanation of its purpose and content. Alternatively the team considered excluding this explanation, instead making it accessible by the user via a help button. The user clicks in a check box to indicate that a value will be entered or selected for each type of product selection criteria. Because there are thousands of titles in a BEC store's inventory, the team decides not to use a drop-down menu for entering a title, but rather the user enters the approximate title of the product. The team realizes that in many cases the title entered will not exactly match the title stored in the database. The logic that processes a title will have to search for the best match. All other criteria, if checked by the user, have a more limited set of options, so drop-down selection boxes are used. The team decides that there are four ways for the user to exit the page, the first three of which are shown in the dialogue diagram in BEC Figure 8-1. The first option is to sub-

mit the product selection, which will take the user to page 1.2.1 (BEC Figure 8-3), the Inventory Review, or output 1. The second option is to return to the welcome page (page 0). The third option is to go back to page 1 to consider other product data tasks. Although not shown in BEC Figure 8-1, a fourth option is to clear the values in page 1.2 to consider another selection. This might occur because the user changes his or her mind after indicating some selections, or upon returning from page 1.2.1, the user may want to enter another selection starting with no selection values.

Page 1.2.1 (BEC Figure 8-3) is the Inventory Review system output, thus the title of the page. The team decides to show the selection criteria in a top frame on the page to remind the user of the selections. Because many items may satisfy a selection, the team allows for scrolling through all results in the bottom frame. Besides the three exit options from this page shown on the dialogue diagram of BEC Figure 8-1— go back to page 1.2, return to the Product Information page (page 1), and return to the Welcome Page—a fourth user action is to print the query results. The students had not even considered the need for printing until designing this screen. The team notes this new platform requirement and will ask Carrie about this at their next meeting. The cost of the printer will not be that great, but the team is more concerned about paper costs, keeping a supply of paper in the printer, and the extra effort to keep the printer free of paper jams, with full ink cartridges, and otherwise in working order. The use of a printer requires more active involvement of store staff than the team had previously explained to Carrie would be necessary.

```
┌─────────────────────────────────────────────────────────────────────────┐
│ INVENTORY REVIEW                                                           │
│  ┌─────────────┐  Title:              ┌──────────────┐ ┌──────────────┐   │
│  │These are the│  Category:           │Return to Product│ │  Return to  │  │
│  │criteria you │  Publisher:          │  Information  │ │ Welcome Page │   │
│  │  selected:  │  Release Date:       └──────────────┘ └──────────────┘   │
│  └─────────────┘                      ┌──────────────┐ ┌──────────────┐   │
│                                       │     Back     │ │    Print     │   │
│                                       └──────────────┘ └──────────────┘   │
│  ┌────────────────────────────────────────────────────────────────┬──┐   │
│  │ Title:                      Artist:                              │▲ │   │
│  │                                                                  │  │   │
│  │ Category:          Publisher:              Release Date:         │  │   │
│  │                                                                  │  │   │
│  │ Description:                                                     │  │   │
│  │                                                                  │  │   │
│  │                                                                  │  │   │
│  │ Price—Sale:        Rental:                                       │  │   │
│  │                                                                  │▼ │   │
│  └────────────────────────────────────────────────────────────────┴──┘   │
└─────────────────────────────────────────────────────────────────────────┘
```

BEC FIGURE 8-3
INTERFACE DESIGN FOR PAGE 1.2.1

Case Summary

The student team feels that the design of the user dialogue and two system pages represents a suitable project deliverable that should be reviewed by someone outside the team. Fortunately, Professor Tann has scheduled project status report presentations for the next session of the team's project class. The MyBroadway student team will present the pencil-and-paper prototype shown in BEC Figures 8-1, 8-2, and 8-3 to obtain an initial reaction to the design. Because the team has not invested a great deal of time in the initial design, the team members believe they can be open to, not defensive in response to, constructive suggestions. Because other teams will also likely walk through human interface designs, the BEC team will see some other creative designs, which will give additional ideas for improvement.

Case Questions

1. Using guidelines from this chapter and other sources, evaluate the usability of the dialogue design depicted in BEC Figure 8-1. Specifically, consider the overall organization, grouping of pages, navigation paths between pages, and depth of the dialogue diagram and how this depth might affect user efficiency.
2. Are there any missing pages in BEC Figure 8-1? Can you anticipate the need for additional pages in the customer interface for MyBroadway? If so, where do these pages come from if not from the list of system inputs and outputs listed in BEC Table 8-1?
3. Using guidelines from this chapter and other sources, evaluate the usability of the two pages shown in BEC Figures 8-2 and 8-3.
4. Chapter 8 encourages the design of a help system early in the design of the human interface. How would you incorporate help into the interface as designed by the St. Claire students?
5. Given the designs for pages 1.2 and 1.2.1, design pages 1.1.2 and 1.3. Assume that the designs for pages 1.2 and 1.2.1 represent the look and feel desired for MyBroadway.
6. The designs for pages 1.2 and 1.2.1 include a Back button. Is this button necessary or desirable?
7. Is the use of drop-down selection lists a good feature of the design of page 1.2? Can you think of a better way to provide these selections?
8. The design for page 1.2.1 includes a Print button. Design the printed version of this page.
9. Are there any other possible navigation paths exiting page 1.2 that are not shown on BEC Figure 8-1? Is page 1.2.1 the only possible result of searching on the selection criteria? If not, design pages for other results.
10. Explain how cookie crumbs could be used in MyBroadway. Are cookie crumbs a desireable navigation aid for this system? Why or why not?

Designing Databases

⊙ **Objectives**

After studying this chapter, you should be able to:

⊙ Concisely define each of the following key database design terms: *relation, primary key, functional dependency, foreign key, referential integrity, field, data type, null value, denormalization, file organization, index,* and *secondary key.*

⊙ Explain the role of designing databases in the analysis and design of an information system.

⊙ Transform an entity-relationship (E-R) diagram into an equivalent set of well-structured (normalized) relations.

⊙ Merge normalized relations from separate user views into a consolidated set of well-structured relations.

⊙ Choose storage formats for fields in database tables.

⊙ Translate well-structured relations into efficient database tables.

⊙ Explain when to use different types of file organizations to store computer files.

⊙ Describe the purpose of indexes and the important considerations in selecting attributes to be indexed.

Chapter Preview . . .

In Chapter 6 you learned how to represent an organization's data graphically using an entity-relationship (E-R) diagram and Microsoft Visio. In this chapter, you learn guidelines for clear and efficient data files and about logical and physical database design. It is likely that the human interface and database design steps will happen in parallel, as illustrated in the SDLC in Figure 9-1.

Logical and physical database design has five purposes:

1. Structure the data in stable structures that are not likely to change over time and that have minimal redundancy.

2. Develop a logical database design that reflects the actual data requirements that exist in the forms (hard copy and computer displays) and reports of an information system. This is why database design is often done in parallel with the design of the human interface of an information system.

3. Develop a logical database design from which we can do physical database design. Because most information systems today use relational database management systems, logical database design usually uses a relational database model, which represents data in simple tables with common columns to link related tables.

4. Translate a relational database model into a technical file and database design.

5. Choose data storage technologies (such as floppy disk, CD-ROM, or optical disk) that will efficiently, accurately, and securely process database activities.

The implementation of a database (i.e., creating and loading data into files and databases) is done during the next phase of the systems development life cycle. Because implementation is very technology specific, we address implementation issues only at a general level in Chapter 10.

FIGURE 9-1
SYSTEMS DEVELOPMENT LIFE CYCLE
Systems analysts design databases during the systems design phase. Database design typically occurs in parallel with other design steps.

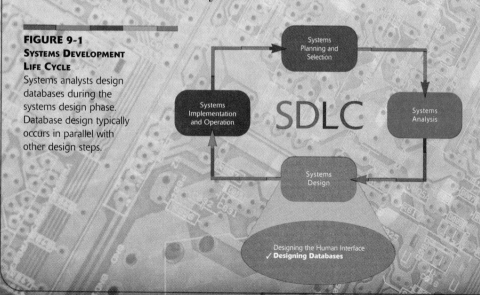

Database Design

File and database design occurs in two steps. You begin by developing a logical database model, which describes data using a notation that corresponds to a data organization used by a database management system. This is the system software responsible for storing, retrieving, and protecting data (such as Microsoft Access, Oracle, or SQL Server). The most common style for a logical database model is the relational database model. Once you develop a clear and precise logical database model, you are ready to prescribe the technical specifications for computer files and databases in which to store the data ultimately. A physical database design provides these specifications.

You typically do logical and physical database design in parallel with other systems design steps. Thus, you collect the detailed specifications of data necessary for logical database design as you design system inputs and outputs. Logical database design is driven not only from the previously developed E-R data model for the application but also from form and report layouts. You study data elements on these system inputs and outputs and identify interrelationships among the data. As with conceptual data modeling, the work of all systems development team members is coordinated and shared through the project dictionary or repository. The designs for logical databases and system inputs and outputs are then used in physical design activities to specify to computer programmers, database administrators, network managers, and others how to implement the new information system. We assume for this text that the design of computer programs and distributed information processing and data networks are topics of other courses, so we concentrate on the aspect of physical design most often undertaken by a systems analyst—physical file and database design.

The Process of Database Design

Figure 9-2 shows that database modeling and design activities occur in all phases of the systems development process. In this chapter we discuss methods that help you finalize logical and physical database designs during the design phase. In logical database design you use a process called normalization, which is a way to build a data model that has the properties of simplicity, nonredundancy, and minimal maintenance.

In most situations, many physical database design decisions are implicit or eliminated when you choose the data management technologies to use with the application. We concentrate on those decisions you will make most frequently and use Microsoft Access to illustrate the range of physical database design parameters you must manage. The interested reader is referred to Hoffer, Prescott, and McFadden (2002) for a more thorough treatment of techniques for logical and physical database design.

There are four key steps in logical database modeling and design:

1. Develop a logical data model for each known user interface (form and report) for the application using normalization principles.
2. Combine normalized data requirements from all user interfaces into one consolidated logical database model; this step is called view integration.
3. Translate the conceptual E-R data model for the application, developed without explicit consideration of specific user interfaces, into normalized data requirements.
4. Compare the consolidated logical database design with the translated E-R model and produce, through view integration, one final logical database model for the application.

During physical database design, you use the results of these four key logical database design steps. You also consider definitions of each attribute; descriptions of where and when data are entered, retrieved, deleted, and updated; expectations for response time and data integrity; and descriptions of the file and database technologies to be used. These inputs allow you to make key physical database design decisions, including the following:

1. Choosing the storage format (called data type) for each attribute from the logical database model; the format is chosen to minimize storage space and to maximize data quality. Data type involves choosing length, coding scheme, number of decimal places, minimum and maximum values, and potentially many other parameters for each attribute.
2. Grouping attributes from the logical database model into physical records (in general, this is called selecting a stored record, or data, structure).
3. Arranging related records in secondary memory (hard disks and magnetic tapes) so that individual and groups of records can be stored, retrieved, and updated rapidly (called file organizations). You should also consider protecting data and recovering data after errors are found.
4. Selecting media and structures for storing data to make access more efficient. The choice of media affects the utility of different file organizations. The primary structure used today to make access to data more rapid is key indexes, on unique and nonunique keys.

In this chapter we show how to do each of the logical database design steps and discuss factors to consider in making each physical file and database design decision.

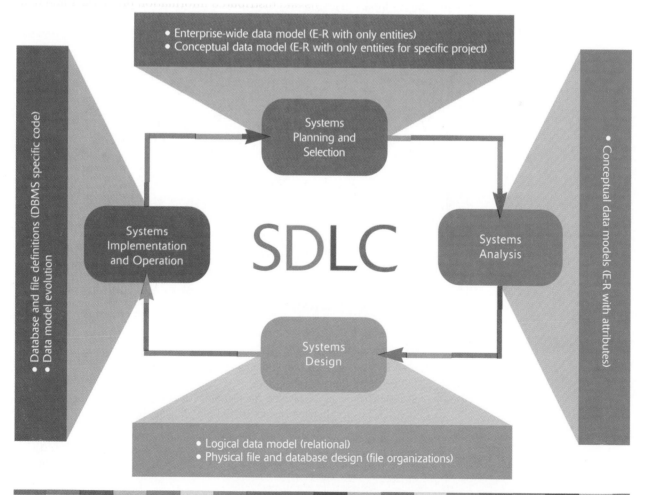

FIGURE 9-2
RELATIONSHIP BETWEEN DATA MODELING AND THE SYSTEMS DEVELOPMENT LIFE CYCLE

A

HIGHEST VOLUME CUSTOMER

ENTER PRODUCT ID.:   M128
START DATE: 11/01/2002
END DATE: 12/31/2002
- - - - - - - - - - - - - - - - - - - -
CUSTOMER ID.:   1256
NAME: Commonwealth Builder
VOLUME: 30

This inquiry screen shows the customer with the largest volume total sales of a specified product during an indicated time period.

Relations:
    CUSTOMER(Customer_ID,Name)
    ORDER(Order_Number,Customer_ID,Order_Date)
    PRODUCT(Product_ID)
    LINE ITEM(Order_Number,Product_ID,Order_Quantity)

B

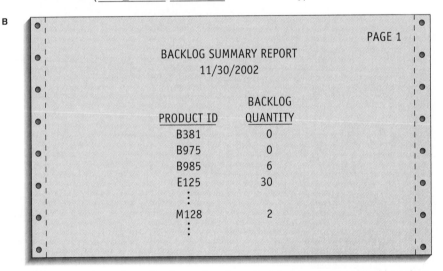

PAGE 1

BACKLOG SUMMARY REPORT
11/30/2002

PRODUCT ID	BACKLOG QUANTITY
B381	0
B975	0
B985	6
E125	30
⋮	
M128	2
⋮	

This report shows the unit volume of each product that has been ordered less than amount shipped through the specified date.

Relations:
    PRODUCT(Product_ID)
    LINE ITEM(Product_ID,Order_Number,Order_Quantity)
    ORDER(Order_Number,Order_Date)
    SHIPMENT(Product_ID,Invoice_Number,Ship_Quantity)
    INVOICE(Invoice_Number,Invoice_Date,Order_Number)

C   CUSTOMER(Customer_ID,Name)
    PRODUCT(Product_ID)
    INVOICE(Invoice_Number,Invoice_Date,Order_Number)
    ORDER(Order_Number,Customer_ID,Order_Date)
    LINE ITEM(Order_Number,Product_ID,Order_Quantity)
    SHIPMENT(Product_ID,Invoice_Number,Ship_Quantity)

D

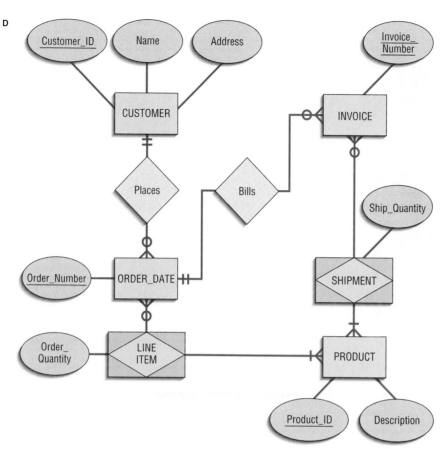

Relations:

    CUSTOMER(Customer_ID,Name,Address)
    PRODUCT(Product_ID,Description)
    ORDER(Order_Number,Customer_ID,Order_Date)
    LINE ITEM(Order_Number,Product_ID,Order_Quantity)
    INVOICE(Invoice_Number,Order_Number)
    SHIPMENT(Invoice_Number,Product_ID,Ship_Quantity)

E   CUSTOMER(Customer_ID,Name,Address)
     PRODUCT(Product_ID,Description)
     ORDER(Order_Number,Customer_ID,Order_Date)
     LINE ITEM(Order_Number,Product_ID,Order_Quantity)
     INVOICE(Invoice_Number,Order_Number,Invoice_Date)
     SHIPMENT(Invoice_Number,Product_ID,Ship_Quantity)

FIGURE 9-3 (*continued*)

Deliverables and Outcomes

During logical database design, you must account for every data element on a system input or output—form or report—and on the E-R model. Each data element (like customer name, product description, or purchase price) must be a piece of raw data kept in the system's database, or in the case of a data element on a system output, the element can be derived from data in the database. Figure 9-3 illustrates the outcomes from the four-step logical database design process. Figures 9-3(A) and 9-3(B) (step 1) contain two sample system outputs for a customer order processing system at Pine Valley Furniture. A description of the associated database requirements, in the form of what we call normalized relations, is listed below each output diagram. Each relation (think of a relation as a table with rows and columns) is named and its attributes (columns) are listed within parentheses. The **primary key** attribute—that

Primary key
An attribute whose value is unique across all occurrences of a relation.

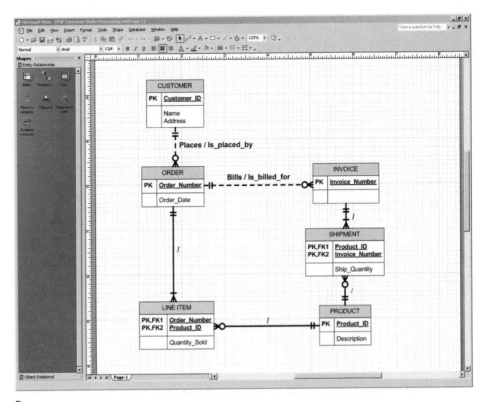

FIGURE 9-3 *(continued)*   F

attribute whose value is unique across all occurrences of the relation—is indicated by an underline, and an attribute of a relation that is the primary key of another relation is indicated by a dashed underline.

In Figure 9-3(A), data are shown about customers, products, and the customer orders and associated line items for products. Each of the attributes of each relation either appears in the display or is needed to link related relations. For example, because an order is for some customer, an attribute of ORDER is the associated Customer_ID. The data for the display in Figure 9-3(B) are more complex. A backlogged product on an order occurs when the amount ordered (Order_Quantity) is less than the amount shipped (Ship_Quantity) for invoices associated with an order. The query refers to only a specified time period, so the Order_Date is needed. The INVOICE Order_Number links invoices with the associated order.

Figure 9-3(C) (step 2) shows the result of integrating these two separate sets of normalized relations. Figure 9-3(D) (step 3) shows an E-R diagram for a customer order processing application that might be developed during conceptual data modeling along with equivalent normalized relations. Figure 9-3(E) (step 4) shows a set of normalized relations that would result from reconciling the logical database designs of Figures 9-3(C) and 9-3(D). Normalized relations like those in Figure 9-3(E) are the primary deliverable from logical database design.

Finally, Figure 9-3(F) shows the E-R diagram drawn in Microsoft Visio. Visio actually shows the tables and relationships between the tables from the normalized relations. Thus, the associative entities, LINE ITEM and SHIPMENT, are shown as entities on the Visio diagram; we do not place relationship names on either side of these entities on the Visio diagram because these represent associative entities. Visio also shows for theses entities the primary keys of the associated ORDER, INVOICE, and PRODUCT entities. Also note that the lines for the Places and Bills relationships are dashed. This is Visio notation to indicate that ORDER and INVOICE have their own primary keys that do not include the primary keys of CUSTOMER and ORDER respectively (what Visio calls non-identifying relationships). Because LINE ITEM and SHIPMENT both include in their primary keys the primary keys of other entities (which is common for associative entities), the relationships around LINE ITEM and SHIPMENT are identifying, and hence the relationship lines are solid.

It is important to remember that relations do not correspond to computer files. In physical database design, you translate the relations from logical database design into specifications for computer files. For most information systems, these files will be tables in a relational database. These specifications are sufficient for programmers and database analysts to code the definitions of the database. The coding, done during systems implementation, is written in special database definition and processing languages, such as Structured Query Language (SQL), or by filling in table definition forms, such as with Microsoft Access. Figure 9-4 shows a possible definition for the SHIPMENT relation from Figure 9-3(E) using Microsoft Access. This display of the SHIPMENT table definition illustrates choices made for several physical database design decisions.

- ❂ All three attributes from the SHIPMENT relation, and no attributes from other relations, have been grouped together to form the fields of the SHIPMENT table.

- ❂ The Invoice Number field has been given a data type of Text, with a maximum length of 10 characters.

- ❂ The Invoice Number field is required because it is part of the primary key for the SHIPMENT table (the value that makes every row of the SHIPMENT table unique is a combination of Invoice Number and Product ID).

- ❂ An index is defined for the Invoice Number field, but because there may be several rows in the SHIPMENT table for the same invoice (different products on the same invoice), duplicate index values are allowed (so Invoice Number is what we will call a secondary key).

- ❂ The Invoice_Number, because it references the Invoice_Number from the INVOICE table, is defined as a Lookup to the first column (Invoice_Number) of the INVOICE table; in this way, all values that are placed in the Invoice_Number field of the SHIPMENT table must correspond to a previously entered invoice.

A

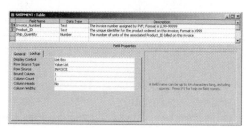

B

FIGURE 9-4
**DEFINITION OF SHIPMENT TABLE
IN MICROSOFT ACCESS
(A) TABLE WITH INVOICE_NUMBER
PROPERTIES
(B) INVOICE_NUMBER LOOKUP
PROPERTIES**

Many other physical database design decisions were made for the SHIPMENT table, but they are not apparent on the display in Figure 9-4. Further, this table is only one table in the PVF Order Entry database, and other tables and structures for this database are not illustrated in this figure.

Relational Database Model

NET SEARCH

Investigate the origins of the relational database model. Visit http://www.prenhall. com/valacich to complete an exercise related to this topic.

Relational database model
Data represented as a set of related tables or relations.

Relation
A named, two-dimensional table of data. Each relation consists of a set of named columns and an arbitrary number of unnamed rows.

Many different database models are in use and are the basis for database technologies. Although hierarchical and network models have been popular in the past, these are not used very often today for new information systems. Object-oriented database models are emerging, but are still not common. The vast majority of information systems today use the relational database model. The **relational database model** represents data in the form of related tables or relations. A **relation** is a named, two-dimensional table of data. Each relation (or table) consists of a set of named columns and an arbitrary number of unnamed rows. Each column in a relation corresponds to an attribute of that relation. Each row of a relation corresponds to a record that contains data values for an entity.

Figure 9-5 shows an example of a relation named EMPLOYEE1. This relation contains the following attributes describing employees: Emp_ID, Name, Dept, and Salary. There are five sample rows in the table, corresponding to five employees.

You can express the structure of a relation by a shorthand notation in which the name of the relation is followed (in parentheses) by the names of the attributes in the relation. The identifier attribute (called the primary key of the relation) is underlined. For example, you would express EMPLOYEE1 as follows:

Employee(Emp_ID,Name,Dept,Salary)

Not all tables are relations. Relations have several properties that distinguish them from nonrelational tables:

1. Entries in cells are simple. An entry at the intersection of each row and column has a single value.
2. Entries in columns are from the same set of values.
3. Each row is unique. Uniqueness is guaranteed because the relation has a nonempty primary key value.
4. The sequence of columns can be interchanged without changing the meaning or use of the relation.
5. The rows may be interchanged or stored in any sequence.

Well-structured relation (or table)
A relation that contains a minimum amount of redundancy and allows users to insert, modify, and delete the rows without errors or inconsistencies.

Well-Structured Relations

What constitutes a **well-structured relation** (or **table**)? Intuitively, a well-structured relation contains a minimum amount of redundancy and allows users to insert, modify, and delete the rows in a table without errors or inconsistencies. EMPLOYEE1 (Figure 9-5) is such a relation. Each row of the table

---

FIGURE 9-5
EMPLOYEE1 RELATION WITH SAMPLE DATA

EMPLOYEE1

Emp_ID	Name	Dept	Salary
100	Margaret Simpson	Marketing	42,000
140	Allen Beeton	Accounting	39,000
110	Chris Lucero	Info Systems	41,500
190	Lorenzo Davis	Finance	38,000
150	Susan Martin	Marketing	38,500

contains data describing one employee, and any modification to an employee's data (such as a change in salary) is confined to one row of the table.

In contrast, EMPLOYEE2 (Figure 9-6) contains data about employees and the courses they have completed. Each row in this table is unique for the combination of Emp_ID and Course, which becomes the primary key for the table. This is not a well-structured relation, however. If you examine the sample data in the table, you notice a considerable amount of redundancy. For example, the Emp_ID, Name, Dept, and Salary values appear in two separate rows for employees 100, 110, and 150. Consequently, if the salary for employee 100 changes, we must record this fact in two rows (or more, for some employees).

The problem with this relation is that it contains data about two entities: EMPLOYEE and COURSE. You will learn to use principles of normalization to divide EMPLOYEE2 into two relations. One of the resulting relations is EMPLOYEE1 (Figure 9-5). The other we will call EMP COURSE, which appears with sample data in Figure 9-7. The primary key of this relation is the combination of Emp_ID and Course (we emphasize this by underlining the column names for these attributes).

Normalization

We have presented an intuitive discussion of well-structured relations; however, we need rules and a process for designing them. **Normalization** is a process for converting complex data structures into simple, stable data structures. For example, we used the principles of normalization to convert the EMPLOYEE2 table with its redundancy to EMPLOYEE1 (Figure 9-5) and EMP COURSE (Figure 9-7).

Normalization
The process of converting complex data structures into simple, stable data structures.

EMPLOYEE2

Emp_ID	Name	Dept	Salary	Course	Date_Completed
100	Margaret Simpson	Marketing	$42,000	SPSS	6/19/2002
100	Margaret Simpson	Marketing	42,000	Surveys	10/7/2002
140	Alan Beeton	Accounting	39,000	Tax Acc	12/8/2002
110	Chris Lucero	Info Systems	41,500	SPSS	1/12/2002
110	Chris Lucero	Info Systems	41,500	C++	4/22/2002
190	Lorenzo Davis	Finance	38,000	Investments	5/7/2002
150	Susan Martin	Marketing	38,500	SPSS	6/19/2002
150	Susan Martin	Marketing	38,500	TQM	8/12/2002

FIGURE 9-6
RELATION WITH REDUNDANCY

EMP COURSE

Emp_ID	Course	Date_Completed
100	SPSS	6/19/2002
100	Surveys	10/7/2002
140	Tax Acc	12/8/2002
110	SPSS	1/22/2002
110	C++	4/22/2002
190	Investments	5/7/2002
150	SPSS	6/19/2002
150	TQM	8/12/2002

FIGURE 9-7
EMP COURSE RELATION

Rules of Normalization

Normalization is based on well-accepted principles and rules. There are many normalization rules, more than can be covered in this text (see Hoffer, Prescott, and McFadden, 2002, for a more complete coverage). Besides the five properties of relations outlined previously, there are two other frequently used rules:

1. *Second normal form (2NF).* Each nonprimary key attribute is identified by the whole key (what we call full functional dependency).
2. *Third normal form (3NF).* Nonprimary key attributes do not depend on each other (what we call no transitive dependencies).

The result of normalization is that every nonprimary key attribute depends upon the whole primary key and nothing but the primary key. We discuss second and third normal form in more detail next.

Functional Dependence and Primary Keys

Normalization is based on the analysis of functional dependence. A **functional dependency** is a particular relationship between two attributes. In a given relation, attribute B is functionally dependent on attribute A if, for every valid value of A, that value of A uniquely determines the value of B. The functional dependence of B on A is represented by an arrow, as follows: A→B (e.g., Emp_ID→Name in the relation of Figure 9-5). Functional dependence does not imply mathematical dependence—that the value of one attribute may be computed from the value of another attribute; rather, functional dependence of B on A means that there can be only one value of B for each value of A. Thus, for a given Emp_ID value, there can be only one Name value associated with it; the value of Name, however, cannot be derived from the value of Emp_ID. Other examples of functional dependencies from Figure 9-3(B) are in ORDER, Order_Number → Order_Date, and in INVOICE, Invoice_Number → Invoice_Date and Order_Number.

An attribute may be functionally dependent on two (or more) attributes, rather than on a single attribute. For example, consider the relation EMP COURSE (Emp_ID, Course,Date_Completed) shown in Figure 9-7. We represent the functional dependency in this relation as follows: Emp_ID, Course→Date_Completed. In this case, Date_Completed cannot be determined by either Emp_ID or Course alone, because Date_Completed is a characteristic of an employee taking a course.

You should be aware that the instances (or sample data) in a relation do not prove that a functional dependency exists. Only knowledge of the problem domain, obtained from a thorough requirements analysis, is a reliable method for identifying a functional dependency. However, you can use sample data to demonstrate that a functional dependency does not exist between two or more attributes. For example, consider the sample data in the relation EXAMPLE(A,B,C,D) shown in Figure 9-8. The sample data in this relation prove that attribute B is not functionally dependent on attribute A because A does not uniquely determine B (two rows with the same value of A have different values of B).

Functional dependency
A particular relationship between two attributes. For a given relation, attribute B is functionally dependent on attribute A if, for every valid value of A, that value of A uniquely determines the value of B. The functional dependence of B on A is represented by A→B.

FIGURE 9-8
EXAMPLE Relation

EXAMPLE

A	B	C	D
X	U	X	Y
Y	X	Z	X
Z	Y	Y	Y
Y	Z	W	Z

Second Normal Form

A relation is in **second normal form (2NF)** if every nonprimary key attribute is functionally dependent on the whole primary key. Thus no nonprimary key attribute is functionally dependent on part, but not all, of the primary key. Second normal form is satisfied if any one of the following conditions apply:

1. The primary key consists of only one attribute (such as the attribute Emp_ID in relation EMPLOYEE1).
2. No nonprimary key attributes exist in the relation.
3. Every nonprimary key attribute is functionally dependent on the full set of primary key attributes.

EMPLOYEE2 (Figure 9-6) is an example of a relation that is not in second normal form. The shorthand notation for this relation is

EMPLOYEE2 (Emp_ID,Name,Dept,Salary,Course,Date_Completed)

The functional dependencies in this relation are the following:

Emp_ID→Name,Dept,Salary
Emp_ID,Course→Date_Completed

The primary key for this relation is the composite key Emp_ID,Course. Therefore, the nonprimary key attributes Name, Dept, and Salary are functionally dependent on only Emp_ID but not on Course. EMPLOYEE2 has redundancy, which results in problems when the table is updated.

To convert a relation to second normal form, you decompose the relation into new relations using the attributes, called *determinants*, that determine other attributes; the determinants are the primary keys of these relations. EMPLOYEE2 is decomposed into the following two relations:

1. EMPLOYEE(Emp_ID,Name,Dept,Salary): This relation satisfies the first second normal form condition (sample data shown in Figure 9-5).
2. EMP COURSE(Emp_ID,Course,Date_Completed): This relation satisfies second normal form condition three (sample data appear in Figure 9-7).

Third Normal Form

A relation is in **third normal form (3NF)** if it is in second normal form and there are no functional dependencies between two (or more) nonprimary key attributes (a functional dependency between nonprimary key attributes is also called a *transitive dependency*). For example, consider the relation SALES(Customer_ID,Customer_Name,Salesperson,Region) (sample data shown in Figure 9-9[A]).

The following functional dependencies exist in the SALES relation:

1. Customer_ID→Customer_Name,Salesperson,Region (Customer_ID is the primary key.)
2. Salesperson→Region (Each salesperson is assigned to a unique region.)

Notice that SALES is in second normal form because the primary key consists of a single attribute (Customer_ID). However, Region is functionally dependent on Salesperson, and Salesperson is functionally dependent on Customer_ID. As a result, there are data maintenance problems in SALES.

1. A new salesperson (Robinson) assigned to the North region cannot be entered until a customer has been assigned to that salesperson (because a value for Customer_ID must be provided to insert a row in the table).
2. If customer number 6837 is deleted from the table, we lose the information that salesperson Hernandez is assigned to the East region.
3. If salesperson Smith is reassigned to the East region, several rows must be changed to reflect that fact (two rows are shown in Figure 9-9[A]).

Phase 3:
Systems
Design

Second normal form (2NF)
A relation is in second normal form if every nonprimary key attribute is functionally dependent on the whole primary key.

Third normal form (3NF)
A relation is in third normal form (3NF) if it is in second normal form and there are no functional (transitive) dependencies between two (or more) nonprimary key attributes.

FIGURE 9-9
REMOVING TRANSITIVE
DEPENDENCIES
(A) RELATION WITH TRANSITIVE
DEPENDENCY
(B) RELATIONS IN 3NF

SALES

Customer_ID	Customer_Name	Salesperson	Region
8023	Anderson	Smith	South
9167	Bancroft	Hicks	West
7924	Hobbs	Smith	South
6837	Tucker	Hernandez	East
8596	Eckersley	Hicks	West
7018	Arnold	Faulb	North

A

SALES1

Customer_ID	Customer_Name	Salesperson
8023	Anderson	Smith
9167	Bancroft	Hicks
7924	Hobbs	Smith
6837	Tucker	Hernandez
8596	Eckersley	Hicks
7018	Arnold	Faulb

SPERSON

Salesperson	Region
Smith	South
Hicks	West
Hernandez	East
Faulb	North

B

These problems can be avoided by decomposing SALES into the two relations, based on the two determinants, shown in Figure 9-9(B). These relations are the following:

SALES1(Customer_ID,Customer_Name,Salesperson)
SPERSON(Salesperson,Region)

Note that Salesperson is the primary key in SPERSON. Salesperson is also a foreign key in SALES1. A **foreign key** is an attribute that appears as a nonprimary key attribute in one relation (such as SALES1) and as a primary key attribute (or part of a primary key) in another relation. You designate a foreign key by using a dashed underline.

A foreign key must satisfy **referential integrity,** which specifies that the value of an attribute in one relation depends on the value of the same attribute in another relation. Thus, in Figure 9-9(B), the value of Salesperson in each row of table SALES1 is limited to only the current values of Salesperson in the SPERSON table. Referential integrity is one of the most important principles of the relational model.

Foreign key
An attribute that appears as a nonprimary key attribute in one relation and as a primary key attribute (or part of a primary key) in another relation.

Referential integrity
An integrity constraint specifying that the value (or existence) of an attribute in one relation depends on the value (or existence) of the same attribute in another relation.

Transforming E-R Diagrams into Relations

Normalization produces a set of well-structured relations that contains all of the data mentioned in system inputs and outputs developed in human interface design. Because these specific information requirements may not represent all future information needs, the E-R diagram you developed in conceptual data modeling is another source of insight into possible data requirements for a new application system. To compare the conceptual data model and the normalized relations developed so far, your E-R diagram must be transformed into relational notation, normalized, and then merged with the existing normalized relations.

Transforming an E-R diagram into normalized relations and then merging all the relations into one final, consolidated set of relations can be accomplished in four steps. These steps are summarized briefly here, and then steps 1, 2, and 4 are discussed in detail in subsequent sections of this chapter.

1 *Represent entities.* Each entity type in the E-R diagram becomes a relation. The identifier of the entity type becomes the primary key of the relation, and other attributes of the entity type become nonprimary key attributes of the relation.

2 *Represent relationships.* Each relationship in an E-R diagram must be represented in the relational database design. How we represent a relationship depends on its nature. For example, in some cases we represent a relationship by making the primary key of one relation a foreign key of another relation. In other cases, we create a separate relation to represent a relationship.

3 *Normalize the relations.* The relations created in steps 1 and 2 may have unnecessary redundancy. So, we need to normalize these relations to make them well structured.

4 *Merge the relations.* So far in database design we have created various relations from both a bottom-up normalization of user views and from transforming one or more E-R diagrams into sets of relations. Across these different sets of relations, there may be redundant relations (two or more relations that describe the same entity type) that must be merged and renormalized to remove the redundancy.

Represent Entities

Each regular entity type in an E-R diagram is transformed into a relation. The identifier of the entity type becomes the primary key of the corresponding relation. Each nonkey attribute of the entity type becomes a nonkey attribute of the relation. You should check to make sure that the primary key satisfies the following two properties:

1. The value of the key must uniquely identify every row in the relation.
2. The key should be nonredundant; that is, no attribute in the key can be deleted without destroying its unique identification.

Some entities may have keys that include the primary keys of other entities. For example, an EMPLOYEE DEPENDENT may have a Name for each dependent, but, to form the primary key for this entity, you must include the Employee_ID attribute from the associated EMPLOYEE entity. Such an entity whose primary key depends upon the primary key of another entity is called a weak entity.

Representation of an entity as a relation is straightforward. Figure 9-10(A) shows the CUSTOMER entity type for Pine Valley Furniture Company. The corresponding CUSTOMER relation is represented as follows:

CUSTOMER(<u>Customer ID</u>,Name,Address,City_State_ZIP,Discount)

In this notation, the entity type label is translated into a relation name. The identifier of the entity type is listed first and underlined. All nonkey attributes are listed after the primary key. This relation is shown as a table with sample data in Figure 9-10(B).

FIGURE 9-10
TRANSFORMING AN ENTITY TYPE TO A RELATION
(A) E-R DIAGRAM
(B) RELATION

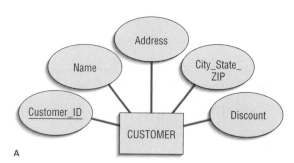

A

CUSTOMER

Customer_ID	Name	Address	City_State_ZIP	Discount
1273	Contemporary Designs	123 Oak St.	Austin, TX 28384	5%
6390	Casual Corner	18 Hoosier Dr.	Bloomington, IN 45821	3%

B

Represent Relationships

The procedure for representing relationships depends on both the degree of the relationship—unary, binary, ternary—and the cardinalities of the relationship.

Binary 1:N and 1:1 Relationships A binary one-to-many (1:N) relationship in an E-R diagram is represented by adding the primary key attribute (or attributes) of the entity on the one side of the relationship as a foreign key in the relation that is on the many side of the relationship.

Figure 9-11(A), an example of this rule, shows the Places relationship (1:N) linking CUSTOMER and ORDER at Pine Valley Furniture Company. Two relations, CUSTOMER and ORDER, were formed from the respective entity types (see Figure 9-11[B]). Customer_ID, which is the primary key of CUSTOMER (on the one side of the relationship) is added as a foreign key to ORDER (on the many side of the relationship).

FIGURE 9-11
REPRESENTING A 1:*N*
RELATIONSHIP
(A) E-R DIAGRAM
(B) RELATIONS

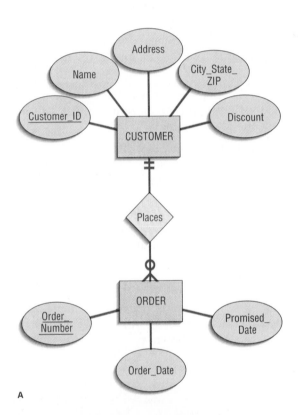

A

CUSTOMER

Customer_ID	Name	Address	City_State_ZIP	Discount
1273	Contemporary Designs	123 Oak St.	Austin, TX 28384	5%
6390	Casual Corner	18 Hoosier Dr.	Bloomington, IN 45821	3%

ORDER

Order_Number	Order_Date	Promised_Date	Customer_ID
57194	3/15/0X	3/28/0X	6390
63725	3/17/0X	4/01/0X	1273
80149	3/14/0X	3/24/0X	6390

B

One special case under this rule was mentioned in the previous section. If the entity on the many side needs the key of the entity on the one side as part of its primary key (this is a so-called weak entity), then this attribute is added not as a nonkey but as part of the primary key.

For a binary or unary one-to-one (1:1) relationship between two entities A and B (for a unary relationship, A and B would be the same entity type), the relationship can be represented by any of the following choices:

1. Adding the primary key of A as a foreign key of B
2. Adding the primary key of B as a foreign key of A
3. Both of the above

Binary and Higher-Degree M:N Relationships Suppose that there is a binary many-to-many (M:N) relationship (or associative entity) between two entity types A and B. For such a relationship, we create a separate relation C. The primary key of this relation is a composite key consisting of the primary key for each of the two entities in the relationship. Any nonkey attributes that are associated with the M:N relationship are included with the relation C.

Figure 9-12(A), an example of this rule, shows the Requests relationship (*M:N*) between the entity types ORDER and PRODUCT for Pine Valley Furniture Company. Figure 9-12(B) shows the three relations (ORDER, ORDER LINE, and PRODUCT) that are formed from the entity types and the Requests relationship. A relation (called ORDER LINE in Figure 9-12[B]) is created for the Requests relationship. The primary key of ORDER LINE is the combination (Order_Number,Product_ID), which is the respective primary keys of ORDER and PRODUCT. The nonkey attribute Quantity_Ordered also appears in ORDER LINE.

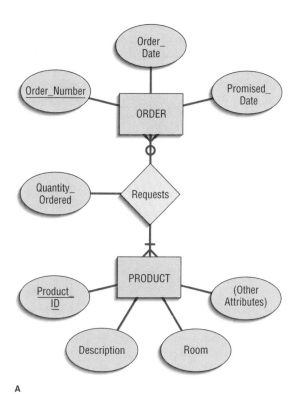

ORDER

Order_Number	Order_Date	Promised_Date
61384	2/17/2002	3/01/2002
62009	2/13/2002	2/27/2002
62807	2/15/2002	3/01/2002

ORDER LINE

Order_Number	Product_ID	Quantity_Ordered
61384	M128	2
61384	A261	1

PRODUCT

Product_ID	Description	(Other Attributes)
M128	Bookcase	—
A261	Wall unit	—
R149	Cabinet	—

A

B

FIGURE 9-12
REPRESENTING AN *M:N* RELATIONSHIP
(A) E-R DIAGRAM
(B) RELATIONS

Occasionally, the relation created from an *M:N* relationship requires a primary key that includes more than just the primary keys from the two related relations. Consider, for example, the following situation:

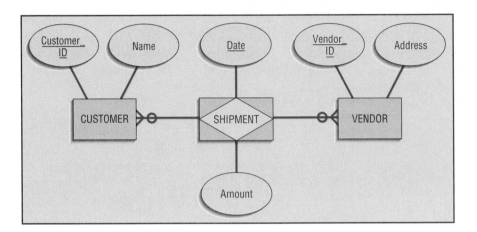

In this case, Date must be part of the key for the SHIPMENT relation to uniquely distinguish each row of the SHIPMENT table, as follows:

SHIPMENT(<u>Customer_ID,Vendor_ID,Date</u>,Amount)

If each shipment has a separate nonintelligent key, say a shipment number, then Date becomes a nonkey and Customer_ID and Vendor_ID become foreign keys, as follows:

SHIPMENT(<u>Shipment_Number</u>,Customer_ID,Vendor_ID,Date,Amount)

In some cases, there may be a relationship among three or more entities. In such cases, we create a separate relation that has as a primary key the composite of the primary keys of each of the participating entities (plus any necessary additional key elements). This rule is a simple generalization of the rule for a binary *M:N* relationship.

Unary Relationships To review, a unary relationship is a relationship between the instances of a single entity type, which are also called recursive relationships. Figure 9-13 shows two common examples. Figure 9-13(A) shows

FIGURE 9-13
TWO UNARY RELATIONS
(A) EMPLOYEE WITH MANAGES RELATIONSHIP (1 : N)
(B) BILL-OF-MATERIALS STRUCTURE (M : N)

A

B

a one-to-many relationship named Manages that associates employees with another employee who is their manager. Figure 9-13(B) shows a many-to-many relationship that associates certain items with their component items. This relationship is called a *bill-of-materials structure.*

For a unary 1:*N* relationship, the entity type (such as EMPLOYEE) is modeled as a relation. The primary key of that relation is the same as for the entity type. Then a foreign key is added to the relation that references the primary key values. A **recursive foreign key** is a foreign key in a relation that references the primary key values of that same relation. We can represent the relationship in Figure 9-13(A) as follows:

> EMPLOYEE(<u>Emp_ID</u>,Name,Birthdate,<u>Manager_ID</u>)

In this relation, Manager_ID is a recursive foreign key that takes its values from the same set of worker identification numbers as Emp_ID.

For a unary *M:N* relationship, we model the entity type as one relation. Then we create a separate relation to represent the *M:N* relationship. The primary key of this new relation is a composite key that consists of two attributes (which need not have the same name) that both take their values from the same primary key. Any attribute associated with the relationship (such as Quantity in Figure 9-13[B]) is included as a nonkey attribute in this new relation. We can express the result for Figure 9-13[B] as follows:

> ITEM(<u>Item_Number</u>,Name,Cost)
> ITEM-BILL(<u>Item_Number,Component_Number</u>,Quantity)

Recursive foreign key
A foreign key in a relation that references the primary key values of that same relation.

Summary of Transforming E-R Diagrams to Relations

We have now described how to transform E-R diagrams to relations. Table 9-1 lists the rules discussed in this section for transforming entity-relationship diagrams into equivalent relations. After this transformation, you should check the resulting relations to determine whether they are in third normal form and, if necessary, perform normalization as described earlier in the chapter.

Merging Relations

As part of the logical database design, normalized relations likely have been created from a number of separate E-R diagrams and various user interfaces. Some of the relations may be redundant—they may refer to the same entities. If so, you should merge those relations to remove the redundancy. This section describes merging relations or *view integration*, which is the last step in logical database design and prior to physical file and database design.

An Example of Merging Relations

Suppose that modeling a user interface or transforming an E-R diagram results in the following 3NF relation:

> EMPLOYEE1(<u>Emp_ID</u>,Name,Address,Phone)

Modeling a second user interface might result in the following relation:

> EMPLOYEE2(<u>Emp_ID</u>,Name,Address,Jobcode,Number_of_Years)

Because these two relations have the same primary key (Emp_ID) and describe the same entity, they should be merged into one relation. The result of merging the relations is the following relation:

> EMPLOYEE(<u>Emp_ID</u>,Name,Address,Phone,Jobcode,Number_of_Years)

Phase 3:
Systems
Design

TABLE 9-1: E-R to Relational Transformation

E-R Structure	Relational Representation
Regular entity	Create a relation with primary key and nonkey attributes.
Weak entity	Create a relation with a composite primary key (which includes the primary key of the entity on which this weak entity depends) and nonkey attributes.
Binary or unary 1:1 relationship	Place the primary key of either entity in the relation for the other entity or do this for both entities.
Binary 1:N relationship	Place the primary key of the entity on the one side of the relationship as a foreign key in the relation for the entity on the many side.
Binary or unary M:N relationship or associative entity	Create a relation with a composite primary key using the primary keys of the related entities, plus any nonkey attributes of the relationship or associative entity.
Binary or unary M:N relationship or associative entity with additional key(s)	Create a relation with a composite primary key using the primary keys of the related entities and additional primary key attributes associated with the relationship or associative entity, plus any nonkey attributes of the relationship or associative entity.
Binary or unary M:N relationship or associative entity with its own key	Create a relation with the primary key associated with the relationship or associative entity, plus any nonkey attributes of the relationship or associative entity and the primary keys of the related entities (as nonkey attributes).

Notice that an attribute that appears in both relations (such as Name in this example) appears only once in the merged relation.

View Integration Problems

When integrating relations, you must understand the meaning of the data and must be prepared to resolve any problems that may arise in that process. In this section, we describe and illustrate three problems that arise in view integration: synonyms, homonyms, and dependencies between nonkeys.

Synonyms In some situations, two or more attributes may have different names but the same meaning, as when they describe the same characteristic of an entity. Such attributes are called **synonyms.** For example, Emp_ID and Employee_Number may be synonyms.

Synonyms
Two different names that are used for the same attribute.

When merging the relations that contain synonyms, you should obtain, if possible, agreement from users on a single standardized name for the attribute and eliminate the other synonym. Another alternative is to choose a third name to replace the synonyms. For example, consider the following relations:

    STUDENT1(Student_ID,Name)
    STUDENT2(Matriculation_Number,Name,Address)

In this case, the analyst recognizes that both the Student_ID and the Matriculation_Number are synonyms for a person's social security number and are identical attributes. One possible resolution would be to standardize on one of the two attribute names, such as Student_ID. Another option is to use a new attribute name, such as SSN, to replace both synonyms. Assuming the latter approach, merging the two relations would produce the following result:

STUDENT(<u>SSN</u>,Name,Address)

Homonyms   In other situations, a single attribute name, called a **homonym,** may have more than one meaning or describe more than one characteristic. For example, the term *account* might refer to a bank's checking account, savings account, loan account, or other type of account; therefore, *account* refers to different data, depending on how it is used.

Homonym
A single attribute name that is used for two or more different attributes.

You should be on the lookout for homonyms when merging relations. Consider the following example:

STUDENT1(<u>Student_ID</u>,Name,Address)
STUDENT2(<u>Student_ID</u>,Name,Phone_Number,Address)

In discussions with users, the systems analyst may discover that the attribute Address in STUDENT1 refers to a student's campus address, whereas in STUDENT2 the same attribute refers to a student's home address. To resolve this conflict, we would probably need to create new attribute names and the merged relation would become:

STUDENT(<u>Student_ID</u>,Name,Phone_Number,Campus_Address, Permanent_Address)

Dependencies between Nonkeys   When two 3NF relations are merged to form a single relation, dependencies between nonkeys may result. For example, consider the following two relations:

STUDENT1(<u>Student_ID</u>,Major)
STUDENT2(<u>Student_ID</u>,Adviser)

Because STUDENT1 and STUDENT2 have the same primary key, the two relations may be merged:

STUDENT(<u>Student_ID</u>,Major,Adviser)

However, suppose that each major has exactly one adviser. In this case, Adviser is functionally dependent on Major:

Major→Adviser

If the above dependency exists, then STUDENT is in 2NF but not 3NF, because it contains a functional dependency between nonkeys. The analyst can create 3NF relations by creating two relations with Major as a foreign key in STUDENT:

STUDENT(<u>Student_ID</u>,Major)
MAJOR ADVISER(<u>Major</u>,Adviser)

Logical Database Design for Hoosier Burger

In Chapter 6 we developed an E-R diagram for a new Inventory Control System at Hoosier Burger (Figure 9-14 repeats the diagram from Chapter 6). In this section we show how this E-R model is translated into normalized relations and how to normalize and then merge the relations for a new report with the relations from the E-R model.

In this E-R model, four entities exist independently of other entities: SALE, PRODUCT, INVOICE, and INVENTORY ITEM. Given the attributes shown in Figure 9-14, we can represent these entities in the following four relations:

SALE(<u>Receipt_Number</u>,Sale_Date)
PRODUCT(<u>Product_ID</u>,Product_Description)
INVOICE(<u>Vendor_ID,Invoice_Number</u>,Invoice_Date,Paid?)

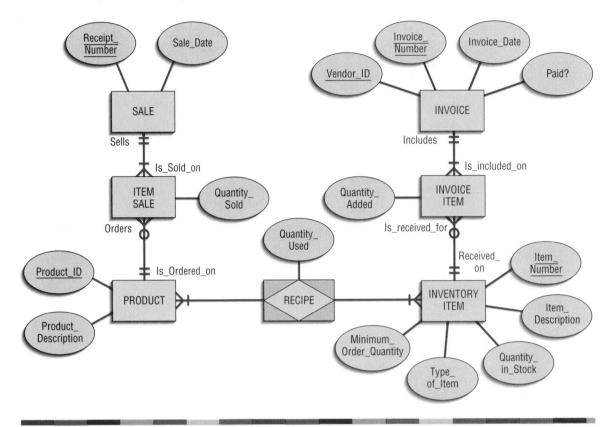

FIGURE 9-14
FINAL E-R DIAGRAM FOR HOOSIER BURGER'S INVENTORY CONTROL SYSTEM

> INVENTORY ITEM(<u>Item_Number</u>,Item_Description,Quantity_in_Stock,_
> Minimum_Order_Quantity,Type_of_Item)

The entities ITEM SALE and INVOICE ITEM as well as the associative entity
RECIPE each have composite primary keys taken from the entities to which they
relate, so we can represent these three entities in the following three relations:

> ITEM SALE(<u>Receipt_Number</u>,<u>Product_ID</u>,Quantity_Sold)
> INVOICE ITEM(<u>Vendor_ID</u>,<u>Invoice_Number</u>,<u>Item_Number</u>,Quantity_ Added)
> RECIPE(<u>Product_ID</u>,<u>Item_Number</u>,Quantity_Used)

Because there are no many-to-many, one-to-one, or unary relationships, we
have now represented all the entities and relationships from the E-R model.
Also, each of the above relations is in 3NF because all attributes are simple, all
nonkeys are fully dependent on the whole key, and there are no dependencies
between nonkeys in the INVOICE and INVENTORY ITEM relations.

Now suppose that Bob Mellankamp wanted an additional report that was not
previously known by the analyst who designed the Inventory Control System
for Hoosier Burger. A rough sketch of this new report, listing volume of purchases from each vendor by type of item in a given month, appears in Figure 9-
15. In this report, the same type of item may appear many times if multiple vendors supply the same type of item.

This report contains data about several relations already known to the analyst, including:

> INVOICE(<u>Vendor_ID</u>,<u>Invoice_Number</u>,Invoice_Date): primary keys, and
> the date is needed to select invoices in the specified month of the report
> INVENTORY ITEM(<u>Item_Number</u>,Type_of_Item): primary key and a non-
> key in the report

Monthly Vendor Load Report
for Month: xxxxx

Page *x* of *n*

Vendor		Type of Item	Total Quantity Added
ID	Name		
V1	V1name	aaa	nnn1
		bbb	nnn2
		ccc	nnn3
V2	V2name	bbb	nnn4
		mmm	nnn5
×			
×			
×			

FIGURE 9-15
HOOSIER BURGER MONTHLY
VENDOR LOAD REPORT

INVOICE ITEM(Vendor_ID,Invoice_Number,Item_Number,Quantity_Added):
primary keys and the raw quantity of items invoiced that are subtotaled by
vendor and type of item in the report

In addition, the report includes a new attribute—Vendor name. After some
investigation, an analyst determines that Vendor_ID→Vendor_Name. Because the
whole primary key of the INVOICE relation is Vendor_ID and Invoice_Number, if
Vendor_Name were part of the INVOICE relation, this relation would violate the
3NF rule. So, a new VENDOR relation must be created as follows:

VENDOR(Vendor_ID,Vendor_Name)

Now, Vendor_ID not only is part of the primary key of INVOICE but also is a
foreign key referencing the VENDOR relation. Hence, there must be a one-to-
many relationship from VENDOR to INVOICE. The systems analyst determines
that an invoice must come from a vendor, and there is no need to keep data about
a vendor unless the vendor invoices Hoosier Burger. An updated E-R diagram,
reflecting these enhancements for new data needed in the monthly vendor load
report, appears in Figure 9-16. The normalized relations for this database are:

SALE(Receipt_Number,Sale_Date)
PRODUCT(Product_ID,Product_Description)
INVOICE(Vendor_ID,Invoice_Number,Invoice_Date,Paid?,Vendor_ID)
INVENTORY ITEM(Item_Number,Item_Description,Quantity_in_Stock,
    Minimum_Order_Quantity,Type_of_Item)
ITEM SALE(Receipt_Number,Product_ID,Quantity_Sold)
INVOICE ITEM(Vendor_ID,Invoice_Number,Item_Number,Quantity_Added)
RECIPE(Product_ID,Item_Number,Quantity_Used)
VENDOR(Vendor_ID,Vendor_Name)

Physical File and Database Design

Designing physical files and databases requires certain information that should
have been collected and produced during prior SDLC phases. This information
includes:

- Normalized relations, including volume estimates
- Definitions of each attribute
- Descriptions of where and when data are used: entered,
  retrieved, deleted, and updated (including frequencies)
- Expectations or requirements for response time and data
  integrity

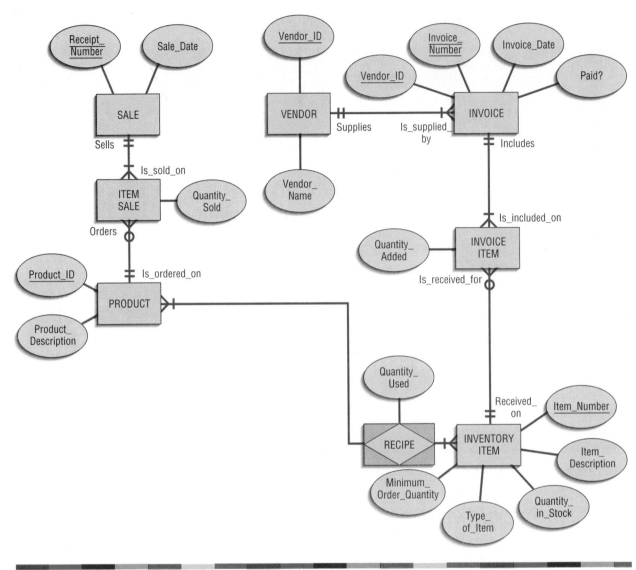

FIGURE 9-16
E-R DIAGRAM CORRESPONDING TO NORMALIZED RELATIONS OF HOOSIER BURGER'S INVENTORY CONTROL SYSTEM

❶ Descriptions of the technologies used for implementing the
files and database so that the range of required decisions
and choices for each is known

Normalized relations are, of course, the result of logical database design.
Statistics on the number of rows in each table as well as the other information
listed above may have been collected during requirements determination in
systems analysis. If not, these items need to be discovered to proceed with
database design.

We take a bottom-up approach to reviewing physical file and database
design. Thus, we begin the physical design phase by addressing the design of
physical fields for each attribute in a logical data model.

Designing Fields

Field
The smallest unit of named
application data recognized by
system software.

A **field** is the smallest unit of application data recognized by system software,
such as a programming language or database management system. An attribute
from a logical database model may be represented by several fields. For exam-

ple, a student name attribute in a normalized student relation might be represented as three fields: last name, first name, and middle initial. Each field requires a separate definition when the application system is implemented.

In general, you will represent each attribute from each normalized relation as one or more fields. The basic decisions you must make in specifying each field concern the type of data (or storage type) used to represent the field and data integrity controls for the field.

Choosing Data Types

A **data type** is a coding scheme recognized by system software for representing organizational data. The bit pattern of the coding scheme is usually immaterial to you, but the space to store data and the speed required to access data are of consequence in the physical file and database design. The specific file or database management software you use with your system will dictate which choices are available to you. For example, Table 9-2 lists the data types available in Microsoft Access 2000.

Data type
A coding scheme recognized by system software for representing organizational data.

Selecting a data type balances four objectives that will vary in degree of importance dependent on the application:

1. Minimize storage space
2. Represent all possible values of the field
3. Improve data integrity for the field
4. Support all data manipulations desired on the field

TABLE 9-2:   Microsoft Access 2000 Data Types

Data Type	Description
Text	Text or combinations of text and numbers, as well as numbers that don't require calculations, such as phone numbers. A specific length is indicated, with a maximum number of characters of 255. One byte of storage is required for each character used.
Memo	Lengthy (up to 65,535 characters) text or combinations of text and numbers. One byte of storage is required for each character used.
Number	Numeric data used in mathematical calculations. Either 1, 2, 4, or 8 bytes of storage space is required, depending on the specified length of the number.
Date/time	Date and time values for the years 100 through 9999. Eight bytes of storage space is required.
Currency	Currency values and numeric data used in mathematical calculations involving data with one to four decimal places. Accurate to 15 digits on the left side of the decimal separator and to 4 digits on the right side. Eight bytes of storage space is required.
Autonumber	A unique sequential (incremented by 1) number or random number assigned by Microsoft Access whenever a new record is added to a table. Typically, 4 bytes of storage is required.
Yes/No	Yes and No values and fields that contain only one of two values (Yes/No, True/False, or On/Off). One bit of storage is required.
OLE object	An object (such as a Microsoft Excel spreadsheet, a Microsoft Word document, graphics, sounds, or other binary data) linked to or embedded in a Microsoft Access table. Up to 1 gigabyte of storage possible.
Hyperlink	Text or combinations of text and numbers stored as text and used as a hyperlink address (typical URL form).
Lookup Wizard	Creates a field that allows you to choose a value from another table (the table's primary key) or from a list of values by using a list box or combo box. Clicking this option starts the Lookup Wizard, which creates a Lookup field. After you complete the wizard, Microsoft Access sets the data type based on the values selected in the wizard. Used for foreign keys to enforce referential integrity. Space requirement depends on length of foreign key or lookup value.

You want to choose a data type for a field that minimizes space, represents every possible legitimate value for the associated attribute, and allows the data to be manipulated as needed. For example, suppose a quantity sold field can be represented by a Number data type. You would select a length for this field that would handle the maximum value, plus some room for growth of the business. Further, the Number data type will restrict users from entering inappropriate values (text), but it does allow negative numbers (if this is a problem, application code or form design may be required to restrict the values to positive).

Be careful—the data type must be suitable for the life of the application; otherwise, maintenance will be required. Choose data types for future needs by anticipating growth. Also, be careful that date arithmetic can be done so that dates can be subtracted or time periods can be added to or subtracted from a date.

Several other capabilities of data types may be available with some database technologies. We discuss a few of the most common of these features next: calculated fields and coding and compression techniques.

Calculated Fields It is common that an attribute is mathematically related to other data. For example, an invoice may include a total due field, which represents the sum of the amount due on each item on the invoice. A field that can be derived from other database fields is called a **calculated** (or **computed** or **derived**) **field** (recall that a functional dependency between attributes does not imply a calculated field). Some database technologies allow you to explicitly define calculated fields along with other raw data fields. If you specify a field as calculated, you would then usually be prompted to enter the formula for the calculation; the formula can involve other fields from the same record and possibly fields from records in related files. The database technology will either store the calculated value or compute it when requested.

Coding and Compression Techniques Some attributes have very few values from a large range of possible values. For example, although a six-digit field (five numbers plus a value sign) can represent numbers –99999 to 99999, maybe only 100 positive values within this range will ever exist. Thus, a Number data type does not adequately restrict the permissible values for data integrity, and storage space for five digits plus a value sign is wasteful. To use space more efficiently (and less space may mean faster access because the data you need are closer together), you can define a field for an attribute so that the possible attribute values are not represented literally but rather are abbreviated. For example, suppose in Pine Valley Furniture each product has a finish attribute, with possible values Birch, Walnut, Oak, and so forth. To store this attribute as Text might require 12, 15, or even 20 bytes to represent the longest finish value. Suppose that even a liberal estimate is that Pine Valley Furniture will never have more than 25 finishes. Thus, a single alphabetic or alphanumeric character would be more than sufficient. We not only reduce storage space but also increase integrity (by restricting input to only a few values), which helps to achieve two of the physical file and database design goals. Codes also have disadvantages. If used in system inputs and outputs, they can be more difficult for users to remember, and programs must be written to decode fields if codes will not be displayed.

Controlling Data Integrity

We have already explained that data typing helps control data integrity by limiting the possible range of values for a field. There are additional physical file and database design options you might use to ensure higher-quality data. Although these controls can be imposed within application programs, it is better to include these as part of the file and database definitions so that the con-

---

Calculated (or computed or derived) field

A field that can be derived from other database fields.

NET SEARCH

Investigate the capabilities of several data compression programs. Visit http://www.prenhall.com/valacich to complete an exercise related to this topic.

trols are guaranteed to be applied all the time as well as uniformly for all programs. There are five popular data integrity control methods: default value, picture control, range control, referential integrity, and null value control.

Phase 3:
Systems
Design

▶ *Default value.* A **default value** is the value a field will assume unless an explicit value is entered for the field. For example, the city and state of most customers for a particular retail store will likely be the same as the store's city and state. Assigning a default value to a field can reduce data entry time (the field can simply be skipped during data entry) and data entry errors, such as typing *IM* instead of *IN* for *Indiana.*

Default value
A value a field will assume unless an explicit value is entered for that field.

▶ *Input Mask.* Some data must follow a specified pattern. An **input mask** (or field **template**) is a pattern of codes that restricts the width and possible values for each position within a field. For example, a product number at Pine Valley Furniture is four alphanumeric characters—the first is alphabetic and the next three are numeric—defined by an input mask of L999 where L means that only alphabetic characters are accepted and 9 means that only numeric digits are accepted. M128 is an acceptable value but 3128 or M12H would be unacceptable. Other types of input masks can be used to convert all characters to uppercase, indicate how to show negative numbers, suppress showing leading zeros, or indicate if entry of a letter or digit is optional.

Input mask
A pattern of codes that restricts the width and possible values for each position of a field.

▶ *Range control.* Both numeric and alphabetic data may have a limited set of permissible values. For example, a field for the number of product units sold may have a lower bound of 0, and a field that represents the month of a product sale may be limited to the values JAN, FEB, and so forth.

▶ *Referential integrity.* As noted earlier in this chapter, the most common example of referential integrity is cross-referencing between relations. For example, consider the pair of relations in Figure 9-17(A). In this case, the values for the foreign key Customer_ID field within a customer order must be limited to the set of Customer_ID values from the customer relation; we would not want to accept an order for a nonexisting or unknown customer. Referential integrity may be useful in other instances. Consider the employee relation example in Figure 9-17(B). In this example, the employee relation has a field of Supervisor_ID. This field refers to the Employee_ID of the employee's supervisor and should have referential integrity on the Employee_ID field within the same relation. Note in this case that because some employees do not have supervisors,

CUSTOMER(**Customer_ID**,Cust_Name,Cust_Address,...)

CUST_ORDER(Order_ID,**Customer_ID**,Order_Date,...)
   and Customer_ID may not be null because every order must be for
   some existing customer

A

EMPLOYEE(**Employee_ID**,**Supervisor_ID**,Empl_Name,...)
   and Supervisor_ID may be null because not all employees have supervisors

B

FIGURE 9-17
EXAMPLES OF REFERENTIAL
INTEGRITY FIELD CONTROLS
(A) REFERENTIAL INTEGRITY
BETWEEN RELATIONS
(B) REFERENTIAL INTEGRITY WITHIN
A RELATION

this is a weak referential integrity constraint because the value of a Supervisor_ID field may be empty.

◗ *Null value control.* A **null value** is a special field value, distinct from 0, blank, or any other value, that indicates that the value for the field is missing or otherwise unknown. It is not uncommon that when it is time to enter data—for example, a new customer—you might not know the customer's phone number. The question is whether a customer, to be valid, must have a value for this field. The answer for this field is probably initially no, because most data processing can continue without knowing the customer's phone number. Later a null value may not be allowed when you are ready to ship product to the customer. On the other hand, you must always know a value for the Customer_ID field. Due to referential integrity, you cannot enter any customer orders for this new customer without knowing an existing Customer_ID value, and customer name is essential for visual verification of correct data entry. Besides using a special null value when a field is missing its value, you can also estimate the value, produce a report indicating rows of tables with critical missing values, or determine whether the missing value matters in computing needed information.

Null value

A special field value, distinct from 0, blank, or any other value, that indicates that the value for the field is missing or otherwise unknown.

Designing Physical Tables

A relational database is a set of related tables (tables are related by foreign keys referencing primary keys). In logical database design you grouped into a relation those attributes that concern some unifying, normalized business concept, such as a customer, product, or employee. In contrast, a **physical table** is a named set of rows and columns that specifies the fields in each row of the table. A physical table may or may not correspond to one relation. Whereas normalized relations possess properties of well-structured relations, the design of a physical table has two goals different from those of normalization: efficient use of secondary storage and data processing speed.

The efficient use of secondary storage (disk space) relates to how data are loaded on disks. Disks are physically divided into units (called pages) that can be read or written in one machine operation. Space is used efficiently when the physical length of a table row divides close to evenly into the length of the storage unit. For many information systems, this even division is very difficult to achieve because it depends on factors, such as operating system parameters, outside the control of each database. Consequently, we do not discuss this factor of physical table design in this text.

A second and often more important consideration when selecting a physical table design is efficient data processing. Data are most efficiently processed when they are stored close to one another in secondary memory, thus minimizing the number of input/output (I/O) operations that must be performed. Typically, the data in one physical table (all the rows and fields in those rows) are stored close together on disk. **Denormalization** is the process of splitting or combining normalized relations into physical tables based on affinity of use of rows and fields. Consider Figure 9-18. In Figure 9-18(A), a normalized product relation is split into separate physical tables with each containing only engineering, accounting, or marketing product data; the primary key must be included in each table. Note, the Description and Color attributes are repeated in both the engineering and marketing tables because these attributes relate to both kinds of data. In Figure 9-18(B), a customer relation is denormalized by putting rows from different geographic regions into separate tables. In both

Physical table

A named set of rows and columns that specifies the fields in each row of the table.

Denormalization

The process of splitting or combining normalized relations into physical tables based on affinity of use of rows and fields.

Phase 3:
Systems
Design

Normalized Product Relation

PRODUCT(<u>Product_ID</u>,Description,Drawing_Number,Weight,Color,Unit_Cost,Burden_Rate,Price,Product_Manager)

Denormalized Functional Area Product Relations for Tables

Engineering: E_PRODUCT(<u>Product_ID</u>,Description,Drawing_Number,Weight,Color)
Accounting: A_PRODUCT(<u>Product_ID</u>,Unit_Cost,Burden_Rate)
Marketing: M_PRODUCT(<u>Product_ID</u>,Description,Color,Price,Product_Manager)

A

Normalized Customer Table

CUSTOMER

Customer_ID	Name	Region	Annual_Sales
1256	Rogers	Atlantic	10,000
1323	Temple	Pacific	20,000
1455	Gates	South	15,000
1626	Hope	Pacific	22,000
2433	Bates	South	14,000
2566	Bailey	Atlantic	12,000

Denormalized Regional Customer Tables

A_CUSTOMER

Customer_ID	Name	Region	Annual_Sales
1256	Rogers	Atlantic	10,000
2566	Bailey	Atlantic	12,000

P_CUSTOMER

Customer_ID	Name	Region	Annual_Sales
1323	Temple	Pacific	20,000
1626	Hope	Pacific	22,000

S_CUSTOMER

Customer_ID	Name	Region	Annual_Sales
1455	Gates	South	15,000
2433	Bates	South	14,000

B

FIGURE 9-18
EXAMPLES OF DENORMALIZATION
(A) DENORMALIZATION BY COLUMNS
(B) DENORMALIZATION BY ROWS

cases, the goal is to create tables that contain only the data used together in programs. By placing data used together close to one another on disk, the number of disk I/O operations needed to retrieve all the data needed in a program is minimized.

Denormalization can increase the chance of errors and inconsistencies that normalization avoided. Further, denormalization optimizes certain data processing at the expense of others, so if the frequencies of different processing activities change, the benefits of denormalization may no longer exist.

Various forms of denormalization can be done, but there are no hard-and-fast rules for deciding when to denormalize data. Here are three common situations in which denormalization often makes sense (see Figure 9-19 for illustrations):

1 *Two entities with a one-to-one relationship.* Figure 9-19(A) shows student data with optional data from a standard scholarship application a student may complete. In this case, one record could be formed with four fields from the STUDENT and SCHOLARSHIP APPLICATION FORM normalized relations. (*Note:* In this case, fields from the optional entity must have null values allowed.)

FIGURE 9-19
POSSIBLE DENORMALIZATION SITUATIONS
(A) TWO ENTITIES WITH A ONE-TO-ONE RELATIONSHIP
(B) A MANY-TO-MANY RELATIONSHIP WITH NONKEY ATTRIBUTES
(C) REFERENCE DATA

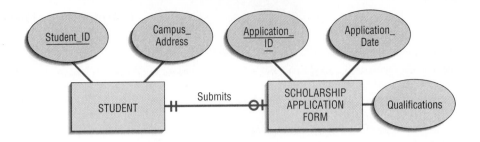

Normalized relations:
   STUDENT(Student_ID,Campus_Address,Application_ID)
   APPLICATION(Application_ID,Application_Date,Qualifications,Student_ID)

Denormalized relation:
   STUDENT(Student_ID,Campus_Address,Application_Date,Qualifications)

   and Application_Date and Qualifications may be null

(*Note:* We assume Application_ID is not necessary when all fields are stored in one record, but this field can be included if it is required application data.)

A

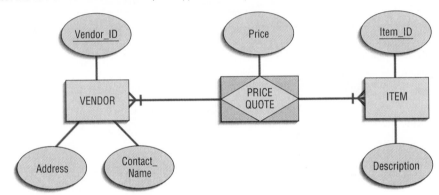

Normalized relations:
   VENDOR(Vendor_ID,Address,Contact_Name)
   ITEM(Item_ID,Description)
   PRICE QUOTE(Vendor_ID,Item_ID,Price)

Denormalized relations:
   VENDOR(Vendor_ID,Address,Contact_Name)
   ITEM-QUOTE(Vendor_ID,Item_ID,Description,Price)

B

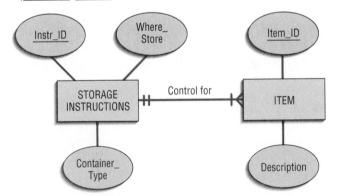

Normalized relations:
   STORAGE(Instr_ID,Where_Store,Container_Type)
   ITEM(Item_ID,Description,Instr_ID)

Denormalized relation
   ITEM(Item_ID,Description,Where_Store,Container_Type)

C

2 *A many-to-many relationship (associative entity) with nonkey attrib-utes.* Figure 9-19(B) shows price quotes for different items from different vendors. In this case, fields from ITEM and PRICE QUOTE relations might be combined into one physical table to avoid having to combine all three tables together. (*Note:* This may create considerable duplication of data—in the example, the ITEM fields, such as Description, would repeat for each price quote—and excessive updating if duplicated data changes.)

3 *Reference data.* Figure 9-19(C) shows that several ITEMs have the same STORAGE INSTRUCTIONS and STORAGE INSTRUCTIONS relate only to ITEMs. In this case, the storage instruction data could be stored in the ITEM table, thus reducing the number of tables to access but also creating redundancy and the potential for extra data maintenance.

Phase 3:
Systems
Design

Arranging Table Rows

The result of denormalization is the definition of one or more physical files. A computer operating system stores data in a **physical file,** which is a named set of table rows stored in a contiguous section of secondary memory. A file con-tains rows and columns from one or more tables, as produced from denormal-ization. To the operating system—like Windows, Linux, or UNIX—each table may be one file or the whole database may be in one file, depending on how the database technology and database designer organize data. The way the operat-ing system arranges table rows in a file is called a **file organization.** With some database technologies, the systems designer can choose among several organizations for a file.

If the database designer has a choice, he or she chooses a file organization for a specific file to provide:

1. Fast data retrieval
2. High throughput for processing transactions
3. Efficient use of storage space
4. Protection from failures or data loss
5. Minimal need for reorganization
6. Accommodation of growth
7. Security from unauthorized use

Often these objectives conflict, and you must select an organization for each file that provides a reasonable balance among the criteria within the resources available.

To achieve these objectives, many file organizations utilize the concept of a pointer. A **pointer** is a field of data that can be used to locate a related field or row of data. In most cases, a pointer contains the address of the associated data, which has no business meaning. Pointers are used in file organizations when it is not possible to store related data next to each other. Because this is often the case, pointers are common. In most cases, fortunately, pointers are hidden from a programmer. Yet, because a database designer may need to decide if and how to use pointers, we introduce the concept here.

Literally hundreds of different file organizations and variations have been created, but we outline the basics of three families of file organizations used in most file management environments: sequential, indexed, and hashed, as illus-trated in Figure 9-20. You need to understand the particular variations of each method available in the environment for which you are designing files.

Sequential File Organizations In a **sequential file organization,** the rows in the file are stored in sequence according to a primary key value (see Figure 9-20[A]). To locate a particular row, a program must normally scan the file from the beginning until the desired row is located. A common example of a sequential file is the alphabetic list of persons in the white pages of a phone directory (ignoring any index that may be included with the directory).

Physical file
A named set of table rows stored in a contiguous section of secondary memory.

File organization
A technique for physically arranging the records of a file.

Pointer
A field of data that can be used to locate a related field or row of data.

Sequential file organization
The rows in the file are stored in sequence according to a primary key value.

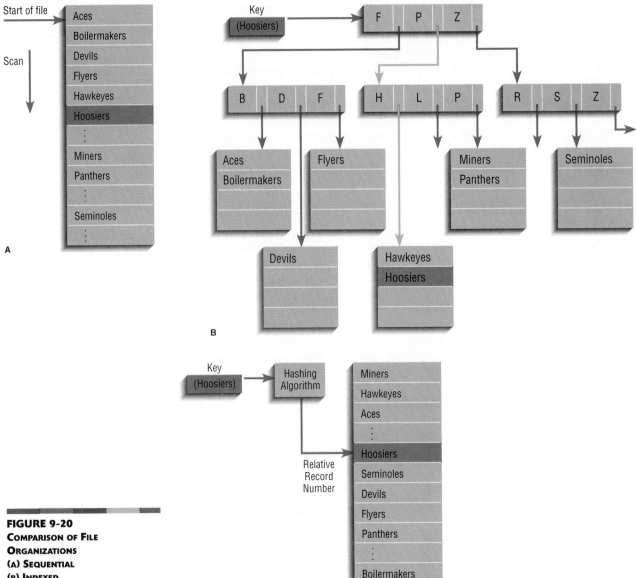

FIGURE 9-20
COMPARISON OF FILE
ORGANIZATIONS
(A) SEQUENTIAL
(B) INDEXED
(C) HASHED

Sequential files are very fast if you want to process rows sequentially, but they are essentially impractical for random row retrievals. Deleting rows can cause wasted space or the need to compress the file. Adding rows requires rewriting the file, at least from the point of insertion. Updating a row may also require rewriting the file, unless the file organization supports rewriting over the updated row only. Moreover, only one sequence can be maintained without duplicating the rows.

Indexed File Organizations In an **indexed file organization**, the rows are stored either sequentially or nonsequentially, and an index is created that allows the application software to locate individual rows (see Figure 9-20[B]). Like a card catalog in a library, an **index** is a structure that is used to determine the rows in a file that satisfy some condition. Each entry matches a key value with one or more rows. An index can point to unique rows (a primary key index, such as on the Product_ID field of a product table) or to potentially more than one row. An index that allows each entry to point to more than one record is called a **secondary key** index. Secondary key indexes are important for supporting many reporting requirements and for providing rapid ad hoc data retrieval. An example would be an index on the Finish field of a product table.

Indexed file organization
The rows are stored either sequentially or nonsequentially, and an index is created that allows software to locate individal rows.

Index
A table used to determine the location of rows in a file that satisfy some condition.

Secondary key
One or a combination of fields for which more than one row may have the same combination of values.

The example in Figure 9-20(B), typical of many index structures, illustrates that indexes can be built on top of indexes, creating a hierarchical set of indexes, and the data are stored sequentially in many contiguous segments. For example, to find the record with key "Hoosiers," the file organization would start at the top index and take the pointer after the entry *P*, which points to another index for all keys that begin with the letters *G* through *P* in the alphabet. Then the software would follow the pointer after the *H* in this index, which represents all those records with keys that begin with the letters *G* through *H*. Eventually, the search through the indexes either locates the desired record or indicates that no such record exists. The reason for storing the data in many contiguous segments is to allow room for some new data to be inserted in sequence without rearranging all the data.

The main disadvantages of indexed file organizations are the extra space required to store the indexes and the extra time necessary to access and maintain indexes. Usually these disadvantages are more than offset by the advantages. Because the index is kept in sequential order, both random and sequential processing are practical. Also, because the index is separate from the data, you can build multiple index structures on the same data file (just as in the library where there are multiple indexes on author, title, subject, and so forth). With multiple indexes, software may rapidly find records that have compound conditions, such as finding books by Tom Clancy on espionage.

The decision of which indexes to create is probably the most important physical database design task for relational database technology, such as Microsoft Access, Oracle, DB2, and similar systems. Indexes can be created for both primary and secondary keys. When using indexes, there is a trade-off between improved performance for retrievals and degrading performance for inserting, deleting, and updating the rows in a file. Thus, indexes should be used generously for databases intended primarily to support data retrievals, such as for decision support applications. Because they impose additional overhead, indexes should be used judiciously for databases that support transaction processing and other applications with heavy updating requirements.

Here are some guidelines for choosing indexes for relational databases:

1. Specify a unique index for the primary key of each table (file). This selection ensures the uniqueness of primary key values and speeds retrieval based on those values. Random retrieval based on primary key value is common for answering multitable queries and for simple data maintenance tasks.
2. Specify an index for foreign keys. As in the first guideline, this speeds processing multitable queries.
3. Specify an index for nonkey fields that are referenced in qualification and sorting commands for the purpose of retrieving data.

To illustrate the use of these rules, consider the following relations for Pine Valley Furniture Company:

PRODUCT(<u>Product_Number</u>,Description,Finish,Room,Price)
ORDER(<u>Order_Number</u>,<u>Product_Number</u>,Quantity)

You would normally specify a unique index for each primary key: Product_Number in PRODUCT and Order_Number in ORDER. Other indexes would be assigned based on how the data are used. For example, suppose that there is a system module that requires PRODUCT and PRODUCT_ORDER data for products with a price below $500, ordered by Product_Number. To speed up this retrieval, you could consider specifying indexes on the following nonkey attributes:

1. Price in PRODUCT because it satisfies rule 3
2. Product_Number in ORDER because it satisfies rule 2

Because users may direct a potentially large number of different queries against the database, especially for a system with a lot of ad hoc queries, you will probably have to be selective in specifying indexes to support the most common or frequently used queries.

Hashed File Organizations In **hashed file organization,** the address of each row is determined using an algorithm (see Figure 9-20[c]) that converts a primary key value into a row address. Although there are several variations of hashed files, in most cases the rows are located nonsequentially as dictated by the hashing algorithm. Thus, sequential data processing is impractical. On the other hand, retrieval of random rows is very fast. There are issues in the design of hashing file organizations, such as how to handle two primary keys that translate into the same address, but again, these issues are beyond our scope (see Hoffer, Prescott, and McFadden [2002] for a thorough discussion).

Summary of File Organizations The three families of file organizations— sequential, indexed, and hashed—cover most of the file organizations you will have at your disposal as you design physical files and databases. Table 9-3 summarizes the comparative features of these file organizations. You can use this table to help choose a file organization by matching the file characteristics and file processing requirements with the features of the file organization.

Designing Controls for Files

Two of the goals of physical table design mentioned earlier are protection from failures or data loss and security from unauthorized use. These goals are achieved primarily by implementing controls on each file. Data integrity controls, a primary type of control, was mentioned earlier in the chapter. Two other important types of controls address file backup and security.

TABLE 9-3: Comparative Features of Sequential, Indexed, and Hashed File Organizations

Factor	File Organization		
	Sequential	**Indexed**	**Hashed**
Storage space	No wasted space	No wasted space for data, but extra space for index	Extra space may be needed to allow for addition and deletion of records
Sequential retrieval on primary key	Very fast	Moderately fast	Impractical
Random retrieval on primary key	Impractical	Moderately fast	Very fast
Multiple key retrieval	Possible, but requires scanning whole file	Very fast with multiple indexes	Not possible
Deleting rows	Can create wasted space or require reorganizing	If space can be dynamically allocated, this is easy, but requires maintenance of indexes	Very easy
Adding rows	Requires rewriting file	If space can be dynamically allocated, this is easy, but requires maintenance of indexes	Very easy, except multiple keys with same address require extra work
Updating rows	Usually requires rewriting file	Easy, but requires maintenance of indexes	Very easy

It is almost inevitable that a file will be damaged or lost, due to either software or human errors. When a file is damaged, it must be restored to an accurate and reasonably current condition. A file and database designer has several techniques for file restoration, including:

- Periodically making a backup copy of a file
- Storing a copy of each change to a file in a transaction log or audit trail
- Storing a copy of each row before or after it is changed

For example, a backup copy of a file and a log of rows after they were changed can be used to reconstruct a file from a previous state (the backup copy) to its current values. This process would be necessary if the current file were so damaged that it could not be used. If the current file is operational but inaccurate, then a log of before images of rows can be used in reverse order to restore a file to an accurate but previous condition. Then a log of the transactions can be reapplied to the restored file to bring it up to current values. It is important that the information system designer make provisions for backup, audit trail, and row image files so that data files can be rebuilt when errors and damage occur.

An information system designer can build data security into a file by several means, including:

NET SEARCH
Investigate further data security methods and techniques. Visit http://www.prenhall.com/valacich to complete an exercise related to this topic.

- Coding, or encrypting, the data in the file so that they cannot be read unless the reader knows how to decrypt the stored values
- Requiring data file users to identify themselves by entering user names and passwords, and then possibly allowing only certain file activities (read, add, delete, change) for selected users for selected data in the file
- Prohibiting users from directly manipulating any data in the file, and rather requiring programs and users to work with a copy (real or virtual) of the data they need; the copy contains only the data that users or programs are allowed to manipulate, and the original version of the data will change only after changes to the copy are thoroughly checked for validity

Security procedures such as these all add overhead to an information system, so only necessary controls should be included.

Physical Database Design for Hoosier Burger

A set of normalized relations and an associated E-R diagram for Hoosier Burger (Figure 9-16) were presented in the section Logical Database Design for Hoosier Burger earlier in this chapter. The display of a complete design of this database would require more documentation than space permits in this text, so we illustrate in this section only a few key decisions from the complete physical database.

As outlined in this chapter, to translate a logical database design into a physical database design, you need to make the following decisions:

- Create one or more fields for each attribute and determine a data type for each field.
- For each field, decide if it is calculated, needs to be coded or compressed, must have a default value or input mask, or must have range, referential integrity, or null value controls.

- For each relation, decide if it should be denormalized to achieve desired processing efficiencies.
- Choose a file organization for each physical file.
- Select suitable controls for each file and the database.

Remember, the specifications for these decisions are made in physical database design, and then the specifications are coded in the implementation phase using the capabilities of the chosen database technology. These database technology capabilities determine what physical database design decisions you need to make. For example, for Microsoft Access, which we assume is the implementation environment for this illustration, the only choice for file organization is indexed, so the file organization decision becomes on which primary and secondary key attributes to build indexes.

We illustrate these physical database design decisions only for the INVOICE table. The first decision most likely would be whether to denormalize this table. Based on the suggestions for possible denormalization presented in the chapter, the only possible denormalization of this table would be to combine it with the VENDOR table. Because each invoice must have a vendor, and the only additional data about vendors not in the INVOICE table is the Vendor_Name attribute, this is a good candidate for denormalization. Because Vendor_Name is not very volatile, repeating Vendor_Name in each invoice for the same vendor will not cause excessive update maintenance. If Vendor_Name is often used with other invoice data when invoice data are displayed, then, indeed, this would be a good candidate for denormalization. So, the denormalized relation to be transformed into a physical table is:

INVOICE(Vendor_ID,Invoice_Number,Invoice_Date,Paid?,Vendor_Name)

The next decision can be what indexes to create. The guidelines presented in this chapter suggest creating an index for the primary key, all foreign keys, and secondary keys used for sorting and qualifications in queries. So, we create a primary key index on the combined fields Vendor_ID and Invoice_Number. INVOICE has no foreign keys. To determine what fields are used as secondary keys in query sorting and qualification clauses, we would need to know the content of queries. Also, it would be helpful to know query frequency, because indexes do not provide much performance efficiency for infrequently run queries. For simplicity, suppose there were only two frequently run queries that reference the INVOICE table, as follows:

1. Display all the data about all unpaid invoices due this week.
2. Display all invoices ordered by vendor, show all unpaid invoices first, then all paid invoices, and order the invoices of each category in reverse sequence by invoice date.

In the first query, both the Paid? and Invoice_Date fields are used for qualification. Paid?, however, may not be a good candidate for an index because there are only two values for this field. The systems analyst would need to discover what percentage of invoices on file are unpaid. If this value is more than 10 percent, then an index on Paid? would not likely be helpful. Invoice_Date is a more discriminating field, so an index on this field would be helpful.

In the second query, Vendor_ID, Paid?, and Invoice_Date are used for sorting. Vendor_ID and Invoice_Date are discriminating fields (most values occur in less than 10 percent of the rows), so indexes on these fields will be helpful. Assuming less than 10 percent of the invoices on file are unpaid, then it would make sense to create the following indexes to make these two queries run as efficiently as possible:

Primary key index: Vendor_ID and Invoice_Number
Secondary key indices: Vendor_ID,Invoice_Date, and Paid?

Table 9-4 shows the decisions made for the properties of each field, based on reasonable assumptions about invoice data. Figure 9-4 illustrates a table definition screen for Microsoft Access. It is the parameters on this screen that must be specified for each field. Table 9-4 summarizes the data type including size (width), format and input mask (picture), default value, validation rule (integrity control), and whether the field is required or is allowed zero length (null value controls); we have already indicated the indexing decision. Recall from Table 9-2 that the data type of Lookup Wizard implements referential integrity, but there are no foreign keys in the INVOICE table because we combined the VENDOR table into the INVOICE table. We do not specify properties under the Lookup tab, which relates to additional data entry and display properties peculiar to Microsoft Access. Remember, we specify these parameters in physical database design, and it is in implementation that the Access tables would be defined using forms such as in Figure 9-4.

We do not illustrate security and other types of controls because these decisions are very dependent on unique capabilities of the technology and a complex analysis of what data which users have the right to read, modify, add, or delete. This section does illustrate the process of making many key physical database design decisions within the Microsoft Access environment.

ELECTRONIC COMMERCE APPLICATION: DESIGNING DATABASES

Like many other analysis and design activities, designing the database for an Internet-based electronic commerce application is no different than the process followed when designing the database for other types of applications. In the last chapter, you read how Jim Woo and the Pine Valley Furniture development team designed the human interface for the WebStore. In this section, we examine the processes Jim followed when transforming the conceptual data model for the WebStore into a set of normalized relations.

Designing Databases for Pine Valley Furniture's WebStore

The first step Jim took when designing the database for the WebStore was to review the conceptual data model—the entity-relationship diagram—developed during the analysis phase of the SDLC (see Figure 6-13 for a review). Given that

TABLE 9-4:   INVOICE Table Field Design Parameters for Hoosier Burger

Field	Physical Design Parameter				
	Data Type and Size	Format and Input Mask	Default Value	Validation Rule	Required, Zero Length
Vendor_ID	Number	Fixed with 0 decimals, 9999	N/A	<0	Required, not 0 length
Invoice_Number	Text, 10	LL99-99999	N/A	N/A	Required, not 0 length
Invoice_Date	Date/Time	Medium date	=Date( )	>#1/1/2000	Not required
Paid?	Yes/No	N/A	False	N/A	Required
Vendor_Name	Text, 30	N/A	N/A	N/A	Required, may be 0 length

there were no associative entities—many-to-many relationships—in the diagram, he began by identifying four distinct entity types that he named:

CUSTOMER
ORDER
INVENTORY
SHOPPING_CART

Once reacquainted with the conceptual data model, he examined the lists of attributes for each entity. He noted that three types of customers were identified during conceptual data modeling, namely: corporate customers, home office customers, and student customers. Yet, all were simply referred to as a "customer." Nonetheless, because each type of customer had some unique information (attributes) that other types of customers did not, Jim created three additional entity types, or subtypes, of customers:

CORPORATE
HOME_OFFICE
STUDENT

Table 9-5 lists the common and unique information about each customer type. As Table 9-5 implies, there needs to be four separate relations to keep track of customer information without having anomalies. The CUSTOMER relation is used to capture common attributes, whereas the additional relations are used to capture information unique to each distinct customer type. In order to identify the type of customer within the CUSTOMER relation easily, a Customer_Type attribute is added to the CUSTOMER relation. Thus, the CUSTOMER relation consists of:

CUSTOMER(<u>Customer_ID</u>,Address,Phone,E-mail,Customer_Type)

In order to link the CUSTOMER relation to each of the separate customer types—CORPORATE, HOME_OFFICE, and STUDENT—all share the same pri-

TABLE 9-5: Common and Unique Information about Each Customer Type*

Common Information about ALL Customer Types		
Corporate Customer	**Home Office Customer**	**Student Customer**
Customer ID	Customer ID	Customer ID
Address	Address	Address
Phone	Phone	Phone
E-mail	E-mail	E-mail
Unique Information about EACH Customer Type		
Corporate Customer	**Home Office Customer**	**Student Customer**
Corporate name	Customer name	Customer name
Shipping method	Corporate name	School
Buyer name	Fax	
Fax		

*Having multiple "types" of an entity, with all sharing common and each type having unique attributes, is modeled in E-R diagrams as a subclass entity and is commonly referred to as an "is a" relationship (e.g., a customer is a corporate customer, a customer is a home office customer, or a customer is a student customer). Please see a comprehensive database management text such as Hoffer, Prescott, and McFadden (2002) for more information on subclass entities and "is a" relationships.

mary key, Customer_ID, in addition to the attributes unique to each. This results in the following relations:

> CORPORATE(<u>Customer_ID</u>,Corporate_Name,Shipping_Method,Buyer_
> Name, Fax)
> HOME_OFFICE(<u>Customer_ID</u>,Customer_Name,Corporate_Name,Fax)
> STUDENT(<u>Customer_ID</u>,Customer_Name,School)

In addition to identifying all the attributes for customers, Jim also identified the attributes for the other entity types. The results of this investigation are summarized in Table 9-6. As described in Chapter 6, much of the order-related information is captured and tracked within PVF's Purchasing Fulfillment System. This means that the ORDER relation does not need to track all the details of the order because the Purchasing Fulfillment System produces a detailed invoice that contains all order details such as the list of ordered products, materials used, colors, quantities, and other such information. In order to access this invoice information, a foreign key, Invoice_ID, is included in the ORDER relation. Additionally, to easily identify which orders belong to a specific customer, the Customer_ID attribute _is also included in ORDER. Two additional attributes, Return_Code and Order_Status, are also included in ORDER. The Return_Code is used to track the return of an order more easily—or a product within an order—whereas Order_Status is a code used to represent the state of an order as it moves through the purchasing fulfillment process. This results in the following ORDER relation:

> ORDER(<u>Order_ID</u>,<u>Invoice_ID</u>,<u>Customer_ID</u>,Return_Code,Order_Status)

In the INVENTORY entity, two attributes—Materials and Colors—could take on multiple values but were represented as single attributes. For example, Materials represents the range of materials that a particular inventory item could be constructed from. Likewise, Colors is used to represent the range of possible product colors. PVF has a long-established set of codes for representing materials and colors; each of these *complex* attributes is represented as a single attribute. For example, the value "A" in the Colors field represents walnut, dark oak, light oak, and natural pine, whereas the value "B" represents cherry and walnut. Using this coding scheme, PVF can use a single character code to represent numerous combinations of colors. This results in the following INVENTORY relation:

> INVENTORY(<u>Inventory_ID</u>,Name,Description,Size,Weight,Materials,
> Colors,Price,Lead_Time)

TABLE 9-6:   **Attributes for Order, Inventory, and Shopping Cart Entities**

Order	Inventory	Shopping_Cart
<u>Order_ID</u> (primary key)	<u>Inventory_ID</u> (primary key)	<u>Cart_ID</u> (primary key)
<u>Invoice_ID</u> (foreign key)	Name	<u>Customer_ID</u> (foreign key)
<u>Customer_ID</u> (foreign key)	Description	<u>Inventory_ID</u> (foreign key)
Return_Code	Size	Material
Order_Status	Weight	Color
	Materials	Quantity
	Colors	
	Price	
	Lead_Time	

Finally, in addition to Cart_ID, each shopping cart contains the Customer_ID and Inventory_ID attributes so that each item in a cart can be linked to a particular inventory item and to a specific customer. In other words, both the Customer_ID and Inventory_ID attributes are foreign keys in the SHOPPING_CART relation. Recall that the SHOPPING_CART is temporary and is kept only while a customer is shopping. When a customer actually places the order, the ORDER relation is created and the line items for the order—the items in the shopping cart—are moved to the Purchase Fulfillment System and stored as part of an invoice. Because we also need to know the selected material, color, and quantity of each item in the SHOPPING_CART, these attributes are included in this relation. This results in the following:

SHOPPING_CART(Cart_ID,Customer_ID,Inventory_ID,Material,Color, Quantity)

Now that Jim had completed the database design for the WebStore, he shared all the design information with his project team so that the design could be turned into a working database during implementation. We read more about the WebStore's implementation in the next chapter.

Key Points Review

1. **Concisely define each of the following key database design terms: *relation, primary key, functional dependency, foreign key, referential integrity, field, data type, null value, denormalization, file organization, index,* and *secondary key*.**

   A relation is a named, two-dimensional table of data. Each relation consists of a set of named columns and an arbitrary number of unnamed rows. In logical database design, a relation corresponds to an entity or a many-to-many relationship from an E-R data model. One or more columns of each relation compose the primary key of the relation, values for which distinguish each row of data in the relation. A functional dependency is a particular relationship between two attributes. For a given relation, attribute B is functionally dependent on attribute A if, for every valid value of A, that value of A uniquely determines the value of B. The functional dependence of B on A is represented by A→B. The primary goal of logical database design is to develop relations in which all the nonprimary key attributes of a relation functionally depend on the whole primary key and nothing but the primary key. Relationships between relations are represented by placing the primary key of the table on the one side of the relationship as an attribute (also known as a foreign key) in the relation on the many side of the relationship. Foreign keys must satisfy referential integrity, which means that the value (or existence) of an attribute depends on the value (or existence) of the same attribute in another relation. The specifications for a database in terms of relations must be transformed into technology-related terms before the database can be implemented. A field is the smallest unit of stored data in a database and typically corresponds to an attribute in a relation. Each field has a data type, which is a coding scheme recognized by system software for representing organizational data. A null value for a field is a special field value, distinct from 0, blank, or any other value, that indicates that the value for the field is missing or otherwise unknown. Denormalization is an important process in designing a physical database, by which normalized relations are split or combined into physical tables based on affinity of use of rows and fields. A file organization is a technique for physically arranging the records of a physical file. Many types of file organizations utilize an index, which is a table (not related to the E-R diagram for the application) used to determine the location of rows in a file that satisfy some condition. An index can be created on a primary or a secondary key, which is one or a combination of fields for which more than one row may have the same combination of values.

2. **Explain the role of designing databases in the analysis and design of an information system.**

   Databases are defined during the systems design phase of the systems development life cycle. They are designed usually in parallel with the design of system interfaces. To design a database, a systems analyst must understand the conceptual database design for the application, usually specified by an E-R diagram, and the data requirements of each system interface (report, form, screen, etc.). Thus, database design is a combination of top-down (driven by an E-R diagram) and bottom-up (driven by specific information requirements in system interfaces) processes. Besides data requirements, systems analysts must also know physical data characteristics (e.g., length and format), frequency of use of the system interfaces, and the capabilities of database technologies.

3. **Transform an entity-relationship (E-R) diagram into an equivalent set of well-structured (normalized) relations.**

   An E-R diagram is transformed into normalized relations by following well-defined principles summarized in Table 9-1. For example, each entity becomes a relation and each many-to-many relationship or associative entity also becomes a relation. The principles also specify how to add foreign keys to relations to represent one-to-many relationships. You may want to review Table 9-1 at this point.

4. **Merge normalized relations from separate user views into a consolidated set of well-structured relations.**

   Separate sets of normalized relations are merged (this process is called view integration) to create a consolidated logical database design. The different sets of relations come from the conceptual E-R diagram for the application, known human system interfaces (reports, screens, forms, etc.), and known or anticipated queries for data which meet certain qualifications. The result of merging is a comprehensive, normalized set of relations for the application. Merging is not simply a mechanical process. A systems analyst must address issues of synonyms, homonyms, and functional dependencies between nonkeys during view integration.

5. **Choose storage formats for fields in database tables.**

   Fields in the physical database design represent the attributes (columns) of relations in the logical database design. Each field must have a data type, and potentially other characteristics such as a coding scheme to simplify the storage of business data, default value, input mask, range control, referential integrity control, or null value control. A storage format is chosen to balance four objectives: (1) Minimize storage space, (2) represent all possible values of the field, (3) improve data integrity for the field, and (4) support all data manipulations desired on the field.

6. **Translate well-structured relations into efficient database tables.**

   Whereas normalized relations possess properties of well-structured relations, the design of a physical table attempts to achieve two goals different from those of normalization: efficient use of secondary storage and data processing speed. Efficient use of storage means that the amount of extra (or overhead) information is minimized. So, file organizations such as sequential are efficient in the use of storage because little or no extra information, besides the meaningful business data, are kept. Data processing speed is achieved by keeping storage data close together that are used together, and by building extra information in the database, which allows data to be quickly found based on primary or secondary key values, or by sequence.

7. **Explain when to use different types of file organizations to store computer files.**

   Table 9-3 summarizes the performance characteristics of different types of file organizations. The systems analyst must decide which performance factors are most important for each application and the associated database. These factors are storage space, sequential retrieval speed, random row retrieval speed, speed of retrieving data based on multiple key qualifications, and the speed to perform data maintenance activities of row deletion, addition, and updating.

8. **Describe the purpose of indexes and the important considerations in selecting attributes to be indexed.**

   An index is information about the primary or secondary keys of a file. Each index entry contains the key value and a pointer to the row that contains that key value. An index facilitates rapid retrieval to rows for queries that involve AND, OR, and NOT qualifications of keys (e.g., all products with a maple finish and unit cost greater than $500 or all products in the office furniture product line). When using indexes, there is a trade-off between improved performance for retrievals and degrading performance for inserting, deleting, and updating the rows in a file. Thus, indexes should be used generously for databases intended primarily to support data retrievals, such as for decision support applications. Because they impose additional overhead, indexes should be used judiciously for databases that support transaction processing and other applications with heavy updating requirements. Typically, you create indexes on a file for its primary key, foreign keys, and other attributes used in qualification and sorting clauses in queries, forms, reports, and other system interfaces.

Key Terms Checkpoint

Here are the key terms from the chapter. The page where each term is first explained is in parentheses after the term.

1. **Calculated** (or **computed** or **derived**) **field (p. 338)**
2. **Data type (p. 337)**
3. **Default value (p. 339)**
4. **Denormalization (p. 340)**
5. **Field (p. 336)**
6. **File organization (p. 343)**
7. **Foreign key (p. 326)**
8. **Functional dependency (p. 324)**
9. **Hashed file organization (p. 346)**
10. **Homonym (p. 333)**
11. **Index (p. 344)**
12. **Indexed file organization (p. 344)**
13. **Input Mask (p. 339)**
14. **Normalization (p. 323)**
15. **Null value (p. 340)**
16. **Physical file (p. 343)**
17. **Physical table (p. 340)**
18. **Pointer (p. 343)**
19. **Primary key (p. 319)**
20. **Recursive foreign key (p. 331)**
21. **Referential integrity (p. 326)**
22. **Relation (p. 322)**
23. **Relational database model (p. 322)**
24. **Second normal form (2NF) (p. 325)**
25. **Secondary key (p. 344)**
26. **Sequential file organization (p. 343)**
27. **Synonyms (p. 332)**
28. **Third normal form (3NF) (p. 325)**
29. **Well-structured relation (or table) (p. 322)**

Match each of the key terms above the definition that best fits it.

_____ 1. A named, two-dimensional table of data. Each relation consists of a set of named columns and an arbitrary number of unnamed rows.

_____ 2. A relation that contains a minimum amount of redundancy and allows users to insert, modify, and delete the rows without errors or inconsistencies.

_____ 3. The process of converting complex data structures into simple, stable data structures.

_____ 4. A particular relationship between two attributes.

_____ 5. A relation for which every nonprimary key attribute is functionally dependent on the whole primary key.

_____ 6. A relation that is in second normal form and that has no functional (transitive) dependencies between two (or more) non-primary key attributes.

_____ 7. An attribute that appears as a nonprimary key attribute in one relation and as a primary key attribute (or part of a primary key) in another relation.

_____ 8. An integrity constraint specifying that the value (or existence) of an attribute in one relation depends on the value (or existence) of the same attribute in another relation.

_____ 9. A foreign key in a relation that references the primary key values of that same relation.

_____ 10. Two different names that are used for the same attribute.

_____ 11. A single attribute name that is used for two or more different attributes.

_____ 12. The smallest unit of named application data recognized by system software.

_____ 13. A coding scheme recognized by system software for representing organizational data.

_____ 14. A field that can be derived from other database fields.

_____ 15. A value a field will assume unless an explicit value is entered for that field.

_____ 16. A pattern of codes that restricts the width and possible values for each position of a field.

_____ 17. A special field value, distinct from 0, blank, or any other value, that indicates that the value for the field is missing or otherwise unknown.

_____ 18. A named set of rows and columns that specifies the fields in each row of the table.

_____ 19. The process of splitting or combining normalized relations into physical tables based on affinity of use of rows and fields.

_____ 20. A named set of table rows stored in a contiguous section of secondary memory.

_____ 21. A technique for physically arranging the records of a file.

_____ 22. A field of data that can be used to locate a related field or row of data.

_____ 23. The rows in the file are stored in sequence according to a primary key value.

_____ 24. The rows are stored either sequentially or nonsequentially, and an index is created that allows software to locate individual rows.

____ 25. A table used to determine the location of rows in a file that satisfy some condition.

____ 26. One or a combination of fields for which more than one row may have the same combination of values.

____ 27. The address for each row is determined using an algorithm.

____ 28. An attribute whose value is unique across all occurrences of a relation.

____ 29. Data represented as a set of related tables or relations.

Review Questions

1. What is the purpose of normalization?
2. List five properties of relations.
3. What problems can arise during view integration or merging relations?
4. How are relationships between entities represented in the relational data model?
5. What is the relationship between the primary key of a relation and the functional dependencies among all attributes within that relation?
6. How is a foreign key represented in relational notation?
7. Can instances of a relation (sample data) prove the existence of a functional dependency? Why or why not?
8. In what way does the choice of a data type for a field help to control the integrity of that field?
9. What is the difference between how a range control statement and a referential integrity control statement are handled by a file management system?
10. What is the purpose of denormalization? Why might you not want to create one physical table or file for each relation in a logical data model?
11. What factors influence the decision to create an index on a field?
12. Explain the purpose of data compression techniques.
13. What are the goals of designing physical tables?
14. What are the seven factors that should be considered in selecting a file organization?
15. What are the four key steps in logical database modeling and design?
16. What are the four steps in transforming an E-R diagram into normalized relations?

Problems and Exercises

1. Assume that at Pine Valley Furniture products are comprised of components, products are assigned to salespersons, and components are produced by vendors. Also assume that in the relation PRODUCT(Prodname,_Salesperson, Compname,Vendor)Vendor is functionally dependent on Compname, and Compname is functionally dependent on Prodname. Eliminate the transitive dependency in this relation and form 3NF relations.

2. Transform the E-R diagram of Figure 6-3 into a set of 3NF relations. Make up a primary key and one or more nonkeys for each entity.

3. Transform the E-R diagram of Figure 9-21 into a set of 3NF relations.

4. Consider the list of individual 3NF relations that follow. These relations were developed from several separate normalization activities.

    PATIENT(Patient_ID,Room_Number,Admit_ Date,_Address)
    ROOM(Room_Number,Phone,Daily_Rate)
    PATIENT(Patient_Number,Treatment_Description, _Address)
    TREATMENT(Treatment_ID,Description,Cost)
    PHYSICIAN(Physician_ID,Name,Department)
    PHYSICIAN(Physician_ID,Name,Supervisor_ID)

    a. Merge these relations into a consolidated set of 3NF relations. Make and state whatever assumptions you consider necessary to resolve any potential problems you identify in the merging process.

    b. Draw an E-R diagram for your answer to part a.

5. Consider the following 3NF relations about a sorority or fraternity:

    MEMBER(Member_ID,Name,Address,Dues_Owed)
    OFFICE(Office_Name,Officer_ID,Term_Start_ Date,Budget)
    EXPENSE(Ledger_Number,Office_Name,_ Expense_Date,Amt_Owed)
    PAYMENT(Check_Number,_Expense_Ledger_ Number,Amt_Paid)

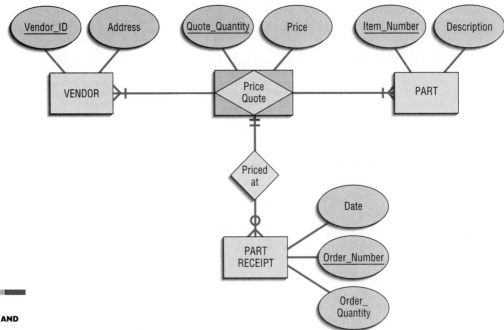

FIGURE 9-21
E-R DIAGRAM FOR PROBLEM AND
EXERCISE 3

RECEIPT(<u>Member_ID</u>,Receipt_Date,Dues_
   Received)
COMMITTEE(<u>Committee_ID</u>,Officer_in_Charge)
WORKERS(<u>Committee_ID</u>,<u>Member_ID</u>)

a. Foreign keys are not indicated in these relations. Decide which attributes are foreign keys
   and justify your decisions.
b. Draw an E-R diagram for these relations, using
   your answer to part a.
c. Explain the assumptions you made about cardinalities in your answer to part b. Explain
   why it is said that the E-R data model is more
   expressive or more semantically rich than the
   relational data model.

6. Consider the following functional dependencies:

   Applicant_ID→Applicant_Name
   Applicant_ID→Applicant_Address
   Position_ID→Position_Title
   Position_ID→Date_Position_Opens
   Position_ID→Department
   Applicant_ID + Position_ID→Date_Applied
   Applicant_ID + Position_ID + Date_Interviewed→

   a. Represent these attributes with 3NF relations.
      Provide meaningful relation names.
   b. Represent these attributes using an E-R dia-
      gram. Provide meaningful entity and relation-
      ship names.

7. Suppose you were designing a file of student
   records for your university's placement office.
   One of the fields that would likely be in this file is

the student's major. Develop a coding scheme for
this field that achieves the objectives outlined in
this chapter for field coding.

8. In Problem and Exercise 3, you developed integrated normalized relations. Choose primary
   keys for the files that would hold the data for
   these relations. Did you use attributes from the
   relations for primary keys or did you design new
   fields? Why or why not?

9. Suppose you created a file for each relation in your
   answer to Problem and Exercise 3. If the following
   queries represented the complete set of accesses
   to this database, suggest and justify what primary
   and secondary key indexes you would build.

   a. For each PART in Item_Number order list in
      Vendor_ID, sequence all the vendors and their
      associated prices for that part.
   b. List all PART RECEIPTs including related
      PART fields for all the parts received on a par-
      ticular day.
   c. For a particular VENDOR, list all the PARTs and
      their associated prices that VENDOR can supply.

10. Suppose you were designing a default value for
    the age field in a student record at your univer-
    sity. What possible values would you consider
    and why? How might the default vary by other
    characteristics about the student, such as school
    within the university or degree sought?

11. Consider Figure 9-19(B). Explain a query that
    would likely be processed more quickly using the
    denormalized relations rather than the normal-
    ized relations.

Discussion Questions

1. Find in the library books or articles with discussions of additional normal forms other than second and third normal forms. Describe each of these additional normal forms and give examples of each. How are these additional normal forms different from those presented in this chapter? What additional benefit does their use provide?
2. Describe the deliverables from file and database design.
3. Discuss what additional information should be collected during requirements analysis that is needed for file and database design and that is not very useful for earlier phases of systems development.
4. Find out what database management systems are available at your university for student use. Investigate which data types these DBMSs support. Compare these DBMSs based upon data types supported, and suggest which types of applications each DBMS is best suited for based on this comparison.
5. Find out what database management systems are available at your university for student use. Investigate what physical file and database design decisions need to be made. Compare this list of decisions to those discussed in this chapter. For physical database and design decisions (or options) not discussed in the chapter, investigate what choices you have and how you should choose among these choices. Submit a report to your instructor with your findings.

Case Problems

1. Pine Valley Furniture

   Development work on Pine Valley Furniture's new Customer Tracking System is proceeding according to plan and is on schedule. The project team has been busy designing the human interfaces, and you have just completed the new tracking system's Customer Profile Form, Products by Demographics Summary Report, and Customer Purchasing Frequency Report.

   Because you are now ready for a new task, Jim Woo asks you to prepare logical data models for the form and two reports that you have just designed and drop them by his office this afternoon. At that time, the two of you will prepare a consolidated database model, translate the Customer Tracking System's E-R data model into normalized relations, and then integrate the logical data models into a final logical data model for the Customer Tracking System.

   a. Develop logical data models for the form and two reports mentioned in the case scenario.
   b. Perform view integration on the logical models developed for part a.
   c. What view integration problems, if any, exist? How should you correct these problems?
   d. Have a fellow classmate critique your logical data model. Make any necessary corrections.

2. Hoosier Burger

   As the lead analyst on the Hoosier Burger project, you have had the opportunity to learn more about the systems development process, work with project team members, and interact with the system's end users, especially with Bob and Thelma. You have just completed the design work for the various forms and reports that will be used by Bob, Thelma, and their employees. Now it is time to prepare logical and physical database designs for the new Hoosier Burger system.

   During a meeting with Hoosier Burger project team members, you review the four steps in logical database modeling and design. It will be your task to prepare the logical models for the Customer Order Form, Customer Account Balance Form, Daily Delivery Sales Report, and Inventory Low-in-Stock Report. At the next meeting, the E-R model will be translated and a final logical model produced.

   a. Develop logical models for each of the interfaces mentioned in the case scenario.
   b. Integrate the logical models prepared for part a into a consolidated logical model.
   c. What types of problems can arise from view integration? Did you encounter any of these problems when preparing the consolidated logical model?
   d. Using your newly constructed logical model, determine which fields should be indexed. Which fields should be designated as calculated fields?

3. Pest Busters

   Pest Busters is a locally owned and operated pest control and lawn maintenance business. Pest Busters offers its customers free termite control estimates, provides insect and rodent control options, and has recently implemented a new lawn care maintenance program. Pest Busters offers its pest control and lawn maintenance services to a large, metropolitan-based clientele, including one-time residential, scheduled maintenance, and commercial customers.

A recent advertising campaign caused an increase in demand for Pest Buster services; this increase in demand is expected to continue. In order to provide faster, more efficient service, Pest Busters hired your consulting company to design, develop, and implement a computer-based system. Your development team is currently preparing the logical and physical database designs for Pest Busters.

a. What are the four steps in logical database modeling and design?
b. Several relations have been identified for this project, including technician, customer, service provided, product inventory, and services offered. What relationships exist among these relations? How should these relationships be represented?
c. Think of the attributes that would most likely be associated with the relations identified in part b. For each data integrity control method discussed in your chapter, provide a specific example.
d. What are the guidelines for choosing indexes? Identify several fields that should be indexed.

CASE: BROADWAY ENTERTAINMENT COMPANY, INC.

Designing the Relational Database for the Customer Relationship Management System

Case Introduction

The students from St. Claire Community College are making good progress in the design of MyBroadway, the Web-based customer relationship management system for Broadway Entertainment Company stores. They recently completed the design for all human interfaces and received tentative approval for these from Carrie Douglass, their client at the Centerville, Ohio, BEC store. This was an important step not only for making progress on the human interfaces but also because this approval validated all the data needed in system inputs and outputs. The entity-relationship diagram the team developed earlier in the project (see the BEC case at the end of Chapter 6) was not grounded in actual system inputs and outputs, but rather from a general understanding of system requirements. The approved inputs and outputs allow the student team to check that the entity types in the E-R diagram can hold all the input data that must be kept and can be used to produce all the outputs from MyBroadway.

Identifying Relations

BEC Figure 6-1 at the end of Chapter 6 identified six entities the students decided were required for the

MyBroadway database. As the students discuss the translation of this diagram into normalized relations, they conclude that the task is fairly straightforward. The translation looks easy because all the relationships are one-to-many. Based on procedures they have been taught in courses at St. Claire, each data entity becomes a relational table. The identifier of each entity can be used at the primary key of the associated relation, and the other attributes of an entity become the nonkey attributes of the associated relation. The relationships are represented as foreign keys, in which the primary key of the entity on the one side of a relationship becomes a foreign key in the relation for the entity on the many side of the relationship. BEC Figure 9-1 shows the team's initial relational data model, based on these rules.

The students are fairly confident that the relations in BEC Figure 9-1 are accurate (both complete and in third normal form), but they see some implementation issues with these relations. First, the primary key of the PRODUCT relation, which is a foreign key in every other relation, is awkward because it has three components. The students are concerned that linking the tables based on three attributes will be inefficient in terms of both storage space and producing pages. The second issue they identify is with the PRODUCT Description, the COMMENT Member_Comment, and the PICK Employee_Comment attributes. These attributes are highly variable in length and may be quite long. These two traits also mean that retrieving

BEC FIGURE 9-1
INITIAL RELATIONS FOR
MYBROADWAY

PRODUCT(Title,Artist,Type,Publisher,Category,Media,Description,Release_Date,Sale_Price,Rental_Price)
COMMENT(Membership_ID,Comment_Time_Stamp,Title,Artist,Type,Parent/Child?,Member_Comment)
REQUEST(Membership_ID,Request_Time_Stamp,Title,Artist,Type)
SALE(Membership_ID,Sale_Time_Stamp,Title,Artist,Type)
RENTAL(Membership_ID,Rental_Time_Stamp,Title,Artist,Type,Due_Date,Refund?)
PICK(Employee_Name,Pick_Time_Stamp,Title,Artist,Rating,Employee_Comment)

and storing data can be time-consuming. The students will have to address these issues before the database is defined.

Designing the Physical Database

The students decide to address the issue of the compound primary key by creating what is called a nonintelligent key. A nonintelligent key is a system-assigned value that has no business meaning. It simply is an artificial attribute that will have a unique value for each row in the PRODUCT table. The three attributes of Title, Artist, and Type then become nonkey attributes, with the nonintelligent key of Product_ID becoming the primary key of PRODUCT and the foreign key in the other relations. Figure 9-2 shows the relations with this modification.

The students address the second issue about the long and variable length of the Description, Member_Comment, and Employee_Comment attributes by defining them each as a memo field. Microsoft Access stores memo fields separately from the other attributes of a relation, which overcomes the problems with long and variable length data.

Because the team members have chosen to use Microsoft Access for the prototype, very few specific physical database design decisions must be made. They will have to select a data type for each attribute. For numeric data, such as PRODUCT Sale_Price, the students will have to decide on a format with number of decimal places, and for text fields, like PRODUCT Category, they will have to establish a maximum length. Access does not allow a designer to choose between file organizations for each table, but the students will need to decide on which attributes to build indexes. They immediately decide to create a primary key index for each table, but they are not sure which secondary key indexes will be best. Other decisions to be made are (1) whether to save storage space by coding fields, like PRODUCT media; (2) whether to define any data integrity controls on each field, such as a default value, input mask, validation rule; and (3) whether a value for the field is required for a new row of the table to be stored (e.g., must there be a specified Release_Date for each PRODUCT?).

Case Summary

The student team has many specific decisions to make in order to finalize the design of the database for MyBroadway. For a prototype, the students remember that some developers will not take the time to make intelligent physical design decisions. However, because the students plan for actual BEC customers to use the MyBroadway prototype, they want the system to be reasonably efficient. Thus, they plan on taking the time to use all the power of Microsoft Access to create an efficient and reliable database. The team decides to analyze further each data input and information output page to understand better how those pages use the database. So, each team member is assigned several pages, and they agree to meet in two days with suggestions for all the physical database design decisions before they begin implementation of the initial prototype.

Case Questions

1. In the questions associated with the BEC case at the end of Chapter 6, you were asked to modify the entity-relationship diagram drawn by the St. Claire student team to include any other entities and the attributes you identified from the BEC cases. Review your answers to these questions, and modify the relations in BEC Figure 9-2 to include your changes.
2. Study your answer to Question 1. Verify that the relations you say represent the MyBroadway database are in third normal form. If they are not, change them so that they are.
3. The E-R diagram you developed in questions in the BEC case at the end of Chapter 6 should have shown minimum cardinalities on both ends of each relationship. Are minimum cardinalities represented in some way in the relations in your answer to Question 2? If not, how are minimum cardinalities enforced in the database?
4. You have probably noticed that the St. Claire students chose to include a time stamp field as part of the primary key for all of relations except PRODUCT. Explain why you think they decided to include this field in each relation and why it is part of the primary key. Are there other alternatives to a time stamp field for creating the primary key of these relations?
5. This BEC case indicated the data types chosen for only a few of the fields of the database. Using your answer to Question 2, select data types, formats, and lengths for each attribute of each relation. Use the data types and formats supported by Microsoft Access. What data type should be

PRODUCT(Product_ID,Title,Artist,Type,Publisher,Category,Media,Description,Release_Date,Sale_Price,Rental_Price)
COMMENT(Membership_ID,Comment_Time_Stamp,Product_ID,Parent/Child?,Member_Comment)
REQUEST(Membership_ID,Request_Time_Stamp,Product_ID)
SALE(Membership_ID,Sale_Time_Stamp,Product_ID)
RENTAL(Membership_ID,Rental_Time_Stamp,Product_ID,Due_Date,Refund?)
PICK(Employee_Name,Pick_Time_Stamp,Product_ID,Rating,Employee_Comment)

BEC FIGURE 9-2
RELATIONS WITH NONINTELLIGENT
PRIMARY KEY FOR PRODUCTS

used for the nonintelligent primary keys? Do you agree with the use of Memo for Description and Member_Comment attributes?

6. This BEC case also mentioned that the students will consider if any fields should be coded. Are any fields good candidates for coding? If so, suggest a coding scheme for each coding candidate field. How would you implement field coding through Microsoft Access?

7. Complete all table and field definitions for the MyBroadway database using Microsoft Access. Besides the decisions you have made in answers to the above questions, fill in all other field definition parameters for each field of each table.

8. The one decision for a relational database that usually influences efficiency the most is index definition. Besides the primary key indexes the students have chosen, what other secondary key indexes do you recommend for this database? Justify your selection of each secondary key index.

9. Using Microsoft Visio, develop an E-R diagram with all the supporting database properties for decisions you made in Case Questions 1–8. Are there database design decisions you made that could not be documented in Visio? Finally, use Visio to generate Microsoft Access table definitions. Did the table generation create the table definitions you would create manually?

10

Systems Implementation and Operation

➲ Objectives

After studying this chapter, you should be able to:

➲ Describe the process of coding, testing, and converting an organizational information system, and outline the deliverables and outcomes of the process.

➲ Apply four installation strategies: direct, parallel, single location, and phased installation.

➲ List the deliverables for documenting the system and for training and supporting users.

➲ Compare the many modes available for organizational information system training, including self-training and electronic performance support systems.

➲ Discuss the issues of providing support for end users.

➲ Explain why systems implementation sometimes fails.

➲ Explain and contrast four types of maintenance.

➲ Describe several factors that influence the cost of maintaining an information system.

Chapter Preview . . .

The implementation and operation phase of the systems development life cycle is the most expensive and time-consuming phase of the entire life cycle. This phase is expensive because so many people are involved in the process. It is time-consuming because of all the work that has to be completed through the entire life of the system. During implementation and operation, physical design specifications must be turned into working computer code. Then the code is tested until most of the errors have been detected and corrected, the system is installed, user sites are prepared for the new system, and users must come to rely on the new system rather than the existing one to get their work done. Even once the system is installed, new features are added to the system, new business requirements and regulations demand system improvements, and corrections are made as flaws are identified from use of the system in new circumstances. These changes will have ripple effects causing rework in many systems development phases.

The seven major activities we are concerned with in this chapter are coding, testing, installation, documentation, training, support, and

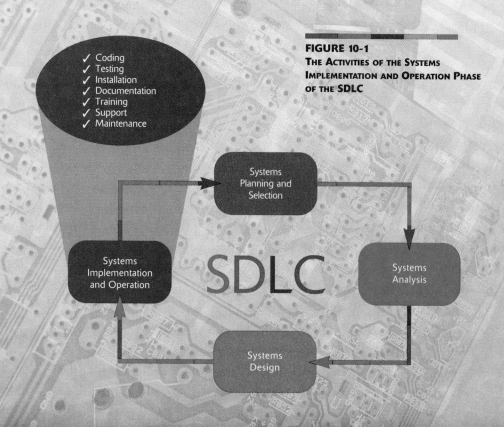

FIGURE 10-1
THE ACTIVITIES OF THE SYSTEMS IMPLEMENTATION AND OPERATION PHASE OF THE SDLC

✓ Coding
✓ Testing
✓ Installation
✓ Documentation
✓ Training
✓ Support
✓ Maintenance

Systems Planning and Selection

Systems Analysis

Systems Design

Systems Implementation and Operation

SDLC

maintenance. These and other activities are highlighted in Figure 10-1. Our intent is not to explain how to program and test systems—most of you have already learned about writing and testing programs in the previous courses you took. Rather, this chapter shows you where coding and testing fit in the overall scheme of implementation and stresses the view of implementation as an organizational change process that is not always successful.

In addition, you will learn about providing documentation about the new system for the information systems personnel who will maintain the system and about providing documentation and conducting training for the system's users. Once training has ended and the system is accepted and used, you must provide a means for users to get answers to their additional questions and to identify needs for further training.

Your first job after graduation may very well be as a maintenance programmer/analyst. Maintenance can begin soon after the system is installed. A question many people have about maintenance relates to how long organizations should maintain a system. Five years? Ten years? Longer? There is no simple answer to this question, but it is most often an issue of economics. In other words, at what point does it make financial sense to discontinue updating an older system and build or purchase a new one? Upper IS management gives significant attention to assessing the trade-offs between maintenance and new development. In this chapter, we describe the maintenance process and the issues that must be considered when maintaining systems. At the end of the chapter, we describe the process of resolving a maintenance request at Pine Valley Furniture.

Systems Implementation and Operation

Systems implementation and operation is made up of seven major activities:

- ◗ Coding
- ◗ Testing
- ◗ Installation
- ◗ Documentation
- ◗ Training
- ◗ Support
- ◗ Maintenance

The purpose of these steps is to convert the final physical system specifications into working and reliable software and hardware, document the work that has been done, and provide help for current and future users and caretakers of the system. Usage of the system leads to changes, so during maintenance, users and others submit maintenance requests, requests are transformed into specific changes to the system, the system is redesigned to accept the changes, and the changes are implemented.

These steps are often done by other project team members besides analysts, although analysts may do some programming and other steps. Often a separate analyst and developer team from those who developed the original system is responsible for testing, documenting, training, and maintenance activities. In any case, analysts are responsible for ensuring that all of these various activities are properly planned and executed. We briefly discuss these activities in three groups:

1. Activities that lead to the system going into operation—coding, testing, and installation
2. Activities that are necessary for successful system operation—documenting the system and training and supporting users
3. Activities that are ongoing and needed to keep the system working and up-to-date—maintenance

The Processes of Coding, Testing, and Installation

Coding, as we mentioned before, is the process whereby the physical design specifications created by the design team are turned into working computer code by the programming team. Depending on the size and complexity of the system, coding can be an involved, intensive activity. Once coding has begun, the testing process can begin and proceed in parallel. As each program module is produced, it can be tested individually, then as part of a larger program, and then as part of a larger system. You learn about the different strategies for testing later in the chapter. We should emphasize that although testing is done during implementation, you must begin planning for testing earlier in the project. Planning involves determining what needs to be tested and collecting test data. This is often done during the analysis phase because testing requirements are related to system requirements.

Installation is the process during which the current system is replaced by the new system. This includes conversion of existing data, software, documentation, and work procedures to those consistent with the new system. Users must give up the old ways of doing their jobs, whether manual or automated, and adjust to accomplishing the same tasks with the new system. Users will sometimes resist these changes and you must help them adjust. However, you cannot control all the dynamics of user-system interaction involved in the installation process.

Deliverables and Outcomes from Coding, Testing, and Installation

Table 10-1 shows the deliverables from the coding, testing, and installation processes. The most obvious outcome is the code itself, but just as important as the code is documentation of the code. Modern programming languages, such as Visual Basic, are said to be largely self-documenting. When standard naming and program design conventions are used, the code itself spells out

TABLE 10-1:   Deliverables from Coding, Testing, and Installation

Action	Deliverable
Coding	Code
	Program documentation
Testing	Test scenarios (test plan) and test data
	Results of program and system testing
Installation	User guides
	User training plan
	Installation and conversion plan
	Hardware and software installation schedule
	Data conversion plan
	Site and facility remodeling plan

much about the program's logic, the meaning of data and variables, and the locations where data are accessed and output. But even well-documented code can be mysterious to maintenance programmers who must maintain the system for years after the original system was written and the original programmers have moved on to other jobs. Therefore, clear, complete documentation for all individual modules and programs is crucial to the system's continued smooth operation. Increasingly, CASE tools are used to maintain the documentation needed by systems professionals.

The results of program and system testing are important deliverables from the testing process because they document the tests as well as the test results. For example, what type of test was conducted? What test data were used? How did the system handle the test? The answers to these questions can provide important information for system maintenance as changes will require retesting and similar testing procedures will be used during the maintenance process.

The next two deliverables, user guides and the user training plan, result from the installation process. User guides provide information on how to use the new system, and the training plan is a strategy for training users so they can quickly learn the new system. The development of the training plan probably began earlier in the project, and some training on the concepts behind the new system may have already taken place. During the early stages of implementation, the training plans are finalized and training on the use of the system begins. Similarly, the installation plan lays out a strategy for moving from the old system to the new. Installation includes installing the system (hardware and software) at central and user sites. The installation plan answers such questions as when and where the new system will be installed, what people and resources are required, which data will be converted and cleansed, and how long the installation process will take. It is not enough that the system is installed; users must actually use it.

As an analyst, your job is to ensure that all of these deliverables are produced and done well, whether by you or others. Coding, testing, and installation work may be done by IS professionals in your organization, contractors, hardware designers, and, increasingly, users. The extent of your responsibilities will vary according to the size and standards of the organization you work for, but your ultimate role includes ensuring that all the coding, testing, and installation work leads to a system that meets the specifications developed in earlier project phases.

The Processes of Documenting the System, Training Users, and Supporting Users

Although the process of documentation proceeds throughout the life cycle, it receives formal attention now because once the system is installed the analysis team's involvement in system development usually ceases. As the team is getting ready to move on to new projects, you and the other analysts need to prepare documents that reveal all of the important information you have learned about this system during its development and implementation. There are two audiences for this final documentation: (1) the information systems personnel who will maintain the system throughout its productive life and (2) the people who will use the system as part of their daily lives.

Larger organizations also tend to provide training and support to computer users throughout the organization, sometimes as part of a corporate university. Some of the training and support is directed to off-the-shelf software packages. For example, it is common to find courses on Microsoft Windows and WordPerfect in organization-wide training facilities. Analysts typically work with corporate trainers to provide training and support tailored to particular computer applications they have helped develop. Centralized information system training facilities tend to have specialized staff who can help with training and support

NET SEARCH
Investigate the career opportunities in the information systems field. Visit http://www. prenhall.com/valacich to complete an exercise related to this topic.

issues. In smaller organizations that cannot afford to have well-staffed centralized training and support facilities, fellow users are the best source of training and support users have, whether the software is customized or off-the-shelf.

Deliverables and Outcomes from Documenting the System, Training Users, and Supporting Users

Table 10-2 shows the deliverables from documenting the system, training users, and supporting users. User documentation can be paper based, but it should also include computer-based modules. For modern information systems, this documentation includes any online help designed as part of the system interface. The development team should think through the user training process: Who should be trained? How much training is adequate for each training audience? What do different types of users need to learn during training? The training plan should be supplemented by actual training modules, or at least outlines of such modules, that at a minimum address the three questions stated previously. Finally, the development team should also deliver a user support plan that addresses such issues as how users will be able to find help once the information system has become integrated into the organization. The development team should consider a multitude of support mechanisms and modes of delivery. Each deliverable is addressed in more detail later in the chapter.

The Process of Maintaining Information Systems

Throughout this book, we have drawn the systems development life cycle as a circle where one phase leads to the next, with overlap and feedback loops. This means that the process of maintaining an information system is the process of returning to the beginning of the SDLC and repeating development steps, focusing on the needs for system change, until the change is implemented.

Four major activities occur within maintenance:

1. Obtaining maintenance requests
2. Transforming requests into changes
3. Designing changes
4. Implementing changes

Obtaining maintenance requests requires that a formal process be established whereby users can submit system change requests. Earlier in the book, we presented a user request document called a System Service Request (SSR). Most companies have some sort of document like an SSR to request new development, to report problems, or to request new system features for an existing system. When developing the procedures for obtaining maintenance requests, organizations must also specify an individual within the organization to collect these requests and manage their dispersal to maintenance personnel. The

TABLE 10-2:   Deliverables from Documenting the System, Training Users, and Supporting Users

Documentation	User training modules
System documentation	Training materials
User documentation	Computer-based training aids
User training plan	User support plan
Classes	Help desk
Tutorials	Online help
	Bulletin boards and other support mechanisms

process of collecting and dispersing maintenance requests is described in much greater detail later in the chapter.

Once a request is received, analysis must be conducted to gain an understanding of the scope of the request. It must be determined how the request will affect the current system and the duration of such a project. As with the initial development of a system, the size of a maintenance request can be analyzed for risk and feasibility (see Chapter 3). Next, a change request can be transformed into a formal design change, which can then be fed into the maintenance implementation phase. Thus, many similarities exist between the SDLC and the activities within the maintenance process. Figure 10-2 equates SDLC phases to the maintenance activities described previously. The figure shows that the first phase of the SDLC—systems planning and selection—is analogous to the maintenance process of obtaining a maintenance request (step 1). The SDLC phase systems analysis is analogous to the maintenance process of transforming requests into a specific system change (step 2). The systems design phase of the SDLC, of course, equates to the designing changes process (step 3). Finally, the SDLC phase implementation and maintenance equates to implementing changes (step 4). This similarity between the maintenance process and the SDLC is no accident. The concepts and techniques used to develop a system initially are also used to maintain it.

Deliverables and Outcomes from Maintaining Information Systems

Because maintenance is basically a subset of the activities of the entire development process, the deliverables and outcomes from the process are the development of a new version of the software and new versions of all design

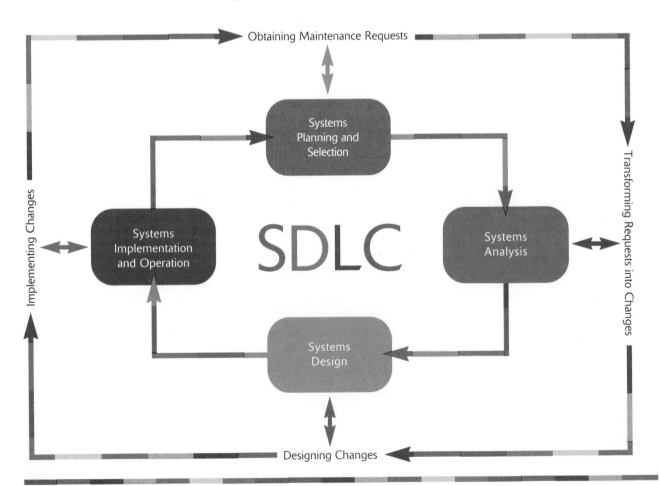

FIGURE 10-2
MAINTENANCE ACTIVITIES IN RELATION TO THE SDLC

documents and training materials developed or modified during the maintenance process. This means that all documents created or modified during the maintenance effort, including the system itself, represent the deliverables and outcomes of the process. Those programs and documents that did not change may also be part of the new system. Because most organizations archive prior versions of systems, all prior programs and documents must be kept to ensure the proper versioning of the system. This enables prior versions of the system to be re-created if needed. A more detailed discussion of configuration management and change control is presented later in the chapter.

Because of the similarities between the steps, deliverables, and outcomes of new development and maintenance, you may be wondering how to distinguish between these two processes. One difference is that maintenance reuses most existing system modules in producing the new system version. Other distinctions are that we develop a new system when there is a change in the hardware or software platform or when fundamental assumptions and properties of the data, logic, or process models change.

Software Application Testing

As we mentioned previously, analysts prepare system specifications that are passed on to programmers for coding. Testing software begins earlier in the systems development life cycle, even though many of the actual testing activities are carried out during implementation. During analysis, you develop an overall test plan. During design, you develop a unit test plan, an integration test plan, and a system test plan. During implementation, these various plans are put into effect and the actual testing is performed.

The purpose of these written test plans is to improve communication among all the people involved in testing the application software. The plan specifies what each person's role will be during testing. The test plans also serve as checklists you can use to determine whether all testing steps have been completed. The overall test plan is not just a single document but a collection of documents. Each of the component documents represents a complete test plan for one part of the system or for a particular type of test.

Some organizations have specially trained personnel who supervise and support testing. Testing managers are responsible for developing test plans, establishing testing standards, integrating testing and development activities in the life cycle, and ensuring that test plans are completed. Testing specialists help develop test plans, create test cases and scenarios, execute the actual tests, and analyze and report test results.

Seven Different Types of Tests

Software application testing is an umbrella term that covers several types of tests. Tests can be done with or without executing the code and they may be manual or automated. Using this framework, we can categorize types of tests as shown in Table 10-3.

TABLE 10-3:   A Categorization of Test Types

	Manual	**Automated**
Without Code Execution	Inspections	Syntax checking
With Code Execution	Walkthroughs	Unit testing
	Desk checking	Integration testing
		System testing
		Stub testing

Inspections
A testing technique in which participants examine program code for predictable language-specific errors.

Let's examine each type of test in turn. **Inspections** are formal group activities where participants manually examine code for occurrences of well-known errors. Syntax, grammar, and some other routine errors can be checked in early stages of coding by automated inspection software, so manual inspection checks are used for more subtle errors. Code inspection participants compare the code they are examining to a checklist of well-known errors for that particular language. Exactly what the code does is not investigated in an inspection. Code inspections have been used by organizations to detect from 60 to 90 percent of all software defects as well as to provide programmers with feedback that enables them to avoid making the same types of errors in future work. The inspection process can also be used to ensure that design specifications are accomplished.

Unlike inspections, what the code does is an important question in a *walkthrough*. Using structured walkthroughs is a very effective method of detecting errors in code. As you saw in Chapter 3, structured walkthroughs can be used to review many systems development deliverables, including design specifications as well as code. Whereas specification walkthroughs tend to be formal reviews, code walkthroughs tend to be informal. Informality makes programmers less apprehensive of criticism and thus helps increase the frequency of walkthroughs. Code walkthroughs should be done frequently when the pieces of work reviewed are relatively small and before the work is formally tested. If walkthroughs are not held until the entire program is tested, the programmer will have already spent too much time looking for errors that the programming team could have found much more quickly. Further, the longer a program goes without being subjected to a walkthrough, the more defensive the programmer becomes when the code is reviewed. Although each organization that uses walkthroughs conducts them differently, there is a basic structure that you can follow that works well (see Figure 10-3).

It should be stressed that the purpose of a walkthrough is to detect errors, not to correct them. It is the programmer's job to correct the errors uncovered in a walkthrough. Sometimes it can be difficult for the reviewers to refrain from suggesting ways to fix the problems they find in the code, but increased experience with the process can help change reviewers' behavior.

Desk checking
A testing technique in which the program code is sequentially executed manually by the reviewer.

What the code does is also important in **desk checking,** an informal process where the programmer or someone else who understands the logic of the program works through the code with a paper and pencil. The programmer executes each instruction, using test cases that may or may not be written down. In one sense, the reviewer acts as the computer, mentally checking each step and its results for the entire set of computer instructions.

Syntax checking is typically done by a compiler. Errors in syntax are uncovered but the code is not executed. For the other three automated techniques, the code is executed.

FIGURE 10-3
GUIDELINES FOR CONDUCTING A CODE WALKTHROUGH
Source: Adapted from Yourdon, 1989.

GUIDELINES FOR CONDUCTING A CODE WALKTHROUGH

1. Have the review meeting chaired by the project manager or chief programmer, who is also responsible for scheduling the meeting, reserving a room, setting the agenda, inviting participants, and so on.
2. The programmer presents his or her work to the reviewers. Discussion should be general during the presentation.
3. Following the general discussion, the programmer walks through the code in detail, focusing on the logic of the code rather than on specific test cases.
4. Reviewers ask to walk through specific test cases.
5. The chair resolves disagreements if the review team cannot reach agreement among themselves and assigns duties, usually to the programmer, for making specific changes.
6. A second walkthrough is then scheduled if needed.

The first such technique is **unit testing,** sometimes called module testing. In unit testing, each module (roughly a section of code that performs a single function) is tested alone in an attempt to discover any errors that may exist in the module's code. Yet because modules coexist and work with other modules in programs and systems, they must be tested together in larger groups. Combining modules and testing them is called **integration testing.** Integration testing is gradual. First you test the highest level, or coordinating module, and only one of its subordinate modules. The process assumes a typical structure for a program, with one highest-level, or main module and various subordinate modules referenced from the main module. Each subordinate module may have a set of modules subordinate to it, and so on, similar to an organization chart. Next, you continue testing subsequent modules at the same level until all subordinate to the highest-level module have been successfully tested together. Once the program has been tested successfully with the high-level module and all of its immediate subordinate modules, you add modules from the next level one at a time. Again, you move forward only when tests are successfully completed. If an error occurs, the process stops, the error is identified and corrected, and the test is redone. The process repeats until the entire program—all modules at all levels—is successfully integrated and tested with no errors.

System testing is a similar process, but instead of integrating modules into programs for testing, you integrate programs into systems. System testing follows the same incremental logic that integration testing does. Under both integration and system testing, not only do individual modules and programs get tested many times, so do the interfaces between modules and programs.

Current practice (as outlined above) calls for a top-down approach to writing and testing modules. Under a top-down approach, the coordinating module is written first. Then the modules at the next level are written, followed by the modules at the next level, and so on, until all of the modules in the system are done. Each module is tested as it is written. Because top-level modules contain many calls to subordinate modules, you may wonder how they can be tested if the lower-level modules haven't been written yet. The answer is **stub testing.** Stubs are two or three lines of code written by a programmer to stand in for the missing modules. During testing, the coordinating module calls the stub instead of the subordinate module. The stub accepts control and then returns it to the coordinating module.

System testing is more than simply expanded integration testing where you are testing the interfaces between programs in a system rather than testing the interfaces between modules in a program. System testing is also intended to demonstrate whether a system meets its objectives. The system test is typically conducted by information systems personnel led by the project team leader, although it can also be conducted by users under the guidance of information systems personnel.

The Testing Process

Up to this point, we have talked about an overall test plan and seven different types of tests for software applications. We haven't said very much about the process of testing itself. Two important things to remember about testing information systems are

1. The purpose of testing is confirming that the system satisfies requirements.
2. Testing must be planned.

Testing is not haphazard. You must pay attention to many different aspects of a system, such as response time, response to extreme data values, response to no input, response to heavy volumes of input, and so on. You must test anything (within resource constraints) that could go wrong or be wrong with a system.

Phase 4:
Implementation and Operation

Unit testing
Each module is tested alone in an attempt to discover any errors in its code; also called *module testing.*

Integration testing
The process of bringing together all of the modules that a program comprises for testing purposes. Modules are typically integrated in a top-down, incremental fashion.

System testing
The bringing together of all the programs that a system comprises for testing purposes. Programs are typically integrated in a top-down, incremental fashion.

Stub testing
A technique used in testing modules, especially where modules are written and tested in a top-down fashion, where a few lines of code are used to substitute for subordinate modules.

At a minimum, you should test the most frequently used parts of the system and as many other paths through the system as time permits. Planning gives analysts and programmers an opportunity to think through all the potential problem areas, list these areas, and develop ways to test for problems. As indicated previously, one part of a test plan is creating a set of test cases, each of which must be carefully documented. See Figure 10-4 for an outline of a test case description and summary.

A test case is a specific scenario of transactions, queries, or navigation paths that represent a typical, critical, or abnormal use of the system. A test case should be repeatable so that it can be rerun as new versions of the software are tested. This is important for all code, whether written in-house, developed by a contractor, or purchased. Test cases need to determine that new software works with other existing software with which it must share data. Even though analysts often do not do the testing, systems analysts, because of their intimate knowledge of applications, often make up or find test data. The people who create the test cases should not be the same people as those who coded and tested the system. In addition to a description of each test case, there must also be a summary of the test results, with an emphasis on how the actual results differed from the expected results. The testing summary will indicate why the results were different and what, if anything, should be done to change the software. Further, this summary will then suggest the need for retesting, possibly introducing new tests necessary to discover the source of the differences.

One important reason to keep such a thorough description of test cases and results is so that testing can be repeated for each revision of an application. Although new versions of a system may necessitate new test data to validate

NET SEARCH
Investigate further tools and methods for software testing. Visit http://www.prenhall.com/valacich to complete an exercise related to this topic.

FIGURE 10-4
TEST CASE DESCRIPTION AND SUMMARY FORM

Pine Valley Furniture Company
Test Case Description and Summary

Test Case Number: Date:
Test Case Description:

Program/Module Name:
Testing State:
Test Case Prepared By:
Test Administrator:
Description of Test Data:

Expected Results:

Actual Results:

Explanation of Differences between Actual and Expected Results:

Suggestions for Next Steps:

new features of the application, previous test data usually can and should be reused. Results from use of the test data with prior versions are compared to new versions to show that changes have not introduced new errors and that the behavior of the system, including response time, is no worse.

Phase 4:
Implementation and Operation

Acceptance Testing by Users

Once the system tests have been satisfactorily completed, the system is ready for **acceptance testing,** which is testing the system in the environment where it will eventually be used. Acceptance refers to the fact that users typically sign off on the system and "accept" it once they are satisfied with it. The purpose of acceptance testing is for users to determine whether the system meets their requirements. The extent of acceptance testing will vary with the organization and with the system in question. The most complete acceptance testing will include **alpha testing,** where simulated but typical data are used for system testing; **beta testing,** in which live data are used in the users' real working environment; and a system audit conducted by the organization's internal auditors or by members of the quality assurance group.

During alpha testing, the entire system is implemented in a test environment to discover whether or not the system is overtly destructive to itself or to the rest of the environment. The types of tests performed during alpha testing include the following:

Acceptance testing
The process whereby actual users test a completed information system, the end result of which is the users' acceptance of it.

Alpha testing
User testing of a completed information system using simulated data.

Beta testing
User testing of a completed information system using real data in the real user environment.

- *Recovery testing.* Forces the software (or environment) to fail in order to verify that recovery is properly performed.
- *Security testing.* Verifies that protection mechanisms built into the system will protect it from improper penetration.
- *Stress testing.* Tries to break the system (e.g., what happens when a record is written to the database with incomplete information or what happens under extreme online transaction loads or with a large number of concurrent users).
- *Performance testing.* Determines how the system performs on the range of possible environments in which it may be used (e.g., different hardware configurations, networks, operating systems, and so on); often the goal is to have the system perform with similar response time and other performance measures in each environment.

In beta testing, a subset of the intended users run the system in their own environments using their own data. The intent of the beta test is to determine whether the software, documentation, technical support, and training activities work as intended. In essence, beta testing can be viewed as a rehearsal of the installation phase. Problems uncovered in alpha and beta testing in any of these areas must be corrected before users can accept the system.

Installation

The process of moving from the current information system to the new one is called **installation.** All employees who use a system, whether they were consulted during the development process or not, must give up their reliance on the current system and begin to rely on the new system. Four different approaches to installation have emerged over the years:

Installation
The organizational process of changing over from the current information system to a new one.

- Direct
- Parallel
- Single location
- Phased

These four approaches are highlighted in Figure 10-5 and Table 10-4. The approach (or combination) an organization decides to use will depend on the scope and complexity of the change associated with the new system and the organization's risk aversion. In practice you will rarely choose a single strategy to the exclusion of all others; most installations will rely on a combination of two or more approaches. For example, if you choose a single location strategy, you have to decide how installation will proceed there and at subsequent sites. Will it be direct, parallel, or phased?

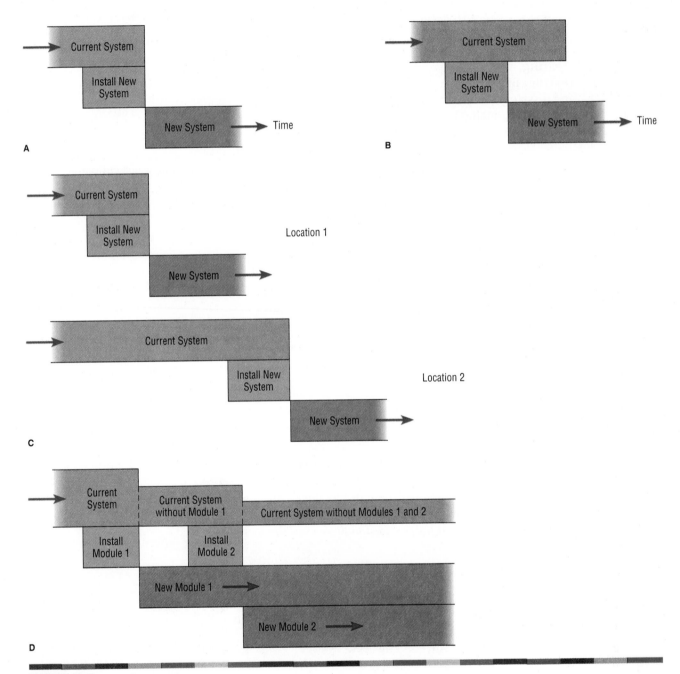

FIGURE 10-5

COMPARISON OF INSTALLATION STRATEGIES

(A) DIRECT INSTALLATION

(B) PARALLEL INSTALLATION

(C) SINGLE LOCATION INSTALLATION (WITH DIRECT INSTALLATION AT EACH LOCATION)

(D) PHASED INSTALLATION

Planning Installation

Each installation strategy involves converting not only software but also data and (potentially) hardware, documentation, work methods, job descriptions, offices and other facilities, training materials, business forms, and other aspects of the system. For example, it is necessary to recall or replace all the current system documentation and business forms, which suggests that the IS department must keep track of who has these items so that they can be notified and receive replacement items.

Phase 4:
Implementation and Operation

TABLE 10-4: Approaches to Information Systems Installation

Characteristics	Positive Aspects	Hazards/Risks
Direct Installation		
• Abrupt	• Low cost	• Operational errors have direct impact on users and organization
• "Cold turkey"	• High interest in making installation a success	• It may take too long to restore old system, if necessary
	• May be the only possible approach if new and existing systems cannot coexist in some form	• Time-consuming, and benefits may be delayed until whole system is installed
Parallel Installation		
• Old and new systems coexist	• New systems can be checked against old system	• Not all aspects of new system can be compared to old system
• Safe	• Impact of operational errors are minimized because old system is also processing all data	• Very expensive due to duplication of effort to run and maintain two systems
		• Can be confusing to users
		• May be a delay until benefits result
		• May not be feasible due to costs or system size
Single Location Installation		
• Pilot approach	• Learning can occur and problems fixed by concentrating on one site	• Burden on IS staff to maintain old and new systems
• Middle-of-the-road approach	• Limits potential harm and costs from system errors or failure to selected pilot sites	• If different sites require data sharing, extra programs need to be written to "bridge" the two systems
• May involve a series of single location installations	• Can use early success to convince others to convert to new system	• Some parts of organization get benefits earlier than other parts
• Each location may be branch office, factory, or department		
Phased Installation		
• Staged, incremental, gradual, based on system functional components	• Allows for system development also to be phased	• Old and new systems must be able to work together and share data, which likely will require extra programming to "bridge" the two systems
• Similar to bringing system out via multiple releases	• Limits potential harm and costs from system error or failure to certain business activities/functions	• Conversion is constant and may extend over a long period, causing frustration and confusion for users
	• Risk spread over time	
	• Some benefits can be achieved early	
	• Each phrase is small and more manageable	

Of special interest in the installation process is the conversion of data. Because existing systems usually contain data required by the new system, current data must be made error-free, unloaded from current files, combined with new data, and loaded into new files. Data may need to be reformatted to be consistent with more advanced data types supported by newer technology used to build the new system. New data fields may have to be entered in large quantities so that every record copied from the current system has all the new fields populated. Manual tasks, such as taking a physical inventory, may need to be done in order to validate data before they are transferred to the new files. The total data conversion process can be tedious. Furthermore, this process may require that current systems be shut off while the data are extracted so that updates to old data, which would contaminate the extract process, cannot occur.

Any decision that requires the current system to be shut down, in whole or in part, before the replacement system is in place must be done with care. Typically, off-hours are used for installations that require a lapse in system support. Whether a lapse in service is required or not, the installation schedule should be announced to users well in advance to let them plan their work schedules around outages in service and periods when their system support might be erratic. Successful installation steps should also be announced, and special procedures put in place so that users can easily inform you of problems they encounter during installation periods. You should also plan for emergency staff to be available in case of system failure so that business operations can be recovered and made operational as quickly as possible. Another consideration is the business cycle of the organization. Most organizations face heavy workloads at particular times of year and relatively light loads at other times. A well-known example is the retail industry, where the busiest time of year is the fall, right before the year's major gift-giving holidays. You wouldn't want to schedule installation of a new point-of-sale system to begin December 1 for a department store.

Planning for installation may begin as early as the analysis of the organization supported by the system. Some installation activities, such as buying new hardware, remodeling facilities, validating data to be transferred to the new system, and collecting new data to be loaded into the new system, must be done before the software installation can occur. Often the project team leader is responsible for anticipating all installation tasks and assigns responsibility for each to different analysts.

Each installation process involves getting workers to change the way they work. As such, installation should be looked at not as simply installing a new computer system, but as an organizational change process. More than just a computer system is involved—you are also changing how people do their jobs and how the organization operates.

Documenting the System

In one sense, every information systems development project is unique and will generate its own unique documentation. In another sense, though, system development projects are probably more alike than they are different. Each project shares a similar systems development life cycle, which dictates that certain activities be undertaken and each of those activities documented. Specific documentation will vary depending on the life cycle you are following, and the format and content of the documentation may be mandated by the organization you work for. Start developing documentation elements early, as the information needed is captured.

We can simplify the situation by dividing documentation into two basic types, system documentation and user documentation. **System documentation** records detailed information about a system's design specifications, its internal workings, and its functionality. System documentation can be further divided into internal and external documentation. **Internal documentation** is part of

---

System documentation
Detailed information about a system's design specifications, its internal workings, and its functionality.

Internal documentation
System documentation that is part of the program source code or is generated at compile time.

the program source code or is generated at compile time. **External documentation** includes the outcome of all of the structured diagramming techniques you have studied in this book, such as data flow and entity-relationship diagrams. **User documentation** is written or other visual information about an application system, how it works, and how to use it. Although not part of the code itself, external documentation can provide useful information to the primary users of system documentation—maintenance programmers. For example, data flow diagrams provide a good overview of a system's structure. In the past, external documentation was typically discarded after implementation, primarily because it was considered too costly to keep up-to-date, but today's CASE environment makes it possible to maintain and update external documentation as long as desired.

Whereas system documentation is intended primarily for maintenance programmers, user documentation is intended mainly for users. An organization may have definitive standards on system documentation, often consistent with CASE tools and the system development process. These standards may include the outline for the project dictionary and specific pieces of documentation within it. Standards for user documentation are not as explicit.

User Documentation

User documentation consists of written or other visual information about an application system, how it works, and how to use it. An excerpt of online user documentation for Microsoft Access 2000 appears in Figure 10-6. Notice that the documentation has hot links to the meaning of important terms. The documentation lists the item necessary to perform the task the user inquired about. The user controls how much of the help is shown by expanding or contracting sections ("About bound, unbound, and calculated controls" is expanded and "About ways to create controls" is contracted). Hypertext techniques, rare in online PC documentation five years ago, are now the rule rather than the exception.

Figure 10-6 represents the content of a reference guide, just one type of user documentation. Other types of user documentation include a quick reference guide, user's guide, release description, system administrator's guide, and acceptance sign-off. The reference guide consists of an exhaustive list of the system's functions and commands, usually in alphabetic order. Most online reference guides allow you to search by topic area or by typing in the first few letters of your keyword. Reference guides are very good for very specific information (as in Figure 10-6) but not as good for the broader picture of how you perform all the steps required for a given task. The quick reference guide provides essential information about operating a system in a short, concise format. Where computer resources are shared and many users perform similar tasks on the same machines (as with airline reservation or mail-order-catalog clerks), quick reference guides

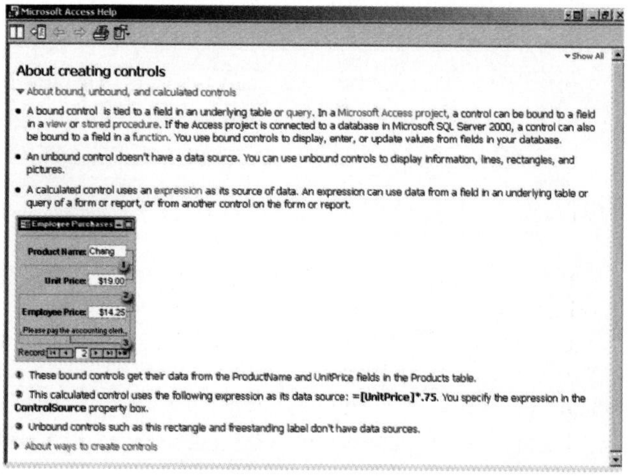

Phase 4:
Implementation
and Operation

External documentation
System documentation that includes the outcome of structured diagramming techniques such as data flow and entity-relationship diagrams.

User documentation
Written or other visual information about an application system, how it works, and how to use it.

FIGURE 10-6
ONLINE USER DOCUMENTATION FOR MICROSOFT ACCESS 2000

are often printed on index cards or as small books and mounted on or near the computer terminal. The purpose of a reference guide is to provide information on how users can use computer systems to perform specific tasks. The information in a user's guide is typically ordered by how often tasks are performed and how complex they are. Increasingly, software vendors are using Web sites to provide additional user guide content. Figure 10-7 shows the Microsoft Access 2000 help page, found by clicking the "Office on the Web" choice within the Help menu, and then selecting Access. Web-based documentation allows the vendor to provide more up-to-date reference material without issuing new software CDs.

Because most software is reissued as new features are added, a release description contains information about a new system release, including a list of complete documentation for the new release, features and enhancements, known problems and how they have been dealt with in the new release, and information about installation. A system administrator's guide is intended primarily for a particular type of user—those who will install and administer a new system—and contains information about the network on which the system will run, software interfaces for peripherals such as printers, troubleshooting, and setting up user accounts. Finally, the acceptance sign-off allows users to test for proper system installation and then signify their acceptance of the new system and its documentation with their signatures.

Preparing User Documentation

NET SEARCH
Investigate further opportunities to improve your documentation skills. Visit http://prenhall. com.valacich to complete an exercise related to this topic.

User documentation, regardless of its content or intended audience, is now most often delivered online in hypertext format. Regardless of format, user documentation is an investment that reduces training and consultation costs. As a future analyst, you need to consider the source of documentation, its quality, and whether its focus is on the information system's functionality or on the tasks the system can be used to perform.

The traditional source of user and system documentation has been the organization's information systems department. Until recently, the bulk of this documentation was system documentation, intended for analysts, programmers, and those who must maintain the system.

In today's end-user information systems environment, users interact directly with many computing resources, users have many options or querying capabilities from which to choose when using a system, and users are able to develop many local applications themselves. Analysts often serve as consultants for these local end-user applications. For end-user applications, the nature and purpose of documentation has changed from documentation intended for the maintenance programmer to documentation for the end user. Application-oriented documentation, whose purpose is to increase user understanding and utilization of the organization's computing resources, has also come to be important.

FIGURE 10-7
STRUCTURE OF AN ONLINE
REFERENCE USER'S GUIDE

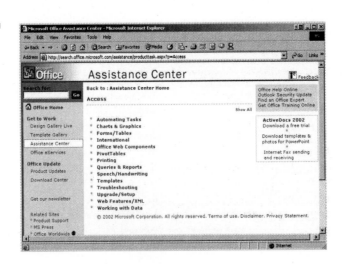

Although some of this user-oriented documentation continues to be supplied by the information systems department, much of it now originates with vendors and with users themselves.

Training and Supporting Users

Training and **support** are critical for the success of an information system. As the person whom the user holds responsible for the new system, you and other analysts on the project team must ensure that high-quality training and support are available. Training and support help people adequately use computer systems to do their primary work. Without proper training and the opportunity to ask questions and gain assistance/consultation when needed, users will misuse, underuse, or not use the information system you develop.

Although training and support can be talked about as if they are two separate things, in organizational practice the distinction between the two is not all that clear, as the two sometimes overlap. After all, both deal with learning about computing. It is clear that support mechanisms are also a good way to provide training, especially for intermittent users of a system. Intermittent or occasional system users are not interested in, nor would they profit from, typical user training methods. Intermittent users must be provided with "point of need support," specific answers to specific questions at the time the answers are needed. A variety of mechanisms, such as the system interface itself and online help facilities, can be designed to provide both training and support at the same time.

Support
Providing ongoing educational and problem-solving assistance to information system users. Support material and jobs must be designed along with the associated information system.

Training Information System Users

Many organizations tend to underinvest in computing skills training. It is true that some organizations institutionalize high levels of information system training, but many others offer no systematic training at all. Some argue that information systems departments are similar to hospitals: Both (1) are high-technology environments, (2) are staffed by well-educated professionals, (3) are capital intensive, and (4) have less than adequate "bedside manners" (Schrage, 1993). It has been shown in many studies that training users to be effective with the systems they have now can be a cost-effective way to increase productivity, even more so than installing hardware and software upgrades.

The type of necessary training will vary by type of system and expertise of users. The list of potential topics from which you must determine if training will be useful include the following:

- Use of the system (e.g., how to enter a class registration request)
- General computer concepts (e.g., computer files and how to copy them)
- Information system concepts (e.g., batch processing)
- Organizational concepts (e.g., FIFO inventory accounting)
- System management (e.g., how to request changes to a system)
- System installation (e.g., how to reconcile current and new systems during phased installation)

As you can see from this partial list, there are potentially many topics that go beyond simply how to use the new system. It may be necessary for you to develop training for users in other areas so that they will be ready, conceptually and psychologically, to use the new system. Some training, like concept training, should begin early in the project because this training can assist in convincing users of the need for system and organizational change.

Each element of training can be delivered in a variety of ways. Table 10-5 lists the most common training methods used by information system departments and

TABLE 10-5:   Types and Frequencies of Training Methods

Method of Training	Relative Frequency
Resident expert	51
Computer-aided instruction	12
Formal courses—several people taught at the same time	10
Software help components	10
Tutorials—one person taught at a time	7
Interactive training manuals—combination of tutorials and computer-aided instruction	5
External sources, such as vendors	5

their approximate relative frequency of use. Users primarily rely on just one of these delivery modes: More often than not, users turn to the resident expert and to fellow users for training. Users are more likely to turn to local experts for help than to the organization's technical support staff because the local expert understands both the users' primary work and the computer systems they use. Given their dependence on fellow users for training, it should not be surprising that end users describe their most common mode of computer training as "self-training."

One conclusion from the experience with user training methods is that an effective strategy for training on a new system is first to train a few key users and then to organize training programs and support mechanisms that involve these users to provide further training, both formal and on-demand. Often, training is most effective if you customize it to particular user groups, and the lead trainers from these groups are in the best position to do this.

Although individualized training is expensive and time-consuming, technological advances and decreasing costs have made this type of training more feasible. Newer training modes include videos, interactive television for remote training, multimedia training, online tutorials, and **electronic performance support systems (EPSS).** These may be delivered via videotapes, CD-ROMs, company intranets, and the Internet.

Electronic performance support systems are online help systems that go beyond simply providing help—they embed training directly into a software package. An EPSS may take on one or more forms: It can be an online tutorial, provide hypertext-based access to context-sensitive reference material, or consist of an expert system shell that acts as a coach. The main idea behind the development of an EPSS is that the user never has to leave the application to get the benefits of training. Users learn a new system or unfamiliar features at their own pace and on their own machines, without having to lose work time to attend remote group training sessions. Furthermore, this learning is on-demand when the user is most motivated to learn, because the user has a task to do. EPSS is sometimes referred to as "just-in-time knowledge."

One example of an EPSS with which you may be familiar is Microsoft's Office Assistant. Office Assistants, illustrated in Figure 10-8, are animated characters that come up on top of such applications as Access and Word. You ask questions, and the Office Assistants return answers that provide educational information, such as about graphics, examples, and procedures, as well as hypertext nodes for jumping to related help topics. Microsoft's Office Assistants communicate with the application you are running to see where you are, so you can determine, by reading the context-sensitive information, if what you want to do is possible from your present location. Some EPSS environments actually walk the user step-by-step through the task, coaching the user on what to do or allowing the user to get additional online assistance at any point.

Electronic performance support system (EPSS)
Component of a software package or application in which training and educational information is embedded. An EPSS may include a tutorial, expert system, and hypertext jumps to reference material.

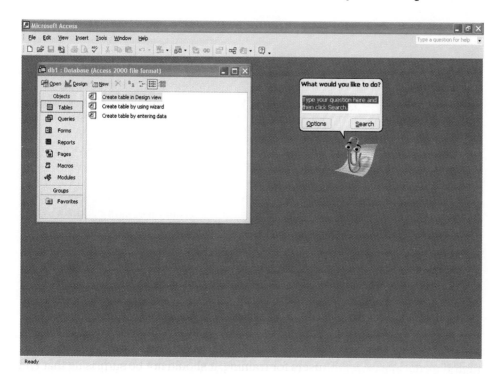

Phase 4:
Implementation
and Operation

FIGURE 10-8
MICROSOFT'S OFFICE ASSISTANT IN ACCESS XP

Training for information systems is increasingly being made available over both company intranets and the Internet. International Data Corporation projects that the online training industry will grow to a $28 billion business by the year 2001. Individual companies may prepare the training and make it available with the help of vendors who convert the content to work on the Internet. Alternatively, an organization may prepare its own training content using course authoring software. Still a third alternative is to access training provided by third-party vendors. Accessing training over the Internet has the potential to save companies, especially global businesses, thousands of dollars each year in training costs. Instead of having to send personnel off site for weeks and pay their travel expenses, companies can gain access to Internet training for lower cost, and personnel can get the training at their desks.

Supporting Information System Users

Users consider support to be extremely important. A 1993 J. D. Power and Associates survey found user support to be the number-one criterion contributing to user satisfaction with personal computing, cited by 26 percent of respondents as the most important factor (Schurr, 1993).

Although more support has been made available online, many organizations also provide institutionalized user support in the form of information centers and help desks. An **information center** is made up of a group of people who can answer questions and assist users with a wide range of computing needs, including the use of particular information systems. Information center staff might do the following:

Information center
An organizational unit whose mission is to support users in exploiting information technology.

- ◗ Install new hardware or software and set up user accounts
- ◗ Consult with users writing programs in fourth-generation languages
- ◗ Extract data from organizational databases onto personal computers
- ◗ Answer basic on-demand questions

- Provide a demonstration site for viewing hardware and software
- Work with users to submit system change requests

When you expect an organizational information center to help support a new system, you will likely want to train the information center staff as soon as possible on the system. Information center staff will need additional training periodically as new system features are introduced, new phases of a system are installed, or system problems and workarounds are identified. Even if new training is not required, information center staff should be aware of system changes, because such changes may result in an increase in demand for information center services. Personnel in an information center or help desk are typically drawn from the ranks of information systems workers or knowledgeable users in functional area departments.

Automating Support Vendors of purchased software may provide an information center (or help desk), but as they have shifted their offerings from primarily expensive mainframe packages to inexpensive off-the-shelf software, they find they can no longer bear the cost of providing the support for free. Most vendors now charge for support, and many have instituted 900 numbers and other automated support mechanisms or sell customers unlimited support for a given monthly or annual charge. Common methods for automating support include online support forums (on private Web sites or public Internet service providers, like America Online), bulletin board systems, on-demand fax, and voice-response systems. Online support forums provide users access to information on new releases, bugs, and tips for more effective usage. On-demand fax allows users to order support information through an 800 number and receive that information instantly over their fax machines. Finally, voice-response systems allow users to navigate option menus that lead to prerecorded messages about usage, problems, and workarounds. Organizations have established similar support mechanisms for systems developed or purchased by the organization. Internal e-mail, group support systems, and office automation can be used to support such capabilities within an organization.

Providing Support through a Help Desk Whether assisted by vendors or going it alone, the center of support activities for a specific information system in many organizations is the help desk. A **help desk** is an information systems department function, staffed by IS personnel, possibly part of the information center unit. The help desk is the first place users should call when they need assistance with an information system. The help desk staff either deals with the users' questions or refers the users to the most appropriate person.

Today, help desks are increasingly common as management comes to appreciate the special combination of technical skills and people skills needed to make good help desk staffers. Many software packages exist to automate the record keeping for a help desk. Records must be kept on each user contact, the content of the question or problem, and the status and resolution of the problem. Help desk managers use the software to track problems with different information systems, assess help desk personnel efficiency and effectiveness, and identify users who require training.

Help desk personnel (as well as the personnel in the more general information center) need to be good at communicating with users, by listening to their problems and intelligently communicating potential solutions. These personnel also need to understand the technology they are helping users with. It is crucial, however, that help desk personnel know when new systems and releases are being implemented and when users are being trained for new systems. Help desk personnel themselves should be well trained on new systems. One sure recipe for disaster is to train users on new systems but not train the help desk personnel these same users will turn to for their support needs.

Help desk
A single point of contact for all user inquiries and problems about a particular information system or for all users in a particular department.

Support Issues for the Analyst to Consider

Support is more than just answering user questions about how to use a system to perform a particular task or about the system's functionality. Support also consists of such tasks as providing for recovery and backup, disaster recovery, and PC maintenance; writing newsletters and offering other types of proactive information sharing; and setting up user groups. It is the responsibility of analysts for a new system to be sure that all forms of support are in place before the system is installed.

For medium to large organizations with active information system units, many of these issues are dealt with centrally. For example, users may be provided with backup software by the central information systems unit and a schedule for routine backup. Policies may also be in place for initiating recovery procedures in case of system failure. Similarly, disaster recovery plans are almost always established by the central IS unit. There may also be IS unit specialists in charge of composing and transmitting newsletters or overseeing automated bulletin boards and organizing user groups.

When all of these (and more) services are provided by central IS, you must follow the proper procedures to include any new system and its users in the lists of those to whom support is provided. You must design training for the support staff on the new system, and you must make sure that system documentation will be available to it. You must make the support staff aware of the installation schedule. You must also keep these people informed as the system evolves. Similarly, any new hardware and off-the-shelf software have to be registered with the central IS authorities.

When there is no official IS support function to provide support services, you must come up with a creative plan to provide as many services as possible. You may have to write backup and recovery procedures and schedules, and the users' departments may have to purchase and be responsible for the maintenance of their hardware. In some cases, software and hardware maintenance may have to be outsourced to vendors or other capable professionals. In such situations, user interaction and information dissemination may have to be more informal than formal: Informal user groups may meet over lunch or over a coffeepot rather than in officially formed and sanctioned forums.

Why Implementation Sometimes Fails

Despite the best efforts of the systems development team to design and build a quality system and to manage the change process in the organization, the implementation effort sometimes fails. Sometimes employees will not use the new system that has been developed for them, or if they do use the system, their level of satisfaction with it is very low.

The conventional wisdom that has emerged over the years is that at least two conditions are necessary for a successful implementation effort: management support of the system under development and the involvement of users in the development process. Yet, despite the support and active participation of management and users, information systems implementation still sometimes fails.

Let's review some insights about the implementation process:

- *Risk.* User involvement in the development process can help reduce the risk of failure when the system is complex, but it can also make failure more likely when there are financial and time constraints in the development process.

- *Commitment to the project.* The system development project should be managed so that the problem being solved is well understood and the system being developed to deal with the problem actually solves it.

▶ *Commitment to change.* Users and managers must be willing to change behaviors, procedures, and other aspects of the organization.

▶ *Extent of project definition and planning.* The more extensive the planning effort, the less likely is implementation failure.

▶ *Realistic user expectations.* The more realistic a user's early expectations about a new system and its capabilities, the more likely it is that the user will be satisfied with the new system and actually use it.

Whether a system implementation fails or succeeds also depends on your definition of success. Although there are many ways to determine if an implementation has been successful, the two most common and trusted are the extent to which the system is used and the user's satisfaction with the system. Whether a user will actually use a new system depends on several additional factors not already mentioned:

1. How relevant the system is to the work the user performs.
2. System ease of use and reliability.
3. User demographics, such as age and degree of computer experience.
4. The more users can do with a system and the more creative ways they can develop to benefit from the system, the more they will use it. Then the more people use the system, the more likely they are to find even more ways to benefit from the system.
5. The more satisfied the users are with the system, the more they will use it. The more they use it, the more satisfied they will be.

It should be clear that, as an analyst and as someone responsible for the successful implementation of an information system, you have more control over some factors than others. For example, you have considerable influence over the system's ease of use and reliability, and you may have some influence over the levels of support that will be provided for users of the system. You have no direct control over a user's demographics, relevance of the system, management support, or the urgency of the problem to the user. However, you can't ignore these factors. You need to understand these factors very well, because you will have to balance them with the factors you can change in your system design and implementation strategy. You may not be able to change a user's demographics or personal stake in a system, but you can help design the system and your implementation strategy with these factors in mind.

The factors mentioned so far are straightforward. For example, a lack of computer experience can make a user hesitant, inefficient, and ineffective with a system, leading to a system's not achieving its full potential benefit. If top management does not seem to care about the system, why should subordinates care? However, additional factors can be categorized as political, and may be more hidden, difficult to effect, and even unrelated to the system you are implementing, yet instrumental to the system's success.

The basis for political factors is that individuals who work in an organization have their own self-interested goals, which they pursue in addition to the goals of their departments and of their organizations. For example, people may act to increase their own power relative to that of their coworkers and, at other times, people will act to prevent coworkers with more power (such as bosses) from using that power or from gaining more. Because information is power, information systems are often seen as instruments of one's ability to influence and exert power. For example, an information system that provides information about the inventory and production capabilities of plant A to other plants may be seen as undesirable to managers in plant A, even if this information makes the company operate more efficiently overall. Users in plant A may

NET SEARCH
Investigate further why information system projects succeed and fail. Visit http:// prenhall. com.valacich to complete an exercise related to this topic.

resist participation in systems development activities, may continue (if possible) to use old systems and ignore the new one, or may initiate delaying tactics to stall the installation of the new system (such as asking for more studies and analysis work to "perfect" the system). Thus, you must attempt to understand the history and politics around an information system, and deal with negative political factors as well as the more objective and operational ones.

Project Closedown

In Chapter 2, you learned about the various phases of project management, from project initiation to closing down the project. If you are the project manager and you have successfully guided your project through all of the phases of the systems development life cycle presented so far in this book, you are now ready to close down your project. Although systems operation is just about to begin, the development project itself is over. As you will see in the following sections, maintenance can be thought of as a series of smaller development projects, each with its own series of project management phases.

As you recall from Chapter 2, your first task in closing down the project involves many different activities, from dealing with project personnel to planning a celebration of the project's ending. You will likely have to evaluate your team members, reassign most to other projects, and perhaps terminate others. As project manager, you will also have to notify all of the affected parties that the development project is ending and that you are now switching to operation and maintenance mode.

Your second task is to conduct postproject reviews with both your management and your customers. In some organizations, these postproject reviews follow formal procedures and may involve internal or electronic data processing (EDP) auditors. The point of a project review is to critique the project, its methods, its deliverables, and its management. You can learn many lessons to improve future projects from a thorough postproject review.

The third major task in project closedown is closing out the customer contract. Any contract that has been in effect between you and your customers during the project (or as the basis for the project) must be completed. This may involve a formal "signing-off" by the clients that your work is complete and acceptable. Maintenance activities will typically be covered under new contractual agreements. If your customer is outside of your organization, you will also likely negotiate a separate support agreement.

Some organizations conduct a postimplementation audit of a system shortly after it goes into operation, during or shortly after project closedown. A system audit may be conducted by a member of an internal audit staff, responsible for checking any data-handling procedure change in the organization. Sometimes a system audit is conducted by an outside organization, such as a management consulting firm or public accounting firm. The purpose of a system audit is to verify that a system works properly by itself and in combination with other systems. A system audit is similar to a system test but is done on a system in operation. A system audit not only checks that the operational system works accurately, but the audit is likely also to review the development process for the system. Such a process audit checks that sound practices were used to design, develop, and test the system. For example, a process audit will review the testing plan and summary of results. Errors found during an audit will generate requests for system maintenance, and in an extreme case, could force a system to cease operation.

As an analyst member of the development team, your job on this particular project ends during project closedown. You will likely be reassigned to another project dealing with some other organizational problem. During your career as a systems analyst, many of your job assignments will be to perform maintenance on existing systems. We cover this important part of the systems implementation and operation phase next.

Conducting Systems Maintenance

A significant portion of an organization's budget for information systems does not go to the development of new systems but to the maintenance of existing systems. We describe various types of maintenance, factors influencing the complexity and cost of maintenance, alternatives for managing maintenance, and the role of CASE during maintenance. Given that maintenance activities consume the majority of information systems–related expenditures, gaining an understanding of these topics will yield numerous benefits to your career as an information systems professional.

Types of Maintenance

Maintenance
Changes made to a system to fix or enhance its functionality.

Corrective maintenance
Changes made to a system to repair flaws in its design, coding, or implementation.

Adaptive maintenance
Changes made to a system to evolve its functionality to changing business needs or technologies.

Perfective maintenance
Changes made to a system to add new features or to improve performance.

Preventive maintenance
Changes made to a system to avoid possible future problems.

There are several types of maintenance that you can perform on an information system, as described in Table 10-6. By **maintenance,** we mean the fixing or enhancing of an information system. **Corrective maintenance** refers to changes made to repair defects in the design, coding, or implementation of the system. For example, if you purchase a new home, corrective maintenance would involve repairs made to things that had never worked as designed, such as a faulty electrical outlet or misaligned door. Most corrective maintenance problems surface soon after installation. When corrective maintenance problems surface, they are typically urgent and need to be resolved to curtail possible interruptions in normal business activities. Some corrective maintenance is due to incompatibilities between the new system and other information systems with which it must exchange data. Corrective maintenance adds little or no value to the organization; it simply focuses on removing defects from an existing system without adding new functionality.

Adaptive maintenance involves making changes to an information system to evolve its functionality to changing business needs or to migrate it to a different operating environment. Within a home, adaptive maintenance might be adding storm windows to improve the cooling performance of an air conditioner. Adaptive maintenance is usually less urgent than corrective maintenance because business and technical changes typically occur over some period of time. Contrary to corrective maintenance, adaptive maintenance is generally a small part of an organization's maintenance effort but does add value to the organization.

Perfective maintenance involves making enhancements to improve processing performance, interface usability, or to add desired, but not necessarily required, system features ("bells and whistles"). In our home example, perfective maintenance would be adding a new room. Many system professionals feel that perfective maintenance is not really maintenance but new development.

Preventive maintenance involves changes made to a system to reduce the chance of future system failure. An example of preventive maintenance might

TABLE 10-6: Types of Maintenance

Type	Description	Approximate Percentage of Maintenance Effort
Corrective	Repair design and programming errors	70
Adaptive	Modify system to environmental changes	10
Perfective	Evolve system to solve new problems or take advantage of new opportunities	15
Preventive	Safeguard system from future problems	5

be to increase the number of records that a system can process far beyond what is currently needed. In our home example, preventive maintenance could be painting the exterior to better protect the home from severe weather conditions. As with adaptive maintenance, both perfective and preventive maintenance are typically a much lower priority than corrective maintenance. Adaptive, perfective, and preventive maintenance activities can lead to corrective maintenance activities if not carefully designed and implemented.

The Cost of Maintenance

Information systems maintenance costs are a significant expenditure. For some organizations, as much as 80 percent of their information systems budget is allocated to maintenance activities (Pressman, 1992). This proportion has risen from roughly 50 percent 10 years ago due to the fact that many organizations have accumulated more and more older systems that require more and more maintenance. This means that you must understand the factors influencing the maintainability of systems. Maintainability is the ease with which software can be understood, corrected, adapted, and enhanced. Systems with low maintainability result in uncontrollable maintenance expenses.

Numerous factors influence the maintainability of a system. These factors, or cost elements, determine the extent to which a system has high or low maintainability. Of these factors, three are most significant: number of latent defects, number of customers, and documentation quality. The others—personnel, tools, and software structure—have noticeable, but less, influence.

- ● *Latent defects.* This is the number of unknown errors existing in the system after it is installed. Because corrective maintenance accounts for most maintenance activity, the number of latent defects in a system influences most of the costs associated with maintaining a system.

- ● *Number of customers for a given system.* In general, the greater the number of customers, the greater the maintenance costs. For example, if a system has only one customer, problem and change requests will come from only one source. Also, training, reporting errors, and support will be simpler. Maintenance requests are less likely to be contradictory or incompatible.

- ● *Quality of system documentation.* Without quality documentation, maintenance effort can increase exponentially. Quality documentation makes it easier to find code that needs to be changed and to understand how the code needs to be changed. Good documentation also explains why a system does what it does and why alternatives were not feasible, which saves wasted maintenance efforts.

- ● *Maintenance personnel.* In some organizations, the best programmers are assigned to maintenance. Highly skilled programmers are needed because the maintenance programmer is typically not the original programmer and must quickly understand and carefully change the software.

- ● *Tools.* Tools that can automatically produce system documentation where none exists can also lower maintenance costs. Also, tools that can automatically generate new code based on system specification changes can dramatically reduce maintenance time and costs.

- ● *Well-structured programs.* Well-designed programs are easier to understand and fix.

Since the mid-1990s, many organizations have taken a new approach to managing maintenance costs. Rather than develop custom systems internally or through contractors, they have chosen to buy packaged application software. Although vendors of packaged software charge an annual maintenance fee for updates, these charges are more predictable and lower than those for custom-developed systems. There still can be internal maintenance work when using packages. One major maintenance task is to make the packaged software compatible with other packages and internally developed systems with which it must cooperate. When new releases of the purchased package appear, maintenance may be needed to make all the packages continue to share and exchange data. Some companies are minimizing this effort by buying comprehensive packages, called enterprise resource planning (ERP) packages, which provide information services for a wide range of organizational functions (from human resources to accounting, manufacturing, and sales and marketing). Although the initial costs to install such ERP packages can be significant, they promise great potential for drastically reducing system maintenance costs.

Measuring Maintenance Effectiveness

Because maintenance can be so costly, it is important to measure its effectiveness. To measure effectiveness, you must measure these factors:

- Number of failures
- Time between each failure
- Type of failure

Mean time between failures (MTBF)
A measurement of error occurrences that can be tracked over time to indicate the quality of a system.

Measuring the number of and time between failures will provide you with the basis to calculate a widely used measure of system quality. This measure is referred to as the **mean time between failures (MTBF).** As its name implies, the MTBF measure shows the average length of time between the identification of one system failure until the next. Over time, you should expect the MTBF value to increase rapidly after a few months of use (and corrective maintenance) of the system. If the MTBF does not rapidly increase over time, it will be a signal to management that major problems exist within the system that are not being adequately resolved through the maintenance process.

A more revealing method of measurement is to examine the failures that are occurring. Over time, logging the types of failures provides a very clear picture of where, when, and how failures occur. For example, knowing that a system repeatedly fails logging new account information to the database when a particular customer is using the system can provide invaluable information to the maintenance personnel. Were the users adequately trained? Is there something unique about this user? Is there something unique about an installation that is causing the failure? What activities were being performed when the system failed?

Tracking the types of failures also provides important management information for future projects. For example, if a higher frequency of errors occurs when a particular development environment is used, such information can help guide personnel assignments, training courses, or the avoidance of a particular package, language, or environment during future development. The primary lesson here is that without measuring and tracking maintenance activities, you cannot gain the knowledge to improve or know how well you are doing relative to the past. To manage effectively and to improve continuously, you must measure and assess performance over time.

Controlling Maintenance Requests

Another maintenance activity is managing maintenance requests. From a management perspective, a key issue is deciding which requests to perform and which to ignore. Because some requests will be more critical than others, some method of prioritizing requests must be determined.

Figure 10-9 shows a flowchart that suggests one possible method for dealing with maintenance change requests. First, you must determine the type of request. If, for example, the request is an error—that is, a corrective maintenance request—then a question related to the error's severity must be asked. If the error is "very severe," then the request has top priority and is placed at the top of a queue of tasks waiting to be performed on the system. If, however, the error is considered "not very severe," then the change request can be categorized and prioritized based upon its type and relative importance. Categorization and prioritization may be done by the same review panel or board that evaluates new system requests.

If the change request is not an error, then you must determine whether the request is to adapt the system to technology changes and/or business requirements or to enhance the system with new business functionality. Adaptation requests will also need to be evaluated, categorized, prioritized, and placed in the queue. Enhancement-type requests must first be evaluated for alignment with future business and information systems' plans. If not aligned, the request will be rejected and the requester will be informed. If the enhancement is

Phase 4:
Implementation and Operation

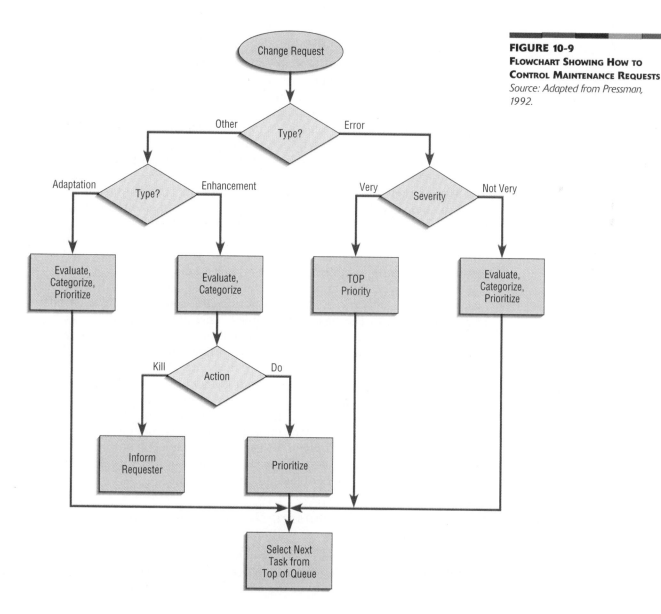

FIGURE 10-9
FLOWCHART SHOWING HOW TO CONTROL MAINTENANCE REQUESTS
Source: Adapted from Pressman, 1992.

aligned with business and information systems' plans, it can then be prioritized and placed into the queue of future tasks. Part of the prioritization process includes estimating the scope and feasibility of the change. Techniques used for assessing the scope and feasibility of entire projects should be used when assessing maintenance requests (see Chapter 3).

Managing the queue of pending tasks is an important activity. The queue of maintenance tasks is dynamic—growing and shrinking based upon business changes and errors. In fact, some lower-priority change requests may never be accomplished, because only a limited number of changes can be accomplished at a given time. In other words, changes in business needs between the time the request was made and when the task finally rises to the top of the queue may result in the request being deemed unnecessary or no longer important given current business directions.

Although each change request goes through an approval process as depicted in Figure 10-9, changes are usually implemented in batches, forming a new release of the software. It is too difficult to manage a lot of small changes. Further, batching changes can reduce maintenance work when several change requests affect the same or highly related modules. Frequent releases of new system versions may also confuse users if the appearance of displays, reports, or data entry screens changes.

Configuration Management

Configuration management
The process of assuring that only authorized changes are made to a system.

Baseline modules
Software modules that have been tested, documented, and approved to be included in the most recently created version of a system.

System librarian
A person responsible for controlling the checking out and checking in of baseline modules when a system is being developed or maintained.

Build routines
Guidelines that list the instructions to construct an executable system from the baseline source code.

A final aspect of managing maintenance is **configuration management**, which is the process of assuring that only authorized changes are made to a system. Once a system has been implemented and installed, the programming code used to construct the system represents the **baseline modules** of the system. These are the software modules for the most recent version of a system where each module has passed the organization's quality assurance process and documentation standards. A **system librarian** controls the baseline source code modules. If maintenance personnel are assigned to make changes to a system, they must first check out a copy of the baseline system modules because no one is allowed to modify the baseline modules directly. Only modules that have been checked out and have gone through a formal check-in process can reside in the library. Before any code can be checked back in to the librarian, the code must pass the quality control procedures, testing, and documentation standards established by the organization.

When various maintenance personnel working on different maintenance tasks complete each task, the librarian notifies those still working that updates have been made to the baseline modules. This means that all tasks being worked on must now incorporate the latest baseline modules before being approved for check-in. Following a formal process of checking modules out and in, a system librarian helps to assure that only tested and approved modules become part of the baseline system. It is also the librarian's responsibility to keep copies of all prior versions of all system modules, including the build routines. **Build routines** are guidelines needed to construct any version of the system that ever existed. It may be important to reconstruct old versions of the system if new ones fail, or to support users that cannot run newer versions on their computer system. Specialized packaged system software exists to support all of the functions of configuration management.

Role of CASE and Automated Development Tools in Maintenance

NET SEARCH
Investigate software for configuration management. Visit http://prenhall.com. valacich to complete an exercise related to this topic.

In traditional systems development, much of the time is spent on coding and testing. When software changes are approved, code is first changed and then tested. Once the functionality of the code is assured, the documentation and specification documents are updated to reflect system changes. Over time, the process of keeping all system documentation current can be a very tedious and time-consuming activity that is often neglected. This neglect makes future maintenance by the same or different programmers difficult.

A primary objective of using CASE and other automated tools for systems development and maintenance is to change radically how code and documentation are modified and updated. When using an integrated development environment, analysts maintain design documents such as data flow diagrams and screen designs, not source code. In other words, design documents are modified and then code generators automatically create a new version of the system from these updated designs. Also, because the changes are made at the design specification level, most documentation changes such as an updated data flow diagram will have already been completed during the maintenance process itself. One of the biggest advantages to using CASE, for example, is its benefits during system maintenance.

In addition to using general automated tools for maintenance, two special-purpose tools, reverse engineering and reengineering tools, are primarily used to maintain older systems that have incomplete documentation or that were developed prior to CASE use. These tools are often referred to as design recovery tools because their primary benefit is to create high-level design documents of a program by reading and analyzing its source code. When original documentation is not available, these tools can save considerable maintenance time by helping maintenance personnel to understand program and data structures.

Web Site Maintenance

All of the discussion on maintenance in this chapter applies to any type of information system, no matter on what platform it runs. There are, however, some special issues anad procedures needed for Web sites, due to their nature and operational status. These issues and procedures include:

- *24X7X365* Most Web sites are never purposely down. In fact an e-commerce Web site has the advantage of continuous operation. Thus, maintenance of pages and the overall site usually must be done without taking the site off-line. However, it may be necessary to lock out use of pages in a portion of a Web site while changes are made to those pages. This can be done by inserting a "Temporarily out of Service" notice on the main page of the section being maintained and disabling all links within that segment. Alternatively, references to the main page of the section can be temporarily rerouted to an alternative location where the current pages are kept while maintenance is performed to create new versions of those pages. The really tricky nuance is keeping the site consistent for a user during a session: that is, it can be confusing to a user to see two different versions of a page within the same online session. Browser caching functions may bring up an old version of a page even when that page changes during the session. One precaution against confusion is locking, as explained above. Another approach is to not lock a page being changed, but to include a date and time stamp of the most recent change. This gives the page visitor an indication of the change, which may reduce confusion.

- *Check for broken links* Arguably the most common maintenance issue for any Web site (besides changing the content of the site) is validating that links from site pages (especially for links that go outside the source site) are still accurate. Periodic checks need to be performed to make sure active pages are found from all links—this can be done via various software such as CyberSpider (www.CyberSpider.com), Doctor HTML (www.imagiware.com), or Linkbot (tetranetrsoftware.com)—note the irony of any URL in a textbook! In addition, periodic human checks need to be performed to make sure that the content found at a still-existing referenced page is still the intended content.

❶ *Re-registration* It may be necessary to re-register a Web site with search engines when the content of your site significantly changes. Re-registration may be necessary for visitors to find your site based on the new or changed content.

❶ *Future editions* One of the most important issues to address to insure effective Web site use is to avoid confusing visitors. Especially frequent visitors can be confused if the site is constantly changing. To avoid confusion, you can post indications of future enhancements to the site and, as with all information systems, you can batch changes to reduce the frequency of site changes.

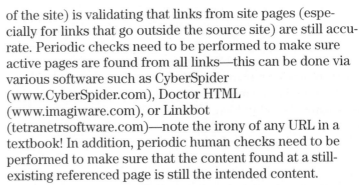

NET SEARCH
Investigate other tools and consulting services for assisting in Web site maintenance. Visit http://prenhall.com/valacich to complete an exercise related to this topic.

Maintaining an Information System at Pine Valley Furniture

Early one Saturday evening, Juanita Lopez, head of the manufacturing support unit of the Purchasing department at Pine Valley Furniture (PVF), was developing a new four-week production schedule to prepare purchase orders for numerous material suppliers. She was working on Saturday evening because she was leaving the next day for a long overdue two-week vacation to the Black Hills of South Dakota. Before she could leave, however, she needed to prepare purchase orders for all material requirements for the next four weeks so that orders could be placed during her absence. She was using the Purchasing Fulfillment System to assist her with this activity.

Midway through the process of developing a new production schedule, the system failed and could not be restarted. When she tried to restart the program, an error message was displayed on the terminal:

Data Integrity Error: Corrupt or missing supplier file.

Given that her plane for Rapid City left in less than 12 hours, Juanita had to figure out some way to overcome this catastrophic system error. Her first thought was to walk over to the offices of the information systems development group within the same building. When she did, she found no one there. Her next idea was to contact Chris Ryan, the project manager for the development and maintenance of the system. She placed a call to Chris's home and found that he was at the grocery store but would be home soon. Juanita left a message for Chris to call her ASAP at the office.

Within 30 minutes, Chris returned the call and was on his way into the office to help Juanita. Although it is not a common occurrence, this is not the first time that Chris has gone into the office to assist users when systems have failed during off-hours. Chris was looking forward to the day when he could handle all of these problems from home using a home PC and secure, high-speed Internet connection; he had been able to do this when all that was required was to scan data files for errors or issue a command to restore a database. Based on

Juanita's explanation of the problem and a few quick inquiries from his home PC, Chris decided he had better make the trip to the office where he had a variety of tools at his disposal.

PVF's systems development methodology for performing system maintenance is a formal process in which a user must first write a Systems Service Request (SSR) before maintenance is performed. After it is reviewed by the project manager, it is then forwarded to the Systems Priority Board. For catastrophic problems requiring instant correction so as not to delay normal business operations, the project manager has the discretion to circumvent the normal request process. After arriving on the scene, reviewing the error messages, and learning of Juanita's pending vacation, Chris believed that the failure with the Purchasing Fulfillment System was an instance where he would circumvent the normal maintenance process. His quick investigation suggested a failure in a new version of a system module that had been installed late on Friday afternoon. Chris noticed that the CASE tool records showed that this replacement module had not been tested against a standard test data set related to the type of work Juanita was doing, which made him suspect that this was the source of the problem. After patching the system to make it run, he would have to go back and document and test his changes so that they conformed to the development standards of PVF.

Over the next two hours, Chris used system backups to rebuild the supplier database. He reinstalled a previous version of the system's potentially faulty module (stored in the CASE library) that seemed more reliable, and then he quickly ran a test data set to check that the patches would hold the system together for now. He had to refresh himself on how to mount a tape cartridge on which the backup supplier data had been archived. Juanita was able to complete her task on time to easily make her flight the next morning. She thanked Chris for "going beyond the call of duty." Her appreciation made Chris feel good, but he was still uneasy. When making the "quick fix" on the system, he did not perform carefully planned testing nor did he confirm what had caused the error. He knew that the system could fail at any time. He did, however, have a copy of all of Juanita's actions just prior to the system failure. He hoped that through a careful review of those actions he would be able to learn why the system failed. But that would be a job for Monday morning.

ELECTRONIC COMMERCE APPLICATION: SYSTEMS IMPLEMENTATION AND OPERATION

Like many other analysis and design activities, systems implementation and operation of an Internet-based electronic commerce application is no different than the processes followed for other types of applications. In the last chapter, you read how Jim Woo and the Pine Valley Furniture development team transformed the conceptual data model for the WebStore into a set of normalized relations. Here we examine how the WebStore system was tested before it was installed and brought online.

Systems Implementation and Operation for Pine Valley Furniture's WebStore

The programming of all WebStore software modules is now complete. The programmers extensively tested each unique module, and it was now time to perform a system-wide test of the WebStore. In this section we examine how test cases were developed, how bugs were recorded and fixed, and how alpha and beta testing was conducted.

Developing Test Cases for the WebStore   To begin the system-wide testing process, Jim and the PVF development team developed test cases to examine every aspect of the system. Jim knew that system testing, like all other

Phase 4:
Implementation
and Operation

aspects of the SDLC, needed to be a very structured and planned process. Before opening the WebStore to the general public, every module and component of the system needed to be tested within a controlled environment. Based upon his experience in implementing other systems, Jim felt that they would need to develop approximately 150 to 200 separate test cases to fully examine the WebStore. To help focus the development of test cases and to assign primary responsibility to members of his team to specific areas of the system, Jim developed the following list of testing categories:

Simple functionality. Add to cart, list section, calculate tax, change personal data
Multiple functionality. Add item to cart and change quantity, create user account, and change address
Function chains. Add item to cart, check out, create user account, purchase
Elective functions. Returned items, lost shipments, item out-of-stock
Emergency/crisis. Missing orders, hardware failure, security attacks

The development group broke into five separate teams, each working to develop an extensive set of cases for each of the testing categories. Each team had one day to develop its test cases. Once developed, each team would lead a walkthrough so that everyone would know the totality of the testing process and to facilitate extensive feedback to each team so that the testing process would be as comprehensive as possible. To make this point, Jim stated, "What happens when a customer repeatedly enters the same product into the shopping cart? Can we handle that? What happens when the customer repeatedly enters and then removes a single product? Can we handle that? Although some of these things are unlikely to ever occur, we need to be confident that the system is robust given to any type of customer interaction. We must develop every test case necessary to give us confidence that the system will operate as intended, 24-7-365!"

A big part of successful system testing is to make sure that no information is lost and that all tests are described in a consistent way. To achieve this, Jim provided all teams with a standard form for documenting each case and for recording the results of each test. This form had the following sections:

Test Case ID
Category/Objective of Test
Description
System Version
Completion Date
Participant(s)
Machine Characteristics (processor, operating system, memory, browser, etc.)
Test Result
Comments

The teams also developed standard codes for each general type of test, and this was used to create the Test Case ID. For example, all tests related to "simple functionality" were given an ID with SF as a prefix and a number as the suffix—for example, SF001. The teams also developed standards for categorizing, listing objectives, and writing other test form contents. Establishing these standards assured that the testing process would be documented consistently.

Bug Tracking and System Evolution An outcome of the testing process is the identification of system bugs. Consequently, in addition to setting a standard method for writing and documenting test cases, Jim and the teams established several other rules to assure a smooth testing process. Experienced developers have long known that an accurate bug tracking process is essential for rapid troubleshooting and repair during the testing process. You can think of bug tracking as creating a "paper trail" that makes it much easier for programmers to find and repair the bug. To make sure that all bugs were documented in a similar way, the team developed a bug tracking form that had the following categories:

Bug Number (simple incremental number)
Test Case ID That Generated the Bug
Is the Bug Replicable?
Effects
Description
Resolution
Resolution Date
Comments

The PVF development team agreed that bug fixes would be made in batches, because all test cases would have to be redone every time the software was changed. Redoing all the test cases each time the software is changed is done to assure that in the process of fixing the bug, no other bugs were introduced into the system. As the system moves along in the testing process—as batches of bugs are fixed—the version number of the software is incremented. During the development and testing phases, the version is typically below the "1.0" first release version.

Alpha and Beta Testing the WebStore   After completing all system test cases and resolving all known bugs, Jim moved the WebStore into the alpha testing phase where the entire PVF development team as well as personnel around the company would put the WebStore through its paces. To motivate employees throughout the company to participate actively in testing the WebStore, several creative promotions and giveaways were held. All employees were given a T-shirt with the motto "I shop at the WebStore, do you?" Additionally, all employees were given $100 to shop at the WebStore and were offered a free lunch for their entire department if they found a system bug while shopping on the system. Also during alpha testing, the development team conducted extensive recovery, security, stress, and performance testing. Table 10-7 provides a sample of the types of tests performed.

After completing alpha testing, PVF recruited several of their established customers to help in beta testing the WebStore. As real-world customers used the system, Jim was able to monitor the system and fine-tune the servers for

Parallel installation
Running the old information system and the new one at the same time until management decides the old system can be turned off.

Direct installation
Changing over from the old information system to a new one by turning off the old system when the new one is turned on.

Phased installation
Changing from the old information system to the new one incrementally, starting with one or a few functional components and then gradually extending the installation to cover the whole new system.

TABLE 10-7:   Sample of Tests Conducted on the WebStore during Alpha Testing

Test Type	Sample of Tests Performed
Recovery	Unplug main server to test power backup system.
Security	Switch off main server to test the automatic switching to backup server.
Stress	Try to purchase without being a customer.
Performance	Try to examine server directory files both within the PVF domain and when connecting from an outside Internet service provider.
	Have multiple users simultaneously establish accounts, process purchases, add to shopping cart, remove from shopping cart, etc.
	Examine response time using different connection speeds, processors, memory, browsers, and other system configurations.
	Examine response time when backing up server data.

optimal system performance. As the system moved through the testing process, fewer and fewer bugs were found. After several days of "clean" usage, Jim felt confident that it was now time to open the WebStore for business.

WebStore Installation Throughout the testing process, Jim kept PVF management aware of each success and failure. Fortunately, because Jim and the development team followed a structured and disciplined development process, there were far more successes than failures. In fact, he was now confident that the WebStore was ready to go online and would recommend to PVF's top management that it was now time to "flip the switch" and let the world enter the WebStore.

Key Points Review

1. **Describe the process of coding, testing, and converting an organizational information system and outline the deliverables and outcomes of the process.**

    Coding is the process whereby the physical design specifications created by the design team are turned into working computer code by the programming team. Once coding has begun, the testing process can begin and proceed in parallel. As each program module is produced, it can be tested individually, then as part of a larger program, and then as part of a larger system. Installation is the process during which the current system is replaced by the new system. This includes conversion of existing data, software, documentation, and work procedures to those consistent with the new system. The deliverables and outcomes from coding, testing, and conversion are program and system code with associated documentation; testing plans, data, and results; and installation user guides, training plan, and conversion plan for hardware, software, data, and facilities.

2. **Apply four installation strategies: direct, parallel, single location, and phased installation.**

    Direct installation is the changing over from the old information system to a new one by turning off the old system when the new one is turned on. **Parallel installation** means running the old information system and the new one at the same time until management decides the old system can be turned off. **Single location installation** is trying out a new information system at one site and using the experience to decide if and how the new system should be deployed throughout the organization. **Phased installation** is changing from the old information system to the new one incrementally, starting with one or a few functional components and then gradually extending the installation to cover the whole new system. Often, a combination or

hybrid of these four strategies is employed for a particular information system installation. The approach (or combination) an organization decides to use depends on the scope and complexity of the change associated with the new system and the organization's risk aversion.

3. **List the deliverables for documenting the system and for training and supporting users.**

    The deliverables are system and user documentation; user training plan for classes and tutorials; user training materials, including computer-based training aids; and a user support plan, including such elements as a help desk, online help materials, bulletin boards, and other support mechanisms.

4. **Compare the many modes available for organizational information system training, including self-training and electronic performance support systems.**

    Training is most frequently done by a resident expert, usually another user in the same or similar department or job function. Other modes of training are computer-aided instruction, formal courses, software help components, tutorials, interactive training manuals, vendors and other external sources, and electronic performance support systems (EPSS). EPSS is the newest training approach, in which a component of a software package or application has training and educational information embedded in it. The EPSS may be a tutorial, an expert system shell, or hypertext jumps to context-sensitive reference material.

5. **Discuss the issues of providing support for end users.**

    Support is more than just answering user questions about how to use a system to perform a particular task or about the system's functionality. Support also consists of such tasks as providing for recovery and backup, disaster recovery, and PC maintenance; writing newsletters and offering other types of proactive information sharing; and setting up user groups. It is the responsibility of analysts for a new system to be sure that all forms of support are in place before the system is installed. For medium to large organizations with active information system units, many of these issues are dealt with centrally. When there is no official IS support function to provide support services, you must come up with a creative plan to provide as many services as possible. You may have to write backup and recovery procedures and schedules, and the users' departments may have to purchase and be responsible for the maintenance of their hardware. In some cases, software and hardware maintenance may have to be outsourced to vendors or other capable professionals.

6. **Explain why systems implementation sometimes fails.**

Even well-executed systems development projects, which have identified the right requirements and designed and installed a sound system, can fail. Research and experience have shown that management support of the system under development and the involvement of users in the development process can be important, but are not sufficient to achieve success. In addition, users must have commitment to the project and commitment to change. Poorly done project definition and planning can set up a project for failure. Users and developers also must have realistic and consistent expectations of the system's capabilities. Of course, the system must be relevant to the work the user performs. Also important are the ease of use and reliability of the system and user demographics, such as age and degree of computer experience. The more users can do with a system and the more creative ways they can develop to benefit from the system, the more they will use it. Then more use leads users to find even more ways to benefit from the system. The more satisfied the users are with the system, the more they will use it. The more they use it, the more satisfied they will be.

7. **Explain and contrast four types of maintenance.**

Corrective maintenance repairs flaws in a system's design, coding, or implementation. Adaptive maintenance implements changes to a system to evolve its functionality to changing business needs or technologies. Perfective maintenance adds new features or improves system performance. Preventive maintenance avoids possible future problems. Corrective maintenance is the most frequent, by far, and should occur primarily shortly after a system release is installed. Corrective maintenance must be made, and usually quickly. Adaptive maintenance also usually must be done. Some adaptive maintenance and all perfective and preventive maintenance are discretionary and must be categorized and prioritized.

8. **Describe several factors that influence the cost of maintaining an information system.**

The factors that influence the cost of maintaining an information system are (1) latent defects, which are unknown errors existing in the system after it is installed, (2) number of customers for a given system, (3) quality of system documentation, (4) maintenance personnel, (5) tools that can automatically produce system documentation where none exists, and (6) well-structured programs. The most influential of these are latent defects, number of customers, and quality of documentation. Also, some companies have adopted a strategy of using packaged application software, especially enterprise resource planning systems, to reduce maintenance costs.

Key Terms Checkpoint

Here are the key terms from the chapter. The page where each term is first explained is in parentheses after the term.

1. **Acceptance testing (p. 373)**
2. **Adaptive maintenance (p. 386)**
3. **Alpha testing (p. 373)**
4. **Baseline modules (p. 390)**
5. **Beta testing (p. 373)**
6. **Build routines (p. 390)**
7. **Configuration management (p. 390)**
8. **Corrective maintenance (p. 386)**
9. **Desk checking (p. 370)**
10. **Direct installation (p. 395)**
11. **Electronic performance support system (EPSS) (p. 380)**
12. **External documentation (p. 377)**
13. **Help desk (p. 382)**
14. **Information center (p. 381)**
15. **Inspections (p. 370)**
16. **Installation (p. 373)**
17. **Integration testing (p. 371)**
18. **Internal documentation (p. 376)**
19. **Maintenance (p. 386)**

20. **Mean time between failures (MTBF) (p. 388)**
21. **Parallel installation (p. 395)**
22. **Perfective maintenance (p. 386)**
23. **Phased installation (p. 395)**
24. **Preventive maintenance (p. 386)**
25. **Single location installation (p. 396)**
26. **Stub testing (p. 371)**
27. **Support (p. 379)**
28. **System documentation (p. 376)**
29. **System librarian (p. 390)**
30. **System testing (p. 371)**
31. **Unit testing (p. 371)**
32. **User documentation (p. 377)**

Match each of the key terms above with the definition that best fits it.

_____ 1. A testing technique in which participants examine program code for predictable language-specific errors.

_____ 2. A testing technique in which the program code is sequentially executed manually by the reviewer.

_____ 3. Component of a software package or application in which training and educational information is embedded. An EPSS may include a tutorial, expert system, and hypertext jumps to reference material.

_____ 4. Written or other visual information about an application system, how it works, and how to use it.

_____ 5. Changing over from the old information system to a new one by turning off the old system when the new one is turned on.

_____ 6. Changes made to a system to evolve its functionality to changing business needs or technologies.

_____ 7. Each module is tested alone in an attempt to discover any errors in its code; also called *module testing*.

_____ 8. The organizational process of changing over from the current information system to a new one.

_____ 9. A measurement of error occurrences that can be tracked over time to indicate the quality of a system.

_____ 10. System documentation that includes the outcome of structured diagramming techniques such as data flow and entity-relationship diagrams.

_____ 11. The process whereby actual users test a completed information system, the end result of which is the users' acceptance of it.

_____ 12. Guidelines that list the instructions to construct an executable system from the baseline source code.

_____ 13. Changes made to a system to avoid possible future problems.

_____ 14. Detailed information about a system's design specifications, its internal workings, and its functionality.

_____ 15. Running the old information system and the new one at the same time until management decides the old system can be turned off.

_____ 16. The process of bringing together all of the modules that a program comprises for testing purposes. Modules are typically integrated in a top-down, incremental fashion.

_____ 17. Changes made to a system to add new features or to improve performance.

_____ 18. An organizational unit whose mission is to support users in exploiting information technology.

_____ 19. A technique used in testing modules, especially modules that are written and tested in a top-down fashion, where a few lines of code are used to substitute for subordinate modules.

_____ 20. Software modules that have been tested, documented, and approved to be included in the most recently created version of a system.

_____ 21. A person responsible for controlling the checking out and checking in of baseline modules when a system is being developed or maintained.

_____ 22. Changing from the old information system to the new one incrementally, starting with one or a few functional components and then gradually extending the installation to cover the whole new system.

_____ 23. The process of assuring that only authorized changes are made to a system.

_____ 24. The bringing together of all the programs that a system comprises for testing purposes. Programs are typically integrated in a top-down, incremental fashion.

_____ 25. Changes made to a system to fix or enhance its functionality.

_____ 26. System documentation that is part of the program source code or is generated at compile time.

_____ 27. Providing ongoing educational and problem-solving assistance to information system users. Support material and jobs must be designed along with the associated information system.

_____ 28. User testing of a completed information system using real data in the real user environment.

_____ 29. Changes made to a system to repair flaws in its design, coding, or implementation.

_____ 30. Trying out a new information system at one site and using the experience to decide if and how the new system should be deployed throughout the organization.

_____ 31. User testing of a completed information system using simulated data.

_____ 32. A single point of contact for all user inquiries and problems about a particular information system or for all users in a particular department.

Review Questions

1. What are the deliverables from coding, testing, and installation?
2. Explain the testing process for code.
3. What are the four approaches to installation? Which is the most expensive? Which is the most risky? How does an organization decide which approach to use?
4. List and define the factors that are important to successful implementation efforts.
5. What is the difference between system documentation and user documentation?
6. List and define the various methods of user training.
7. Describe the delivery methods many vendors employ for providing support.
8. List the steps in the maintenance process and contrast them with the phases of the systems development life cycle.
9. What are the different types of maintenance and how do they differ?
10. Describe the factors that influence the cost of maintenance. Are any factors more important? Why?
11. What types of measurements must be taken to gain an understanding of the effectiveness of maintenance? Why is tracking mean time between failures an important measurement?
12. Describe the process for controlling maintenance requests. Should all requests be handled in the same way or are there situations when you should be able to circumvent the process? If so, when and why?
13. What is meant by configuration management? Why do you think organizations have adopted the approach of using a systems librarian?

Problems and Exercises

1. One of the difficult aspects of using the single location approach to installation is choosing an appropriate location. What factors should be considered in picking a pilot site?
2. You have been a user of many information systems including, possibly, a class registration system at your school, a bank account system, a word processing system, and an airline reservation system. Pick a system you have used and assume you were involved in the beta testing of that system. What criteria would you apply to judge whether this system were ready for general distribution?
3. Why is it important to keep a history of test cases and the results of those test cases even after a system has been revised several times?
4. What is the purpose of electronic performance support systems? How would you design one to support a word processing package? A database package?
5. Discuss the role of a centralized training and support facility in a modern organization. Given advances in technology and the prevalence of self-training and consulting among computing end users, how can such a centralized facility continue to justify its existence?
6. Is it good or bad for corporations to rely on vendors for computing support? List arguments both for and against reliance on vendors as part of your answer.
7. Suppose you were responsible for establishing a training program for users of Hoosier Burger's inventory control system (described in previous chapters). Which forms of training would you use? Why?
8. Your university or school probably has some form of microcomputer center or help desk for students. What functions does this center perform? How do these functions compare to those outlined in this chapter?
9. Suppose you were responsible for organizing the user documentation for Hoosier Burger's inventory control system (described in previous chapters). Write an outline that shows the documentation you would suggest be created, and generate the table of contents or outline for each element of this documentation.

10. In what ways is a request to change an information system handled differently from a request for a new information system?
11. What can a systems analyst do to reduce the frequency of corrective maintenance, the most common form of maintenance?
12. What other information should be collected on a System Service Request for maintenance as opposed to a System Service Request for a new system?
13. Briefly discuss how a systems analyst can manage each of the six cost elements of maintenance.
14. Suppose an information system were developed following a rapid application development approach like prototyping. How might maintenance be different than if the system had been developed following the traditional life cycle? Why?
15. Figure 10-1 shows an arrow going from systems implementation and operation to systems planning and selection. Explain the meaning of this arrow. What causes the transition from implementation and operation to planning and selection? How do maintenance activities within implementation and operation relate to the whole SDLC?

Discussion Questions

1. If possible, ask a systems analyst you know or have access to about implementation. Ask what the analyst believes is necessary for a successful implementation. Compare what the analyst believes are the factors that influence successful implementation to the factors discussed in this chapter.
2. Talk with people you know who use computers in their work. Ask them to get copies of the user documentation they rely on for the systems they use. Analyze the documentation. Would you consider it good or bad? Support your answer. Whether good or bad, how might you improve it?
3. Volunteer to work for a shift at a help desk at your school's computer center. Keep a journal of your experiences. What kind of users did you have to deal with? What kinds of questions did you get? Do you think help desk work is easy or hard? What skills are needed by someone in this position?
4. Let's say your professor has asked you to help him or her train a new secretary on how to prepare class notes for electronic distribution to class members. Your professor uses word processing software and an e-mail package to prepare and distribute the notes. Assume the secretary knows nothing about either package. Prepare a user task guide that shows the secretary how to complete this task.
5. Study an information systems department with which you are familiar or to which you have access. How does this department measure the effectiveness of systems maintenance? What specific metrics are used, and how are these metrics used to effect changes in maintenance practices? If there is a history of measurements over several years, how can changes in the measurements be explained?

Case Problems

1. Pine Valley Furniture

   Pine Valley Furniture's Customer Tracking System is now entering the final phases of the systems development life cycle. This is a busy time for the project team; project team members are busy coding, testing, training end users, and finalizing the system's documentation.

   To enhance your learning experience, Jim Woo has asked you to participate in the implementation process. As a result of this assignment, you have been attending all meetings concerning coding, testing, installation, end-user training, and documentation. During several of these meetings, the installation strategies, necessary end-

user training, and required documentation have been discussed. You recall from your recent systems analysis and design course that several options for each of these areas is available.

a. Locate a technical writing article on the Web. Briefly summarize this article.
b. Which installation options are available for the Customer Tracking System? Which would you recommend?
c. How can you determine if implementation has been successful?
d. What conditions are necessary for a successful implementation effort?

2. Hoosier Burger

The development of Hoosier Burger's information system is nearing completion. At recent project meetings, the types of testing, training, documentation, and installation strategies appropriate for Hoosier Burger have been discussed. The end users have little computer experience, and thus require several types of training and supporting documentation.

Fred Jones, one of the project's team members, has recommended using a direct installation approach. Because Hoosier Burger's information system is relatively small, he feels that the direct approach is the best installation strategy to pursue. The new system could be installed at the beginning of the week and be up and running for the weekend traffic. However, Paula Freeman does not like this idea. She feels that a parallel approach is more appropriate. She worries that if the system crashes it may be difficult to return to the old system.

a. What types of training will Hoosier Burger's end users need?
b. What types of documentation would you recommend for Hoosier Burger's end users?
c. Which installation strategy would you recommend pursuing?
d. What support issues should be considered?

3. Kitchen Plus

Kitchen Plus is one of the nation's top kitchenware producers. The company has several product lines, including cookware, small appliances, cutlery, and tableware. Over the last several years, the company has watched its market share begin to slip. Several information system projects were rushed into development, including an MRP project. Kitchen Plus executives felt that the new MRP system would enable the company to reduce escalating costs, especially in the areas of inventory, labor, and shipping.

The new MRP system has just been installed, and it is now time to close down the project. As project manager, one of your tasks is to evaluate project team members. Most of the team members performed well, and their work is exemplary. However, Joe McIntire's performance is a different story. Joe was asked to complete several tasks for this project, assisting with interviewing, diagramming, testing, and documentation preparation. Several end users called and complained about Joe's interrogation methods. Additionally, his diagrams were incomplete, sloppily done, and not completed by the due date. During the testing phase, Joe took a week off from work; Pauline Applegate was assigned to take over Joe's duties.

a. Identify the tasks involved in project close-down.
b. How would you evaluate Joe's performance?
c. What types of maintenance problems can you expect from this information system?
d. What factors will influence the maintainability of this system?

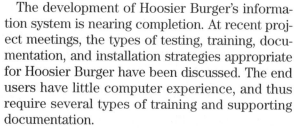

CASE: BROADWAY ENTERTAINMENT COMPANY, INC.

Designing a Testing Plan for the Customer Relationship Management System

Case Introduction

The students from St. Claire Community College are eager to get reactions to the initial prototype of MyBroadway, the Web-based customer relationship management system for Carrie Douglass, manager of the Centerville, Ohio, Broadway Entertainment Company (BEC) store. Based on the user dialogue design (see BEC Figure 8-1 at the end of Chapter 8), the team divided up the work of building the prototype. Tracey Wesley accepted responsibility for defining the database, starting from the normalized relations they developed (see the BEC case at the end of Chapter 9), and then populating with sample data the portions of the database that in production would come from the BEC store and corporate systems. Because John Whitman and Aaron Sharp had the most Microsoft Access experience on the team, they were responsible for developing the menus, forms, and displays for specific subsets of the customer pages. Missi Davies accepted the role of developing and managing the process of testing the system. The team decided that it would be desirable to have someone who was not directly involved in developing the system take responsibility for all aspects of testing. Testing would include tests conducted by Missi herself as well as use of the prototypes by BEC store employees and customers. While Tracey, John, and Aaron were developing the prototype, Missi began organizing the testing plan.

Preparing the Testing Plan

Now that the database and human interface elements of the system had been reasonably well outlined, Missi has a general understanding of the likely functionality and operation of MyBroadway. Because MyBroadway has a natural modular design, Missi believes that a top-down, modular approach can be used as a general process for testing.

Missi decides that the testing plan must involve a sequence of related steps, in which separate modules and then combinations of modules are used. Missi will test the individual modules and will initially test combinations of modules. But, once she has tested all the customer pages in major categories of functionality, and is reasonably sure they work, then it will be time to test the prototype with store employees and customers.

After studying the dialogue design for MyBroadway (BEC Figure 8-1 at the end of Chapter 8), she determines that there are five major modules that could be independently tested by employees and customers. These testing modules correspond to the five pages on the third level of the dialogue diagram: pages 1.1, 1.2, 1.3, 2.1, and 2.2. Missi decides, however, that such a piecemeal testing will be too confusing for employees and customers, but these modules can drive the internal testing process.

Before her independent tests of pages and modules can be done, she knows that she will have to test Tracey's work on building the database. Missi pulls out the E-R diagram the team developed for MyBroadway (see BEC Figure 6-1 at the end of Chapter 6). Data for the Comment, Pick, and Request entities will be entered by customers of MyBroadway. Product, Sale, and Rental data are fed from in-store BEC systems. So, Missi sees the steps of the testing plan emerging.

Missi determines that the first step is to have Tracey build the database and populate it with sample data simulating the feeds from in-store systems. For the prototype, the team won't actually build the feeds. Missi contacts Carrie to request printouts of data on products and sales and rental history. Missi asks Tracey to be prepared to test the loading of Product data first. It would appear that only Product data from the in-store data feeds are needed to test pages 1.1, 1.2, and 1.3. This approach will allow Missi to test Tracey's work on loading Product data before John needs a stable database for the Product Information module he is developing. Then Missi will test John's work on the Product Information module of MyBroadway. She can separately test Tracey's work of loading sales and rental history and Aaron's work to build the Rental Information module.

Missi will select some of the data Carrie provides to give to Tracey for use in Tracey's testing. Once she sees the data, Missi may create some other fictitious data to cover special circumstances (e.g., products that have missing field values or extreme field values). Missi will keep some of the data Carrie sends her for use in her own testing of the procedures Tracey builds for loading data. Missi decides that John and Aaron should do their own testing of pages until they believe that the pages are working properly. She wants them to use their own test data for this purpose, and she will develop separate test data when she looks at their pages.

Missi puts her ideas for a testing plan into a rough outline (BEC Figure 10-1). At this point, the outline does not show a time line or sequence of steps. Missi knows that she has to develop this time line before she can present the testing plan to her team members. After she reviews this outline and time line with the whole team, Aaron, who is maintaining the project schedule, will use Microsoft Project to enter these activities into the official project schedule Professor Tann, the instructor of the information systems project course at St. Claire, requires each team to maintain.

Preparing a Test Case

Among all the work Missi must do to manage the testing process, she must develop a detailed test case for each of the testing steps assigned to her in the overall testing plan in BEC Figure 10-1. Having not tested a new system before, Missi believes that she should develop one case so that she can get feedback from Professor Tann before proceeding to develop the rest of the cases. Missi decides to develop one of the easiest test cases first, and a suitable candidate for this appears to be the test for page 1.1.1, the entry by a customer of a new comment on a product.

Missi reviews the normalized relations the team developed for the database (BEC Figure 9-2 at the end of Chapter 9). Page 1.1.1 deals with entry of data in the Comment table. The Comment table in the database will contain data for:

- *Membership_ID.* Indicates who is entering the comment. Missi assumes this datum will be collected in a prior page, so this datum will not be an integral part of testing page 1.1.1.

- *Comment_Time_Stamp.* Indicates when the comment is entered. John's procedures for the page will have to get this computer system value and store it in the table, but the user won't deal with the data. Whether the time is captured correctly can be tested during the integration testing of pages 1.1.1 and 1.1.2, because comments on a product are reviewed in chronological order.

Collect Product, Sales, and Rental data from Carrie
    Select subset for Tracey
    Create extra sample data for Tracey's tests
    Design test documentation format for Tracey
    Give data to Tracey
    Create extra sample data for Missi's tests
Design test documentation format for John and Aaron
Design test documentation format for Missi's testing
Conduct module tests
    Conduct walkthrough with Tracey on Product data entry procedures
    Test Product data entry procedures from Tracey
        Provide Tracey with feedback on testing
    Test Product Information navigation pages from John and Aaron
    Test page 1.1.1
    Test page 1.1.2
    Do integration test of pages 1.1.1 and 1.1.2
    Test page 1.2.1
    Test page 1.3
    Conduct walkthrough with Tracey on Rental and Sales history data entry procedures
    Test Rental and Sales history data entry procedures from Tracey
        Provide Tracey with feedback on testing
    Test Rental Information navigation pages from John and Aaron
    Test pages 2.1.1 and 2.1.1.1
    Test page 2.2
        Provide John and Aaron with feedback on testing
Conduct tests of revisions made from initial module tests
    Cycle revising and retesting until system is ready for client testing
Conduct tests with employees
    Design employee feedback forms to collect test results
    Test pages 1 and below
    Test pages 2 and below
Review employee test results
    Cycle revising and retesting internally until system addresses employee concerns
Conduct tests with customers
    Design customer feedback forms to collect test results
    Test whole system with customers
Review customer test results
    Cycle revising and retesting internally until system addresses customer concerns
Summarize results of testing for inclusion in final report to professor and client

BEC FIGURE 10-1
OUTLINE OF TESTING PLAN FOR
MYBROADWAY

● *Product_ID.* Indicates on which product the customer is entering a comment. This value will be entered or selected on page 1.1, so pages 1.1 and 1.1.1 should be tested together. Missi asks John how he is designing page 1.1. John has decided to have users select the product through a series of questions so that only a valid product is used in pages 1.1.1 and 1.1.2. Thus, tests on this field will check that only existing products appear among the values for a customer to select.

● *Parent/Child?* Indicates whether the comment is entered by a parent or a child. This field has two values, and John says he will use a pair of radio buttons for entry of this field, and the choice of "Parent" will be the default on the page. Thus, there is no meaningful data entry test of this, but when doing the integration testing with page 1.1.2, Missi will check that the proper value was recorded. Missi makes a note to tell Tracey that John assumes that "Parent" is the default value for this field in the database.

● *Member_Comment.* Free-form textual comment entered by the customer about the selected product. Special cases of this field are that the customer submits the comment before entering a value for this field or enters a comment longer than can be stored. Because the team chose to use the Memo data type for this field, a truncated field is unlikely.

Besides making sure MyBroadway can handle alternative values for each field, Missi considers other important tests she learned about in her classes at St. Claire. Because the prototype will be used on only one PC in the store, there are no issues of concurrent use of the Comment data. Also, stress testing is of no concern for the same reason. Because Carrie has never indicated that security was a concern, no security testing is to be done. A type of recovery test would be to turn off the power to the PC during the entry of data. Performance testing is also of no concern with the limited usage prototype the team is building.

Case Summary

Missi is fairly confident that she has a good start on detailing the testing plan for MyBroadway. Once she can put her ideas on the example test case into a form for Professor Tann to review along with the testing outline, Missi will be ready to set a time line for testing. She needs to check with her team members to see when they think each module of the system will be ready for testing, and when they will need the instructions from her on how they should do their individual alpha testing.

Case Questions

1. Using Figure 10-4 as a guide, develop a test case description and summary form for the test Missi has designed for page 1.1.1.
2. Critically evaluate the outline of a testing plan Missi has developed (BEC Figure 10-1). Can you think of missing steps? Are there too many steps, and should some steps be combined?
3. The testing outline in BEC Figure 10-1 does not show sequencing of steps and what steps could be done in parallel. Develop a testing schedule from this figure, using Microsoft Project or other charting tool, that shows how you would suggest sequencing the testing steps. Make assumptions for the length of each testing step.

4. One element of the testing plan outline is cycling the alpha testing of pages as Missi tests and finds problems, and then the other team members have to rewrite the code for the problematic module. What guideline would you use to determine when to stop the alpha testing and release the module for beta testing with employees?
5. Design the test documentation format that Tracey, John, or Aaron is to use to explain how that student tested his or her code and the results of that testing.
6. Design the customer feedback form to be used to capture comments from customers during their use of the MyBroadway prototype. What measures of usability should be established for MyBroadway, and is a customer feedback form a sufficient means to capture all the usability measures you believe should be collected?
7. How would you suggest that the beta testing with customers be conducted? For example, should users use the system directly or through someone else at the keyboard and mouse? Should the customer be observed while he or she is using the system, either by a student team member watching or by videotape?

Appendix A
Object-Oriented Analysis and Design

After studying this appendix, you should be able to:

- ○ Define the following key terms: *association, class diagram, event, object, object class, operation, sequence diagram, state, state transition, Unified Modeling Language,* and *use case.*
- ○ Describe the concepts and principles underlying the object-oriented approach.
- ○ Develop a simple requirements model using use-case diagrams.
- ○ Develop a simple object model of the problem domain using class diagrams.
- ○ Develop simple requirements models using state and sequence diagrams.

The Object-Oriented Modeling Approach

In Chapters 1 through 10, you learned about traditional methods of systems analysis and design. You also learned how to use data flow diagrams and entity-relationship diagrams to model your system. There are environments where an object-oriented rather than a traditional approach is needed. This appendix covers the techniques and graphical diagrams that systems analysts use for object-oriented analysis and design. As with the traditional modeling techniques, the deliverables from project activities using object-oriented modeling are data flow and entity-relationship diagrams and repository descriptions. A major characteristic of these diagrams in object-oriented modeling is how tightly they are linked with each other. The object-oriented modeling approach provides several benefits, including:

1. The ability to tackle more challenging problem domains
2. Improved communication among users, analysts, designers, and programmers
3. Reusability of analysis, design, and programming results
4. Increased consistency among the models developed during object-oriented analysis, design, and programming

An object-oriented systems development life cycle consists of a progressively developing representation of a system component (what we will call an object) through the phases of analysis, design, and implementation. In the early stages of development, the model built is abstract, focusing on external qualities of the application system such as data structures, timing and sequence of processing operations, and how users interact with the system. As the model evolves, it becomes more and more detailed, shifting the focus to how the system will be built and how it should function.

In the analysis phase, a model of the real-world application is developed showing its important properties. It abstracts concepts from the application domain and describes what the intended system must do, rather than how it will be done. The model specifies the functional behavior of the system, independent of concerns relating to the environment in which it is to be finally

The original version of this appendix was written by Professor Atish P. Sinha.

implemented. In the design phase, the application-oriented analysis model is adapted and refined to suit the target implementation environment. That is followed by the implementation phase, where the design is implemented using a programming language and/or a database management system. The techniques and notations that are incorporated into a standard object-oriented language are called the **Unified Modeling Language,** or **UML.**

The techniques and notations within UML include:

- ▶ *Use cases*, which represent the functional requirements or the "what" of the system
- ▶ *Class diagrams*, which show the static structure of data and the operations that act on the data
- ▶ *State diagrams*, which represent dynamic models of how objects change their states in response to events
- ▶ *Sequence diagrams*, which represent dynamic models of interactions between objects

The Unified Modeling Language (UML) allows the modeler to specify, visualize, and construct the artifacts of software systems, as well as business models. It builds upon and unifies the semantics and notations of leading object-oriented methods, and has been adopted as an industry standard.

The UML notation is useful for graphically depicting object-oriented analysis and design models. It not only allows you to specify the requirements of a system and capture the design decisions but also promotes communication among key persons involved in the development effort. A developer can use an analysis or design model expressed in the UML notation to communicate with domain experts, users, and other stakeholders.

To represent a complex system effectively, the model developed needs to have a small set of independent views of the system. UML allows you to represent multiple views of a system using a variety of graphical diagrams, such as the use-case diagram, class diagram, state diagram, sequence diagram, and collaboration diagram. The underlying model integrates those views so that the system can be analyzed, designed, and implemented in a complete and consistent fashion.

We first show how to develop a use-case model during the requirements analysis phase. Next, we show how to model the static structure of the system using class and object diagrams. You then learn how to capture the dynamic aspects using state and sequence diagrams. Finally, we provide a brief description of component diagrams, which are generated during the design and implementation phases.

Use-Case Modeling

Use-case modeling is applied to analyze the functional requirements of a system. Use-case modeling is done in the early stages of system development, during the analysis phase, to help developers understand the functional requirements of the system without worrying about how those requirements will be implemented. The process is inherently iterative; developers need to involve the users in discussions throughout the model development process and finally come to an agreement on the requirements specification.

A use-case model consists of actors and use cases. An **actor** is an external entity that interacts with the system (similar to an external entity in data flow diagramming). It is someone or something that exchanges information with the system. A **use case** represents a sequence of related actions initiated by an actor; it is a specific way of using the system. An actor represents a role that a user can play. The actor's name should indicate that role. Actors help you to identify the use cases they carry out.

Unified Modeling Language (UML)
A notation that allows the modeler to specify, visualize, and construct the artifacts of software systems, as well as business models.

Actor
An external entity that interacts with the system (similar to an external entity in data flow diagramming).

Use case
A complete sequence of related actions initiated by an actor; it represents a specific way to use the system.

During the requirements analysis stage, the analyst sits down with the intended users of the system and makes a thorough analysis of what functions they desire from the system. These functions are represented as use cases. For example, a university registration system has a use case for class registration and another for student billing. These use cases, then, represent the typical interactions the system has with its users.

In UML, a use-case model is depicted diagrammatically as in Figure A-1. This **use-case diagram** is for a university registration system, which is shown as a box. Outside the box are four actors—Student, Registration Clerk, Instructor, and Bursar's Office—that interact with the system (shown by the lines touching the actors). An actor is shown using a stickman symbol with its name below. Inside the box are four use cases—*Class registration, Registration for special class, Prereq courses not completed,* and *Student billing*—which are shown as ellipses with their names inside. These use cases are performed by the actors outside the system.

A use case is always initiated by an actor. For example, *Student billing* is initiated by the Bursar's Office. A use case can interact with actors other than the one that initiated it. The *Student billing* use case, although initiated by the Bursar's Office, interacts with the Students by mailing them tuition invoices. Another use case, *Class registration,* is carried out by two actors, Student and Registration Clerk. This use case performs a series of related actions aimed at registering a student for a class.

The numbers on each end of the interaction lines indicate the number of instances of the use case with which the actor is associated. For example, the Bursar's Office causes many (*) *Student billing* use case instances to occur, each one for exactly one student.

A use case represents a complete functionality. You should not represent an individual action that is part of an overall function as a use case. For example, although submitting a registration form and paying tuition are two actions performed by users (students) in the university registration system, we do not show them as use cases because they do not specify a complete course of events; each of these actions is executed only as part of an overall function or use case. You can think of Submit registration form as one of the actions of the *Class registration* use case, and Pay tuition as one of the actions of the *Student billing* use case.

A use case may participate in relationships with other use cases. An extends relationship, shown in Microsoft Visio as a line with a hollow triangle pointing

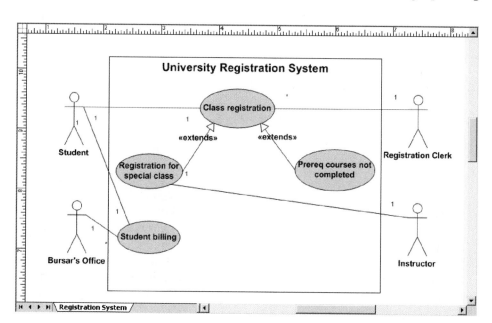

FIGURE A-1
USE-CASE DIAGRAM FOR A UNIVERSITY REGISTRATION SYSTEM DRAWN USING MICROSOFT VISIO

toward the extended use case and labeled with the << extends>> symbol, extends a use case by adding new behaviors or actions. In Figure A-1, for example, the *Registration for special class* use case extends the *Class registration* use case by capturing the additional actions that need to be performed in registering a student for a special class. Registering for a special class requires prior permission of the instructor, in addition to the other steps carried out for a regular registration. You may think of *Class registration* as the basic course, which is always performed—independent of whether the extension is performed or not—and *Registration for special class* as an alternative course, which is performed only under special circumstances.

Another example of an extends relationship is that between the *Prereq courses not completed* and *Class registration* use cases. The former extends the latter in situations where a student registering for a class has not taken the prerequisite courses.

(Note that Microsoft Visio does not use the latest, standard notation for an extends relationship. The current UML standard has a dashed line and a regular arrow head pointing in the opposite direction as shown in Figure A-1.)

Figure A-2 shows a use-case diagram for Hoosier Burger. The Customer actor initiates the *Order food* use case; the other actor involved is the Service Person. A specific scenario would represent a customer placing an order with a service person.

So far you have seen one kind of relationship, extends, between use cases. Another kind of relationship is included, which arises when one use case references another use case. An include relationship is also shown diagrammatically as a dashed line with a regular head pointed toward the use case that is being used; the line is labeled with the <<include>> symbol. In Figure A-2, for example, the include relationship between the *Reorder supplies* and *Track sales and inventory data* use cases implies that the former uses the latter while executing. Simply put, when a manager reorders supplies, the sales and inventory data are tracked. The same data are also tracked when management reports are produced, so there is another include relationship between the *Produce management reports* and *Track sales and inventory data* use cases.

FIGURE A-2
USE-CASE DIAGRAM FOR A
HOOSIER BURGER SYSTEM

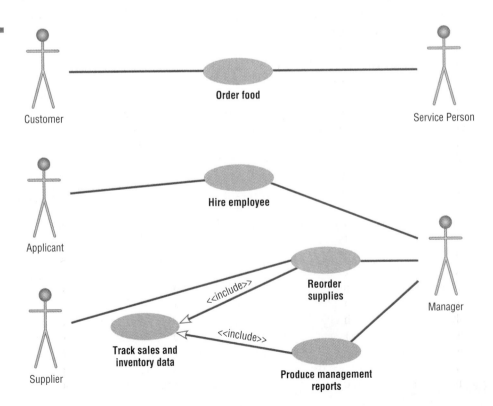

The *Track sales and inventory data* is a generalized use case, representing the common behavior among the specialized use cases, *Reorder supplies* and *Produce management reports*. When *Reorder supplies* or *Produce management reports* is performed, the entire *Track sales and inventory data* is used.

Object Modeling: Class Diagrams

In the object-oriented approach, we model the world in objects. An **object** is an entity that has a well-defined role in the application domain, and has state, behavior, and identity. An object is a concept, abstraction, or thing that makes sense in an application context. An object could be a tangible or visible entity (e.g., a person, place, or thing); it could be a concept or event (e.g., Department, Performance, Marriage, Registration, etc.); or it could be an artifact of the design process (e.g., User Interface, Controller, Scheduler, etc.).

An object has a state and exhibits behavior through operations that can examine or affect its state. The **state** of an object encompasses its properties (attributes and relationships) and the values those properties have, and its **behavior** represents how an object acts and reacts. An object's state is determined by its attribute values and links to other objects. An object's behavior depends on its state and the operation being performed. An operation is simply an action that one object performs upon another in order to get a response.

Consider the example of a student, Mary Jones, represented as an object. The state of this object is characterized by its attributes, say, name, date of birth, year, address, and phone, and the values these attributes currently have. For example, name is "Mary Jones," year is "junior," and so on. Its behavior is expressed through operations such as calc-gpa, which is used to calculate a student's current grade point average. The Mary Jones object, therefore, packages both its state and its behavior together.

All objects have an identity; that is, no two objects are the same. For example, if there are two Student instances with the same name and date of birth, they are essentially two different objects. Even if those two instances have identical values for all the attributes, the objects maintain their separate identities. At the same time, an object maintains its own identity over its life. For example, if Mary Jones gets married and changes her name, address, and phone, she will still be represented by the same object.

You can depict an **object class,** a set of objects that shares a common structure and a common behavior, graphically in a class diagram as in Figure A-3(A). A **class diagram** shows the static structure of an object-oriented model: the object classes, their internal structure, and the relationships in which they participate. In UML, a class is represented by a rectangle with three compartments separated by horizontal lines. The class name appears in the top compartment, the list of attributes in the middle compartment, and the list of operations in the bottom compartment of a box. The figure shows two classes, Student and Course, along with their attributes and operations.

Objects belonging to the same class may also participate in similar relationships with other objects; for example, all students register for courses and, therefore, the Student class can participate in a relationship called "registers-for" with another class called Course.

An **object diagram,** also known as instance diagram, is a graph of instances that are compatible with a given class diagram. In Figure A-3(B), we have shown an object diagram with two instances, one for each of the two classes that appear in Figure A-3(A). A static object diagram is an instance of a class diagram, providing a snapshot of the detailed state of a system at a point in time.

In an object diagram, an object is represented as a rectangle with two compartments. The names of the object and its class are underlined and shown in

Object
An entity that has a well-defined role in the application domain, and has state, behavior, and identity.

State
A condition that encompasses an object's properties (attributes and relationships) and the values those properties have.

Behavior
A manner that represents how an object acts and reacts.

Object class
A set of objects that shares a common structure and a common behavior.

Class diagram
A diagram that shows the static structure of an object-oriented model: the object classes, their internal structure, and the relationships in which they participate.

Object diagram
A graph of instances that are compatible with a given class diagram; also called an *instance diagram*.

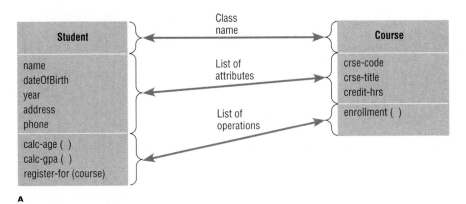

A

FIGURE A-3
UML CLASS AND OBJECT
DIAGRAMS
(A) CLASS DIAGRAM SHOWING
TWO CLASSES
(B) OBJECT DIAGRAM WITH TWO
INSTANCES

Mary Jones: Student
name=Mary Jones
dateOfBirth=4/15/1978
year=junior
. . .

:Course
crse-code=MIS385
crse-title=Database Mgmt
credit-hrs=3

B

Operation
A function or a service that is provided by all the instances of a class.

Encapsulation
The technique of hiding the internal implementation details of an object from its external view.

Association
A relationship between object classes.

Association role
The end of an association where it connects to a class.

Multiplicity
An indication of how many objects participate in a given relationship.

the top compartment using the following syntax: objectname:classname. The object's attributes and their values are shown in the second compartment. For example, we have an object called Mary Jones, who belongs to the Student class. The values of the name, dateOfBirth, and year attributes are also shown.

An **operation,** such as calc-gpa in Student (see Figure A-3[A]), is a function or a service that is provided by all the instances of a class. It is only through such operations that other objects can access or manipulate the information stored in an object. The operations, therefore, provide an external interface to a class; the interface presents the outside view of the class without showing its internal structure or how its operations are implemented. This technique of hiding the internal implementation details of an object from its external view is known as **encapsulation** or information hiding.

Representing Associations

An **association** is a relationship among object classes. As in the E-R model, the degree of an association relationship may be one (unary), two (binary), three (ternary), or higher (*n*-ary), as shown in Figure A-4. An association is depicted as a solid line between the participating classes. The end of an association where it connects to a class is called an **association role.** A role may be explicitly named with a label near the end of an association (see the "manager" role in Figure A-4[A]). The role name indicates the role played by the class attached to the end near which the name appears. For example, the manager role at one end of the Manages relationship implies that an employee can play the role of a manager.

Each role has a **multiplicity,** which indicates how many objects participate in a given association relationship. In a class diagram, a multiplicity specification is shown as a text string representing an interval (or intervals) of integers in the following format: lower-bound..upper-bound. The interval is considered to be closed, which means that the range includes both the lower and upper bounds. In addition to integer values, the upper bound of a multiplicity can be a star character (*), which denotes an infinite upper bound. If a single integer value is specified, it means that the range includes only that value.

The most common multiplicities in practice are 0..1, *, and 1. The 0..1 multiplicity indicates a minimum of 0 and a maximum of 1 (optional one), whereas *

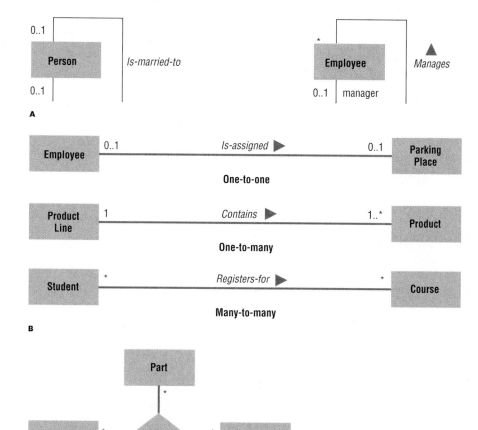

FIGURE A-4
EXAMPLES OF ASSOCIATION RELATIONSHIPS OF DIFFERENT DEGREES
(A) UNARY
(B) BINARY
(C) TERNARY

(or equivalently, 0..*) represents the range from 0 to infinity (optional many). A single 1 stands for 1..1, implying that exactly one object participates in the relationship (mandatory one).

Figure A-4(B) shows three binary relationships: Is-assigned (one-to-one), Contains (one-to-many), and Registers-for (many-to-many). A solid triangle next to an association name shows the direction in which the association is read. For example, the Contains association is read from Product Line to Product.

Figure A-4(C) shows a ternary relationship called Supplies among Vendor, Part, and Warehouse. As in an E-R diagram, we represent a ternary relationship using a diamond symbol and place the name of the relationship there.

The class diagram in Figure A-5(A) (known as a static structure chart in Microsoft Visio) shows binary associations between Student and Faculty, between Course and Course Offering, between Student and Course Offering, and between Faculty and Course Offering. The diagram shows that a student may have an adviser, whereas a faculty member may advise up to a maximum of 10 students. Also, although a course may have multiple offerings, a given course offering is scheduled for exactly one course. Notice that a faculty member can play two roles: instructor and adviser. Figure A-5(B) shows another example of a class diagram, that for a customer order, using standard UML notation.

Representing Generalization

In the object-oriented approach, you can abstract the common features (attributes and operations) among multiple classes, as well as the relationships they participate in, into a more general class. This is known as *generalization*. The

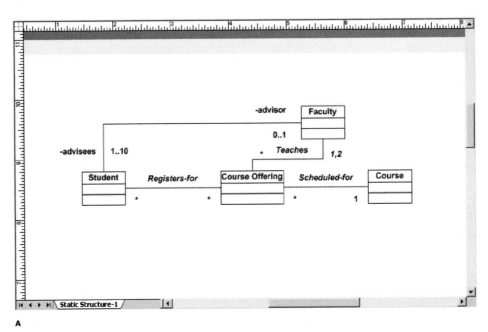

FIGURE A-5
EXAMPLES OF BINARY ASSOCIATION
RELATIONSHIPS
(A) UNIVERSITY EXAMPLE
(A STATIC STRUCTURE CHART IN
MICROSOFT VISIO)
(B) CUSTOMER ORDER EXAMPLE

classes that are generalized are called *subclasses*, and the class they are generalized into is called a *superclass*.

Consider the example shown in Figure A-6(A). There are three types of employees: hourly employees, salaried employees, and consultants. The features that are shared by all employees: empName, empNumber, address, dateHired, and printLabel, are stored in the Employee superclass, whereas the features peculiar to a particular employee type are stored in the corresponding subclass (e.g., hourlyRate and computeWages of Hourly Employee). A generalization path is shown as a solid line from the subclass to the superclass, with a hollow triangle at the end of, and pointing toward, the superclass. You can show a group of generalization paths for a given superclass as a tree with multiple branches connecting the individual subclasses, and a shared segment with a hollow triangle pointing toward the superclass. In Figure A-6(B), for instance, we have combined the generalization paths from Outpatient to Patient, and from Resident Patient to Patient, into a shared segment with a triangle pointing toward Patient.

You can indicate the basis of a generalization by specifying a discriminator next to the path. A discriminator shows which property of an object class is being abstracted by a particular generalization relationship. For example, in Figure A-6(A), we discriminate the Employee class on the basis of employment type (hourly, salaried, consultant).

A subclass inherits all the features from its superclass. For example, in addition to its own special features—hourlyRate and computeWages—the Hourly Employee subclass inherits empName, empNumber, address, dateHired, and printLabel from Employee. An instance of Hourly Employee will store values

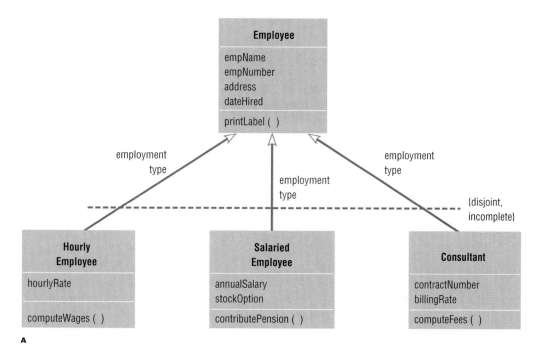

A

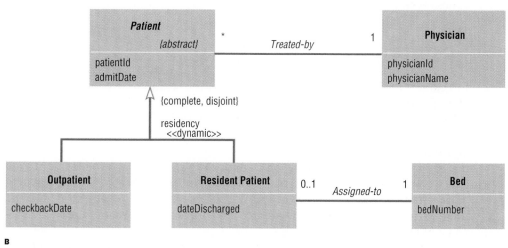

B

FIGURE A-6
EXAMPLES OF GENERALIZATION, INHERITANCE, AND CONSTRAINTS
(A) EMPLOYEE SUPERCLASS WITH THREE SUBCLASSES
(B) ABSTRACT PATIENT CLASS WITH TWO CONCRETE SUBCLASSES

for the attributes of Employee and Hourly Employee and, when requested, will apply the printLabel and computeWages operations.

Inheritance is one of the major advantages of using the object-oriented model. It allows code reuse: There is no need for a programmer to write code that has already been written for a superclass. The programmer writes only code that is unique to the new, refined subclass of an existing class. Proponents of the object-oriented model claim that code reuse results in productivity gains of several orders of magnitude.

Notice that in Figure A-6(B) the Patient class is in italics, implying that it is an abstract class. An **abstract class** is a class that has no direct instances, but whose descendants may have direct instances. A class that can have direct instances (e.g., Outpatient or Resident Patient) is called a **concrete class.**

The Patient abstract class participates in a relationship called Treated-by with Physician, implying that all patients, outpatients and resident patients

Abstract class
A class that has no direct instances, but whose descendants may have direct instances.

Concrete class
A class that can have direct instances.

alike, are treated by physicians. In addition to this inherited relationship, the Resident Patient class has its own special relationship called Assigned-to with Bed, implying that only resident patients may be assigned to beds. Semantic constraints among the subclasses can be expressed using the *complete*, *incomplete*, *disjoint*, and *overlapping* keywords. *Complete* means that every instance must be an instance of some subclass, whereas *incomplete* means that an instance may be an instance of the superclass only. *Disjoint* means that no instance can be an instance of more than one subclass at the same time, whereas *overlapping* allows concurrent participation in multiple subclasses.

Representing Aggregation

Aggregation
A *part-of* relationship between a component object and an aggregate object.

An **aggregation** expresses a Part-of relationship between a component object and an aggregate object. It is a stronger form of association relationship (with the added "part-of" semantics) and is represented with a hollow diamond at the aggregate end. For example, Figure A-7 shows a personal computer as an aggregate of CPU (up to four for multiprocessors), hard disks, monitor, keyboard, and other objects. Note that aggregation involves a set of distinct object instances, one of which contains or is composed of the others. For example, a Personal Computer object is related to (consists of) CPU objects, one of its parts. In contrast, generalization relates object classes: An object (e.g., Mary Jones) is simultaneously an instance of its class (e.g., Graduate Student) and its superclass (e.g., Student).

Dynamic Modeling: State Diagrams

In this section, we show you how to model the dynamic aspects of a system from the perspective of state transitions. In UML, state transitions are shown using state diagrams. A state diagram depicts the various state transitions or changes an object can experience during its lifetime, along with the events that cause those transitions.

A *state* is a condition during the life of an object during which it satisfies some condition(s), performs some action(s), or waits for some event(s). The state changes when the object receives some event; the object is said to undergo a **state transition.** The state of an object depends on its attribute values and links to other objects.

State transition
The changes in the attributes of an object or in the links an object has with other objects.

An **event** is something that takes place at a certain point in time. It is a noteworthy occurrence that triggers a state transition. Some examples of events are a customer places an order, a student registers for a class, a person applies for a loan, and a company hires a new employee. For the purpose of modeling, an event is considered to be instantaneous, though, in reality, it might take some time. A state, on the other hand, spans a period of time. An object remains in a particular state for some time before transitioning to another state. For exam-

Event
Something that takes place at a certain point in time; it is a noteworthy occurrence that triggers a state transition.

FIGURE A-7
EXAMPLE OF AGGREGATION

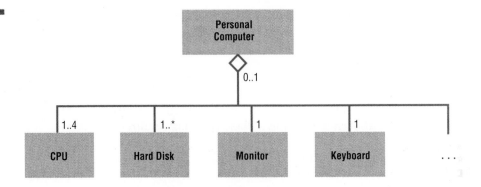

ple, an Employee object might be in the Part-time state (as specified in its employment-status attribute) for a few months, before transitioning to a Full-time state, based on a recommendation from the manager (an event).

In UML, a state is shown as a rectangle with rounded corners. In Figure A-8, for example, we have shown different states of a Student object, such as Inquiry, Applied, Approved, Rejected, and so on. This state diagram shows how the object transitions from an initial state (shown as a small, solid, filled circle) to other states, when certain events occur or when certain conditions are satisfied. When a new Student object is created, it is in its initial state. The event that created the object, Inquires, transitions it from the initial state to the Inquiry state. When a student in the Inquiry state submits an application for admission, the object transitions to the Applied state. The transition is shown as a solid arrow from Inquiry (the source state) to Applied (the target state), labeled with the name of the event, Submits application.

A transition may be labeled with a string consisting of the event name, parameters of the event, guard condition, and action expression. A transition, however, does not have to show all the elements; it shows only those relevant to the transition. For example, we label the transition from Inquiry to Applied with simply the event name. But, for the transition from Applied to Approved, we show the event name (evaluate), the guard condition (acceptable), and the action taken by the transition (mail approval letter). It simply means that an applicant is approved for admission if the admissions office evaluates the application and finds it acceptable. If acceptable, a letter of approval is mailed to the student. The guard condition is shown within square brackets, and the action is specified after the "/" symbol.

If the evaluate event results in a not-acceptable decision (another guard condition), a rejection letter is mailed (an action) and the Student object undergoes a state transition from the Applied to the Rejection state. It remains in that

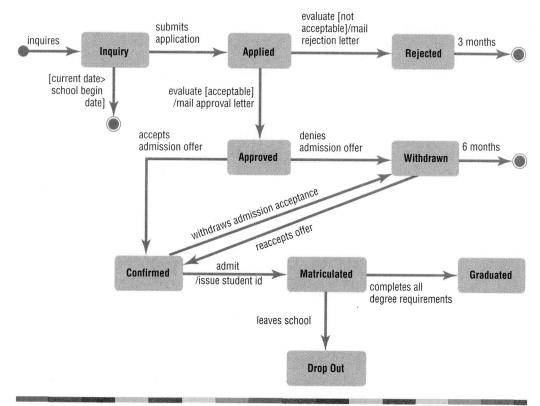

FIGURE A-8
State Diagram for the Student Object

state for three months, before reaching the final state. In the diagram, we have shown an elapsed-time event, three months, indicating the amount of time the object waits in the current state before transitioning. The final state is shown as a bull's eye: a small, solid, filled circle surrounded by another circle. After transitioning to the final state, the Student object ceases to exist.

Notice that the Student object may transition to the final state from two other states: Inquiry and Withdrawn. For the transition from Inquiry, we have not specified any event name or action, but we have shown a guard condition, current date > school begin date. This condition implies that the Student object ceases to exist beyond the first day of school unless, of course, the object has moved in the meantime from the Inquiry state to some other state.

The state diagram shown in Figure A-8 captures all the possible states of a Student object, the state transitions, the events or conditions that trigger those transitions, and the associated actions. For a typical student, it captures the student's sojourn through college, right from the time when he or she expressed an interest in the college until graduation.

Dynamic Modeling: Sequence Diagrams

In this section we show how to design some of the use cases we identified earlier in the analysis phase. A use-case design describes how each use case is performed by a set of communicating objects. In UML, an interaction diagram is used to show the pattern of interactions among objects for a particular use case. There are two types of interaction diagrams: sequence diagrams and collaboration diagrams. We show you how to design use cases using sequence diagrams.

A **sequence diagram** depicts the interactions among objects during a certain period of time. Because the pattern of interactions varies from one use case to another, each sequence diagram shows only the interactions pertinent to a specific use case. It shows the participating objects by their lifelines, and the interactions among those objects—arranged in time sequence—by the messages they exchange with one another.

Figure A-9 shows a sequence diagram for a scenario of the *Class registration* use case in which a student registers for a course that requires one or more prerequisite courses. The vertical axis of the diagram represents time and the horizontal axis represents the various participating objects. Time increases as we go down the vertical axis. The diagram has six objects, from an instance of Registration Window on the left, to an instance of Registration called "a New Registration" on the right. Each object is shown as a vertical dashed line called the lifeline; the lifeline represents the object's existence over a certain period of time. An object symbol—a box with the object's name underlined—is placed at the head of each lifeline.

A thin rectangle, superimposed on the lifeline of an object, represents an activation of the object. An **activation** shows the time period during which the object performs an operation, either directly or through a call to some subordinate operation. The top of the rectangle, which is at the tip of an incoming message, indicates the initiation of the activation, and the bottom its completion.

Objects communicate with one another by sending messages. A message is shown as a solid arrow from the sending object to the receiving object. For example, the checkIfOpen message is represented by an arrow from the Registration Entry object to the Course Offering object.

Messages could be of different types. Each type is indicated in a diagram by a particular type of arrowhead. A **synchronous message,** shown as a full, solid arrowhead, is one where the caller has to wait for the receiving object to complete executing the called operation before it itself can resume execution. An example of a synchronous message is checkIfOpen. When a Registration Entry

Sequence diagram
A depiction of the interactions among objects during a certain period of time.

Activation
The time period during which an object performs an operation.

Synchronous message
A type of message in which the caller has to wait for the receiving object to finish executing the called operation before it can resume execution itself.

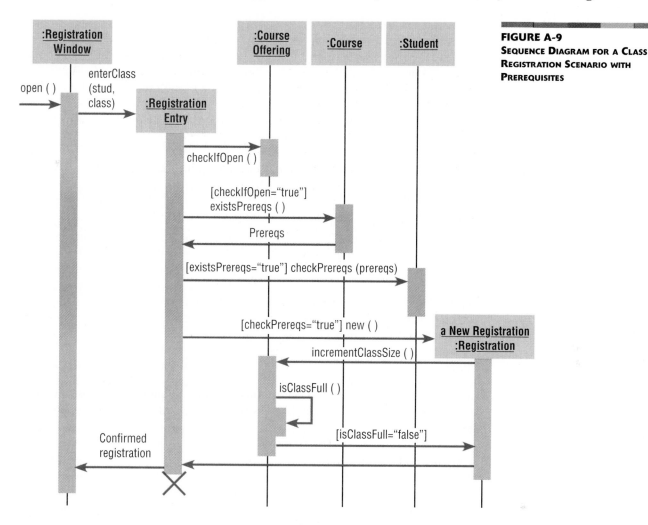

FIGURE A-9
SEQUENCE DIAGRAM FOR A CLASS REGISTRATION SCENARIO WITH PREREQUISITES

object sends this message to a Course Offering object, the latter responds by executing an operation called checkIfOpen (same name as the message). After the execution of this operation is completed, control is transferred back to the calling operation within Registration Entry with a return value, "true" or "false."

A synchronous message always has an associated return message. The message may provide the caller with some return value(s) or simply acknowledge to the caller that the operation called has been successfully completed. We have not shown the return for the checkIfOpen message; it is implicit. We have explicitly shown the return for the existsPrereqs message from Registration Entry to Course. The tail of the return message is aligned with the base of the activation rectangle for the existsPrereqs operation.

A **simple message** simply transfers control from the sender to the recipient without describing the details of the communication. In a diagram, the arrowhead for a simple message is drawn as a transverse tick mark. As we have seen, the return of a synchronous message is a simple message. The "open" message in Figure A-9 is also a simple message; it simply transfers control to the Registration Window object.

Simple message
A message that transfers control from the sender to the recipient without describing the details of the communication.

Designing a Use Case with a Sequence Diagram

Let's see how we can design use cases. We will draw a sequence diagram for an instance of the Class registration use case, one in which the course has prerequisites. Here's a description of this scenario:

1. Registration Clerk opens the registration window and enters the registration information (student and class).
2. Check if the class is open.
3. If the class is open, check if the course has any prerequisites.
4. If the course has prerequisites, then check if the student has taken all of those prerequisites.
5. If the student has taken those prerequisites, then register the student for the class, and increment the class size by one.
6. Check if the class is full; if not, do nothing.
7. Display the confirmed registration in the registration window.

In response to the "open" message from Registration Clerk (external actor), the Registration Window pops up on the screen and the registration information is entered. This creates a new Registration Entry object, which then sends a checkIfOpen message to the Course Offering object (representing the class the student wants to register for). There are two possible return values: true or false. In this scenario, the assumption is that the class is open. We have therefore placed a guard condition, checkIfOpen = "true," on the message existsPrereqs. The guard condition ensures that the message will be sent only if the class is open. The return value is a list of prerequisites; the return is shown explicitly in the diagram.

For this scenario, the fact that the course has prerequisites is captured by the guard condition, existsPrereqs = "true." If this condition is satisfied, the Registration Entry object sends a checkPrereqs message, with "prereqs" as an argument, to the Student object to determine if the student has taken those prerequisites. If the student has taken all the prerequisites, the Registration Entry object creates an object called a New Registration, which denotes a new registration.

Next, a New Registration sends a message called incrementClassSize to Course Offering in order to increase the class size by one. The incrementClassSize operation within Course Offering then calls upon isClassFull, another operation within the same object. Assuming that the class is not full, the isClassFull operation returns control to the calling operation with a value of "false." Next, the incrementClassSize operation completes and relinquishes control to the calling operation within a New Registration.

Finally, on receipt of the return message from a New Registration, the Registration Entry object destroys itself (the destruction is shown with a large X) and sends a confirmation of the registration to the Registration Window. Note that Registration Entry is not a persistent object; it is created on the fly to control the sequence of interactions and is deleted as soon as the registration is completed. In between, it calls several other operations within other objects by sequencing the following messages: checkIfOpen, existsPrereqs, checkPrereqs, and new.

Apart from the Registration Entry object, "a New Registration" is also created during the time period captured in the diagram. The messages that created these objects are represented by arrows pointing directly toward the object symbols. For example, the arrow representing the message called "new" is connected to the object symbol for "a New Registration." The lifeline of such an object begins when the message that creates it is received (the dashed vertical line is hidden behind the activation rectangle).

Moving to Design

When you move to design, you start with the existing set of analysis models and keep adding technical details. For example, you might add several interface classes to your class diagrams to model the windows that you will later implement using a GUI development tool such as Visual Basic or Visual C++. You would define all the operations in detail, specifying the procedures, signatures, and return values completely. If you decide to use a relational DBMS, you need to map the object classes and relationships to tables, primary keys, and foreign keys. The models generated during the design phase will therefore be much more detailed than the analysis models.

Figure A-10 shows a three-layered architecture, consisting of a User Interface package, a Business Objects package, and a Database package. The packages represent different generic subsystems of an information system. The dashed arrows represent the dependencies among the packages. For example, the User Interface package depends on the Business Objects package; the packages participate in a client-supplier relationship. If you make changes to some of the business objects, the interface (e.g., screens) might change.

A package consists of a group of classes. Classes within a package are cohesive. That is, they are tightly coupled. The packages themselves should be loosely coupled so that changes in one package do not affect the other packages a great deal. In the architecture of Figure A-10, the User Interface package contains all the windows, the Business Objects package contains the problem domain objects that you identified during analysis, and the Database package contains a Persistence class for data storage and retrieval. In the university registration system that we considered earlier, the User Interface package could include Microsoft Windows class libraries for developing different types of Windows. The Business Objects package would include all the domain classes, such as Student, Course, Course Offering, Registration, and so on. If you are using an SQL server, the classes in the Database package would contain operations for data storage, retrieval, and update (all using SQL commands).

During design, you would also refine the other analysis models. For example, you may need to show the interaction between a new window object you introduced during design and the other existing objects in a sequence diagram. Also, once you have selected a programming language for each of the operations shown in the sequence diagram, you should provide the exact names that you will be using in the program, along with the names of all the arguments.

In addition to the types of diagrams you have seen so far, two other types of diagrams—component diagrams and deployment diagrams—are pertinent during the design phase. A **component diagram** shows the software components or modules and their dependencies. For example, you can draw a component diagram showing the modules for source code, binary code, and executable code and their dependency relationships. Figure A-11 shows a component diagram for the university registration system. In this figure, three software components have been identified: Class Scheduler, Class Registration, and GUI (graphical user interface). The small circles in the diagram represent interfaces. The registration interface, for example, is used to register a student for a class, and the schedule update interface is used for updating a class schedule.

Another type of diagram, a deployment diagram (not illustrated), shows how the software components, processes, and objects are deployed into the physical architecture of the system. It shows the configuration of the hardware units (e.g., computers, communication devices, etc.) and how the software (components, objects, etc.) is distributed across the units. For example, a deployment diagram for the university registration system might show the topology of nodes in a client/server architecture, and the deployment of the Class Registration component to a Windows NT Server and of the GUI component to client workstations.

When the design phase is complete, you move on to the implementation phase where you code the system. If you are using an object-oriented programming language, translating the design models to code should be relatively straightforward. Programming of the system is followed by testing. The system is developed after going through multiple iterations, with each new iteration providing a better version of the system. The models that you developed during analysis, design, and implementation are navigable in both directions.

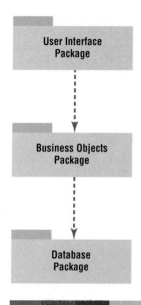

FIGURE A-10
AN EXAMPLE OF UML PACKAGES AND DEPENDENCIES

Component diagram
A diagram that shows the software components or modules and their dependencies.

FIGURE A-11
A COMPONENT DIAGRAM FOR
CLASS REGISTRATION

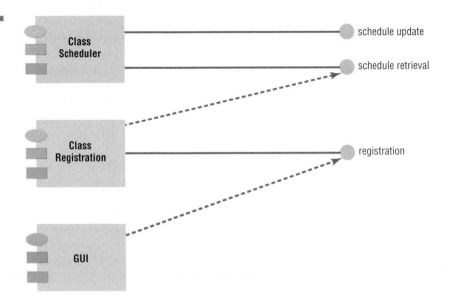

Key Points Review

1. **Define the following key terms: *association, class diagram, event, object, object class, operation, sequence diagram, state, state transition, Unified Modeling Language, and use case.***

   An association is a relationship between object classes. A class diagram shows the static structure of an object-oriented model: the object classes, their internal structure, and the relationships in which they participate. An event is something that takes place at a certain point in time; it is a noteworthy occurrence that triggers a state transition. An object is an entity that has a well-defined role in the application domain, and has state, behavior, and identity; and an object class is a set of objects that share a common structure and a common behavior. An operation is a function or a service that is provided by all the instances of a class. A sequence diagram depicts the interactions among objects during a certain period of time. A state encompasses an object's properties (attributes and relationships) and the values those properties have, and a state transition changes the attributes of an object or the links an object has with other objects. The Unified Modeling Language (UML) is a notation that allows the modeler to specify, visualize, and construct the artifacts of software systems, as well as business models. A use case is a complete sequence of related actions initiated by an actor; it represents a specific way to use the system.

2. **Describe the concepts and principles underlying the object-oriented approach.**

   The fundamental concept of the object-oriented approach is that we can model the world as a set of related objects with their associated states—attributes and behaviors. Through different uses of an object, the object's state changes. The internal implementation details of an object can be hidden from external view by the technique of encapsulation. A class of objects may be a superset or subset of other classes of objects, forming a generalization hierarchy or network of object classes. In this way an object may inherit the properties of the superclasses to which it is related. An object may also be a part of another more aggregate object.

3. **Develop a simple requirements model using use-case diagrams.**

   A use-case diagram consists of a set of related actions initiated by actors. A use case represents a complete functionality, not an individual action. A use case may extend another use case by adding new behaviors or actions. A use case may use another use case when one use case calls on another use case.

4. **Develop a simple object model of the problem domain using class diagrams.**

   A class diagram shows the static structure of object classes, their internal structure, and the relationships in which they participate. The structure of a class includes its name, attributes, and operations. Each object has an object identifier separate from its attributes. An object class can be either abstract (having no direct instances) or concrete (having direct instances). Object classes may have associations similar to relationships in the entity-relationship notation with multiplicity and degree. The end of an association where it connects to a class can be labeled with an association role. A class diagram can also show the generalization relationships between object classes, and subclasses can be complete or incomplete and disjoint or overlapping. In addition, a class diagram may show the aggregation association among object classes.

5. **Develop simple requirements models using state and sequence diagrams.**

   State and sequence diagrams show the dynamic behavior of a system. A state diagram shows all the possible states of an object and the events that trigger an object to transition from one state to another. A state transition occurs by changes in the attributes of an object or in the links an object has with other objects. An object begins in an initial state and ends in a final state. A state may have a guard condition, which checks that certain object properties exist before the transition may occur. When a state transition occurs, specified actions may take place. A sequence diagram depicts the interactions among objects during a certain period of time. The vertical axis of the diagram represents time (going down the axis), and the horizontal axis represents the various participating objects. Each object has a lifeline, which represents the object's existence over a certain period. Objects communicate with each other by sending messages. Among the different types of messages are synchronous (for which the caller has to wait for the receiving object to complete the called operation before the caller can resume execution) and simple (for which control is transferred from the sender to the recipient).

Key Terms Checkpoint

Here are the key terms from the appendix. The page where each term is first explained is in parentheses after the term.

1. **Abstract class** (p. 413)
2. **Activation** (p. 416)
3. **Actor** (p. 406)
4. **Aggregation** (p. 414)
5. **Association** (p. 410)
6. **Association role** (p. 410)
7. **Behavior** (p. 409)
8. **Class diagram** (p. 409)
9. **Component diagram** (p. 419)

10. **Concrete class** (p. 413)
11. **Encapsulation** (p. 410)
12. **Event** (p. 414)
13. **Multiplicity** (p. 410)
14. **Object** (p. 409)
15. **Object class** (p. 409)
16. **Object diagram** (p. 409)
17. **Operation** (p. 410)
18. **Sequence diagram** (p. 416)

19. **Simple message** (p. 417)
20. **State** (p. 409)
21. **State transition** (p. 414)
22. **Synchronous message** (p. 416)
23. **Unified Modeling Language (UML)** (p. 406)
24. **Use case** (p. 406)
25. **Use-case diagram** (p. 407)

Match each of the key terms with the definition that best fits it.

_____ 1. A diagram that depicts the use cases and actors for a system.

_____ 2. Something that takes place at a certain point in time; it is a noteworthy occurrence that triggers a state transition.

_____ 3. The time period during which an object performs an operation.

_____ 4. A set of objects that share a common structure and a common behavior.

_____ 5. A notation that allows the modeler to specify, visualize, and construct the artifacts of software systems, as well as business models.

_____ 6. A type of message in which the caller has to wait for the receiving object to finish executing the called operation before it can resume execution itself.

_____ 7. The end of an association where it connects to a class.

_____ 8. A graph of instances that are compatible with a given class diagram; also called an *instance diagram.*

_____ 9. The changes in the attributes of an object or in the links an object has with other objects.

_____ 10. An entity that has a well-defined role in the application domain, and has state, behavior, and identity.

_____ 11. A diagram that shows the software components or modules and their dependencies.

_____ 12. An external entity that interacts with the system (similar to an external entity in data flow diagramming).

_____ 13. A complete sequence of related actions initiated by an actor; it represents a specific way to use the system.

_____ 14. A "part-of" relationship between a component object and an aggregate object.

_____ 15. An indication of how many objects participate in a given relationship.

_____ 16. The technique of hiding the internal implementation details of an object from its external view.

_____ 17. A function or a service that is provided by all the instances of a class.

_____ 18. A condition that encompasses an object's properties (attributes and relationships) and the values those properties have.

_____ 19. A manner that represents how an object acts and reacts.

_____ 20. A diagram that shows the static structure of an object-oriented model: the object

classes, their internal structure, and the relationships in which they participate.

_____ 21. A relationship between object classes.

_____ 22. A class that has no direct instances, but whose descendants may have direct instances.

_____ 23. A class that can have direct instances.

_____ 24. A depiction of the interactions among objects during a certain period of time.

_____ 25. A message that transfers control from the sender to the recipient without describing the details of the communication.

Review Questions

1. Compare and contrast the object-oriented analysis and design models with structured analysis and design models.

2. Give an example of an abstract use case. Your example should involve at least two other use cases and show how they are related to the abstract use case.

3. Explain the use of association role for an association on a class diagram.

4. Give an example of generalization. Your example should include at least one superclass and three subclasses, and a minimum of one attribute and

one operation for each of the classes. Indicate the discriminator and specify the semantic constraints among the subclasses.

5. Give an example of aggregation. Your example should include at least one aggregate object and three component objects. Specify the multiplicities at each end of all the aggregation relationships.

6. Give an example of state transition. Your example should show how the state of the object undergoes a transition based on some event.

Problems and Exercises

1. The use-case diagram shown in Figure A-1 captures the Student billing function, but does not contain any function for accepting tuition payment from students. Revise the diagram to capture this functionality. Also, express some common behavior among two use cases in the revised diagram by using "include" relationships.

2. Suppose that the employees of the university are not billed for tuition. Their spouses do not get a full-tuition waiver, but pay for only 25 percent of the total tuition. Extend the use-case diagram of Figure A-1 to capture these situations.

3. Draw a class diagram, showing the relevant classes, attributes, operations, and relationships for the following situation (if you believe that you need to make additional assumptions, clearly state them for each situation).

A laboratory has several chemists who work on one or more projects. Chemists also may use certain kinds of equipment on each project. Attributes of Chemist include name and phoneNo. Attributes

of Project include projectName and startDate. Attributes of Equipment include serialNo and cost. The organization wishes to record assignDate—that is, the date when a given equipment item was assigned to a particular chemist working on a specified project—as well as totalHours—that is, the total number of hours the chemist has used the equipment for the project. The organization also wants to track the usage of each type of equipment by a chemist. It does so by computing the average number of hours the chemist has used that equipment on all assigned projects. A chemist must be assigned to at least one project and one equipment item. A given equipment item need not be assigned, and a given project need not be assigned either a chemist or an equipment item.

4. An organization has been entrusted with developing a Registration and Title system that maintains information about all vehicles registered in a particular state. For each vehicle that is regis-

tered with the office, the system has to store the name, address, telephone number of the owner, the start date and end date of the registration, plate information (issuer, year, type, and number), sticker (year, type, and number), and registration fee. In addition, the following information is maintained about the vehicles themselves: the number, year, make, model, body style, gross weight, number of passengers, diesel powered (yes/no), color, cost, and mileage. If the vehicle is a trailer, diesel powered and number of passengers are not relevant. For travel trailers, the body number and length must be known. The system needs to maintain information on the luggage capacity for a car, maximum cargo capacity and maximum towing capacity for a truck, and horsepower for a motorcycle. The system issues registration notices to owners of vehicles whose registrations are due to expire after two months. When the owner reviews the registration, the system updates the registration information on the vehicle.

a. Develop a static object model by drawing a class diagram that shows all the object classes, attributes, operations, relationships, and multiplicities. For each operation, show its argument list.

b. Draw a state diagram that captures all the possible states of a Vehicle object, right from the time the vehicle was manufactured until it goes to the junkyard. In drawing the diagram, you may make any necessary assumptions, as long as they are realistic.

c. Select any state or event from the high-level state diagram that you have drawn and show its fine structure (substates and their transitions) in a lower-level diagram.

d. One of the use cases for this application is "Issue registration renewal notice," which is performed once every day. Draw a sequence diagram, in generic form, showing all possible object interactions for this use case.

Appendix B
Rapid Application Development and CASE Tools

After studying this appendix, you should be able to:

- Explain the Rapid Application Development (RAD) approach and how it differs from traditional approaches to information systems development.
- Describe the systems development components essential to RAD.
- List and describe the typical components of a comprehensive CASE environment.
- Describe how CASE tools can be used to support RAD.
- Describe visual and emerging development tools and how they can be used to support RAD.
- Discuss the conceptual pillars that support the RAD approach.
- Explain the advantages and disadvantages of RAD as an exclusive systems development methodology.

Rapid Application Development

As a systems analyst, you may find yourself at a company that needs to install an information system quickly. Although the traditional approaches to design that we've talked about are proven, they may not meet all needs all the time. In this appendix, we cover Rapid Application Development, a technique for developing systems quickly. **Rapid Application Development (RAD)** is an approach to developing information systems that promises better and cheaper systems and more rapid deployment. Five key factors work together in the RAD process. We discuss each of these factors in this appendix:

1. Extensive user involvement
2. Joint Application Design sessions
3. Prototyping
4. Integrated CASE tools
5. Code generators

RAD grew out of the convergence of two trends: the increased speed and turbulence of doing business in the late 1980s and early 1990s, and the ready availability of high-powered computer-based tools to support systems development and easy maintenance. As the conditions of doing business in a changing, competitive global environment became more turbulent, management in many organizations began to question if it made sense to wait two to three years to develop systems that would be obsolete upon completion. On the other hand, CASE tools and prototyping software were diffusing throughout organizations, making it relatively easy for end users to see what their systems would look like before they were completed. Why not use these tools to address the problems of developing systems more productively in a rapidly changing business environment? So RAD was born.

The creation of RAD did not lead to its immediate adoption, however. To some, RAD was not an approach that could be taken seriously for enterprise-wide information systems. But increasing disenchantment with traditional systems development methods and their long development times led more and more

> **Rapid Application Development (RAD)** Systems development methodology created to radically decrease the time needed to design and implement information systems. RAD relies on extensive user involvement, Joint Application Design sessions, prototyping, integrated CASE tools, and code generators.

firms to consider RAD seriously. The ready availability of increasingly powerful software tools created to support RAD also increased interest in this approach. RAD is becoming more and more a legitimate way to develop information systems.

The Process of Developing an Application Rapidly

As you will see in this appendix, RAD is not a single methodology but a general strategy of developing information systems. As such, there are many different approaches to developing applications rapidly. Some are special life cycles, such as the Martin RAD life cycle. Others focus more on specific software tools and visual development environments that enable the process of rapidly developing and deploying applications. Whatever strategy is taken, however, the goal remains the same: to analyze a business problem rapidly, to design a viable system solution through intense cooperation between users and developers, and to get the finished application into the hands of users quickly, saving time, money, and other resources in the process.

As Figure B-1 shows, the same phases followed in the traditional SDLC are also followed in RAD, but the phases are combined to produce a more streamlined development technique. Planning and design phases in RAD are shortened by focusing work on system functional and user interface requirements at the expense of detailed business analysis and concern for system performance issues. Also, usually RAD looks at the system being developed in isolation from other systems, thus eliminating the time-consuming activities of coordinating with existing standards and systems during design and development. The emphasis in RAD is generally less on the sequence and structure of processes in the life cycle and more on doing different tasks in parallel with each other and on using prototyping extensively. Notice also that the iteration in the RAD life cycle is limited to the design and development phases. This is where the bulk of the work in a RAD approach takes place. Although it is possible in RAD for there to be a return to planning once design has begun, it is rarely done. Similarly, although it is possible to return to development from the cutover phase (when the system is turned over to the user), RAD is designed to minimize iteration at

FIGURE B-1
RAD SYSTEMS DEVELOPMENT LIFE CYCLE COMPARED TO STANDARD SDLC

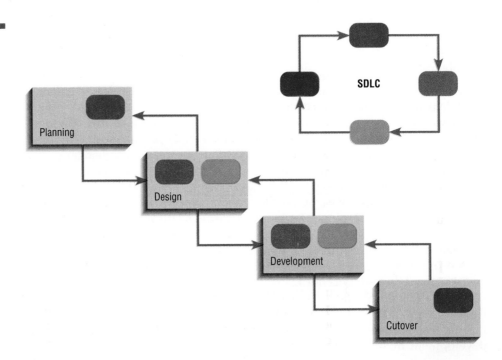

this point in the life cycle. The high level of user commitment and involvement throughout RAD implies that the system that emerges should be more readily accepted by the user community (and hence more easily implemented during cutover) than would be a system developed using traditional techniques.

Components of RAD

To succeed, RAD relies on bringing together several systems development components found in this book. As you might have gathered from the definition, RAD depends on extensive user involvement. End users are involved from the beginning of the development process, where they participate in application planning; through requirements determination, where they work with analysts in system prototyping; and design and implementation, where they work with system developers to validate final elements of the system's design. Much of end-user involvement takes place in the prototyping process, where users and analysts work together to design interfaces and reports for new systems.

The prototyping is conducted in sessions that resemble traditional Joint Application Design (JAD) sessions. The primary difference is that in RAD, the prototype becomes the basis for the new system—the screens designed during prototyping become screens in the production system. This is accomplished through reliance on integrated CASE tools, which contain prototyping tools that allow users and analysts to build prototypes during analysis and code generators for creating code from the designs end users and analysts create. Alternatively, RAD may employ visual development environments instead of CASE tools with code generators, but the benefits from rapid prototyping are the same. In many cases, the basis for the production system is being built even as users are talking about the system during development workshops. In many cases, end users can get hands-on experience with the developing system before the design workshops are over. To help speed the process further, the reuse of templates, components, or previous systems described in the CASE tool repository is strongly encouraged. To help you understand how CASE and visual development tools can be used in RAD, we introduce you to the basics about them in the next section.

CASE and Visual Development Environments and Their Role in RAD

Computer-aided software engineering (CASE) refers to automated software tools used by systems analysts to develop information systems. These tools can be used to automate or support activities throughout the systems development process with the objective of increasing productivity and improving the overall quality of systems. CASE helps provide an engineering-type discipline to software development and to the automation of the entire software life cycle process, sometimes with a single family of integrated software tools. In general, CASE assists systems builders in managing the complexities of information system projects and helps assure that high-quality systems are constructed on time and within budget.

Components of CASE

CASE tools are used to support a wide variety of SDLC activities. CASE tools can be used to help in the project identification and selection, project initiation and planning, analysis, and design phases (**upper CASE**) and/or in the implementation and maintenance phases (**lower CASE**) of the SDLC (Figure B-2). A

Computer-aided software engineering (CASE)
Software tools that provide automated support for some portion of the systems development process.

Upper CASE
CASE tools designed to support the systems planning and selection, systems analysis, and systems design phases of the systems development life cycle.

Lower CASE
CASE tools designed to support the systems implementation and operation phase of the systems development life cycle.

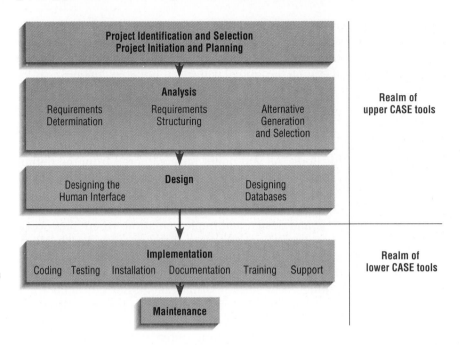

FIGURE B-2
THE RELATIONSHIP BETWEEN CASE TOOLS AND THE SYSTEMS DEVELOPMENT LIFE CYCLE

Cross life cycle CASE
CASE tools designed to support activities that occur across multiple phases of the systems development life cycle.

Repository
A centralized database that contains all diagrams, forms and report definitions, data structure, data definitions, process flows and logic, and definitions of other organizational and system components; it provides a set of mechanisms and structures to achieve seamless data-to-tool and data-to-data integration.

I-CASE
An automated systems development environment that provides numerous tools to create diagrams, forms, and reports; provides analysis, reporting, and code generation facilities; and seamlessly shares and integrates data across and between tools.

third category of CASE, **cross life cycle CASE,** consists of tools used to support activities that occur *across* multiple phases of the SDLC. For example, tools used to assist in ongoing activities, such as managing the project, developing time estimates for activities, and creating documentation, are often considered cross life cycle tools.

Over the past several years, vendors of upper, lower, and cross life cycle CASE products have "opened up" their systems through the use of standard databases and data conversion utilities to share information across products and tools more easily. An integrated and standard database called a **repository** is the common method for providing product and tool integration and has been a key factor in enabling CASE to manage larger, more complex projects more easily and to seamlessly integrate data across various tools and products. Integrated CASE or **I-CASE** will be described in more detail later in this appendix. User interface standards such as Microsoft's Windows have also greatly eased the integration and deployment of these systems. The general types of CASE tools are listed below:

- Diagramming tools that enable system process, data, and control structures to be represented graphically.
- Computer display and report generators that help prototype how systems "look and feel" to users. Display (or form) and report generators also make it easier for the systems analyst to identify data requirements and relationships.
- Analysis tools that automatically check for incomplete, inconsistent, or incorrect specifications in diagrams, forms, and reports.
- A central repository that enables the integrated storage of specifications, diagrams, reports, and project management information.
- Documentation generators that help produce both technical and user documentation in standard formats.
- Code generators that enable the automatic generation of program and database definition code directly from the design documents, diagrams, forms, and reports.

Besides providing an array of tools, most CASE products also support ad hoc inquiry into and extraction from the repository. Security features, which may be important in some development environments, are also widely available. For example, if you contract with a custom software developer, you may expect it to secure your system specifications so that other project teams may not access your system requirements, design, and code. Some more advanced CASE products also support version control, which allows one repository to contain the description of several versions or releases of the same application system. Also, some CASE products provide import and export facilities to move data automatically between the CASE repository and other software development tools such as word processors, software libraries, and testing environments. Finally, as a shared development database, CASE environments should provide facilities for backup and recovery, user account management, and usage accounting. In other words, to provide the greatest benefits, CASE should be used to support all activities within the SDLC. However, two components of CASE in particular are most useful in the RAD process: CASE form and report generators and CASE code generators. Visual development environments are useful to RAD for the same reasons: They allow rapid development of prototypes and ease the process of moving from design to code.

CASE Form and Report Generator Tools

Automated tools for developing computer displays (forms) and reports help the systems analyst design how the user will interact with the new system. These tools, referred to as **form and report generators,** are most commonly used for two purposes: (1) to create, modify, and test prototypes of computer display forms and reports and (2) to identify which data items to display or collect for each form or report. Figure B-3 shows an example of a form layout design using Oracle's Developer form design facility. This facility has numerous features to help you quickly design forms and windows that look and feel consistent to your users. For example, Developer has a template feature that allows you to define common headings, footers, and function key assignments. Once defined, all forms in the system inherit these template definitions. Also, if a common change is desired for all forms, you can simply change the template definition and all system forms will automatically inherit this change. Once you are happy with the design, you can quickly test its usability by converting the design template (Figure B-3) into a working prototype.

You will find that using automated tools for developing forms and reports is useful for both you and the eventual users of your system. For users, interacting

Form and report generators
CASE tools that support the creation of system forms and reports in order to prototype how systems will "look and feel" to users.

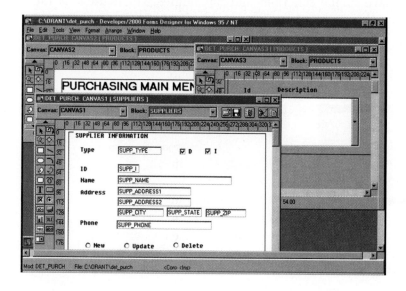

FIGURE B-3
FORM DESIGN TOOL FROM ORACLE'S DEVELOPER CASE ENVIRONMENT
Source: Screen display reproduced with the permission of Oracle Corporation.

with you during the early stages of the SDLC as forms and reports are outlined may help to ease system implementation. Involved users will be more familiar with the system when it is completed. As a result, these users may require less training than uninvolved users. Additionally, they may feel more positive that the system will meet their needs. From your perspective as the analyst, close interaction with users will help you develop a common frame of reference and enable you to better understand their data and processing requirements.

CASE Code Generation Tools

Code generators are automated systems that produce high-level program source code from diagrams and forms used to represent the system. As target environments vary on several dimensions, such as hardware and operating system platforms, many code generators are designed to be special-purpose systems that produce source code for a particular environment in a particular programming language. Most CASE tools that generate source code take a more flexible approach by producing standard source code and database definitions. Using standard language conventions, CASE-generated code can typically be compiled and executed on numerous hardware and operating system platforms with no, or very minor, changes. Yet standard code and definitions may not take advantage of special hardware or operating system features of specific environments.

> **Code generators**
> CASE tools that enable the automatic generation of program and database definition code directly from the design documents, diagrams, forms, and reports stored in the repository.

Visual Development Tools

Visual development tools are a relatively new and extremely powerful way to rapidly develop systems. Visual Basic by Microsoft, PowerBuilder by Sybase, ColdFusion by Allaire Corporation, Delphi by Borland International, and others are some of the most widely used development environments today. These visual tools allow systems developers to build new user interfaces, reports, and other features into new and existing systems in a fraction of the time previously required. Instead of building a screen, report, or menu by typing crude commands, designers use visual tools to "draw" the design using predefined objects. For example, to build a menu system in a visual programming environment is simple. Analysts can quickly list the order of menu commands in a development module called the Menu Editor (see Figure B-4[A]) and instantly test the look of their design (see Figure B-4[B]). In fact, once designed, these systems convert the design into the appropriate computer instructions, but all these details are hidden from the developer.

Although the popularity and the capabilities of most visual programming environments continue to expand, each environment has its own way of doing things. Yet, all have similar functions and capabilities. The market for these tools is very competitive; therefore, if one tool adds a new feature, most others quickly follow. Consequently, each tool has its strengths and weaknesses; it would be impossible to generalize that one system or approach to visual development is best. Visual Basic is arguably the most popular visual development

FIGURE B-4
BUILDING A MENU WITH A VISUAL DEVELOPMENT TOOL
(A) MENU EDITOR COMMANDS TO BUILD A MENU IN VISUAL BASIC
(B) MENU SYSTEM CREATED USING VISUAL BASIC'S MENU EDITOR

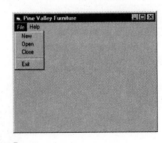

tool and is very powerful and easy to use (see Figure B-5 for an example of this development environment). In fact, Visual Basic is the macro programming language in the Microsoft software suite of tools such as Word and Excel. This means that Visual Basic has a wide range of users, from individuals streamlining their home expense budget in Excel to professional systems developers in organizations such as Xerox and the Vancouver Port Authority.

PowerBuilder is another visual development environment that is extremely powerful and somewhat more expensive than Visual Basic (see Figure B-6 for

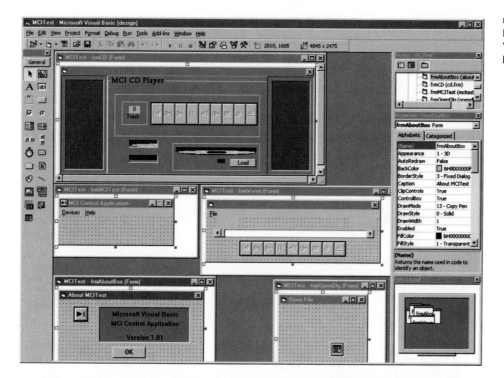

FIGURE B-5
VISUAL BASIC DEVELOPMENT ENVIRONMENT

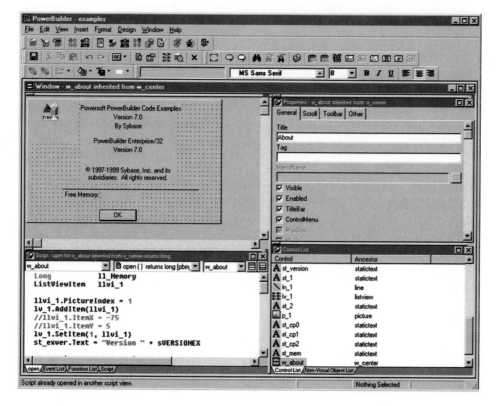

FIGURE B-6
POWERBUILDER DEVELOPMENT ENVIRONMENT
Source: Sybase, Inc., 1998.

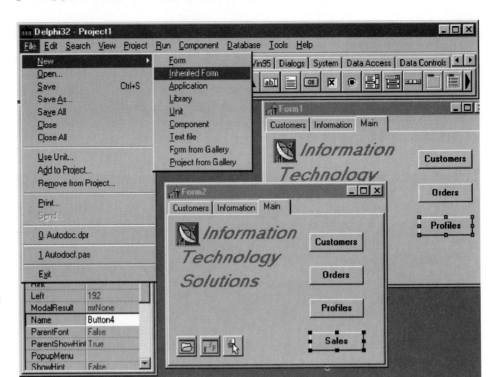

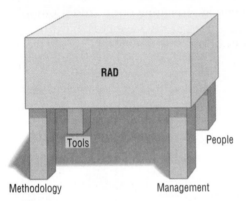

FIGURE B-8
MARTIN'S FOUR NECESSARY PILLARS FOR THE RAD APPROACH: TOOLS, PEOPLE, METHODOLOGY, AND MANAGEMENT.

an example of this development environment). PowerBuilder's greatest strength is that it runs and develops applications for a broad range of hardware, computing platforms, and databases. PowerBuilder is primarily used by professional systems developers in larger organizations such as American Express, Fox Broadcasting, Nissan Motors, and Sega Video Games. Delphi is another powerful visual programming tool that has a large and loyal following among professional developers (see Figure B-7). Delphi, like Visual Basic and PowerBuilder, has extensive object-oriented capabilities that developers can quickly and easily include in their systems. Organizations such as American Airlines, BMW, Coors, and NASA have reported great success when using Delphi (see www.inprise.com/delphi).

Approaches to RAD

James Martin, one of the best-known information systems authorities in the world and the man who invented RAD, suggests four important supporting components or "pillars" of RAD, as shown in Figure B-8:

1. Tools
2. People

3. Methodology
4. Management

The first pillar of RAD consists of software development tools. Tools, necessary as they are, are not enough to make RAD work. According to Martin, the other three pillars of RAD are people, who must be trained in the right skills; a coherent methodology, which spells out the proper tasks to be done in the proper order; and the support and facilitation of management.

Martin suggests that no information systems organization should be converted to the RAD approach overnight. Instead, a small group of well-trained and dedicated professionals, called a RAD cell, should be created to demonstrate the viability of RAD through pilot projects. Over time, the cell can grow, gradually adding more people, skills, and projects, until RAD is the predominant approach of the information systems unit.

Software Tools for RAD

There are many different software tools that can be successfully used to support the RAD approach. Remember that there is no silver bullet for software development, and that includes software tools that support RAD. Some tools will work better for certain applications than others, and developers need to have a good understanding of the limits as well as the possibilities of any development environment they decide to use.

We mentioned earlier how integrated CASE tools can be used to support RAD. Some CASE tools that can support RAD have excellent project management and diagramming capabilities, such as Advantage Gen from Computer Associates (Figure B-9). Other CASE tools used in RAD need to have prototyping facilities as well as code generators in order to provide the best level of support. CASE tools with these capabilities include Oracle's Developer/(see Figure B-3) and Borland's Delphi (see Figure B-7). Both of these CASE tools allow developers and users to work together closely to design windows and navigation paths between them. Developer generates code in several languages, including C and C++.

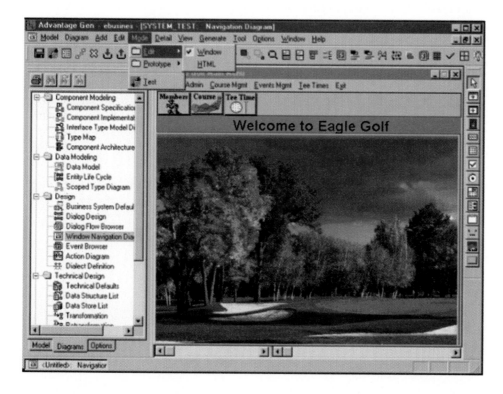

FIGURE B-9
AN EXAMPLE OF A CASE TOOL THAT CAN BE USED DURING RAD, FROM COMPUTER ASSOCIATES ADVANTAGE GEN
Used by permission.

Martin's RAD Life Cycle

Recall that Martin listed tools as just one of his four pillars necessary for RAD. Another pillar is methodology. For methodology, Martin developed a specific RAD life cycle, which contains the basic phases of any RAD life cycle (see Figure B-1)—planning, design, development, and cutover—although he calls the phases requirements planning, user design, construction, and cutover (see Figure B-10).

Requirements planning incorporates elements of the traditional planning and analysis phases. During this phase, high-level managers, executives, and knowledgeable end users determine system requirements, but the determination is done in the context of a discussion of business problems and business areas. Once specific systems have been identified for development, users and analysts conduct a joint requirements planning workshop to reach agreement on system requirements. The overall planning process is not all that much different from planning in the traditional SDLC.

During user design, end users and information systems professionals participate in JAD workshops, where they use integrated CASE tools to support the rapid prototyping of system design. Users and analysts work closely and quickly to create prototypes that capture system requirements and that become the basis for the physical design of the system being developed. Users sign off on the CASE-based design—there are no paper-based specifications. Because user design ends with agreement on a computer-based design, the gap between the end of design and the handing over of the new system to users might only take three months instead of the usual 18.

In construction, the same information systems professionals who created the design now generate code using the CASE tools' code generator. End users also participate, validating screens and other aspects of the design as the application system is being built. Construction can be combined with user design into one phase for smaller systems.

Cutover means delivery of the new system to its end users. Because the RAD approach is so fast, planning for cutover must begin early in the RAD process. Cutover involves many of the traditional activities of implementation, including testing, training users, dealing with organizational changes, and running the new and old systems in parallel, but all of these activities occur on an accelerated basis.

FIGURE B-10
MARTIN'S FOUR-PHASE RAD LIFE CYCLE

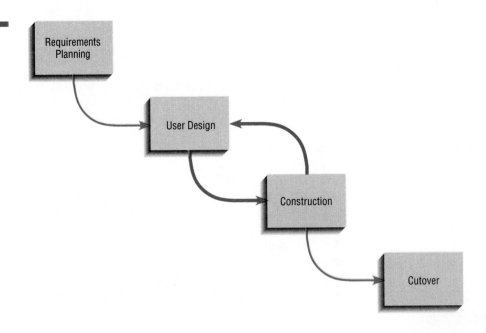

According to Martin, RAD can produce a system in six months that would take 24 months to produce using the traditional systems development life cycle.

RAD Success Stories

When used successfully, the results of RAD can be dramatic. Here are three stories about successful use of the RAD approach. The first two stories focus on the use of a particular software tool suite, Borland's Delphi. Both of the Delphi stories are based on postings to the Delphi Web pages. The third story is based on the use of another RAD tool, IBM's VisualAge for Java.

RAD Success with Borland's Delphi in the U.S. Navy and First National Bank of Chicago

Two RAD success stories with Delphi, one from the U.S. government and one from the financial business sector, suggest the broad applicability of the RAD approach.

U.S. Navy Fleet Modernization   The U.S. Navy has as an ongoing challenge: the modernization of its fleet as well as its maintenance. Modernization has become especially crucial as the Navy has downsized from a fleet of almost 600 ships in the 1980s to just 260 in 1997. The Navy wanted to upgrade its fleet modernization management system at the same time it was renovating the fleet. The requirement was to move from three character-based systems to a unified, GUI-based system based on a single database. The Fleet Modernization program chose Delphi to build the new system. Two aspects of Delphi that were attractive to the Navy were its support of rapid prototyping and the promise of reuse of components in the object-oriented Pascal environment. The final system was developed with Delphi using a RAD approach in six months, an estimated savings of 50 percent. The Navy estimates the new system resulted in immediate savings of 20 percent per year due to reduced maintenance costs.

First National Bank of Chicago   First National Bank of Chicago is one of the 10 largest banks in the United States, with assets of over $122 billion. In October 1994, the bank was awarded a bid to develop a new Electronic Federal Tax Payment System, intended to foster faster tax withholdings and eliminate all paper in the process. The contract, signed by a joint venture of the bank with Mercantile Bank of St. Louis, specified exacting system requirements and very short deadlines. The venture chose Delphi for the job. The use of Delphi enabled rapid development and rapid prototyping of a GUI front end to the system.

At one point, over 125 programmers organized into several teams were working on the project. Fifteen to twenty developers worked on the GUI front end. Rapid prototyping for requirements elicitation allowed coding to begin in May 1995, only seven months after the contract was awarded. The system was completed and ready for testing in March 1996, only 10 months later. After three months of certification testing, the system went live in July 1996. The size of the database used in the Electronic Federal Tax Payment System is not trivial at 225 tables, with 250 million rows and 55 gigabytes in size. By the year 2000, 500 million rows were maintained online. RAD techniques, with Delphi's support, allowed the project to come in on time and on budget, despite the extensive requirements and short deadlines specified in the contract.

RAD Success with VisualAge for Java

Comdata is the leading provider of financial and information services to the North American transportation industry. Over 3,000 trucking companies rely on Comdata's MOTRS (Modular Over The Road System) to help manage their businesses and to monitor the spending of their truck drivers. These drivers use Comdata's Comcheck card to purchase fuel, food, and lodging while on the road.

The original version of MOTRS was PC-based, written in VisualBasic. Comdata customers accessed the system through dial-up, uploaded account data, and downloaded reports. Long distance customers ran different versions of MOTRS, written for different computer platforms, Comdata also incurred large costs for updating the system so it would run on different platforms. Comdata decided that it needed a new version of MOTRS by the end of 1999. The new version had to be Y2K compliant, eliminate the need for long distance dial-up access, and there had to be a single version of the system. These requirements pointed to an e-business solution. Comdata decided to deploy MOTRS on the Internet, and they chose IBM Global Services as the vendor to help them migrate their system.

To develop the Web-based MOTRS, IBM used VisualAge for Java and the IBM WebSphere Application Server software to create servlets. Servlets are programming modules, written in Java, that reside on the server and expand the functions of the Web server. Visual Age for Java was also used to create Java applets. Applets are embedded in the HTML code on Web pages and are downloaded to the customer client machine and executed by the client Web browser. The WebSphere Application Server software was used to manage and control the servlets.

The MOTRS Internet project took only nine months to complete. This included three months for research and design, three months for coding, and three months for testing: one month each for quality assurance, acceptance testing by Comdata, and beta testing by customers. The entire project cost $500,000, including the server software and hardware, development, and everything else. The system became generally available in September 1998, long before the deadline for deployment set for the end of 1999. After the redevelopment effort, MOTRS was Y2K compliant, there was a single PC version of it, and customers now pay only $20 per month for an Internet Service Provider account through which to access the system.

Advantages and Disadvantages of RAD

As these success stories show, the primary advantage of RAD is obvious: information systems developed in as little as one-quarter the usual time. But shorter development cycles also mean cheaper systems, as fewer organizational resources need to be devoted to develop any particular system. Martin points out that RAD also involves smaller development teams, which results in even more savings. Finally, because there is less time between the end of design and conversion, the new system is closer to current business needs and is therefore of higher quality than would be a similar system developed the traditional way.

Others point out, however, that although RAD works, it only works well for systems that have to be developed quickly. Such systems include those that are developed in response to quickly changing business conditions, perceived opportunities to gain a competitive edge by moving quickly in a market or industry, or new government regulations. If speed is a goal, however, then other aspects of an application must often be sacrificed. In general, in a systems development effort, you can obtain two of the three key characteristics of the development effort—speed, cost, and quality—but never all three. If hard-and-fast goals for speed and cost are set, quality of the final application will suffer. Similarly, if you must achieve a certain level of quality and you must do so in a fixed amount of time, the cost will be greater than it might otherwise have been. Finally, if you have set goals for cost and quality, you will take longer to develop the application than you might have wanted originally.

It should also be pointed out that taking too much time to develop a system can actually have a negative effect on the quality of the system. Too much time between the initial requirements determination and the delivery of the completed system adds many complications to the overall development process. It will be more difficult to estimate requirements accurately at the time the system will be delivered because business conditions will have changed so much during the intervening time. Although predicting the future is always difficult, predicting the near-term future is more certain than predicting a future three to five years away. Changing business conditions will also make it easier for system owners and developers to become sidetracked during the development process. Finally, too much time between the beginning and end of a project almost guarantees problems with feature creep, again due to changing business conditions. The key, then, is finding the right balance between time to completion and quality. Too much time can be just as bad as too little.

Although RAD may be preferred for all systems development due to its emphasis on speed of development, RAD's emphasis on what a system does and how it does it implies that essential information about the business models and processes is not included. In its highly accelerated analysis and design phases, RAD leaves little room for understanding the business area: what it does, what its functions are and who performs them, or what the people in the business area do in their jobs. The greater the reliance on RAD, the greater the risk that many systems will be out of alignment with the business.

Another drawback of RAD has to do with the very thing that makes it so attractive as a systems development methodology: its speed. Because the RAD process puts such an emphasis on speed, many important software engineering concepts, such as interface consistency, programming standards, module reuse, scalability, and systems administration, are overlooked. For example:

- *Consistency*. In their efforts to paint screens quickly, RAD analysts often ignore the need to be consistent both within an application and across a suite of related applications. Areas of concern include window size and color, consistent format masks, and using the same error message for the same offense.

- *Programming standards*. Documentation standards and data naming standards should be established early in RAD or it may be difficult to implement them later.

- *Module reuse*. Many times during prototyping, analysts forget that similar screen or report designs may have already been created. Often in RAD there is no mechanism in place that allows analysts to easily determine whether modules that can be reused are already in existence.

- *Scalability*. If the system designed during RAD is useful, its use will gradually spread beyond those initial users who helped build it. Developers should build such growth into their initial system. Scalability also applies to hardware; system scope; the number, type, and users of reports; growth in the software team developing and maintaining the system as it grows; user training as system use expands; and security.

> *Systems administration.* System administration is typically ignored altogether during RAD, as the emphasis is usually on the excitement of seeing a new application system develop before users' eyes. Important system administration issues include database maintenance and reorganization, backup and recovery, distribution or installation of application updates, and scheduling and implementing planned system downtime and restarts.

Although RAD can reduce a project's cost, because fewer resources and less time go into the overall development effort, there are ways RAD can end up increasing costs. RAD is reliant on high levels of user commitment and participation, so it is crucial that key members of the user community are involved during the entire RAD process. The problem is that key users have their own work to do. Pulling key individuals away from their own work can come at a high cost to the organization. If the key user can't do his or her job because of a RAD effort, the organization must either wait for the user to get back to the job or find another way to accomplish the same work. Neither solution is optimal and the organization may ultimately suffer. Again, a suitable balance has to be found between a key user's commitment to his or her job and to the RAD effort.

Clearly, RAD is a powerful approach to information systems development, with the potential for many payoffs. Just as clearly, there are risks associated with using RAD, especially for every systems development project in an organization's portfolio of projects. The relative advantages and disadvantages of the RAD approach are summarized in Table B-1.

TABLE B-1: Advantages and Disadvantages of RAD

Advantages	Disadvantages
Dramatic time savings during the systems development effort	More speed and lower cost may lead to lower overall system quality (e.g., due to lack of attention to internal controls)
Can save time, money, and human effort	Danger of misalignment of system developed via RAD with the business due to missing information on underlying business processes
Tighter fit between user requirements and system specifications	May have inconsistent internal designs within and across systems
Works especially well where speed of development is important, as with rapidly changing business conditions or where systems can capitalize on strategic opportunities	Possible violation of programming standards related to inconsistent naming conventions and insufficient documentation
Ability to rapidly change system design as demanded by users	Difficulties with module reuse for future systems
System optimized for users involved in RAD process	Lack of scalability designed into the system
Concentrates on essential system elements from user viewpoint	Lack of attention to later systems administration built into the system (e.g., not integrated into overall enterprise data model and missing system recovery features)
Strong user stake and ownership of system	High costs of commitment on the part of key user personnel

Key Points Review

1. **Explain the Rapid Application Development (RAD) approach and how it differs from traditional approaches to information systems development.**

    Rapid Application Development (RAD) is an alternative to the traditional systems development life cycle. All of the phases of the traditional life cycle are included in RAD, but the phases are executed at an accelerated rate. The abbreviated RAD life cycle, as defined by James Martin, begins with requirements planning, followed by user design, construction, and cutover.

2. **Describe the systems development components essential to RAD.**

    RAD relies on heavy user involvement, Joint Application Development sessions, prototyping, integrated CASE tools, and code generators to design and implement systems quickly.

3. **List and describe the typical components of a comprehensive CASE environment.**

    The components of a comprehensive CASE system are divided into upper, lower, and cross life cycle CASE tools, covering different segments of the SDLC. Upper CASE tools—diagramming tools, form and report generators, and analysis tools—are used primarily to support project identification and selection, project initiation and planning, analysis, and design. Lower CASE tools—code generators—are used primarily to support system implementation and maintenance. Cross life cycle tools—project management tools—coordinate project activities. The repository and documentation generators are used across multiple life cycle phases to support project management, activity estimation, and documentation creation.

4. **Describe how CASE tools can be used to support RAD.**

    The two CASE components that can provide the most support to the RAD process are form and report generators and code generators. Form and report generators can be used by analysts and end users to develop working prototypes quickly that allow users to get an idea of the look and feel of the system being developed. Code generators can be used to build working prototypes very quickly, with few errors.

5. **Describe visual and emerging development tools and how they can be used to support RAD.**

    Visual development tools allow systems developers to build new user interfaces, reports, and other features into new and existing systems in a fraction of the time previously required. Through the use of development tools like Visual Basic by Microsoft, PowerBuilder by Powersoft, or Delphi by Borland, new systems can be constructed in a fraction of the time previously required.

6. **Discuss the conceptual pillars that support the RAD approach.**

    Martin lists the four conceptual pillars of RAD as tools, people, methodology, and management. Tools like CASE are necessary for RAD to work, but tools alone are not enough. The right methodology is key, as is employing the right people in the RAD effort with active support from the managers of the information technology unit in the organization.

7. **Explain the advantages and disadvantages of RAD as an exclusive systems development methodology.**

    The primary advantage of RAD is the quick development of systems, but quick development may also lead to cost savings and higher-quality systems. RAD does have drawbacks: With its emphasis on developing systems quickly, the detailed business models that underlie information systems are often neglected, leading to the risk that systems may be out of alignment with the overall business. Similarly, the speed of development may lead to analysts' overlooking systems engineering concepts such as consistency, programming standards, module reuse, scalability, and systems administration. If applied successfully, however, RAD may result in dramatic savings and improved performance. Systems may be designed and implemented in one-quarter to one-half the time needed for the traditional life cycle approach.

Key Terms Checkpoint

Here are the key terms from the appendix. The page where each term is first explained is in parentheses after the term.

1. **Code generators (p. 430)**
2. **Computer-aided software engineering (CASE) (p. 427)**
3. **Cross life cycle CASE (p. 428)**
4. **Form and report generators (p. 429)**
5. **I-CASE (p. 428)**
6. **Lower CASE (p. 427)**
7. **Rapid Application Development (RAD) (p. 425)**
8. **Repository (p. 428)**
9. **Upper CASE (p. 427)**

Match each of the key terms above with the definition that best fits it.

_____ 1. CASE tools designed to support activities that occur *across* multiple phases of the systems development life cycle.

_____ 2. CASE tools that support the creation of system forms and reports in order to prototype how systems will "look and feel" to users.

_____ 3. Systems development methodology created to radically decrease the time needed to design and implement information systems. RAD relies on extensive user involvement, Joint Application Design sessions, prototyping, integrated CASE tools, and code generators.

_____ 4. CASE tools that enable the automatic generation of program and database definition code directly from the design documents, diagrams, forms, and reports stored in the repository.

_____ 5. CASE tools designed to support the systems implementation and operation phase of the systems development life cycle.

_____ 6. An automated systems development environment that provides numerous tools to create diagrams, forms, and reports; provides analysis, reporting, and code generation facilities; and seamlessly shares and integrates data across and between tools.

_____ 7. CASE tools designed to support the systems planning and selection, systems analysis, and systems design phases of the systems development life cycle.

_____ 8. Software tools that provide automated support for some portion of the systems development process.

_____ 9. A centralized database that contains all diagrams, forms and report definitions, data structure, data definitions, process flows and logic, and definitions of other organizational and system components; it provides a set of mechanisms and structures to achieve seamless data-to-tool and data-to-data integration.

Review Questions

1. List and briefly define the four phases of RAD, as defined by James Martin.
2. Describe each major component of a comprehensive CASE system.
3. Describe how CASE is used to support each phase of the SDLC.
4. List and briefly explain two different views for the four pillars of RAD.
5. Explain the systems development components essential to RAD.
6. What trends in information systems encouraged the invention of the RAD approach to systems development?
7. Explain the concept of scalability and its influence on systems development.
8. Explain the advantages and disadvantages of RAD.
9. Given the relative advantages and disadvantages of RAD, why would a systems development group follow the RAD approach for a project? When is the best time to use RAD and when is the best time to use a more traditional method?
10. What is the role of JAD in Rapid Application Development?

Problems and Exercises

1. Compare RAD and prototyping. How are these methodologies different from one another? In what ways are they similar?
2. How might RAD be used in conjunction with the structured systems development life cycle? Are RAD and the SDLC totally different approaches or could they complement each other?
3. What types of tools are necessary to do RAD? Is it possible to do RAD without fourth-generation languages and other tools?
4. One of the criticisms of RAD is that it may cause a system to be out of alignment with the direction of the business. Explain how this may occur and suggest what might be done to overcome this potential hazard of RAD.
5. Consider Figure B-1. From what you read in this appendix, answer the following questions:

   a. How does the RAD planning phase differ from the corresponding SDLC phases?
   b. How does the RAD design phase differ from the corresponding SDLC phases?
   c. How does the RAD development phase differ from the corresponding SDLC phase?
   d. How does the RAD cutover phase differ from the corresponding SDLC phases?

6. Describe the characteristics of a situation requiring a new information system that would be ideal for the use of the RAD approach.
7. Review the sample computer forms in Figure B-3. How is a form generator different from a standard graphics package like, say, Microsoft Windows Paintbrush or PowerPoint? Why not simply use one of these graphics packages instead of using a CASE tool? What is gained by using a CASE tool rather than a graphics package? To answer these questions adequately, you may need to call a CASE tool vendor directly or find CASE tool product evaluations in the popular press.
8. A goal stated by many vendors of CASE products is to have CASE ultimately be able to automatically generate (and regenerate to any platform and with any changes) 100 percent of the code, error-free, for a new or modified information system. This goal is considered important in order to achieve systems development productivity gains necessary to deal with systems backlog, to improve system quality, and to enhance our ability to maintain systems. Do you think this goal is possible? Why or why not?

References

Chapter 1: The Systems Development Environment

Aktas, A. Z. *Structured Analysis and Design of Information Systems.* Upper Saddle River, NJ: Prentice Hall, 1987.

Alavi, M., and I. R. Weiss. "Managing the Risks Associated with End-User Computing." *Journal of MIS* 2 (Winter 1985): 5–20.

Bohm, C., and I. Jacopini. "Flow Diagrams, Turing Machines, and Languages with only Two Formation Rules." *Communications of the ACM* 9 (May 1966): 366–71.

Bourne, K. C. "Putting Rigor Back in RAD." *Database Programming & Design* 7 (8) (August 1994): 25–30.

Davis, G. B. "CAUTION: User-Developed Systems Can Be Dangerous to Your Organization." In *End-user computing: Concepts, issues and applications.* Edited by R. R. Nelson, 209–28. New York: Wiley, 1989.

DeMarco, T. *Structured Analysis and System Specification.* Upper Saddle River, NJ: Prentice Hall, 1979.

International Technology Association of America. "Bouncing Back: Jobs, Skills and the Continuing Demand for IT Workers." Accessed 2002, www.it.org.

Martin, E. W., D. W. DeHayes, J. A. Hoffer, and W. C. Perkins. *Managing Information Technology: What Managers Need to Know.* 2d ed. New York: Macmillan, 1994.

Mumford, E. "Participative Systems Design: A Structure and Method." *Systems, Objectives, Solutions* 1 (1) (1981): 5–19.

Naumann, J. D., and A. M. Jenkins. "Prototyping: The New Paradigm for Systems Development. *MIS Quarterly* 6 (3) (1982): 29–44.

Yourdon, E., and L. L. Constantine. *Structured Design.* Upper Saddle River, NJ: Prentice Hall, 1979.

Chapter 2: Managing the Information Systems Project

Abdel-Hamid, T. K. "Investigating the Impacts of Managerial Turnover/Succession on Software Project Performance." *Journal of Management Information Systems* 9 (2) (1992): 127–44.

Boehm, B. W. *Software Engineering Economics.* Upper Saddle River, NJ: Prentice Hall, 1981.

Butler, J. "Automating Process Trims Software Development Fat." *Software Magazine* 14 (8) (August 1994): 37–46.

Grupe, F. H., and D. F. Clevenger. "Using Function Point Analysis as a Software Development Tool." *Journal of Systems Management* 42 (December 1991): 23–26.

Hoffer, J. A., M. B. Prescott, and F. R. McFadden. *Modern Database Management.* 6th ed. Upper Saddle River, NJ: Prentice Hall, 2002.

Kettelhut, M. C. "Avoiding Group-Induced Errors in Systems Development." *Journal of Systems Management* 42 (December 1991): 13–17.

Miller, R. W. "How to Plan and Control with PERT." *Managing Projects and Programs,* Harvard Business School Press. Boston: Harvard, 1989.

Nicholas, J. *Project Management for Business and Technology: Principles and Practice.* Upper Saddle River, NJ: Prentice Hall, 2001.

Page-Jones, M. *Practical Project Management.* New York: Dorset House, 1985.

Pressman, R. S. *Software Engineering.* 3d ed. New York: McGraw-Hill, 1992.

Rettig, M. "Software Teams." *Communications of the ACM* 33 (10) (1990): 23–27.

Thé, L. "IS-Friendly Project Management." *Datamation* (April 1, 1996): 79–82.

Thomsett, R. Foreword to *Practical project management,* by M. Page-Jones. New York: Dorset House, 1985.

"Top 10 Signs of IS Project Failure." *Datamation* 1 (19) (June 1996).

Zachary, G. P. "Agony and Ecstasy of 200 Code Writers Beget Windows NT." *Wall Street Journal* (May 26, 1993): A1–A6.

Chapter 3: Systems Planning and Selection

Atkinson, R. A. "The Motivations for Strategic Planning." *Journal of Information Systems Management* 7 (4) (1990): 53–56.

Atkinson, R. A., and J. Montgomery. "Reshaping IS Strategic Planning." *Journal of Information Systems Management* 7 (4) (1990): 9–15.

Carlson, C. K., E. P. Gardner, and S. R. Ruth. "Technology-Driven Long-Range Planning." *Journal of Information Systems Management* 6 (3) (1989): 24–29.

Cash, J. I., F. W. McFarlan, J. L. McKenney, and L. M. Applegate. *Corporate Information Systems Management.* 3d ed. Boston: Irwin, 1992.

IBM. "Business Systems Planning." In *Advanced System Development/ Feasibility Techniques.* Edited by J. D. Couger, M. A. Colter, and R. W. Knapp, 236–314. New York: Wiley, 1982.

Kerr, J. "The Power of Information Systems Planning." *Database Programming & Design* 3 (December 1990): 60–66.

King, J. L., and E. Schrems. "Cost Benefit Analysis in Information Systems Development and Operation." *ACM Computing Surveys* 10 (1) (1978): 19–34.

Koory, J. L., and D. B. Medley. *Management Information Systems: Planning and Decision Making.* Cincinnati, OH: South-Western, 1987.

Lederer, A. L., and J. Prasad. "Nine Management Guidelines for Better Cost Estimating." *Communications of the ACM* 35 (2) (1992): 51–59.

Martin, J. *Information Engineering.* Upper Saddle River, NJ: Prentice Hall, 1990.

McKeen, J. D., T. Guimaraes, and J. C. Wetherbe. "A Comparative Analysis of MIS Project Selection Mechanisms." *Data Base* 25 (February 1994): 43–59.

Moriarty, T. "Framing Your System." *Database Programming & Design* 4 (June 1991): 57–59.

Morton, C. "Information Competition: Can OLTP and DSS Peacefully Coexist?" *Data Base Management* 2 (June 1992): 24–28.

Parker, M. M., and R. J. Benson. *Information Economics.* Upper Saddle River, NJ: Prentice Hall, 1988.

Parker, M. M., and R. J. Benson. "Enterprisewide Information Management: State-of-the-Art Strategic Planning." *Journal of Information Systems Management* 6 (Summer 1989): 14–23.

Porter, M. *Competitive Strategy: Techniques for Analyzing Industries and Competitors.* New York: Free Press, 1980.

Porter, M. *Competitive Advantage.* New York: Free Press, 1985.

Pressman, R. S. *Software Engineering.* 3d ed. New York: McGraw-Hill, 1992.

Radosevich, L. "Can You Measure Web ROI?" *Datamation* (July 1996): 92–96.

Reingruber, M. J., and D. L. Spahr. "Putting Data Back in Database Design." *Data Base Management* 2 (March 1992): 19–21.

Shank, J. K., and V. Govindarajan. *Strategic Cost Management.* New York: Free Press, 1993.

Sowa, J. F., and J. A. Zachman. "Extending and Formalizing the Framework for Information Systems Architecture." *IBM Systems Journal* 31 (3) (1992): 590–616.

Yourdon, E. *Structured Walkthroughs.* 4th ed. Upper Saddle River, NJ: Prentice Hall, 1989.

Zachman, J. A. "A Framework for Information Systems Architecture." *IBM Systems Journal* 26 (March 1987): 276–92.

Chapter 4: Determining System Requirements

Carmel, E. "Supporting Joint Application Development with Electronic Meeting Systems: A Field Study." Unpublished doctoral dissertation, University of Arizona, 1991.

Carmel, E., J. F. George, and J. F. Nunamaker, Jr. "Supporting Joint Application Development (JAD) with Electronic Meeting Systems: A Field Study." *Proceedings of the thirteenth international conference on information systems.* Dallas, TX, December 1992: 223–32.

Carmel, E., R. Whitaker, and J. F. George. "Participatory Design and Joint Application Design: A Transatlantic Comparison." *Communications of the ACM* 36 (June 1993): 40–48.

Davenport, T. H. *Process Innovation: Reengineering Work Through Information Technology.* Boston: Harvard Business School Press, 1993.

Dennis, A. R., J. F. George, L. Jessup, J. F. Nunamaker, Jr., and D. R. Vogel. "Information Technology to Support Electronic Meetings." *MIS Quarterly* 12 (December 1988) : 591–624.

Dobyns, L., and C. Crawford-Mason. *Quality or Else.* Boston: Houghton Mifflin, 1991.

Hammer, M., and J. Champy. *Reengineering the Corporation.* New York: Harper Business, 1993.

Lucas, M. A. "The Way of JAD." *Database Programming & Design* 6 (July 1993): 42–49.

Mintzberg, H. *The Nature of Managerial Work.* New York: Harper & Row, 1973.

Moad, J. "After Reengineering: Taking Care of Business." *Datamation* 40 (20) (1994): 40–44.

Wood, J., and D. Silver. *Joint Application Design.* New York: Wiley, 1989.

Chapter 5: Structuring System Requirements: Processing Modeling

Celko, J. "I. Data Flow Diagrams." *Computer Language* 4 (January 1987): 41–43.

DeMarco, T. *Structured Analysis and System Specification.* Upper Saddle River, NJ: Prentice Hall, 1979.

Gane, C., and T. Sarson. *Structured Systems Analysis.* Upper Saddle River, NJ: Prentice Hall, 1979.

Gibbs, W. W. "Software's Chronic Crisis." *Scientific American* 271 (September 1994): 86–95.

Hammer, M., and J. Champy. *Reengineering the Corporation.* New York: Harper Business, 1993.

Yourdon, E. *Managing the Structured Techniques.* 4th ed. Upper Saddle River, NJ: Prentice Hall, 1989.

Yourdon, E., and L. L. Constantine. *Structured Design.* Upper Saddle River, NJ: Prentice Hall, 1979.

Chapter 6: Structuring System Requirements: Conceptual Data Modeling

Bruce, T. A. *Designing Quality Databases with IDEF1X Information Models.* New York: Dorset House Publications, 1992.

Chen, P. P-S. "The Entity-Relationship Model—Toward a Unified View of Data." *ACM Transactions on Database Systems* 1 (March 1976): 9–36.

Hoffer, J. A., M. B. Prescott, and F. R. McFadden. *Modern Database Management.* 6th ed. Upper Saddle River, NJ: Prentice Hall, 2002.

Moody, D. "The Seven Habits of Highly Effective Data Modelers." *Database Programming & Design* 9 (October 1996): 57, 58, 60–62, 64.

Teorey, T. J., D. Yang, and J. P. Fry. "A Logical Design Methodology for Relational Databases Using the Extended Entity-Relationship Model." *Computing Surveys* 18 (2) (June 1986): 197–221.

Chapter 7: Selecting the Best Alternative Design Strategy

Applegate, L. M., and R. Montealegre. "Eastman Kodak Company: Managing Information Systems Through Strategic Alliances." Harvard Business School case 9-192-030. Cambridge, MA: President and Fellows of Harvard College, 1991.

Harper, D. "Seek a Partner, Not a Vendor." *Industrial Distribution* 83 (April 1994): 97.

Hartmann, C. R. "How to Write a Proposal." *D & B Reports* 42 (March–April 1993): 62.

King, R. T., Jr., and J. E. Rigdon. "Hewlett Prints Computer Services in Big Capital Letters." *Wall Street Journal* (June 3, 1994): B8.

"Microcomputer Procurement Guidelines." *Public Works* 125 (April 15, 1994): G23 1.

Mikulski, F. A. "*Managing Your Vendors: The Business of Buying Technology.*" Upper Saddle River, NJ: Prentice Hall, 1993.

Moad, J. "Inside an Outsourcing Deal." *Datamation* 39 (February 15, 1993): 20–27.

"More Companies Are Chucking Their Computers." *BusinessWeek* (June 19, 1989): 72–74.

Semich, J. W. "Is it Bye-Bye Borland?" *Datamation* 40 (June 1994, 12): 52–54.

Stein, M. "Don't Bomb Out when Preparing RFPs." *Computerworld* 27 (February 15, 1993): 102.

Zachary, G. P. "How Ashton-Tate Lost its Leadership in PC Software Arena." *Wall Street Journal* (April 1, 1990): A1–A2.

Zachary, G. P., and W. M. Bulkeley. "Ashton-Tate Loses Flagship Software's Copyright Shield." *Wall Street Journal* (December 14, 1990): B1, B4.

Chapter 8: Designing the Human Interface

Alter, S. *Information Systems: A Management Perspective.* 3d ed. Reading, MA: Addison-Wesley, 1999.

Apple. *Human Interface Guidelines: The Apple Desktop Interface.* Reading, MA: Addison-Wesley, 1987.

Benbasat, I., A. S. Dexter, and P. Todd. "The Influence of Color and Graphical Information Presentation in a Managerial Decision Simulation." *Human-Computer Interaction* 2 (1986): 65–92.

Blattner, M., and E. Schultz. "User Interface Tutorial." Presented at the 1988 Hawaii International Conference on System Sciences, Kona, Hawaii, January 1988.

Bloombecker, J. J. "Short-Circuiting Computer Crime." *Datamation* 55 (Oct. 1, 1989): 71–72.

Carroll, J. M. *Designing Interaction.* Cambridge: Cambridge University Press, 1991.

Castro, E. (2001). *XML for the World Wide Web.* Berkley, CA: Peachpit Press.

Dumas, J. S. *Designing User Interfaces for Software.* Upper Saddle River, NJ: Prentice Hall, 1988.

Fernandez, E. B., R. C. Summers, and C. Wood. *Database Security and Integrity.* Reading, MA: Addison-Wesley, 1981.

Flanders, V. and Willis, M. (1998). *Web Pages That Suck: Learn Good Design by Looking at Bad Design*. Alameda, CA: Sybex Publishing.

Hoffer, J. A., M. B. Prescott, and F. R. McFadden. *Modern Database Management*. 6th ed. Upper Saddle River, NJ: Prentice Hall, 2002.

Huff, D. *How to Lie with Statistics*. New York: W. W. Norton, 1954.

IBM. *Systems Application Architecture: Common User Access Guide to User Interface*. IBM Document SC34-4289-00, October 1991.

Jarvenpaa, S. L., and G. W. Dickson. "Graphics and Managerial Decision Making: Research Based Guidelines." *Communications of the ACM* 31 (6) (1988): 764–74.

Johnson, J. (2000). *GUI Bloopers: Don'ts and Do's for Software Developers and Web Designers*. San Diego, CA: Academic Press.

May, J. C. *Extending the Macintosh Toolbox: Programming Menus, Windows, Dialogues, and More*. Reading, MA: Addison-Wesley, 1991.

McKay, E.N. (1999). *Developing User Interfaces for Microsoft Windows*. Redmond, WA: Microsoft Press.

Nielsen, J. (1996). Marginalia of Web Design. November; www.useit.com/alertbox/9611.html.

Nielsen, J. (1997). Loyalty on the Web. August 1; www.useit.com/alertbox/9708a.html.

Nielsen, J. (1998a). Using Link Titles to Help Users Predict Where They Are Going. January 11; www.useit.com/alertbox/980111.html.

Nielsen, J. (1998b). Personalization is Over-Rated. October 4; www.useit.com/alertbox/981004.html.

Nielsen, J. (1998c). Web Pages Must Live Forever. November 29; www.useit.com/alertbox/981129.html.

Nielsen, J. (1999a). "User Interface Directions for the Web." *Communications of the ACM*, 42(1), 65–71.

Nielsen, J. (1999b). Trust or Bust: Communicating Trustworthiness in Web Design. March 7; www.useit.com/alertbox/990307.html.

Nielsen, J. (2000). *Designing Web Usability: The Practice of Simplicity*. Indianapolis, IN: New Riders Publishing.

Norman, K. L. *The Psychology of Menu Selection*. Norwood, NJ: Ablex, 1991.

Porter, A. *C++ Programming for Windows*. Berkeley, CA: Osborne McGraw-Hill, 1993.

Shneiderman, B. (1997). *Designing the User Interface: Strategies for Effective Human-Computer Interaction*. 3rd Edition. Reading, MA: Addison-Wesley.

Sun Microsystems (1999). Java Look and Feel Design Guidelines. Reading, MA: Addison-Wesley.

Wagner, R. "A GUI Design Manifesto." *Paradox Informant* 5 (June 1994): 36–42.

Chapter 9: Designing Databases

Babad, Y. M., and J. A. Hoffer. "Even No Data has a Value." *Communications of the ACM* 27 (August 1984): 748–56.

Codd, E. F. "A Relational Model of Data for Large Relational Databases." *Communications of the ACM* 13 (June 1970): 77–87.

Gibson, M., C. Hughes, and W. Remington. "Tracking the Trade-Offs with Inverted Lists." *Database Programming & Design* 2 (January 1989): 28–34.

Hoffer, J. A., M. B. Prescott, and F. R. McFadden. *Modern Database Management*. 6th ed. Upper Saddle River, NJ: Prentice Hall, 2002.

Navathe, S., R. Elmasri, and J. Larson. "Integrating User Views in Database Design." *Computer* (January 1986): 50–62.

Rodgers, U. "Denormalization: Why, What, and How?" *Database Programming & Design* 2 (12) (December 1989): 46–53.

Viehman, P. "24 Ways to Improve Database Performance." *Database Programming & Design* 7 (2) (February 1994): 32–41.

Chapter 10: Systems Implementation and Operation

Bell, P., and C. Evans. *Mastering Documentation*. New York: Wiley, 1989.

Bloor, R. "The Disappearing Programmer." *DBMS* 7 (9) (August 1994): 14–16.

Brooks, F. P., Jr. *The Mythical Man-Month*. Anniversary edition. Reading, MA: Addison-Wesley, 1995.

Cole, K., O. Fischer, and P. Saltzman. "Just-in-Time Knowledge Delivery." *Communications of the ACM* 40 (7) (1997): 49–53.

Crowley, A. "The Help Desk Gains Respect." *PC Week* 10 (November 15, 1993): 138.

Dillon, N. "Internet-Based Training Passes Audit." *Computerworld* (November 3, 1997): 47–48.

Eason, K. *Information Technology and Organisational Change*. London: Taylor & Francis, 1988.

Ginzberg, M. J. "Early Diagnosis of MIS Implementation Failure: Promising Results and Unanswered Questions." *Management Science* 27 (4) (1981): 459–78.

Ginzberg, M. J. "Key Recurrent Issues in the MIS Implementation Process." *MIS Quarterly* 5 (2) (June 1981): 47–59.

Hanna, M. "Using Documentation as a Life-Cycle Tool." *Software Magazine* (December 1992): 41–46.

Henderson, J. C., and M. E. Treacy. "Managing End-User Computing for Competitive Advantage." *Sloan Management Review* (Winter 1986): 3–14.

Ives, B., and M. H. Olson. "User Involvement and MIS Success: A Review of Research." *Management Science* 30 (5) (1984): 586–603.

Jones, C. "How to Measure Software Costs." *Application Development Trends* (May 1997): 32–36.

Kling, R., and S. Iacono. "Desktop Computerization and the Organization of Work." *Computers in the Human Context*. Edited by T. Forester, 335–56. Cambridge, MA: MIT Press, 1989.

Lee, D. M. S. "Usage Pattern and Sources of Assistance for Personal Computer Users." *MIS Quarterly*, 10 (December 1986): 313–25.

Markus, M. L. "Implementation Politics: Top Management Support and User Involvement." *Systems/Objectives/Solutions* 1 (4) (1981): 203–15.

Martin, E. W., C. V. Brown, D. W. DeHayes, J. A. Hoffer, and W. C. Perkins. *Managing Information Technology: What Managers Need to Know*. 4th ed. Upper Saddle River, NJ: Prentice Hall, 2002.

Mosley, D. J. *The Handbook of MIS Application Software Testing*. Englewood Cliffs, NJ: Yourdon Press, 1993.

Nelson, R. R., and P. H. Cheney. "Training End Users: An Exploratory Study. *MIS Quarterly* 11 (December 1987): 547–59.

Pressman, R. S. *Software Engineering: A Practitioner's Approach*. 2nd ed. New York: McGraw-Hill, 1992.

Schrage, M. "Unsupported Technology: A Prescription for Failure." *Computerworld* 27 (May 10, 1993): 31.

Schurr, A. "Support Is No. 1." *PC Week* 10 (November 15, 1993): 126.

Tait, P., and I. Vessey. "The Effect of User Involvement on System Success: A Contingency Approach." *MIS Quarterly* 12 (1) (March 1988): 91–108.

Torkzadeh, G., and W. J. Doll. "The Place and Value of Documentation in End-User Computing." *Information & Management* 24 (3) (1993): 147–58.

Yourdon, E. *Managing the Structured Techniques*. 4th ed. Upper Saddle River, NJ: Prentice Hall, 1989.

Appendix A: Object-Oriented Analysis and Design

Booch, G. *Object-Oriented Analysis and Design with Applications*. 2d ed. Redwood City, CA: Benjamin/Cummings, 1994.

Coad, P., and E. Yourdon. *Object-Oriented Analysis.* 2d ed. Upper Saddle River, NJ: Prentice Hall, 1991.

Coad, P., and E. Yourdon. *Object-Oriented Design.* Upper Saddle River, NJ: Prentice Hall, 1991.

Eriksson, H., and M. Penker. *UMLToolkit.* New York: Wiley, 1998.

Fowler, M. *UML Distilled: A Brief Guide to the Object Modeling Language.* Reading, MA: Addison-Wesley, 2000.

Jacobson, I., M. Christerson, P. Jonsson, and G. Overgaard. *Object-Oriented Software Engineering: A Use-Case Driven Approach.* Reading, MA: Addison-Wesley, 1992.

Rumbaugh, J., M. Blaha, W. Premerlani, F. Eddy, and W. Lorensen. *Object-Oriented Modeling and Design.* Upper Saddle River, NJ: Prentice Hall, 1991.

UML Document Set. Version 1.0 Santa Clara: Rational Software Corp, January 1997.

UML Notation Guide. Version 1.0 Santa Clara: Rational Software Corp, January 1997.

Appendix B: Rapid Application Development and CASE Tools

Advantage Gen. www.ca.com/solutions. Information viewed on May 29, 2002.

Bourne, K. C. "Putting Rigor Back in RAD." *Database Programming & Design* 7 (8) (1994): 25–30.

Brooks, F. P., Jr. *The Mythical Man-Month* Anniversary edition. Reading, MA: Addison-Wesley, 1995.

Classe, A. "Rapid Reaction." *ComputerWeekly* (January 25, 1996), www.computerweekly.co.uk/cwhome.asp.

Delphi corporate Web site. Accessed on May 29, 2002, www.inprise.com/delphi.

Gibson, M. L., and C. T. Hughes. *Systems Analysis and Design: A Comprehensive Methodology with CASE.* Danvers, MA: Boyd & Fraser, 1994.

Harrar, G. "Welcome to IS Boot Camp." *Forbes ASAP* 152 (10) (1993): 112–18.

Martin, J. *Rapid Application Development.* New York: Macmillan, 1991.

McConnell, S. *Rapid Development.* Redmond, WA: Microsoft Press, 1996.

Glossary of Acronyms

Note: Some acronyms are abbreviations for entries in the Glossary of Terms. For these and some other acronym entries, we list in parentheses the chapter or appendix in which the associated term is defined or introduced. Other acronyms are generally used in the information systems field and are included here for your convenience.

1:1   One-to-One

1:*M*   One-to-Many

1NF   First Normal Form (9)

2NF   Second Normal Form (9)

3NF   Third Normal Form (9)

4GL   Fourth-Generation Language

ACM   Association for Computing Machinery

AITP   Association of Information Technology Professionals

API   Application Program Interface

ASM   Association for Systems Management

ASP   Active Server Pages

ATM   Automated Teller Machine

AT&T   American Telephone & Telegraph

BEC   Broadway Entertainment Company (2)

BPP   Baseline Project Plan (3)

BPR   Business Process Reengineering (4)

BSP   Business Systems Planning

CASE   Computer-Aided Software Engineering (B)

CATS   Customer Activity Tracking System (3)

CD   Compact Disk

CDP   Certified Data Processing

CD-ROM   Compact Disk-Read Only Memory

CGI   Common Gateway Interface

COBOL   COmmon Business Oriented Language

COCOMO   COnstruction COst MOdel

CRT   Cathode Ray Tube

C/S   Client/Server

CUA   Common User Access (8)

DB2   Data Base 2

DBMS   Database Management System

DCR   Design Change Request

DFD   Data Flow Diagram (5)

DPMA   Data Processing Managers Association

DSL   Digital Subscriber Loop

DSS   Decision Support System (1)

E-business   Electronic Business

E-commerce   Electronic Commerce (3)

EDI   Electronic Data Interchange (3)

EDS   Electronic Data Systems

EFT   Electronic Funds Transfer

EIS   Executive Information System (1)

E-mail   Electronic Mail

EPSS   Electronic Performance Support System (20)

E-R   Entity-Relationship

ERD   Entity-Relationship Diagram (6)

ERP   Enterprise Resource Planning (7)

ES   Expert System (1)

ESS   Executive Support System

ET   Estimated Time (2)

EUC   End-User Computing

FORTRAN   FORmula TRANslator

FTP   File Transfer Protocol

GDSS   Group Decision Support System

GSS   Group Support System

GUI   Graphical User Interface (8)

HTML   Hypertext Markup Language

HTTP   Hypertext Transfer Protocol

IBM   International Business Machines

I-CASE   Integrated Computer-Aided Software Engineering (13)

IDSS   Integrated Development Support System

IE   Information Engineering

IEF   Information Engineering Facility

IMAP   Internet Message Access Protocol

I/O   Input/Output

IS   Information System (1)

ISA   Information Systems Architecture

ISAM   Indexed Sequential Access Method

ISDN   Integrated Services Digital Network

ISP   Information Systems Planning

ISPs   Internet Service Providers

IT   Information Technology

IU   Indiana University

JAD   Joint Application Design (1)

KLOC   Thousand Lines of Code

LAN   Local Area Network (3)

MIS   Management Information System (1)

M:N   Many-to-Many

MRP   Material Requirements Planning

MTBF   Mean Time Between Failures (10)

MTTR   Mean Time to Repair Defect

NPV   Net Present Value (3)

OA   Office Automation

OO   Object-Oriented

OOAD   Object-Oriented Analysis and Design (1)

ODBC   Open Database Connectivity

PC   Personal Computer

PD   Participatory Design (1)

PERT   Project Evaluation and Review Technique (2)

PIP   Project Initiation and Planning (1)

POP   Post Office Protocol

POS   Point-of-Sale

POTS   Plain Old Telephone Services

PTR   Problem Tracking Report

PV   Present Value (3)

PVF   Pine Valley Furniture (1)

RAD   Rapid Application Development (1)

RAM   Random Access Memory

R&D   Research and Development

RFP   Request for Proposal (7)

RFQ   Request For Quote (7)

ROI   Return on Investment (3)

ROM   Read Only Memory

SDLC   Systems Development Life Cycle (1)

SDM   Systems Development Methodology

SGML   Standard Generalized Markup Language

SMTP   Simple Mail Transfer Protocol

SNA   System Network Architecture

SOW   Statement of Work (3)

SPTS   Sales Promotion Tracking System

SQL   Structure Query Language

SSR   System Service Request (3)

TCP/IP   Transmission Control Protocol/Internet Protocol

TPS   Transaction Processing System (1)

TQM   Total Quality Management

TVM   Time Value of Money (3)

UML   Unified Modeling Language (A)

URL   Uniform Resource Locator

VAN   Value Added Network

VB   Visual Basic

VBA   Visual Basic for Applications

VPN   Virtual Private Network

VSAM   Virtual Sequential Access Method

WIP   Work in Progress

XML   Extensible Markup Language

Glossary of Terms

Abstract class A class that has no direct instances, but whose descendants may have direct instances.

Acceptance testing The process whereby actual users test a completed information system, the end result of which is the users' acceptance of it.

Action stubs That part of a decision table that lists the actions that result for a given set of conditions.

Activation The time period during which an object performs an operation.

Actor An external entity that interacts with the system (similar to an external entity in data flow diagramming).

Adaptive maintenance Changes made to a system to evolve its functionality to changing business needs or technologies.

Aggregation A *part-of* relationship between a component object and an aggregate object.

Alpha testing User testing of a completed information system using simulated data.

Application independence The separation of data and the definition of data from the applications that use these data.

Application server A "middle-tier" software and hardware combination that lies between the Web server and the corporate network and systems.

Application software Software designed to process data and support users in an organization. Examples of application software include spreadsheets, word processors, and database management systems.

Association A relationship between object classes.

Association role The end of an association where it connects to a class.

Associative entity An entity type that associates the instances of one or more entity types and contains attributes that are peculiar to the relationship between those entity instances.

Attribute A named property or characteristic of an entity that is of interest to the organization.

Audit trail A record of the sequence of data entries and the date of those entries.

Balancing The conservation of inputs and outputs to a data flow diagram process when that process is decomposed to a lower level.

Baseline modules Software modules that have been tested, documented, and approved to be included in the most recently created version of a system.

Baseline Project Plan (BPP) The major outcome and deliverable from the project initiation and planning phase that contains an estimate of the project's scope, benefits, costs, risks, and resource requirements.

Behavior Represents how an object acts and reacts.

Beta testing User testing of a completed information system using real data in the real user environment.

Binary relationship A relationship between instances of two entity types.

Boundary The line that marks the inside and outside of a system, and that sets off one system from other systems in the organization.

Break-even analysis A type of cost-benefit analysis to identify at what point (if ever) benefits equal costs.

Build routines Guidelines that list the instructions to construct an executable system from the baseline source code.

Business case A written report that outlines the justification for an information system. The report highlights economic benefits and costs and the technical and organizational feasibility of the proposed system.

Business process reengineering (BPR) The search for, and implementation of, radical change in business processes to achieve breakthrough improvements in products and services.

Calculated (or computed or derived) field A field that can be derived from other database fields.

Candidate key An attribute (or combination of attributes) that uniquely identifies each instance of an entity type.

Cardinality The number of instances of entity B that can (or must) be associated with each instance of entity A.

Class diagram Shows the static structure of an object-oriented model: the object classes, their internal structure, and the relationships in which they participate.

Closed-ended questions Questions in interviews and on questionnaires that ask those responding to choose from among a set of specified responses.

Code generators CASE tools that enable the automatic generation of program and database definition code directly from the design documents, diagrams, forms, and reports stored in the repository.

Cohesion The extent to which a system or subsystem performs a single function.

Component An irreducible part or aggregation of parts that makes up a system; also called a *subsystem*.

Component diagram Shows the software components or modules and their dependencies.

Computer-aided software engineering (CASE) Software tools that provide automated support for some portion of the systems development process.

Conceptual data model A detailed model that shows the overall structure of organizational data while being independent of any database management system or other implementation considerations.

Concrete class A class that can have direct instances.

Condition stub That part of a decision table that lists the conditions relevant to the decision.

Configuration management The process of ensuring that only authorized changes are made to a system.

Constraint A limit to what a system can accomplish.

Context diagram A data flow diagram of the scope of an organizational system that shows the system boundaries, external entities that interact with the system, and the major information flows between the entities and the system.

Cookie crumbs A technique for showing a user where they are in a Web site by placing a series of "tabs" on a Web page that show a user where they are and where they have been.

Corrective maintenance Changes made to a system to repair flaws in its design, coding, or implementation.

Coupling The extent to which subsystems depend on each other.

449

Critical path The shortest time in which a project can be completed.

Critical path scheduling A scheduling technique whose order and duration of a sequence of task activities directly affect the completion date of a project.

Cross life cycle CASE CASE tools designed to support activities that occur *across* multiple phases of the systems development life cycle.

Data Raw facts about people, objects, and events in an organization.

Database A shared collection of logically related data designed to meet the information needs of multiple users in an organization.

Data flow Data in motion, moving from one place in a system to another.

Data flow diagram (DFD) A graphic that illustrates the movement of data between external entities and the processes and data stores within a system.

Data-oriented approach A strategy of information systems development that focuses on the ideal organization of data rather than on where and how they are used.

Data store Data at rest, which may take the form of many different physical representations.

Data type A coding scheme recognized by system software for representing organizational data.

Decision table A matrix representation of the logic of a decision, which specifies the possible conditions for the decision and the resulting actions.

Decomposition The process of breaking the description of a system down into small components; also called *functional decomposition.*

Default value The value a field will assume unless an explicit value is entered for that field.

Degree The number of entity types that participate in a relationship.

Deliverable An end product in a phase of the SDLC.

Denormalization The process of splitting or combining normalized relations into physical tables based on affinity of use of rows and fields.

Design strategy A particular approach to developing an information system. It includes statements on the system's functionality, hardware and system software platform, and method for acquisition.

Desk checking A testing technique in which the program code is sequentially executed manually by the reviewer.

DFD completeness The extent to which all necessary components of a data flow diagram have been included and fully described.

DFD consistency The extent to which information contained on one level of a set of nested data flow diagrams is also included on other levels.

Dialogue The sequence of interaction between a user and a system.

Dialogue diagramming A formal method for designing and representing human-computer dialogues using box and line diagrams.

Discount rate The interest rate used to compute the present value of future cash flows.

Disruptive technologies Technologies that enable the breaking of long-held business rules that inhibit organizations from making radical business changes.

Economic feasibility A process of identifying the financial benefits and costs associated with a development project.

Electronic commerce (EC) Internet-based communication to support day-to-day business activities.

Electronic data interchange (EDI) The use of telecommunications technologies to transfer business documents directly between organizations.

Electronic performance support system (EPSS) Component of a software package or application in which training and educational information is embedded. An EPSS may include a tutorial, expert system, and hypertext jumps to reference material.

Encapsulation The technique of hiding the internal implementation details of an object from its external view.

End users Non-information-system professionals in an organization. End users often request new or modified software applications, test and approve applications, and may serve on project teams as business experts.

Enterprise solutions software or enterprise resource planning (ERP) system A system that integrates individual traditional business functions into a series of modules so that a single transaction occurs seamlessly within a single information system rather than in several separate systems.

Entity A person, place, object, event, or concept in the user environment about which the organization wishes to maintain data.

Entity instance (instance) A single occurrence of an entity type.

Entity-relationship diagram (E-R diagram) A detailed, logical, and graphical representation of the entities, associa-

tions, and data elements for an organization or business area.

Entity type A collection of entities that share common properties or characteristics.

Environment Everything external to a system that interacts with the system.

Event Something that takes place at a certain point in time; it is a noteworthy occurrence that triggers a state transition.

External documentation System documentation that includes the outcome of structured diagramming techniques such as data flow and entity-relationship diagrams.

Extranet Internet-based communication to support business-to-business activities.

Feasibility study Determines if the information system makes sense for the organization from an economic and operational standpoint.

Field The smallest unit of named application data recognized by system software.

File organization A technique for physically arranging the records of a file.

Foreign key An attribute that appears as a nonprimary key attribute in one relation and as a primary key attribute (or part of a primary key) in another relation.

Form A business document that contains some predefined data and may include some areas where additional data are to be filled in. An instance of a form is typically based on one database record.

Formal system The official way a system works as described in organizational documentation.

Form and report generators CASE tools that support the creation of system forms and reports in order to prototype how systems will "look and feel" to users.

Functional dependency A particular relationship between two attributes. For a given relation, attribute B is functionally dependent on attribute A if, for every valid value of A, that value of A uniquely determines the value of B. The functional dependence of B on A is represented by A'B.

Gantt chart A graphical representation of a project that shows each task as a horizontal bar whose length is proportional to its time for completion.

Gap analysis The process of discovering discrepancies between two or more sets of data flow diagrams or discrepancies within a single DFD.

Hashed file organization The address for each row is determined using an algorithm.

Help desk A single point of contact for all user inquiries and problems about a particular information system or for all users in a particular department.

Homonym   A single attribute name that is used for two or more different attributes.

I-CASE   An automated systems development environment that provides numerous tools to create diagrams, forms, and reports; provides analysis, reporting, and code generation facilities; and seamlessly shares and integrates data across and between tools.

Identifier   A candidate key that has been selected as the unique, identifying characteristic for an entity type.

Incremental commitment   A strategy in systems analysis and design in which the project is reviewed after each phase and continuation of the project is rejustified in each of these reviews.

Index   A table used to determine the location of rows in a file that satisfy some condition.

Indifferent condition   In a decision table, a condition whose value does not affect which actions are taken for two or more rules.

Informal system   The way a system actually works.

Information   Data that have been processed and presented in a form suitable for human interpretation, often with the purpose of revealing trends or patterns.

Information center   An organizational unit whose mission is to support users in exploiting information technology.

Information systems analysis and design   The process of developing and maintaining an information system.

Inspections   A testing technique in which participants examine program code for predictable language-specific errors.

Installation   The organizational process of changing over from the current information system to a new one.

Intangible benefit   A benefit derived from the creation of an information system that cannot be easily measured in dollars or with certainty.

Intangible cost   A cost associated with an information system that cannot be easily measured in terms of dollars or with certainty.

Integration testing   The process of bringing together all of the modules that a program comprises for testing purposes. Modules are typically integrated in a top-down, incremental fashion.

Interface   Point of contact where a system meets its environment or where subsystems meet each other.

Internal documentation   System documentation that is part of the program source code or is generated at compile time.

Internet   A large worldwide network of networks that use a common protocol to communicate with each other; a global computing network to support business-to-consumer electronic commerce.

Interrelated components   Dependence of one part of the system on one or more other system parts.

Intranet   Internet-based communication to support business activities within a single organization.

JAD session leader   The trained individual who plans and leads Joint Application Design sessions.

Joint Application Design (JAD)   A structured process in which users, managers, and analysts work together for several days in a series of intensive meetings to specify or review system requirements.

Key business processes   The structured, measured set of activities designed to produce a specific output for a particular customer or market.

Legal and contractual feasibility   The process of assessing potential legal and contractual ramifications due to the construction of a system.

Level-0 diagram   A data flow diagram that represents a system's major processes, data flows, and data stores at a high level of detail.

Level-*n* diagram   A DFD that is the result of *n* nested decompositions of a series of subprocesses from a process on a level-0 diagram.

Lightweight graphics   The use of small simple images to allow a Web page to be more quickly displayed.

Lower CASE   CASE tools designed to support the systems implementation and operation phase of the systems development life cycle.

Maintenance   Changes made to a system to fix or enhance its functionality.

Mean time between failures (MTBF)   A measurement of error occurrences that can be tracked over time to indicate the quality of a system.

Modularity   Dividing a system into chunks or modules of equal size.

Multiplicity   Indicates how many objects participate in a given relationship.

Multivalued attribute   An attribute that may take on more than one value for each entity instance.

Normalization   The process of converting complex data structures into simple, stable data structures.

Network diagram   A diagram that depicts project tasks and their interrelationships.

Null value   A special field value, distinct from 0, blank, or any other value, that indicates that the value for the field is missing or otherwise unknown.

Object   An entity that has a well-defined role in the application domain and has state, behavior, and identity.

Object class   A set of objects that share a common structure and a common behavior.

Object diagram   A graph of instances that are compatible with a given class diagram.

One-time cost   A cost associated with project start-up and development, or system start-up.

Open-ended questions   Questions in interviews and on questionnaires that have no prespecified answers.

Operation   A function or a service that is provided by all the instances of a class.

Operational feasibility   The process of assessing the degree to which a proposed system solves business problems or takes advantage of business opportunities.

Outsourcing   The practice of turning over responsibility of some or all of an organization's information systems applications and operations to an outside firm.

Participatory Design (PD)   A systems development approach that originated in northern Europe in which users and the improvement of their work lives are the central focus.

Perfective maintenance   Changes made to a system to add new features or to improve performance.

PERT chart   A technique that uses optimistic, pessimistic, and realistic time to calculate the expected time for a particular task.

Physical file   A named set of table rows stored in a contiguous section of secondary memory.

Physical table   A named set of rows and columns that specifies the fields in each row of the table.

Picture (or template)   A pattern of codes that restricts the width and possible values for each position of a field.

Pointer   A field of data that can be used to locate a related field or row of data.

Political feasibility   The process of evaluating how key stakeholders within the organization view the proposed system.

Present value The current value of a future cash flow.

Preventive maintenance Changes made to a system to avoid possible future problems.

Primary key An attribute whose value is unique across all occurrences of a relation.

Primitive DFD The lowest level of decomposition for a data flow diagram.

Process The work or actions performed on data so that they are transformed, stored, or distributed.

Processing logic The steps by which data are transformed or moved and a description of the events that trigger these steps.

Process modeling Graphically representing the processes that capture, manipulate, store, and distribute data between a system and its environment and among components within a system.

Process-oriented approach An approach to developing information systems that focuses on how and when data are moved and changed.

Project A planned undertaking of related activities to reach an objective that has a beginning and an end.

Project closedown The final phase of the project management process, which focuses on bringing a project to an end.

Project execution The third phase of the project management process, in which the plans created in the prior phases (project initiation and planning) are put into action.

Project initiation The first phase of the project management process, in which activities are performed to assess the size, scope, and complexity of the project and to establish procedures to support later project activities.

Project management A controlled process of initiating, planning, executing, and closing down a project.

Project manager A systems analyst with a diverse set of skills—management, leadership, technical, conflict management, and customer relationship—who is responsible for initiating, planning, executing, and closing down a project.

Project planning The second phase of the project management process, which focuses on defining clear, discrete activities and the work needed to complete each activity within a single project.

Project workbook An online or hardcopy repository for all project correspondence, inputs, outputs, deliverables, procedures, and standards that is used for performing project audits, orientation of new team members, communication with management and customers, identifying future projects, and performing postproject reviews.

Prototyping Building a scaled-down version of the desired information system.

Purpose The overall goal or function of a system.

Rapid Application Development (RAD) Systems development methodology created to radically decrease the time needed to design and implement information systems. RAD relies on extensive user involvement, Joint Application Design sessions, prototyping, integrated CASE tools, and code generators.

Recurring costs A cost resulting from the ongoing evolution and use of a system.

Recursive foreign key A foreign key in a relation that references the primary key values of that same relation.

Referential integrity An integrity constraint specifying that the value (or existence) of an attribute in one relation depends on the value (or existence) of an attribute in the same or another relation.

Relation A named, two-dimensional table of data. Each relation consists of a set of named columns and an arbitrary number of unnamed rows.

Relational database model Data represented as a set of related tables or relations.

Relationship An association between the instances of one or more entity types that is of interest to the organization.

Repeating group A set of two or more multivalued attributes that are logically related.

Report A business document that contains only predefined data; it is a passive document used only for reading or viewing. A form typically contains data from many unrelated records or transactions.

Repository A centralized database that contains all diagrams, forms and report definitions, data structure, data definitions, process flows and logic, and definitions of other organizational and system components; it provides a set of mechanisms and structures to achieve seamless data-to-tool and data-to-data integration.

Request for proposal (RFP) A document provided to vendors to ask them to propose hardware and system software that will meet the requirements of a new system.

Resources Any person, group of people, piece of equipment, or material used in accomplishing an activity.

Rules That part of a decision table that specifies which actions are to be followed for a given set of conditions.

Scalable The ability to upgrade seamlessly the capabilities of the system through either hardware upgrades, software upgrades, or both.

Schedule feasibility The process of assessing the degree to which the potential time frame and completion dates for all major activities within a project meet organizational deadlines and constraints for affecting change.

Scribe The person who makes detailed notes of the happenings at a Joint Application Design session.

Secondary key One or a combination of fields for which more than one row may have the same combination of values.

Second normal form (2NF) A relation is in second normal form if every nonprimary key attribute is functionally dependent on the whole primary key.

Sequence diagram Depicts the interactions among objects during a certain period of time.

Sequential file organization The rows in the file are stored in sequence according to a primary key value.

Simple message A message that transfers control from the sender to the recipient without describing the details of the communication.

Slack time The amount of time that an activity can be delayed without delaying the project.

Source/Sink The origin and/or destination of data; also called *external entities*.

Statement of Work (SOW) A document prepared for the customer during project initiation and planning that describes what the project will deliver and outlines generally and at a high level all work required to complete the project.

State transition Changes in the attributes of an object or in the links an object has with other objects.

Structured English Modified form of the English language used to specify the logic of information system processes. Although there is no single standard, Structured English typically relies on action verbs and noun phrases and contains no adjectives or adverbs.

Stub testing A technique used in testing modules, especially where modules are written and tested in a top-down fashion, where a few lines of code are used to substitute for subordinate modules.

Support Providing ongoing educational and problem-solving assistance to information system users. Support material and jobs must be designed along with the associated information system.

Synchronous message   A type of message in which the caller has to wait for the receiving object to finish executing the called operation before it can resume execution itself.

Synonyms   Two different names that are used for the same attribute.

System   A group of interrelated procedures used for a business function, with an identifiable boundary, working together for some purpose.

System documentation   Detailed information about a system's design specifications, its internal workings, and its functionality.

System librarian   A person responsible for controlling the checking out and checking in of baseline modules when a system is being developed or maintained.

Systems analysis   Phase of the SDLC in which the current system is studied and alternative replacement systems are proposed.

Systems analyst   The organizational role most responsible for the analysis and design of information systems.

Systems design   Phase of the SDLC in which the system chosen for development in systems analysis is first described independent of any computer platform (logical design) and is then transformed into technology-specific details (physical design) from which all programming and system construction can be accomplished.

Systems development life cycle (SDLC)   The series of steps used to mark the phases of development for an information system.

Systems development methodology   A standard process followed in an organization to conduct all the steps necessary to analyze, design, implement, and maintain information systems.

Systems implementation and operation   Final phase of the SDLC, in which the information system is coded, tested, and installed in the organization, and in which the information system is systematically repaired and improved.

Systems planning and selection   The first phase of the SDLC, in which an organization's total information system needs are analyzed and arranged, and in which a potential information systems project is identified and an argument for continuing or not continuing with the project is presented.

System testing   The bringing together of all the programs that a system comprises for testing purposes. Programs are typically integrated in a top-down, incremental fashion.

Tangible benefit   A benefit derived from the creation of an information system that can be measured in dollars and with certainty.

Tangible cost   A cost associated with an information system that can be easily measured in dollars and with certainty.

Technical feasibility   The process of assessing the development organization's ability to construct a proposed system.

Template-based HTML   Templates to display and process common attributes of higher-level, more abstract items.

Ternary relationship   A simultaneous relationship among instances of three entity types.

Third normal form (3NF)   A relation is in third normal form if it is in second normal form and there are no functional (transitive) dependencies between two (or more) nonprimary key attributes.

Time value of money (TVM)   The process of comparing present cash outlays to future expected returns.

Unary relationship (recursive relationship)   A relationship between the instances of one entity type.

Unified Modeling Language (UML)   A notation that allows the modeler to specify, visualize, and construct the artifacts of software systems, as well as business models.

Unit testing   Each module is tested alone in an attempt to discover any errors in its code.

Upper CASE   CASE tools designed to support the systems planning and selection, systems analysis, and systems design phases of the systems development life cycle.

Use case   A complete sequence of related actions initiated by an actor; it represents a specific way to use the system.

Use-case diagram   A diagram that depicts the use cases and actors for a system.

User documentation   Written or other visual information about an application system, how it works, and how to use it.

Walkthrough   A peer group review of any product created during the systems development process; also called a *structured walkthrough*.

Web server   A computer that is connected to the Internet and stores files written in HTML (hypertext markup language) that are publicly available through an Internet connection.

Well-structured relation (table)   A relation that contains a minimum amount of redundancy and allows users to insert, modify, and delete the rows without errors or inconsistencies.

Work breakdown structure   The process of dividing the project into manageable tasks and logically ordering them to ensure a smooth evolution between tasks.

Index

A

Acceptance testing, 373
Action stubs, 177
Adaptive maintenance, 386
Alpha testing, 373, 395–396
Alternative system. *See* Design strategy
Analysis and design. *See* Systems analysis and design
Analysis paralysis, 122
Apple, 104
Application independence, 13
Application server/object framework, 258–259
Application software, development of, 4
Assessment of project feasibility
    benefits, determination of project, 89–90
    categories, 86
    consideration factors, 96
    cost-benefit analysis, 89
    costs, determination of project, 90–95
    economic feasibility, definition of, 89
    legal and contractual feasibility, 95
    operational feasibility, 95
    political feasibility, 95
    schedule feasibility, 95
    technical feasibility, 95
    time value of money, 92–93
Assessment of usability, 300
Associative entities, 214–215, 343
Attributes, 208–209
Attributive entity, 210
Audit trail, creation of, 292

B

Balancing, 166–168
Baseline modules, 390
Baseline project plan, 85
    building the, 96–100
    change, initiating, 51–52
    development of, 49–50
    execution of, 50
    feasibility assessment, 97, 99
    introduction, 96, 97
    management issues, 97, 99–100
    progress, monitoring, 50
    reviewing the, 100–103
    reviewing the plan, 100–103
    roles during walkthrough, 100
    scope of, 96–97

    system description, 97, 98–99
    updating the, 251–253
BEA. *See* Break-even analysis
Beta testing, 373, 395–396
Binary 1:1 relationships, 328–329, 332, 342
Binary 1:N relationships, 328–329
Binary and higher-degree M:N relationships, 329–330
Binary relationship, 213
Bottom-up approach, 206
Boundary, 6, 7
BPP. *See* Baseline project plan
BPR. *See* Business process reengineering
Break-even analysis, 93–94
Broadway Entertainment Company, Inc.
    background, 70–71
    conceptual data modeling case study, 230–232
    corporate systems and applications, 75
    design strategy, formulating a, 267–269
    designing the human interface application, 309–312
    determining requirements application, 150–153
    history, corporate, 71–72
    in-store systems, 73–74
    information systems, development of, 72–73
    initiating and planning application, 113–117
    introduction, 70
    organization, 71–72
    process modeling application, 192–196
    relational database, designing the, 358–360
    summary, 76
    testing plan, development of, 401–404
Budget, development of preliminary project, 49
Bug tracking, 394–395
Build routines, 390
Business case, 85
Business documents, use of, 122, 131–134
Business process reengineering, 140–141, 171–172
Business rules, 212

C

C++, 16
Calculated fields, 338
Candidate keys, 209

CASE. *See* Computer-aided software engineering tools
Case statements, 174
Champy, James, 141, 171–172
Classification and ranking development projects, 82
Client/server model, 10
Closed-end questions, 126
Closing out the customer contract, 385
Clothing Shack case problems, 149
Coding techniques, 338
Coding, testing, and installation, process of, 365–366
Cohesion, 9
Communication methods, 52
Communication plan, development of a, 48
Compaq, 241
Completeness, DFD, 169
Completes, 211
Compression techniques, 338
Computed fields, 338
Computer-aided software engineering tools, 5, 26, 169, 390–391
Conceptual data modeling
    bottom-up approach, 206
    definition, 200
    deliverables and outcomes, 202
    electronic commerce application, 219–224
    entity-relationship (E-R) data models, 200, 202, 204, 212–216 (*See also* Entity-relationship (E-R) data models)
    goals of, 200
    information gathering, 205–206
    procedures, 201–202
    questions for system users and business managers, sample, 205
    relationship between data modeling and systems development life cycle, 201
    top-down approach, 205
    Visio E-R notation, 203
Condition stubs, 177
Conditional statements, 174
Configuration management, 390
Consistency, DFD, 169
Constraints, 7, 120
Constructs, 206
Context diagram, 160
Contract, completion of the customer, 53

Photo Credits